MW00991479

PTEROSAURS

PTERO

Natural History,

MARK P.

PRINCETON UNIVERSITY PRESS

SAURS

Evolution, Anatomy

WITTON

PRINCETON AND OXFORD

Published by Princeton University Press,
41 William Street,
Princeton, New Jersey 08540

In the United Kingdom:
Princeton University Press,
6 Oxford Street, Woodstock, Oxfordshire OX20 1TW

press.princeton.edu

Library of Congress Cataloging-in-Publication Data

Witton, Mark P., 1984–
Pterosaurs : natural history, evolution, anatomy /
Mark P. Witton.
pages cm
ISBN 978-0-691-15061-1 (hardback)
1. Pterosauria—Evolution. 2. Pterosauria—
Anatomy. I. Title.
QE862.P7W58 2013
567.918–dc23 2012042596

British Library Cataloging-in-Publication Data
is available

This book has been composed in ITC New Baskerville
and Trade Gothic

Printed on acid-free paper. ∞

Printed in China

1 3 5 7 9 10 8 6 4 2

To my family, who can finally be content that I took their advice and wrote and illustrated a book

and

to Georgia, for putting up with me and the funny old dead reptiles I'm always going on about.

Contents

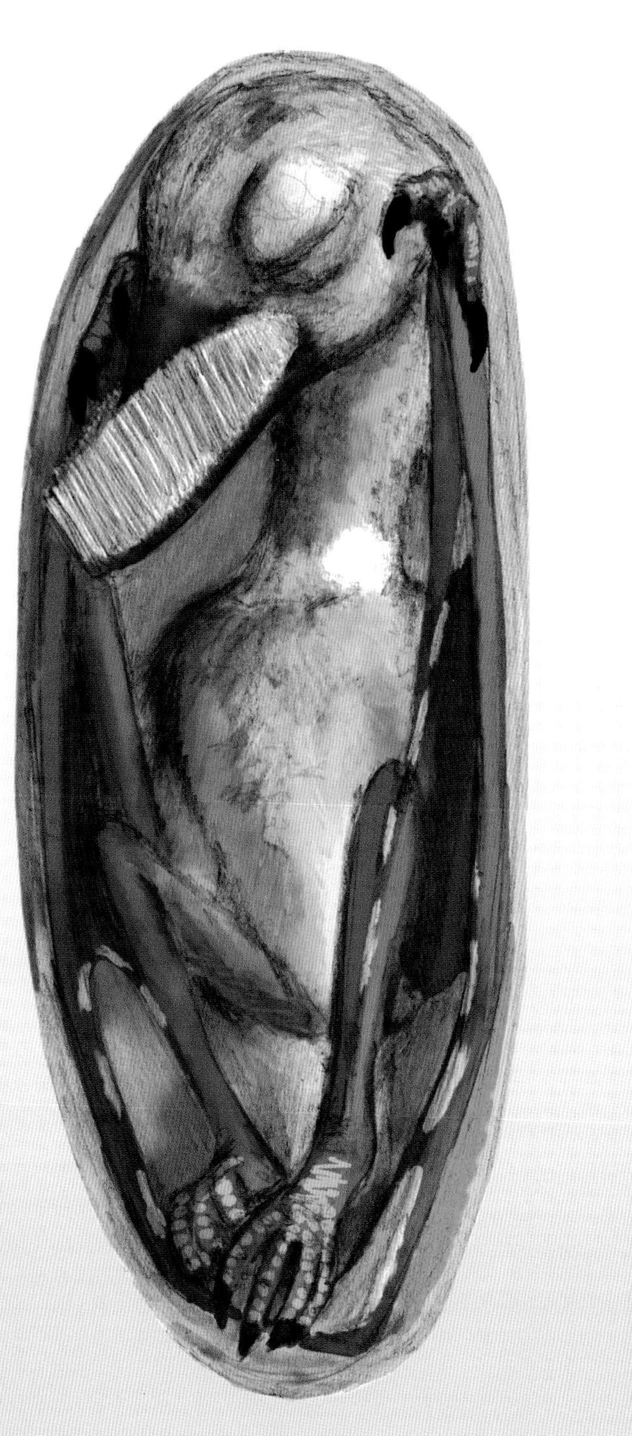

ix *Preface*

xi *Acknowledgments*

1 **1. Leathery-Winged Harpies**

4 **2. Understanding the Flying Reptiles**

12 **3. Pterosaur Beginnings**

23 **4. The Pterosaur Skeleton**

39 **5. Soft Bits**

56 **6. Flying Reptiles**

64 **7. Down from the Skies**

74 **8. The Private Lives of Pterosaurs**

90 **9. The Diversity of Pterosaurs**

95 **10. Early Pterosaurs and Dimorphodontidae**

104 **11. Anurognathidae**

113 **12. "Campylognathoidids"**

123 **13. Rhamphorhynchidae**

135 **14. Wukongopteridae**

143 **15. Istiodactylidae**

152 **16. Ornithocheiridae**

164 **17. Boreopteridae**

170 **18. Pteranodontia**

183 **19. Ctenochasmatoidea**

201 **20. Dsungaripteroidea**

211 **21. Lonchodectidae**

216 **22. Tapejaridae**

228 **23. Chaoyangopteridae**

234 **24. Thalassodromidae**

244 **25. Azhdarchidae**

259 **26. The Rise and Fall of the Pterosaur Empire**

265 *References*

283 *Index*

Preface

Picture this scene. I'm currently somewhere above Asia—Russia, probably—in a cramped seat on a Boeing 767 on my way to Beijing. I left London at some point yesterday evening, but have skipped through so many time zones I'm no longer really aware what the time is. The chap in front of me has his seat reclined back so far that my laptop is practically embedded in my stomach. I've got my carry-on baggage wedged between my feet because my fellow passengers have stowed piles of additional bags into my overhead compartment. (Seriously, what happened to being allowed *one* carry-on item?) For in-flight escapism, I can either watch a Russian-dubbed version of the Eddie Murphy kiddie movie *Dr. Doolittle 2* or listen to some numbingly similar dance tracks on the headphones provided by the airline. After experiencing this Europop haunting my every move during an eight-hour wait at Moscow airport—no matter how hard I tried to escape—I'm understandably reluctant to plug myself into them again.

This situation is combined with the fact that, in the last 36 hours or so, I've slept for about an hour. Maybe 90 minutes, tops. Between late night flights, uncomfortable airport seats, and a stifling heat wave during my Moscow transfer, grabbing some z's has not been easy at all. There is the promise of some sleep later on the flight, but only if my hindquarters recover from being shoved against a seat that, according to the few nerve endings still working down there, is as soft as chrome steel.

Now, you're probably asking yourself, what, if anything, does this have to do with the beasties adorning the cover of this book? My current discomfort stems from the fact that Beijing is hosting Flugsaurier 2010, the latest in a semiregular academic conference in which all the pterosaur experts of the world unite to present new finds, debate nuances in long-running arguments, view important pterosaur specimens, and visit key pterosaur fossil localities. It is, if you like, a celebration of all things pterosaur and not to be missed by any budding pterosaurophile. To some, this event may seem somewhat strange. Unlike their dinosaur cousins, pterosaurs have never really made a sensational popular or academic splash. They have several dozen dedicated academics researching them, and they cameo in the odd book, movie, or documentary from time to time, but they've never really made it into the same leagues as woolly mammoths, *Tyrannosaurus rex,* or the other prehistoric A-listers. Still, there has to be a reason why 60- to 70-odd pterosaurologists from around the world endure the exorbitant costs and traveling discomforts to talk pterosaurs every few years. Sheer eccentricity may explain this activity for some, but it's unlikely that we're *all* mad, so one wonders just why pterosaurs hold such appeal for this dedicated few. Why do perfectly respectable, intelligent men and women dedicate their lives to these long-dead animals and the related, comparatively low-paid, labor-demanding profession of paleontology when they could be making better livings as medical doctors, stockbrokers, or lawyers?

With a little luck, the book in your hands will indirectly provide an answer to this question, providing an overview of the diversity and sheer awesomeness of everyone's favorite leathery-winged reptiles before we reach the end of the story, or my 100,000 word limit, whichever arrives first. If I'm really fortunate, a number of you will even want to delve deeper into the primary literature on these animals, for which an extensive reference list is provided. It should be stressed that the people behind the literature comprising this list—many of whom I'm on my way to meet in Beijing—have already contributed far more to this book than I will, and I offer my sincere thanks to them for providing shoulders to stand on. I apologize in advance for any mistakes I inadvertently make in relaying their findings.

We'd best get cracking, then. I hope you enjoy what follows.

Mark Witton (just north of some cumulus clouds, topically listening to David Bowie's "China Girl," somewhere above Asia, August 2010).

Acknowledgments

A whole host of people have helped steer this book to completion, and I'm too close to my word count to thank them in the manner they deserve. Each of the following contributed to the process of assembling this book, and I'm extremely grateful to all of them. In no particular order, tremendous thanks to Brian Andres, Chris Bennett, Richard Butler, Luis Chiappe, Laura Codorniú, Stephen Czerkas, Fabio Dalla Vecchia, Ross Elgin, Erno Endenberg, Mike Habib, Dave Hone, Elaine Hyder, Martin Lockley, Bob Loveridge, Lü Junchang, Carl Mehling, Markus Moser, Darren Naish, Attila Ősi, Kevin Padian, Ryan Ridgely, Rico Stecher, Jose Sanz, Lorna Steel, David Unwin, Andre Veldmeijer, Larry Witmer, Ewan Wolff, and the nice people at the London Natural History Museum Picture Library, for supplying advice, thoughts, or imagery for use here. The folks at the University of Portsmouth are thanked for giving me a work space and access to technical literature. Kudos to my editor, Robert Kirk, and the other staff at Princeton University Press for running with this project, and the excellent Sheila Dean for turning my flippant prose into something that reads like science. I owe every one of you several beers, and sincere apologies if I've overlooked anyone else. If I did, feel free to slug me in the arm next time we meet, and then demand your promised beverage.

A few people deserve a special moment in the spotlight. Helmut Tischlinger not only allowed me to reproduce some of his excellent work—easily some of the most striking images in this book—but went the extra mile by arranging permission for photograph use from different pterosaur collections. Dave Martill, my PhD supervisor, was instrumental in realizing my interest in pterosaurs and, without him, this book—and a great number of the other cool things I've done over the last few years—would never have happened. My office buddies, old and new (too many to list here, but they know who they are), are not only great company in the office, but are also some of the best friends I could ask for; they continually remind me that there is more to life than work. My parents, sisters, and the rest of my family have always been supportive of my crazy ambition to work in the loopiest profession in the world, and I hope they're proud of this cumulation of my career so far. (Don't worry, Dad: I'll be reliably employed one day, you'll see.) A great number of online friends and acquaintances have also been extremely enthusiastic and supportive of this project. It's extremely flattering that people I've never met face-to-face have been saying how much they're looking forward to this, and I hope it lives up to their expectations.

Disacknowledgment

Finally, although customary in book acknowledgements to honor those who help steer projects to completion, it seems unfair to not mention the tremendous *negative* impact on this project made by Georgia Maclean-Henry. As the single most destructive force against this work, she took my attention from this project so frequently that we ended up moving in with each other halfway through the writing process and have ended up making some sort of home together. She continues to distract me from all kinds of work to this very day and, frankly, I could not be happier about it.

1
Leathery-Winged Harpies

If TV, film, and overzealous internet users have taught me anything, it's that the prehistoric world was harsh and brutal, and everyday existence was a life-or-death struggle. These terrible landscapes would be unrecognizable to our modern eyes, and only the biggest, nastiest animals survived. Consider, for example, the giant birds that stalked the Earth as recently as two million years ago. Taller than basketball players, they kicked and stabbed small, defenseless mammals to death. The ancestors of our pet cats and dogs wielded sabre teeth and bone-crushing jaws that they used to hunt giant elephants and rhinoceros, themselves armed with tusks and horns that would shame their mightiest modern relatives. The world was even more ferocious before these birds and mammals existed. During the span of time known as the Mesozoic (245–65 million years ago, or "Ma"), terrible reptiles ruled the day and predator-prey arms races were more intense than the Cold War. Gangs of carnivorous dinosaurs attacked their enormous herbivorous relatives, contesting their switchblade claws and armor-piercing teeth against the spikes, clubs, shields, and armored hides of their quarry. The Mesozoic oceans were just as deadly, teeming with giant, snaggle-toothed marine reptiles that render Moby Dick as intimidating as Flipper. Even the planet itself had a bad attitude in this Age of Reptiles. Angry volcanoes perpetually smoked, continents ripped themselves to pieces, and gigantic meteorites collided with the Earth, thowing enough dust and ash into the skies to block out the sun and cause cataclysmic extinction events.

The skies of this terrible age were no less formidable. They were dominated by a group of lanky grotesques, strange hybrids of birds and bats with a decidedly reptilian flavor. Their oversize heads were bristling with ferocious teeth or else bore savage beaks, each used to spear fish from primordial seas. Their outstretched membranous wings attained dimensions rivaling the wingspans of small aircraft, but were supported by lank, undermuscled limbs and tiny bodies. They were weak, flimsy, and pathetic animals, barely able to power their own locomotion and reliant on cliffs and headwinds to achieve flight. They were virtually helpless when grounded, barely able to push or drag themselves about, and completely at the mercy of any carnivorous reptile that fancied chewing on their hollow, twiglet-like bones. These creaky beasts were an archaic first attempt by vertebrate animals to achieve flight before graceful birds and nimble bats inherited the skies later in Earth's history. Given their obvious physical ineptitude, it's hardly surprising these creatures, the pterosaurs, collectively bought the farm at the end of the Mesozoic, along with any dinosaur that was not lucky enough to have evolved into a bird.

But That's All Hokum

Of course, the world and animals described above are nothing but caricatures of reality, the sort of landscape you may expect to find in Arthur Conan Doyle's *The Lost World* or similar-grade fiction. The popular view of primordial Earth as violent and totally unfamiliar is probably entirely untrue, a construct of poor scientific communication, overdramatic media representation, and romantic storytellers. In reality, ancient animals were no less sophisticated or intelligent, nor more freakish and savage, than those alive today. Paleontological research has probed deeply into the exotic and strange natures of many ancient animals to reveal that they merely represent "extreme" variants of anatomies and behaviors we see in our modern, familiar species. Such research has not made animals like the long-necked sauropods or giant theropod dinosaurs any less spectacular, but they are certainly not as mysterious and enigmatic as they once were.

Pterosaurs, which translates from Greek to "winged lizards," have suffered more than most in their depiction as ancient savages. All that remains of these animals are their fossil bones, oddly proportioned skeletons that have proved difficult to comprehend and continue to cause frequent controversies among those who study them. The pterosaur's ability to fly, combined with bizarre anatomy, often gigantic size, and an old-fashioned attitude that extinct animals were inherently inferior to modern species, resulted in them being perceived as crude, biological hang gliders, which were rather useless at

FIG. 1.1. Two *Rhamphorhynchus*, fish-eating pterosaurs of the Late Jurassic, doing what pterosaurs do best.

FIG. 1.2. The major pterosaur bauplans of the 130–150 species currently known. See chapters 9–25 for the identities of each animal.

everything but remaining airborne. Constant, often unwarranted, comparisons with birds and bats has cast pterosaurs as evolutionary also-rans, vertebrates that took the bold first stab at powered flight but were ultimately only the warm-up act for later, more sophisticated fliers.

Thankfully, these attitudes have slowly changed. Most modern pterosaurologists perceive pterosaurs as successful, diverse animals with interesting and intricate life histories, and in this book we'll discover the evidence for this change in attitude. We'll see that, while pterosaurs were undeniably very well adapted for powered flight (fig. 1.1), they were also skilled walkers, runners, and swimmers. They lived in diverse habitats across the globe and fueled their active lifestyles with prey caught in distinct, dynamic ways. They grew from precocial beginnings to adulthood and invested heavily in social and sexual display before becoming old and, in some cases, sick and arthritic. We'll meet numerous pterosaur groups (fig. 1.2) and over one hundred species spread across a dynasty spanning almost the entire Mesozoic, beginning in the Triassic period (245–205 Ma), thriving in the Jurassic (205–145 Ma), before ending at the very end of the Cretaceous, 65 Ma.

Our overview of pterosaurs is split into three parts. We'll start with an assessment of their general paleobiology, looking at their anatomy, locomotion, and other generalities of their lifestyles. Then, beginning with chapter 9, we'll meet, chapter by chapter, the diverse array of pterosaur groups currently recognized by pterosaur researchers, or "pterosaurologists." Finally, in the last chapter, we'll ponder their evolutionary story and try to ascertain why the skies of modern times are not full of membranous, reptilian wings in the way they once were.

2
Understanding the Flying Reptiles

FIG. 2.1. A modern reconstruction of the first pterosaur known to science, the ctenochasmatoid *Pterodactylus antiquus*. Although known since 1784, many details of its anatomy, including its headcrest, keratinous beak tips, and webbed feet, were not found in fossils until this century.

Pterosaurs have a track record of being a rather difficult group of animals to study. Since their discovery well over two centuries ago, they have frequently confounded attempts to comprehend their relationships to other animals (chapter 3), their terrestrial locomotion (chapter 7), or even simply parts of their anatomy (chapters 4 and 5). The history of these specific aspects of pterosaur research will be covered in subsequent chapters but, before we get to these, we will take a moment to familiarize ourselves with a broad outline of the first 230 years of pterosaur studies. In addition to the subsequent chapters offered here, readers particularly interested in the history of pterosaur research may enjoy the excellent discourses on this topic by the godfather of modern pterosaurology, Peter Wellnhofer (1991a, 2008). As we'll see below, Wellnhofer almost single-handedly revolutionized pterosaur research and laid the foundations for our modern understanding of these animals. He also provided a critical link between the three main ages of pterosaur paleontology, which can be roughly divided into the fruitful 1800s, a slump in the middle of the twentieth century, and our current golden age of pterosaurology.

1700–1900: Discovery After Discovery, Revelation After Revelation

The first pterosaur fossil, an exquisitely preserved, complete skeleton of an animal that would later be called *Pterodactylus* (figs. 2.1 and 2.2; also see chapter 19), was found at some point between 1767 and 1784 in the world famous Jurassic Solnhofen Limestone of Germany, the same deposit that would later yield the famous fossil bird *Archaeopteryx* (Wellnhofer

Fig. 2.2. The first pterosaur fossil known to science, a complete specimen of the Jurassic species *Pterodactylus antiquus*. Photograph courtesy of Helmut Tischlinger (specimen housed in the Bavarian State Collection of Palaeontology and Geology, Munich; used with permission).

1991a). The skeleton was brought to the attention of the Italian naturalist Cosimo Collini. Despite careful study, he remained unsure about the nature of the fossil. He thought the specimen represented a petrified amphibious creature, and suggested that the long fourth digit on each hand may have represented the spar of a flipper. Collini published his illustration and description of the fossil in 1784, the first documentation of a pterosaur in scientific literature. His work was soon noticed by the most eminent natural historian and comparative anatomist of the time, Baron Georges Cuvier. Cuvier's tremendous knowledge of animal form enabled him to see through the alien nature of the remains, and he realized from Collini's illustration that the elongate fourth finger was not a flipper at all, but the supporting strut of a membranous wing on an ancient, flighted, and reptilian creature (Cuvier 1801; see Taquet and Padian 2004 for details).

It is hard for us to imagine how significant Collini's *Pterodactylus* was to those originally studying it. Far from just revealing the existence of pterosaurs, this animal was also strong evidence for a concept that was once considered radical by even the most eminent nineteenth century researchers: extinction. Most fossils known at that point were the remains of marine

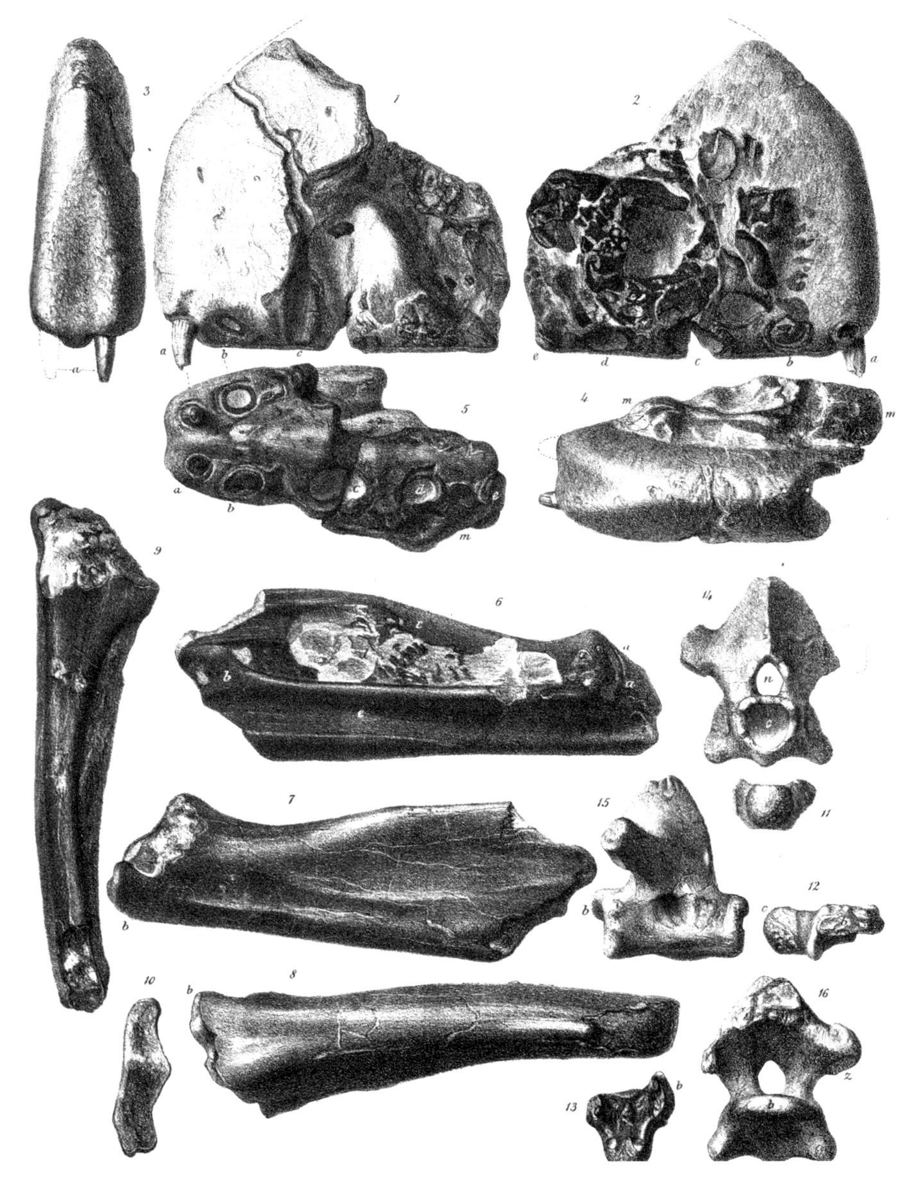

Fig. 2.3. One of many exceptional hand-drawn plates from Richard Owen's 1861 Palaeontographical Society monograph on Cretaceous reptiles, portraying the fragmentary pterosaur remains from the British Cambridge Greensand (see chapter 16 for more on these deposits). Owen authored numerous pterosaur monographs in his career, all of them illustrated with similarly excellent draftsmanship.

animals that, although never seen by humans, may have still existed in unexplored depths of the oceans. Collini's pterosaur, by contrast, was a relatively large, conspicuous creature that would live in our own realm if it existed today. This distinctive creature, as well as several other newly discovered, large, terrestrial fossil vertebrates, had never been witnessed in the surveyed parts of the world, forcing scholars like Collini and Cuvier to grapple with, and eventually accept, the concepts of life before human history and extinction (Taquet and Padian 2004). These ideas were only some of the heretical concepts proposed in what we now term the Age of Enlightenment, a century-long period in which eighteenth-century researchers began to place empirical observations and data before the religious doctrine that had previously been used to explain the natural world. Thus, our introduction to pterosaurs coincided with a rich period of scientific discovery, making them an integral part of the foundations of paleontology and geology.

Spurred on by a new understanding of the world, nineteenth-century investigators were prolific in their documentation and analysis of pterosaur fossils, although it was the middle of the nineteenth century before Cuvier's reptilian identification of pterosaurs was fully accepted. Fossils of new pterosaurs were being found across southern Germany and in other parts of Europe (e.g., Buckland 1829) and by the late 1800s, they were uncovered in the newly opened and fossil-rich deposits of North America (Marsh 1871). Much of the pterosaur material available to these Victorian investigators was rather scrappy, and following the once fashionable idea that every slight difference among fossil specimens was of taxonomic significance, dozens of pterosaur species were named that are of dubious validity to our modern eyes. Pterosaurologists are still struggling with the fallout of this overzealous naming, but by way of redeeming themselves, many Victorian paleontologists were also extremely skilled at describing and illustrating their pterosaur specimens (fig. 2.3). Thus, over one hundred years on, their work remains relevant and valuable to modern researchers, and is still cited heavily in modern literature.

1900–1970: Midlife Crisis

The turn of the twentieth century started well for pterosaur scientists. The British paleontologist Harry Seeley published the world's first popular pterosaur book *Dragons of the Air* in 1901 (fig. 2.4), and the first aeronautical studies of pterosaur flight were performed shortly after (Hankin and Watson 1914). The chimneys of the pterosaur research factory continued to smoke until the 1930s, when pterosaur research inexplicably almost ground to a halt for three decades. Only a handful of contributions were made to technical pterosaur literature during this time (fig. 2.5), and many of these were simply brief descriptions of scrappy new pterosaur fossils. As the seeds of our modern age were sown by men walking on the Moon, the advent of digital technology, and Jimi Hendrix playing guitar with his teeth, our perception of pterosaurs gathered dust and was no more advanced than it was at the end of the Victorian era.

1970–Present: The New Golden Age

Interest in pterosaurs picked up again in the 1970s, largely thanks to the work of the aforementioned father of modern pterosaurology, Peter Wellnhofer. Dedicating his research almost entirely to pterosaurs, Wellnhofer's first publications on these animals were landmark works on the taxonomy, growth, and functional

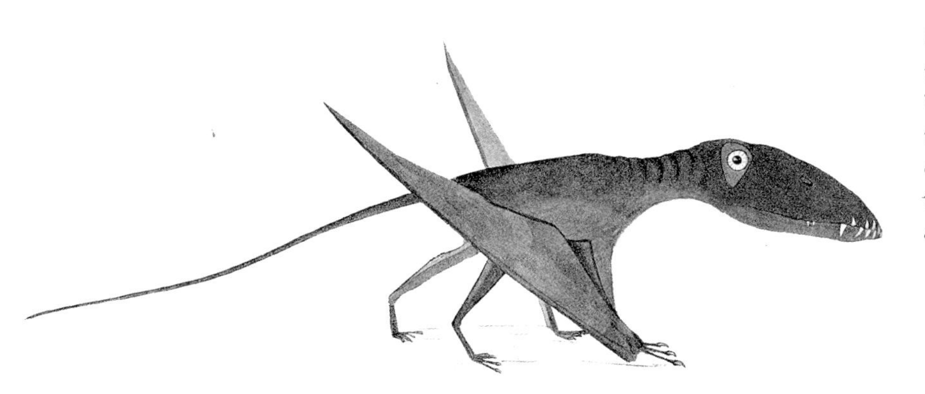

FIG. 2.4. Harry Seeley's (1901) reconstruction of the early pterosaur *Dimorphodon*. While many of Seeley's notions of pterosaur taxonomy and relationships have fallen out of favor with modern workers, his restorations of erect, quadrupedal pterosaurs were ahead of their time.

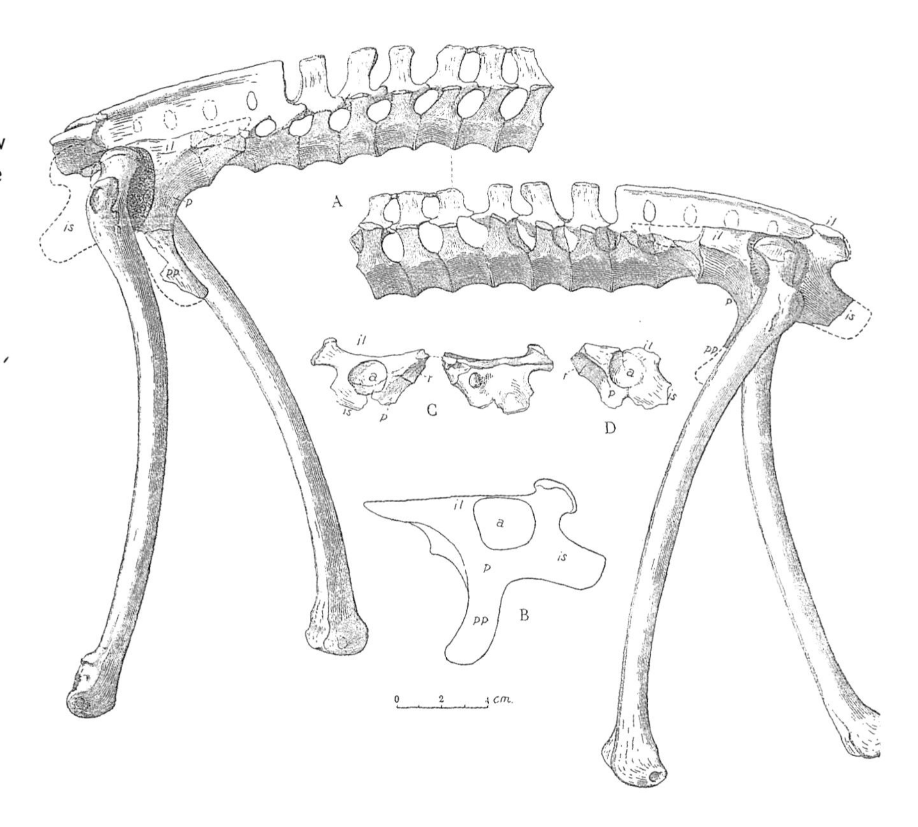

Fig. 2.5. C. C. Young's (1964) illustration of the posterior dorsal vertebrae, pelvis, and femora of the Cretaceous Chinese pterosaur *Dsungaripterus weii*. Young's work on this animal was one of the few significant contributions made to pterosaur science between 1930 and 1970.

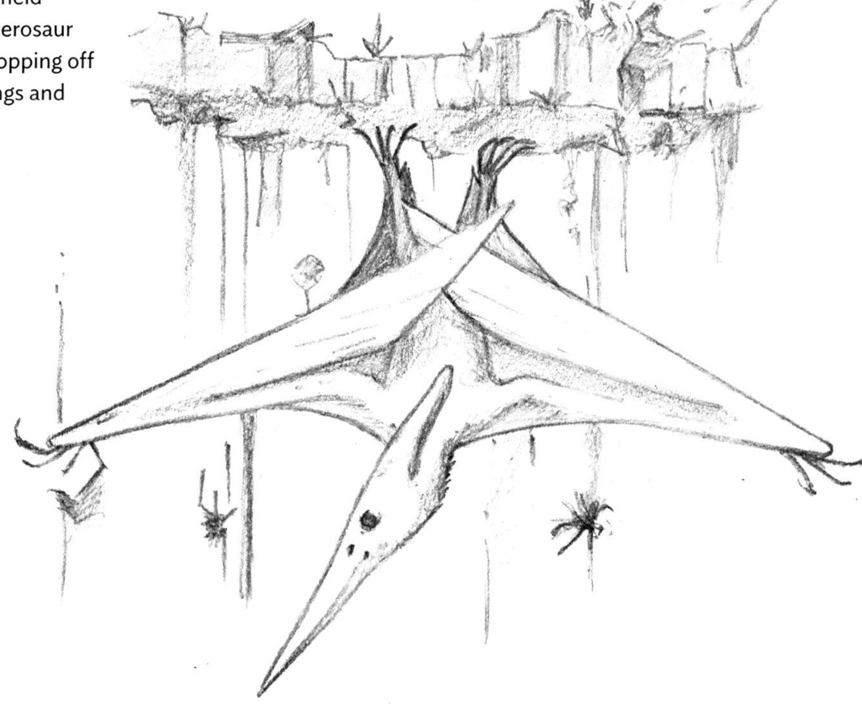

Fig. 2.6. Restoration of the Cretaceous pterosaur *Pteranodon longiceps* in a "resting" position, as imagined by early pterosaur flight researchers Cherrie Bramwell and George Whitfield (1974). At this point in history, the only way a large pterosaur was thought capable of becoming airborne was by dropping off a cliff (as demonstrated here) or by opening their wings and catching headwinds, akin to giant kites.

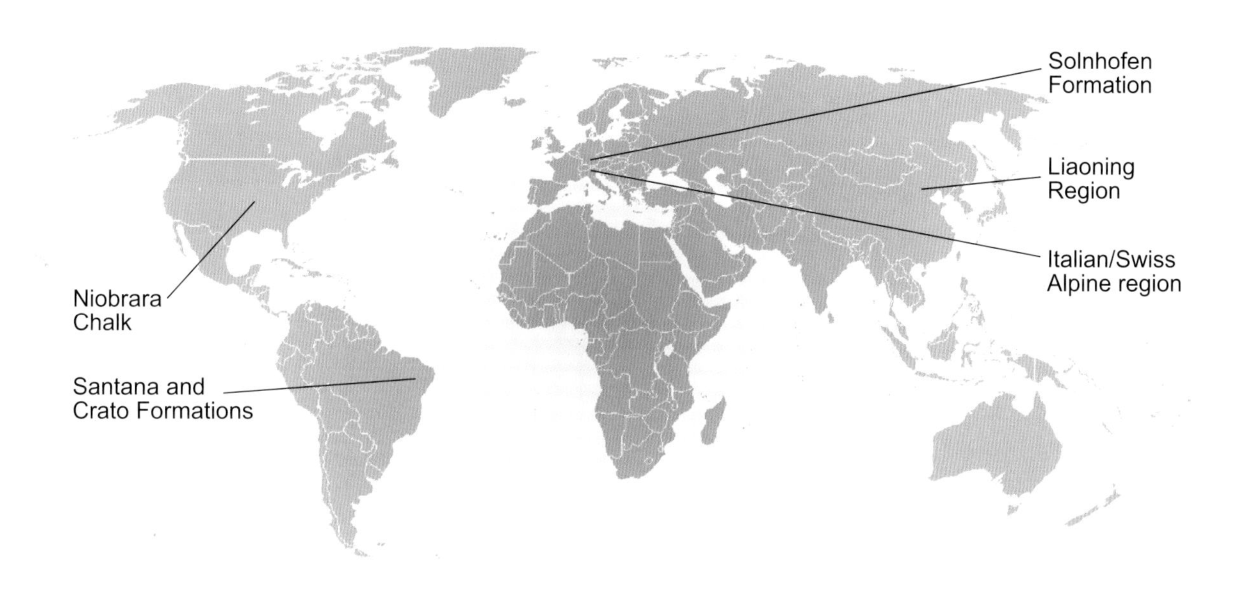

FIG. 2.7. Sites of pterosaur Lagerstätten, some of the most important fossil deposits to modern pterosaur researchers.

morphology of the pterosaurs of the Solnhofen Limestone, the same deposit that yielded the first pterosaur remains almost two hundred years earlier (Wellnhofer 1970, 1975). In 1978, he summarized pterosaur knowledge known at that time in the *Handbuch der Paläoherpetologie, Teil 19: Pterosauria* (translated from German to *Handbook of Paleoherpetology, Volume 19: Pterosauria*, or the *Pterosaur Handbook* for short) and documented, in the superb detail typical of his work, the anatomy of newly discovered, exquisitely preserved pterosaurs from Brazil (Wellnhofer 1985, 1987a, 1991b). Rounding off his achievements was the incredibly successful *Encyclopedia of Pterosaurs* (Wellnhofer 1991a), the first popular pterosaur book since Seeley's 1901 effort, which remains an incredibly useful reference tool for academics and laymen alike. His substantial body of work, only touched on here, not only established a modern framework for subsequent pterosaur researchers to work with but, as importantly, made pterosaurs an attractive topic to investigate again. By the early 1980s, pterosaur research had surpassed the intensity of its pre-1930 heyday with new discussions of pterosaur taxonomy, the description of numerous new species and a keen interest in pterosaur flight (e.g., Bramwell and Whitfield 1974; Stein 1975; also see fig. 2.6). Although dated now, these early flight studies paved the way for modern research into pterosaur flight biomechanics and are still routinely discussed by modern pterosaur workers.

As if rewarding this renewed interest in flying reptiles, the 1980s and 1990s saw fossil sites in Brazil and China termed "Lagerstätten" (fossil sites of exceptionally high preservation quality) yielding the best pterosaur fossils anyone had ever seen (fig. 2.7). The Brazilian Santana Formation was the first to do so, providing fantastically preserved, three-dimensional remains of large pterosaurs (e.g., Unwin 1988a), while the neighboring Crato Formation provided crushed pterosaur material with extensive soft-tissue preservation (reviewed by Unwin and Martill 2007). New discoveries in the Liaoning Province of China revealed similarly crushed pterosaur fossils in the late 1990s, but unlike the Crato Formation, these remains were often complete (fig. 2.8). These deposits have revolutionized our appreciation of pterosaur diversity by revealing several entirely new pterosaur groups, and have allowed for a greater understanding of pterosaur anatomy and systematics than ever before. In turn, this has provided a significant window into the nature of the more fragmentary pterosaur material common to non-Lagerstätten deposits, so a greater understanding of the group's evolutionary history has been gained overall.

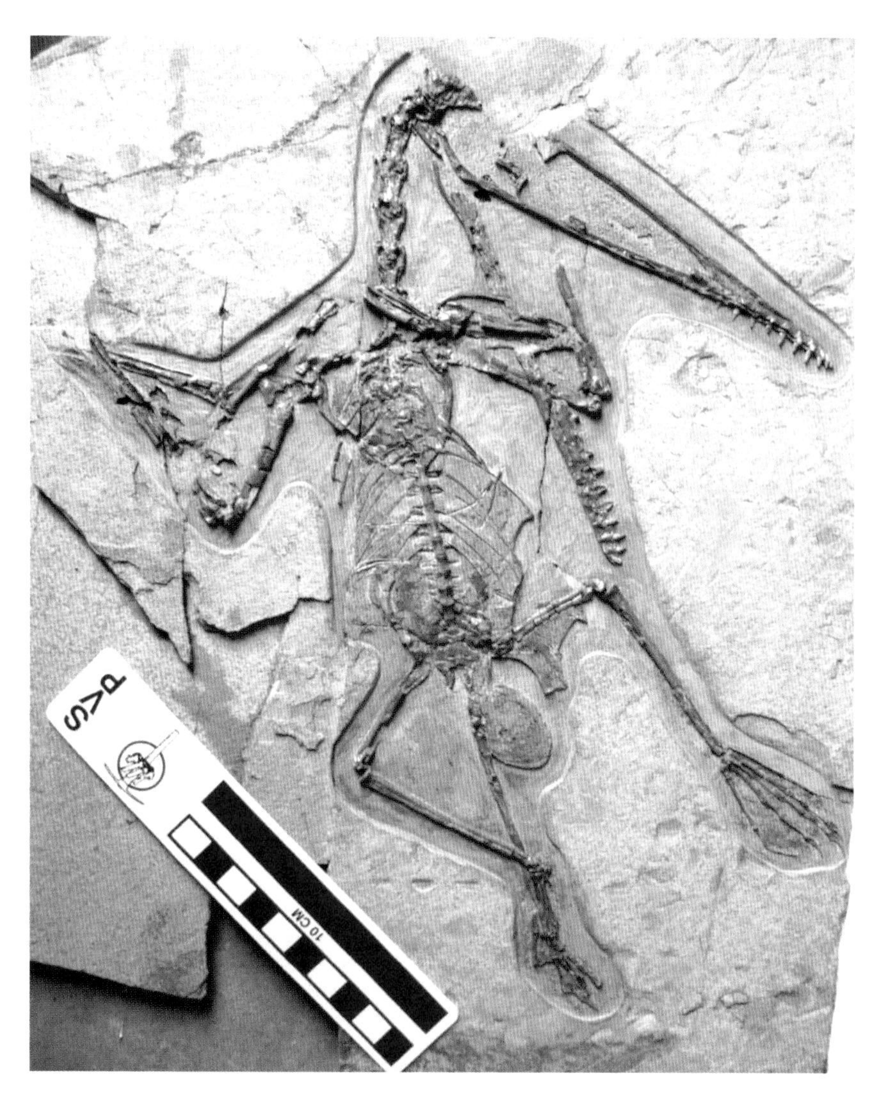

FIG. 2.8. The only pterosaur specimen definitively known to represent a female individual, a *Darwinopterus modularis* known as "Mrs. T," preserved alongside her unlaid egg. This fossil from the Jurassic Tiaojishan Formation of China's Liaoning Province is one of hundreds from Lagerstätte deposits that have changed our perception of pterosaurs in recent years. Photograph courtesy of David Unwin.

The monumental explosion in technology that marked the end of the twentieth century has also changed how we understand pterosaurs. Scanning electron microscopes (SEMs) allow observation of pterosaur remains at nanometer scales, and viewing specimens under ultraviolet (UV) light reveals soft-tissue structures that are invisible to the naked eye (fig. 2.9). Computed tomography (CT) scanning of pterosaur fossils has revealed brain casts and the complicated internal structure of pterosaur bones that were previously only accessible in broken fossils or through destructive sectioning techniques (fig. 2.10). Pterosaurology has also benefited, as has much of paleontology, by the availability of powerful computing power. A simple laptop can hold two centuries of technical pterosaur literature, predict relationships between pterosaur species, calculate scaling regimes joint mechanics, and provide a platform to communicate ideas between research institutes, as well as describe and illustrate fossil specimens for publication. Pterosaurology has never progressed so rapidly or been so accessible, and is only set to grow as a paleontological discipline as more and more students are drawn to it.

This third century of pterosaur research, then, is bound to be the most productive yet, where every specimen, new and old, offers greater insight into more aspects of pterosaur paleobiology than ever before. Alas, all this productivity means we've got a lot of ground to cover and necessitates the brevity of this overview. We had best tuck away our nostalgia and move on to our next stop: Just what *are* pterosaurs?

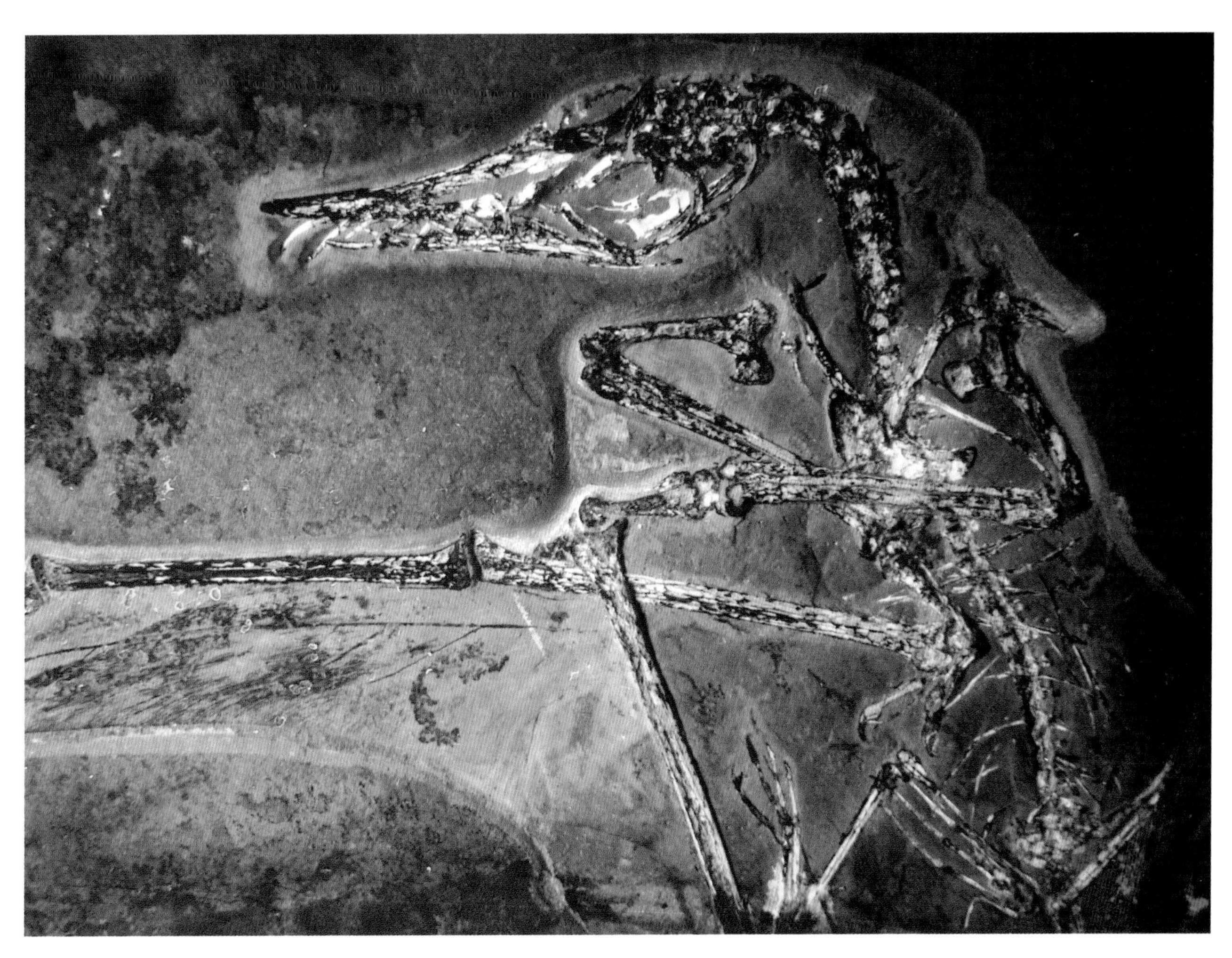

Fig. 2.9. A *Rhamphorhynchus muensteri* skeleton from the Jurassic Solnhofen Limestone under ultraviolet light. Note how the tissues of the wing membranes are clearly visible. Photograph courtesy of Helmut Tischlinger.

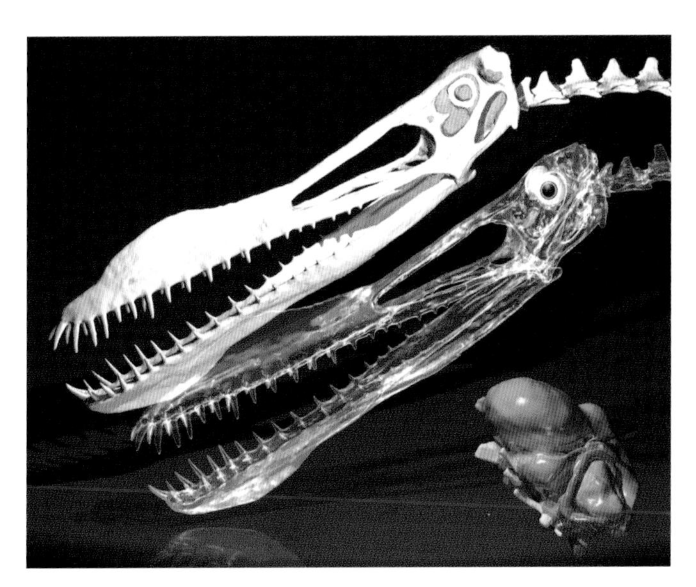

Fig. 2.10. CT visualization of the skull of *Anhanguera santanae*, revealing details of the brain endocast (bottom right) without a hint of damage to the fossil specimen (see chapter 5 for more on pterosaur brains). Image provided and updated from those of Witmer et al. (2003) by Larry Witmer and Ryan Ridgely, Ohio University.

3
Pterosaur Beginnings

Fig. 3.1. Stage C of the Hypothetical Pterosaur Ancestor, or HyPtA C. The fossil record has yet to reveal an "intermediate" form between fully fledged pterosaurs and their possible ancestors, meaning we can only speculate on their anatomy and appearance.

Trying to ascertain what sort of reptile pterosaurs are is one of the most fundamental goals of pterosaurologists, but pinning down their ancestry has proved very challenging. Much of their anatomy is so significantly different from that of other reptiles that their specific evolutionary origins are obscured. In addition, we have yet to find any "protopterosaur" species that bridge the gap between them and their reptilian ancestors (fig. 3.1). This is a problem that pterosaur researchers simply have to tackle despite its difficulties. Because pterosaurs are extinct, we are reliant on their extant relatives for insights into their likely muscle structure, respiratory physiology, and other aspects of their paleobiology that could not fossilize. Thus, if we want to truly understand pterosaurs, we must ascertain what grade of reptiles they were. Thankfully, some significant advances in resolving this problem have been made in recent years, and although some questions are still debated, pterosaur origins are perhaps not as murky as they once were.

Flying Reptiles

The broadest components of pterosaur classification are easy enough to work through. Pterosaurs are quite obviously members of Vertebrata, animals with backbones. Their possession of four well-developed limbs grants them membership to the clade Tetrapoda (animals that have, or had at some point in their evolutionary history, four limbs), and the development of their limb girdles and ankles tells us they are part of Amniota, animals that can lay eggs outside of water. We can quickly ascertain, therefore, that pterosaurs have something to do with the principle amniote groups of mammals, reptiles, and birds, though figuring out which one took early pterosaurologists some time. A number of early pterosaur workers considered them to be mammals (e.g., Soemmerring 1812; Newman 1843) and even reconstructed them with bat-like skeletons, lobed ears, and very mammalian-like external genitalia (see Taquet and Padian 2004). Others thought pterosaur skeletons showed an avian affinity, suggesting they were intermediaries between dinosaurs and birds. The main propoent of this idea, Harry Seeley, proposed the alternative name "ornithosaurs" (bird-lizards), though this term never caught on (Seeley 1864, 1901). An even stranger interpretation held pterosaurs as real-life echoes of griffins, the mammal-bird hybrids of mythology (Wagler 1830).

Close examination of the pterosaur skeleton demonstrates that their origins lie some distance from both birds and mammals, however. Although pterosaur bones are highly modified compared to other amniotes, they still retain numerous hallmarks of a reptilian ancestry. As we mentioned in chapter 2, this is not a new discovery. Georges Cuvier highlighted the reptilian characteristics of pterosaur skeletons as early as 1801 and spent the next few decades arguing his case against advocates of the mammalian pterosaur

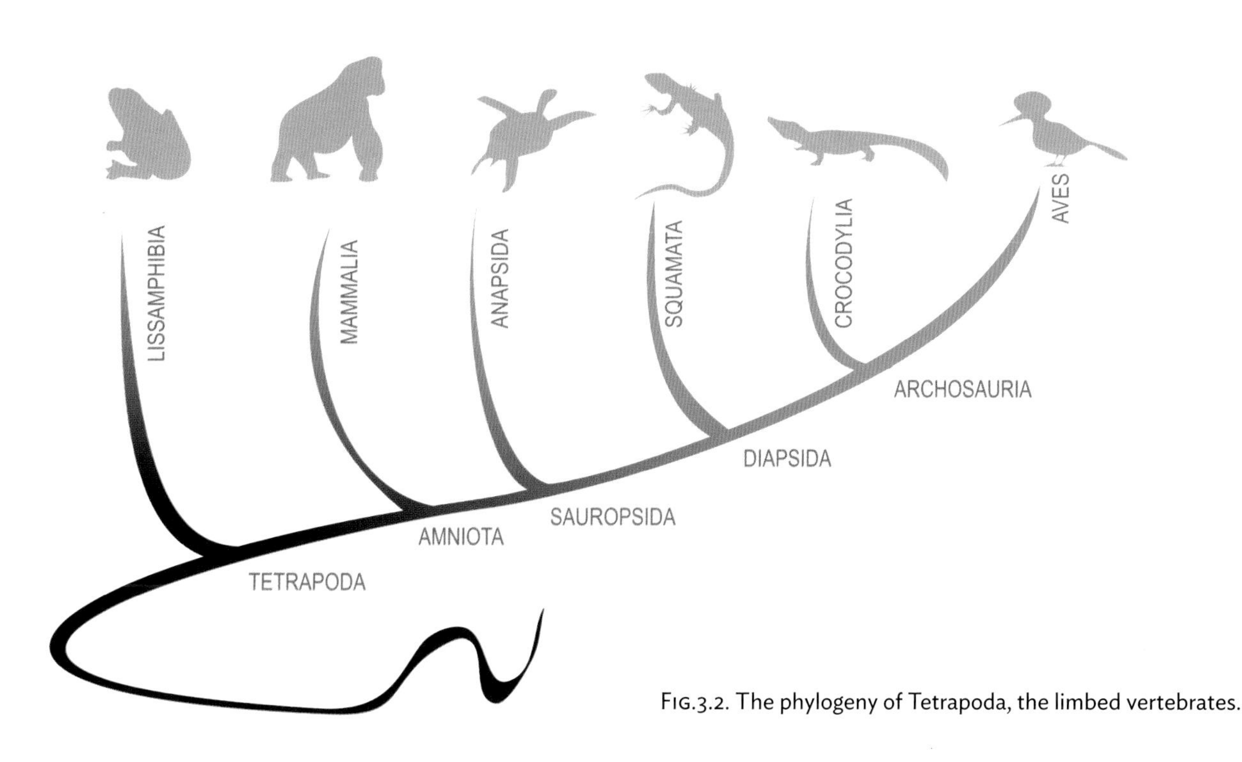

FIG.3.2. The phylogeny of Tetrapoda, the limbed vertebrates.

hypothesis. In particular, Cuvier noted that pterosaur quadrates, strut-like bones found at the back of the skull, articulate with the lower jaw in a manner unlike those of mammals, but very much like those of reptiles. He also noted that pterosaur teeth were far simpler than those of most mammals and were continually replaced throughout their lives; both of these characteristics are reptilian. Additional features of their skulls (see below), trunk skeleton, neck vertebrae, and feet also supported his reptilian identification. Eventually, these ideas won out over those proposing mammalian or avian links for pterosaurs, and nowadays, pterosaurs are incontrovertibly considered to be of reptilian descent.

Modern reptiles (or Sauropsida, to give them their formal title) include turtles and tortoises (Testudines), lizards and snakes (Squamata), the tuatara (the sole survivor of a once widespread group, the Sphenodontia), and Archosauria (crocodiles and birds) (fig. 3.2). This is a very limited view of reptile diversity compared to what we see in the fossil record, however. Reptilian evolution has generally been considered to hold two major branches, one that led to testudines, and the other that begat squamates, sphenodontians, and archosaurs, and we know from fossil remains that these branches gave rise to many more lineages than are alive today (fig. 3.3).[1] The testudine lineage spawned not only our extant shelly friends but also numerous extinct forms like the mighty, armored pareiasaurs; lithe aquatic mesosaurs; and numerous, superficially lizard-like species. The squamate + archosaur branch holds a great menagerie of marine reptiles (including the famous plesiosaurs, ichthyosaurs, and mosasaurs); a host of bizarre protorosaurs (sometimes referred to as "prolacertiforms"); a group of superficially crocodile-like forms that begat Archosauria (so-called Archosauriformes); the tremendous diversity of true crocodiles and their ancestors (the Crurotarsi); and of course the nonavian dinosaurs. (For excellent overviews of sauropsid and other vertebrate evolution, highly recommended texts are Mike Benton's 2005 *Vertebrate Palaeontology* and Liem et al.'s 2001 *Functional Anatomy of the Vertebrates*).

With so many branches sprouting from the reptile tree, identifying pterosaurs as reptiles gives no more than the loosest address for their origins, so further precision is needed if we hold any hope of understanding their paleobiology. We can easily rule out the testudine lineage because these reptiles have a skull configuration quite unlike that of pterosaurs. Testudines and their relatives lack temporal fenestrae, holes behind the eye socket that facilitate anchorage and bulging of jaw muscles. We call animals without these holes *anapsids*, but pterosaurs are *diapsids*, with two temporal fenestrae behind the eye, one above the other. (You, like all mammals, have another configuration with a single opening behind each eye—the *synapsid* condition. This distinction is one of the many reasons that pterosaurs are not considered members of Mammalia.) Ruling out an anapsid identification only adds limited resolution to pterosaur ancestry however, as all nontestudine reptiles are diapsids. This leaves us with a great many places where pterosaurs could plug into the reptile tree, and as we'll see below, figuring out which other diapsids they nestle closest to has proved the most contentious part of unraveling pterosaur relationships.

The Plot Thickens

At one time or another, pterosaurs have been considered members of most major diapsid groups (fig. 3.3). Peters (2000) argued that a suite of features across pterosaur skeletons placed them among a diverse group of Permian-Triassic (300–245, and 245–205 Ma, respectively) reptiles known as protorosaurs.2 In particular, Peters identified affinities with the strange Triassic hindlimb-glider *Sharovipteryx* (fig. 3.4A), a species which was also linked to pterosaur ancestry in several works by Lambert Halstead (e.g., Halstead 1975) because of extensive flight membranes stretching between its body and hindlimbs (Gans et al. 1987; Unwin, Alifanov, et al. 2000). Given that that pterosaurs presumably had a gliding stage at one point in their ancestry (see below), this hypothesis has obvi-

[1] We should note that this "classic" interpretation of reptile phylogeny has been strongly challenged in recent years. Numerous studies suggest that Testudines may be more deeply nested within reptiles than previously realized. Exactly where turtles and their kin plot among reptile evolution is very contentious, but support for the "classic" testudine position in the reptile tree is waning among modern researchers (see Rieppel 2008 for a review). Fortunately, this debate does not seriously affect our consideration of pterosaur origins, but readers should be aware that some of the statements made regarding testudine phylogeny here are by no means certain.

[2] There is some suggestion that established "Protorosauria" may not be a united group. Instead, "protorosaur" species may cluster as successive offshoots from the archosauromorph line (see Dilkes 1998; Modesto and Sues 2004; and Hone and Benton 2007 for different protorosaur phylogenies). There is no apparent consensus on this issue at the time of writing, so I've stuck with the term "protorosaurs" here for the sake of brevity and readability. Readers are invited to mentally substitute something like "nonarchosauriform archosauromorphs" for every use of "protorosaur" in the text if they feel like being more taxonomically savvy, however.

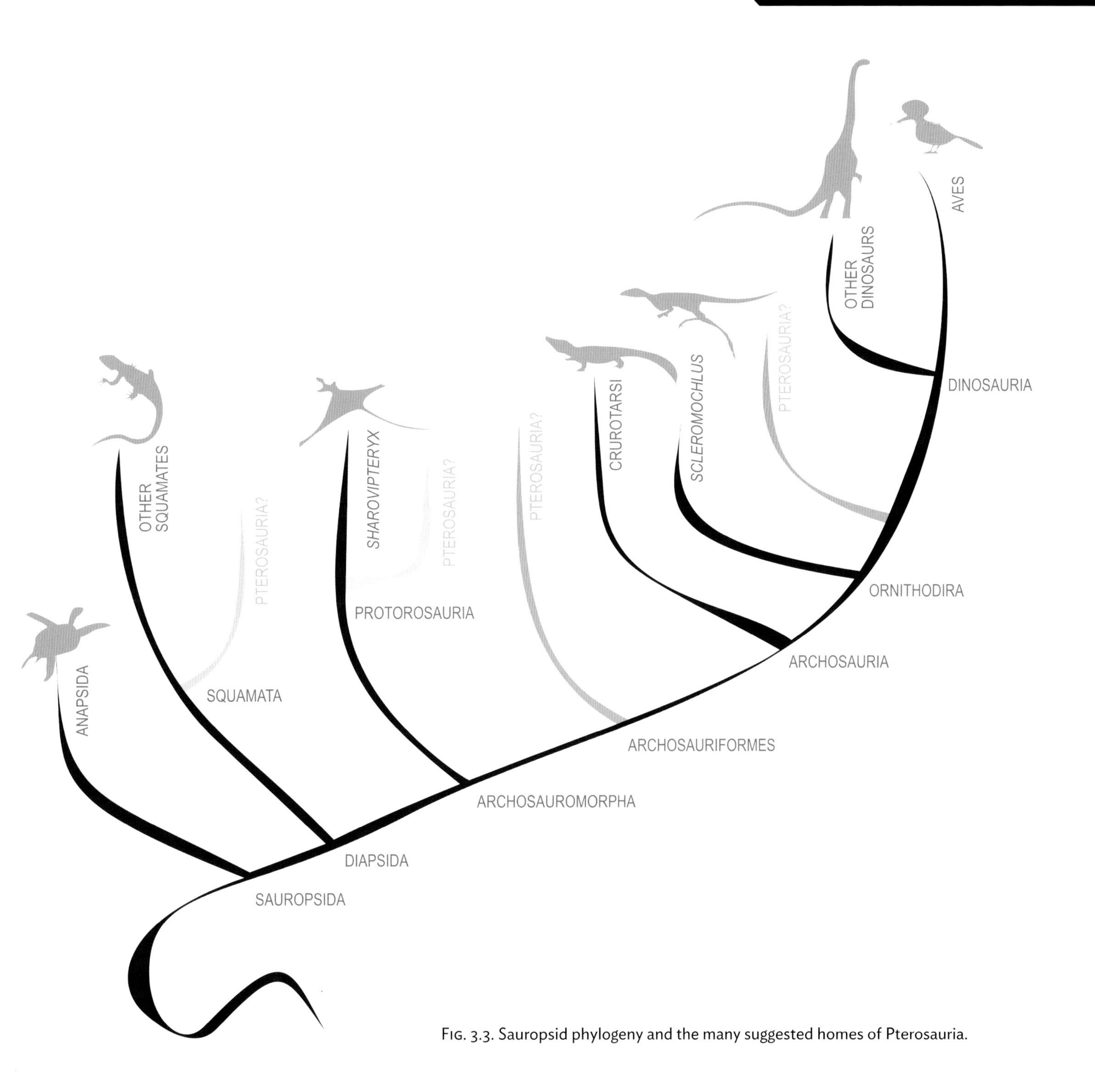

Fig. 3.3. Sauropsid phylogeny and the many suggested homes of Pterosauria.

ous immediate appeal. A position among protorosaurs would place pterosaurs at the base of Archosauromorpha, a group that includes protorosaurs and the archosauriforms. Radical new work by Peters (2008) has sent pterosaurs and the protorosaurs tumbling to the base of the diapsid tree, nestling them among Squamata. This idea has proved highly controversial, contradicting not only other interpretations of pterosaur origins but an otherwise very well-established overview of diapsid evolution (Gower and Wilkinson 1996).

Most authors have placed pterosaurs in more derived positions within Archosauromorpha, among the Archosauriformes. Bennett (1996a) and Unwin (2005) have suggested that features of the pterosaur skull, vertebrae, pelvis, and hindlimbs are indicative of such a placement alongside animals like *Euparkeria*,

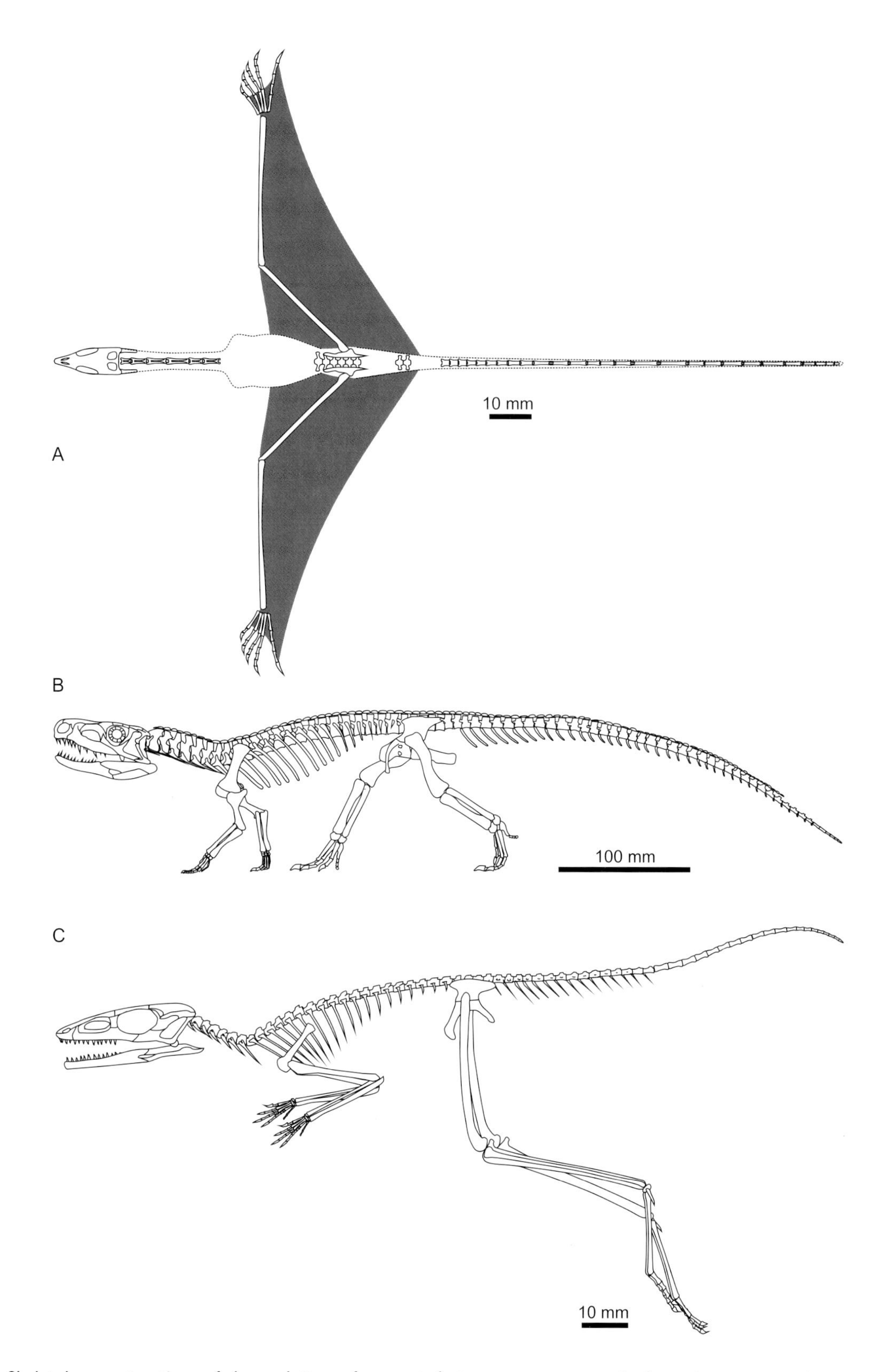

Fig. 3.4. Skeletal reconstructions of close relatives of purported pterosaur ancestors. A, the archosauromorph protorosaur *Sharovipteryx mirabilis*; B, the archosauriform *Euparkeria capensis*; C, the ornithodiran *Scleromochlus taylori*. The trunk and forelimb anatomy of *Sharovipteryx* is poorly known and has not been reconstructed here, though recent preparation of the holotype has revealed some details of its short forelimbs (Dyke et al. 2006). A, based on Unwin, Alifanov, and Benton (2000), Gans et al. (1987), and Peters (2000); B, based on Paul (2002); C, based on Benton (1999).

a small, somewhat crocodile-like Triassic quadruped (fig. 3.4B). A more popular idea is that pterosaurs lie within Archosauria as the major sister lineage to dinosaurs, comprising a group known as Ornithodira (or sometimes the synonymous term "Avemetatarsalia" is used) (e.g., Padian 1984a, Gauthier 1986; Sereno 1991; Benton 1999; Hone and Benton 2007, 2008; Nesbitt et al. 2010; Nesbitt and Hone 2010; Nesbitt 2011). In this scenario, the long-legged archosaur *Scleromochlus* (fig. 3.4C) is considered their closest known relative. The ornithodiran hypothesis was founded on characteristics of pterosaur hindlimbs that indicate, like dinosaurs, they had the ability to stand upright. Recent research has proposed that a great number of features from other regions of the pterosaur skeleton also support this idea (Hone and Nesbitt 2010; Nesbitt 2011).

Each of these positions within Diapsida has supporting and opposing arguments. That said, Peter's (2008) proposal that pterosaurs are housed within Squamata has received little following among pterosaur workers and is perhaps the most unlikely hypothesis currently under consideration. This is largely due to the highly controversial techniques used in his analyses and anatomical interpretations (see Bennett 2005; and Hone and Benton 2007 for critiques). There seems little similarity between the skulls of pterosaurs and the highly modified, mobile skulls of squamates, or any similarity between the trunk and limb skeletons of each group. When these differences are considered, along with the fact that other proposed relationships are far better supported, it seems very unlikely that squamate and pterosaur evolution is intertwined in any way.

The notion of pterosaurs stemming from protorosaurs, and thus making them early offshoots of Archosauromorpha, may have more credence. Protorosaurs are best characterized by their neck vertebrae (the "cervical" vertebrae), which are much longer than those of the trunk skeleton; their tall, blocky neural spines (projections of bone rising above the rest of the vertebra); and long cervical ribs (typically thin, rodlike bones articulating with the neck vertebra—see chapter 4 for more on osteological nomenclature) (e.g., Evans and King 1993; Dilkes 1998). Their lower jaws also lack openings (fenestrae) in their posterior regions, which contrasts with the perforate lower jaw bones of archosauriformes. Pterosaurs also have elongate necks compared to many other archosauromorphs and, like protorosaurs, some lack openings in the side of their lower jaws. These similarities are not without complication, however. The cervicals of early pterosaur may be long, but they lack the blocky neural spines and long ribs characteristic of protorosaurs. Similarly, recent research shows that early pterosaurs possessed fenestrate lower jaws and that only later, relatively derived species had the imperforate variant (Nesbitt and Hone 2010). Further analysis reveals far more differences than similarities between these groups. Pterosaurs, for example, have completely formed lower temporal fenestrae and an opening between their eye and nose openings—the antorbital fenestra—that are quite unlike the same features of protorosaurs. Their postcranial proportions are also very contrasting; most protorosaurs are fairly elongate, squat animals, but pterosaurs have very compact torsos and long limbs.

With the case for pterosaurs as basal archosauromorphs not appearing particularly strong, some burden is placed on the gliding protorosaur *Sharovipteryx* to defend this hypothesis. A cursory glance at *Sharovipteryx* suggests it may be a suitable pterosaur relative, with its very elongate, pterosaur-like legs where the shins are the longest elements; hollow bones; and flight membranes with radiating internal fibers (Gans et al. 1987; Unwin, Alifanov, et al. 2000; Unwin 2005). It is hard to find other features that reliably link this animal with pterosaurs however, and their anatomy is generally so different that close ancestry is unlikely (Sereno 1991; Bennett 1996a; Hone and Benton 2007; Nesbitt et al. 2010). The skull of *Sharovipteryx* is proportionally small, the neck is relatively long compared to the body and head, and while the legs are long, the (poorly known) arms are of a typically short protorosaur fashion. Early pterosaurs show the opposite of each of these conditions. *Sharovipteryx* also lacks an antorbital opening in its skull and has a foot skeleton quite unlike that of any pterosaur. The fact both groups possess membranous flight surfaces is also of little consequence, because flight membranes have evolved independently in dozens of unrelated animal groups, including many extinct ones. In sum, it seems unlikely that *Sharovipteryx* has anything to do with the evolution of pterosaurs, but instead represents another independent reptilian experiment with gliding flight. Indeed, the manner in which *Sharovipteryx* glided with its hindlimbs seems very unique among flying animals, and also contrasts with the way flight was achieved in pterosaurs.

It seems, therefore, that we have to look further up the reptile tree to find a likely origin point for pterosaurs. A placement for pterosaurs among Archosauriformes initially seems likely because pterosaurs share a number of features with the basal members of this group. Like pterosaurs, archosauriforms have skulls with antorbital fenestrae, reduced bone counts in their fifth toes, perforated lower jaws, and many other anatomical similarities (Nesbitt and Hone 2010). This suggests that pterosaurs are almost certainly members of this group but only one, rather controversial analysis has found Pterosauria to lie at the base of Archosauriformes rather than among the more derived archosaurs, which also bear all the features listed above (Bennett 1996a). Pterosaurs only seem to plot among early archosauriformes when their adaptations to cursoriality (that is, being able to run with upright limbs) are ignored as they were by Bennett (1996a) because of an assumption that these traits developed convergently with archosaurs. Bennett's analysis was criticized for this approach, suggesting he disregarded useful morphological data and biased the results (e.g., Benton 1999; Hone and Benton 2007). Interestingly, if Bennett's assessment is conducted without excluding hindlimb characteristics, pterosaurs are whisked up the archosauriform tree to plot within Archosauria as ornithodirans, the dinosaur sister group.

This brings us to what is probably the most compelling argument for pterosaur ancestry currently available, the notion that they are true archosaurs united with dinosaurs in the group Ornithodira. Not only do the slender proportions of *Scleromochlus* and dinosaur ancestors bear greater resemblance to pterosaur anatomy than the squat, long-bodied basal archosauriform or protorosaur species discussed above (Sereno 1991), but many aspects of their detailed anatomy are also very similar. Ornithodirans have classically been united by a set of hindlimb characters including fusion of the two proximal ankle bones to the shin; the reduction of the outer shin bone (the fibula); the structure of the foot; several limb and hip proportions; and a lack of bony scales along their backs (Gauthier 1986; Sereno 1991; Benton 1999). More recently, this list was expanded to up to 26 possible characters pertaining to the entire skeleton, including features of the neural spines; proportions of the hand, hindlimb, and trunk skeleton; construction of the skull vertebral proportions; and bone wall thicknesses. These findings counter previous criticisms that ornithodirans were merely united by features of locomotion and, to my mind, sets this hypothesis up as the one to beat. With this interpretation, it would seem that crocodiles and birds are the closest modern pterosaur relatives, and that these archosaurs provide our best insight into aspects of their soft-tissue anatomy and physiology.

The End of the Debate?

This is not to say, however, that no resistance to the ornithodiran hypothesis exisits. Bennett (1996a) argued that many of the characters purported to unite Ornithodira are only superficial, and Unwin (2005) pointed that the shield-like pelves were quite different from those of other ornithodirans. This may not be surprising, however, given that pterosaur hindlimbs were, uniquely among ornithodirans, used to support the wing in flight (see chapters 4 and 5) and their pelves may have needed broad areas to anchor strong muscles that stabilized the wing. Arguments by the same authors that ornithodirans were digitigrade (that is, standing and walking on tiptoes) bipeds and not plantigrade (standing and walking with the ankle touching the ground, as we do) quadrupeds like pterosaurs, may be flawed too, because some recently discovered ornithodirans were almost certainly quadrupeds (e.g., Nesbitt et al. 2010). Other ornithodirans, including the pterosaur relative *Scleromochlus*, are suspected of hopping about on plantigrade feet (Sereno and Arcucci 1993, 1994; Benton 1999). This latter scenario is quite intriguing given that pterosaur hindlimbs show several adaptations for powerful leaping, suggesting that this behavior may have played a role in the early development of pterosaur flight (see below; chapter 7; and Bennett 1997a).

Although strength and support for the ornithodiran hypothesis seems to be increasing as more ornithodirans and early dinosaur relatives are discovered and more complex and sophisticated analyses are performed, contention over pterosaur origins is likely to continue until the discovery of a "protopterosaur" species that fills the evolutionary distance between pterosaurs and other reptiles. Until then, the scrappy nature of the early pterosaur fossil record, the propensity of animals to evolve convergent features, and our own inability to agree on interpretations of fossils will probably continue to fuel this debate.

In the Absence of Proper Data, Speculate Wildly

The absence of a protopterosaur has not deterred pterosaurologists from making comments and devising speculations on the likely form of the pterosaur ancestor. Because their flight adaptations are much of what make pterosaurs so distinctive, such exercises in imagination are closely tied to predicting how pterosaurs learned to fly. To date, Rupert Wild has made the most thorough attempt at this in a series of papers on some of the geologically oldest known pterosaurs from Triassic rocks of Italy (Wild 1978, 1983, 1984a). Wild imagined his "protopterosaur" (dubbed "*Protopterosaurus*" by Wellnhofer [1991a]) as a very lizard-like, tree-climbing animal with membranes between its limbs and body that acted as a parachute when

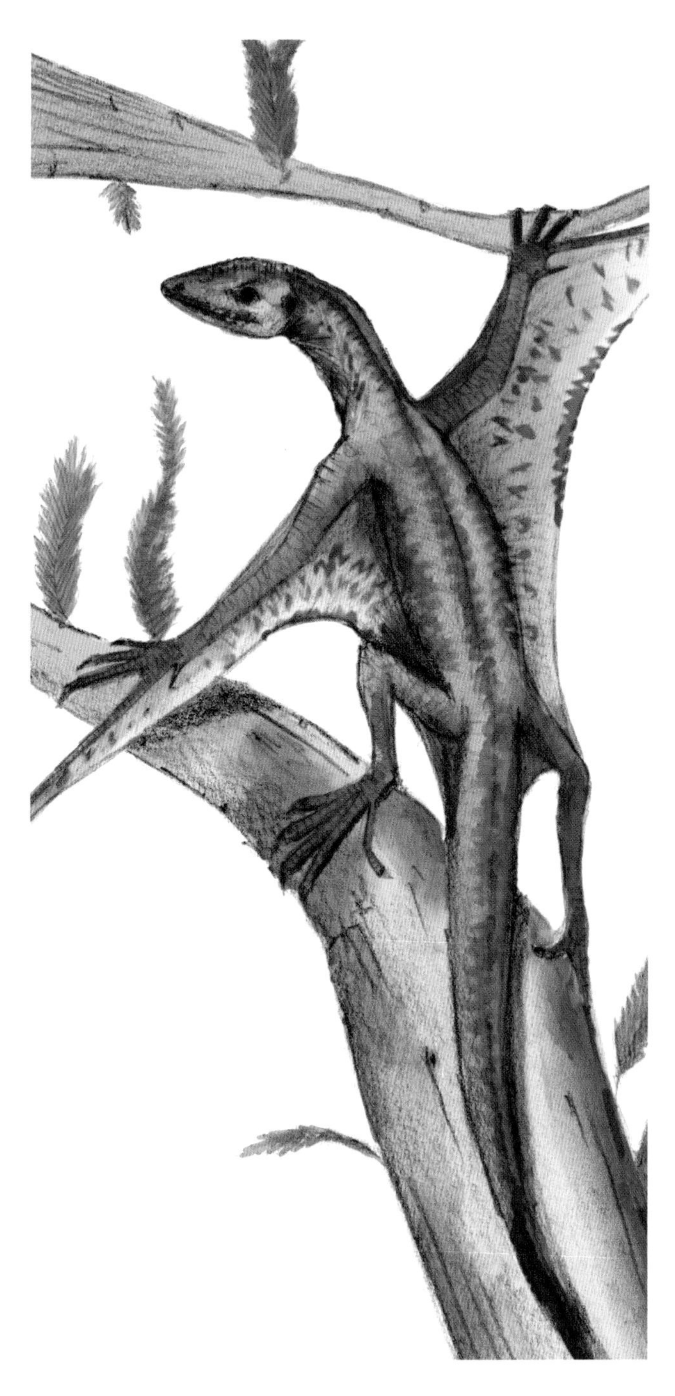

Fig. 3.5. The "*Protopterosaurus*" animal imagined to represent an early stage of pterosaur evolution by Rupert Wild (1978, 1983, 1984a). Based on Wellnhofer's (1991a) reconstruction of the same concept.

falling from trees (fig. 3.5). Over time, these limbs and membranes became larger and modest gliding abilities developed. This promoted the growth of a more efficient airfoil, and eventually the forelimb anatomy developed to the point where the protopterosaur could flap and power its own flight. Wellnhofer (1991a) and Unwin (2005) broadly agreed with this concept, though Unwin detailed that flying insects were the impetus that pulled pterosaurs into the air (see chapter 6).

Decades on, though, Wild's protopterosaur concept seems rather dated. There was little idea at all as to what sort of reptiles pterosaurs were in Wild's day, leading to the rather generalized lizard-like anatomy of his protopterosaur. We now have a better handle on possible pterosaur ancestors, and as no strong contenders for this role are particularly lizard like, protopterosaurs probably also were not. We must also consider that very few of the gliding animal lineages we know of have started *actively* flying. The distinction between gliding flight and active flight is an important one. Gliding is entirely gravity assisted, requiring only a suitably sized and shaped airfoil and an elevated starting position. There is no power input from the glider itself, and subsequently no significant lift or thrust generated that will prolong the flight path. Active, or powered, flight is far more sophisticated, with propelling wing beats that generate both lift and thrust to resist gravity and prolong flight time. Such flight is a demanding activity, the remit of energetic species that can sustain long periods of intense activity without tiring. The evolution of powered flight in birds and bats seems to have followed prior development of active lifestyles and elevated metabolisms, so the same may be true for pterosaurs. The lizard-like visage of Wild's protopterosaur does not suggest it was especially active; its sprawling limbs, heavy tail, long torso, and atrophied limb muscles indicate a more sluggish, sessile existence that may make for a fine parachutist or glider, but not for the development of active flight. Thus, we may need to reimagine the pterosaur ancestor as something that already possessed some fundamental requirements for flight. We also have to acknowledge that flight was not likely to evolve in a snap of evolutionary fingers. In all likelihood, many different stages of evolution were involved that produced a number of distinctive anatomical forms. We may have just about enough information regarding pterosaur ancestry to speculate what these stages may have been like, documenting the transition from a small ornithodiran to fully developed pterosaur. To separate this concept from that of Wild's, however, we won't refer to it as the "protopterosaur" or "*Protopterosaurus*." This is, after all, the

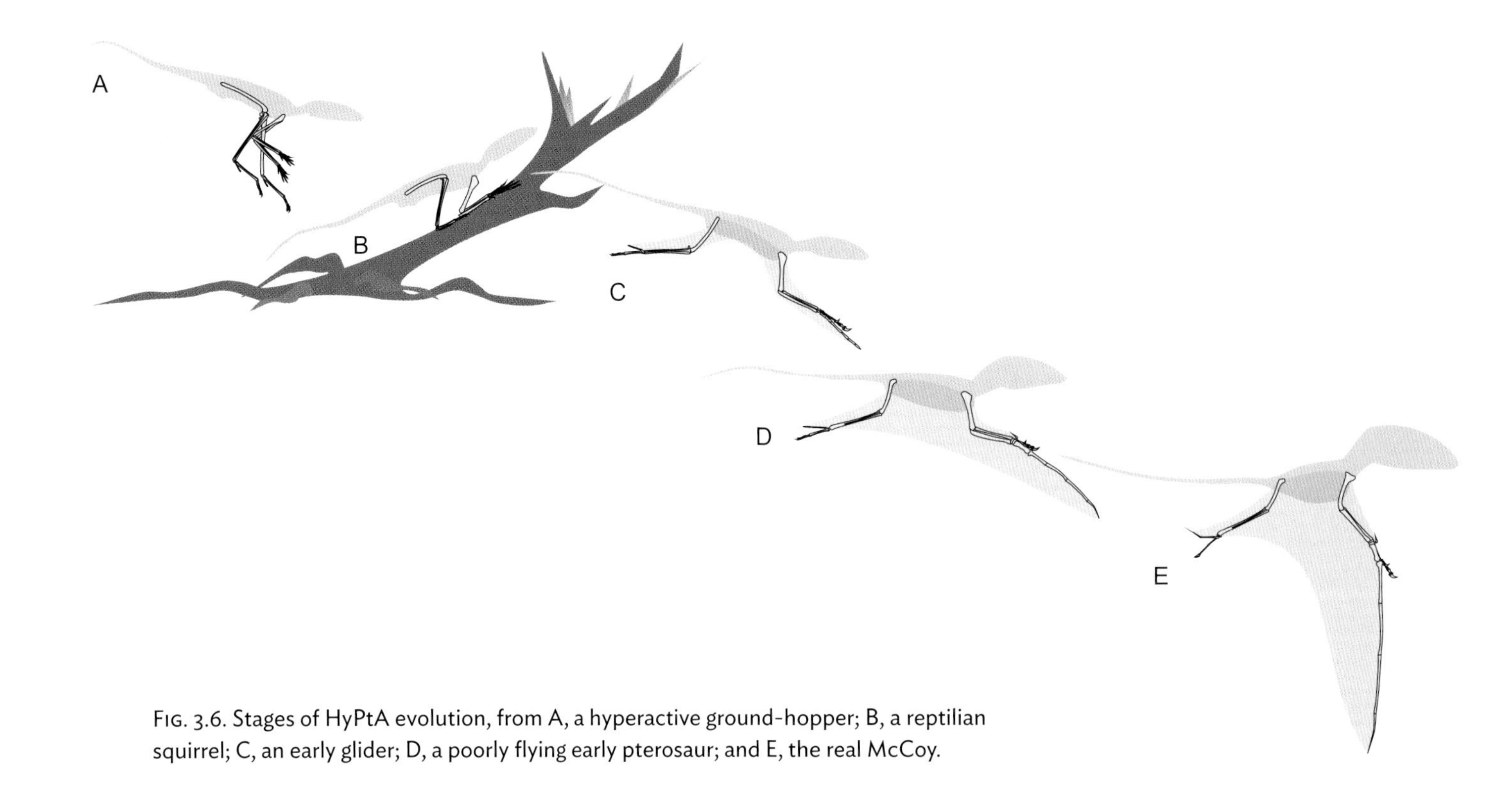

Fig. 3.6. Stages of HyPtA evolution, from A, a hyperactive ground-hopper; B, a reptilian squirrel; C, an early glider; D, a poorly flying early pterosaur; and E, the real McCoy.

twenty-first century, the time of unnecessary acronyms and cute product names like "Google" and "iPad," and txt spk wiv the kidz. With that in mind, meet the HyPtAs, the Hypothetical Pterosaur Ancestors (figs. 3.1, 3.6, 3.7).

Sketching HyPtA

Our HyPtAs are small animals, perhaps only 200–300 mm long from their snouts to tail tips. This is not only in keeping with the sizes of early pterosaurs and their likely ornithodiran ancestors but brings an immediate advantage to our goal of evolving flight: small body sizes seem to lubricate the wheels of evolution. Smaller animals reproduce and adapt faster, and are more resistant to mechanical forces relative to body mass than larger ones. These characteristics provide the useful ecological and functional plasticity necessary for rapidly developing flight. The first HyPtA (stage A in fig. 3.6) somewhat resembles *Scleromochlus* in its appearance and already has a fully erect posture in which the legs and arms extend directly beneath the body. HyPtA A bounds along when moving at speed, leaping with alternating actions of its long arms and legs. In concert with its derived stance, HyPtA A has developed a fuzzy coat in an effort to manage its body temperature. As an active animal, it burns up energy rapidly to sustain a high body temperature and attempts to avoid losing heat to the environment with its insulating coat. This metabolism is fueled with an invertebrate diet found at ground level on woodland floors, and as a primary ground-dwelling animal, it only occasionally scrabbles up low vegetation or rocks in search of additional food or safety. Like most of its reptilian ancestors, HtPtA A has five fingers, the fourth being the longest, and five toes.

HyPtA B is a lithe, agile tree-climber and, compared to HyPtA A, has longer, more robust forelimbs for this purpose. It still forages on the ground, but also finds much of its invertebrate prey in tree canopies and rocky crags. Its shoulder and hip joints are more mobile to aid climbing and it bears long, crampon-like claws on its fingers and toes, as did early pterosaurs (see Unwin 1988b; and chapters 7 and 10 for further details). Its long forelimbs make the animal rather front heavy, forcing it into a habitually quadrupedal stance. HyPtA B has maintained the bounding gait of its ancestor and uses powerful leaps to not only move quickly, but also to cross long distances between elevated positions (a behavior postulated for early pterosaurs in Bennett 1997a). The fifth finger, which was already small in its ancestors, has proved of little use in climbing and has become vestigial. It will be lost

Fig. 3.7. Life restoration of HyPtA D. Living in complex, tiered habitats like rocky canyons (as shown here) or forest canopies may have been a driving force in the development of pterosaur flight. It is much easier and safer to jump, glide, or fly across a gap than it is to climb down, run across, and climb up the other side again, after all.

entirely in HyPtA B's descendants. The fifth toe, however, is a useful stabilizing tool and has become more mobile and elongate, while the tail is now a useful balance when climbing, and has become lighter and longer.

By the next stage, HyPtA C has not only further elaborated the climbing adaptations seen in its stage B predecessor but shows a significant step toward flight by bearing flaps of elastic skin between its body and limbs (fig. 3.1). These membranes, located between the shoulder and wrist, between the legs (and spanning each side of the body between the legs, including the fifth toe), body, and arms (including the hand and fourth finger), provide HyPtA C with crude gliding and parachuting abilities that extend its leaping distances between elevated positions and cushion falls. The frequent use of the membranes to prolong forward motion has refined their shape to a more winglike guise, which is quite different in profile to parachutes suited to merely slowing vertical descent. The fourth finger is the primary agent in making the membranes more winglike, elongating them to produce a relatively narrow distal wing portion. Such a large finger has required that its supporting bone in the hand, the fourth metacarpal, must also be robust, and it now dominates HyPtA C's hand anatomy. The fifth toe, another membrane-supporting digit, has become more mobile and elongate to support the trailing edge of the posterior membrane. The mobile limbs developed by HyPtA C's ancestors have paid off here, as they free both limb sets to rotate out sideways and create efficient airfoils.

A far more pterosaur-like appearance is achieved in HyPtA D (figs. 3.6D and 3.7), thanks to further elaboration of the gliding apparatus seen in HyPtA C. The forelimb, fifth toe, and fourth finger are now even longer and more robust, but the fourth finger has become so large that it is unwieldy in terrestrial locomotion and must be stowed away when the animal is grounded. Accordingly, the grooves in the end of the fourth metacarpal that dictate the articulatory range of this "wing finger" have extended and now permit rotation of their distal airfoil behind the hand, parallel to the long axis of the body. In this position, the wing finger is less prone to damage and allows the animal to move more easily in cluttered spaces (Bennett 2008). The claw on the fourth finger has also been lost, since it bore no function and only snagged on vegetation as the animal moved around (Bennett 2008). Modified wrist bones have begun to tension the anteriormost membrane and, in time, will develop an ability to change the camber of this forewing. The flight membranes have also become more sophisticated, tightening as elastic filaments and thick fibers stiffen their trailing, unsupported regions, thereby reducing fluttering of the membranes and increasing aerodynamic performance.

Perhaps the most progressive anatomical changes in HyPtA D, however, concern its shoulders. The long evolutionary history of climbing in the HyPtA lineage has granted HyPtA D a strong set of pectoral muscles that not only help it to grapple trees and rock faces, but are now useful for manipulating its wings and controlling its glide path. At some point, manipulation of these wings in the vertical plane produced flapping, and self-propellant flight was achieved. The heightened metabolism ancestral to HyPtA D is critical here because only this could power sustained flapping, and thus power bouts of flight. Accordingly, the shoulders and chest of HyPtA D are relatively larger and deeper than its ancestors, reflecting development of early flight musculature. The increasing strength of the shoulders and forelimbs eventually allowed them to leap from a quadrupedal start into the air from flat, unelevated positions (see chapter 7) and to perform fully flapping flight. At this point, the flight of our pterosaur ancestor would probably be relatively crude, short-lived affairs, but this would be honed and refined in its descendants until they were capable of efficient, maneuverable, and prolonged flight strategies. Flight for these early pterosaurs is only a means of transportation as it is too clumsy for chasing agile insects on the wing (see chapter 6), so HyPtA D still forages terrestrially. Nevertheless, this ability to fly may see us classifying HyPtA D as a bona fide pterosaur, because the attainment of powered flight is often used as a defining characteristic for birds, and could be equally applied to pterosaurs. In time, the membrane-supporting wrist bones would become the characteristic pteroid bone (see next chapter), the shoulders would deepen to allow for even larger flight muscles, and the wings would become longer, more efficient airfoils. With no immediate competitors for domination of Mesozoic skies, we may imagine that the first pterosaurs diversified and spread rapidly, perhaps quickly developing into the early pterosaurs we find in Late Triassic rocks today.

4
The Pterosaur Skeleton

The heart of pterosaur research beats through our interpretations of their skeletal anatomy. Aside from rarely preserved soft tissues and pterosaur footprints, virtually everything we know about pterosaurs has been derived from studying their fossil bones (figs. 4.1 and 4.2). Accordingly, anyone with a keen interest in these animals must become familiar with their osteology, so we will devote the next few pages to this task. There are two main aspects to cover here: the bones that comprise the pterosaur skeleton, and the unusual construction of the bones themselves. In the interests of brevity and readability, I've not cited specific literature for much of this review, nor listed particular reference specimens, but more detail can be gleaned from the comprehensive osteologies of several pterosaur species provided by Wellnhofer (1970, 1975, 1978, 1985, 1991a, 1991b), Bennett (2001), Unwin (2003, 2005), and Kellner (2003). Readers seeking anatomical details of specific pterosaur bauplans can find them in the discussions of each pterosaur group in chapters 10–25.

Seeking Direction

Before we begin, we should ensure that we are familiar with the biological terms we will use to describe the pterosaur skeleton. Anatomists use a series of standardized phrases for directing their way around vertebrate bodies, and although this does involve learning a little jargon, I assure you this is far simpler than tying ourselves in grammatical knots trying to avoid them! Beginning with the obvious, *anterior* and *posterior* refer to the head and tail ends of the body respectively. *Ventral* refers to the underside of the body, *dorsal* is the opposite. *Lateral* projects away from the spinal column at a perpendicular angle, and *medial* is the inverse. *Proximal* and *distal* dictate being closer to or further away from the center of the body, respectively. These terms are often combined to describe anatomy more accurately. For example, a feature may be said to be *dorsoventrally* flattened, or projecting *anterolaterally*. Though perhaps a little daunting at first to the uninitiated, these terms should quickly become second nature, and when talking about animal anatomy in the future, you'll wonder how you ever got by without them.

The Skull

Pterosaur skulls and jaws are probably more diverse than any other parts of their bodies (fig. 4.3). All pterosaurs have proportionally large skulls, and increasingly so throughout their evolutionary history. As with all vertebrate skulls, pterosaurs comprise their skulls from a series of bones that are mirrored along the body midline and arranged a little differently in each species. In juvenile pterosaurs, the divisions between each bone, known as sutures, are clear and most bones can be easily identified. In adult pterosaurs, the skull bones are fused together so extensively that all traces of their sutures are indiscernible, and individual bones are difficult to identify.

The first pterosaurs bore relatively tall, stunted skulls (fig. 4.3A), but as the group evolved, their skulls were generally elongated so that most forms bore low, pincerlike jaw tips (fig. 4.3B). The rostrum, defined here as the portion of skull in front of the nasal opening (as in Martill and Naish 2006), typically extends in line with the rest of the jaw but may be deflected dorsally, ventrally, or is sometimes strangely curved. All pterosaurs possess a nasal opening that housed their nostrils. An antorbital fenestra, the characteristic skull opening of Archosauriformes discussed in chapter 3, is located behind the nasal opening. A large number of pterosaurs combine their nasal and antorbital openings into one single perforation, the nasoantorbital fenestra. Many pterosaur fossils show a partial division of this opening toward the back of the fenestra, the existence of which shows that the "nasal" portion of the nasoantorbital fusion was much larger than the antorbital.

The orbit, or eye socket, sits toward the back of the skull. These are generally circular, oval, or pear-shaped, and are large when compared to body size, but are dwarfed by the inflated skull proportions of many pterosaurs. (As we'll see below, many oversize aspects of pterosaur skeletons are "inflated" by air sacs, or pneumatic tissues, creating the illusion that

FIG. 4.1. The petrified remains of rotting, decaying pterosaur bodies are now the only evidence of their existence. Although other pterosaur tissues are known to paleontologists, their fossilized bones, such as this *Dsungaripterus* skull, are their most commonly preserved components.

their skeletons are comprised of much heavier bones than they actually were. Much of the skull, for example, would be full of air in life and very lightweight for its size.) A sclerotic ring, a series of tiny bone plates arranged in a circle, support the eyeball. Two further skull openings mentioned in chapter 3, the upper and lower temporal fenestra, are seen behind the orbit. The upper is typically larger than the lower, and in some forms the lower is almost closed. The braincase is located between the eye sockets and the upper temporal fenestra, and as with the orbits, is dwarfed by the size of the skull but generally quite generous compared to comparably sized reptiles.

The medial margin of the upper temporal fenestra is often confluent with the base of a supraoccipital cranial crest, a variably shaped cranial ornament that emerges from the posterodorsal region of some pterosaur skulls. Other crests may emerge further forward on the skull, sometimes projecting dorsally from the rostrum and extending along the skull midline until they reach the relatively bulbous cranial regions (fig. 4.4). Numerous variations on pterosaur crest structure are now known, and it seems that many—perhaps most—pterosaurs bore cranial crests of one description or another. Recent pterosaur discoveries have revealed that many pterosaurs crests are comprised of both bone and soft tissues, with the bony element often reduced to a low ridge of fibrous bone that supported much larger soft-tissue expanses. Despite their often expansive lateral areas, the crests are typically only millimeters thick and project abruptly from the skull midline when viewed in anterior or posterior profile.

The posterior, or occipital, face of the pterosaur skull is vertical in some early pterosaurs, but as the group evolved, it reclined to an almost subhorizontal

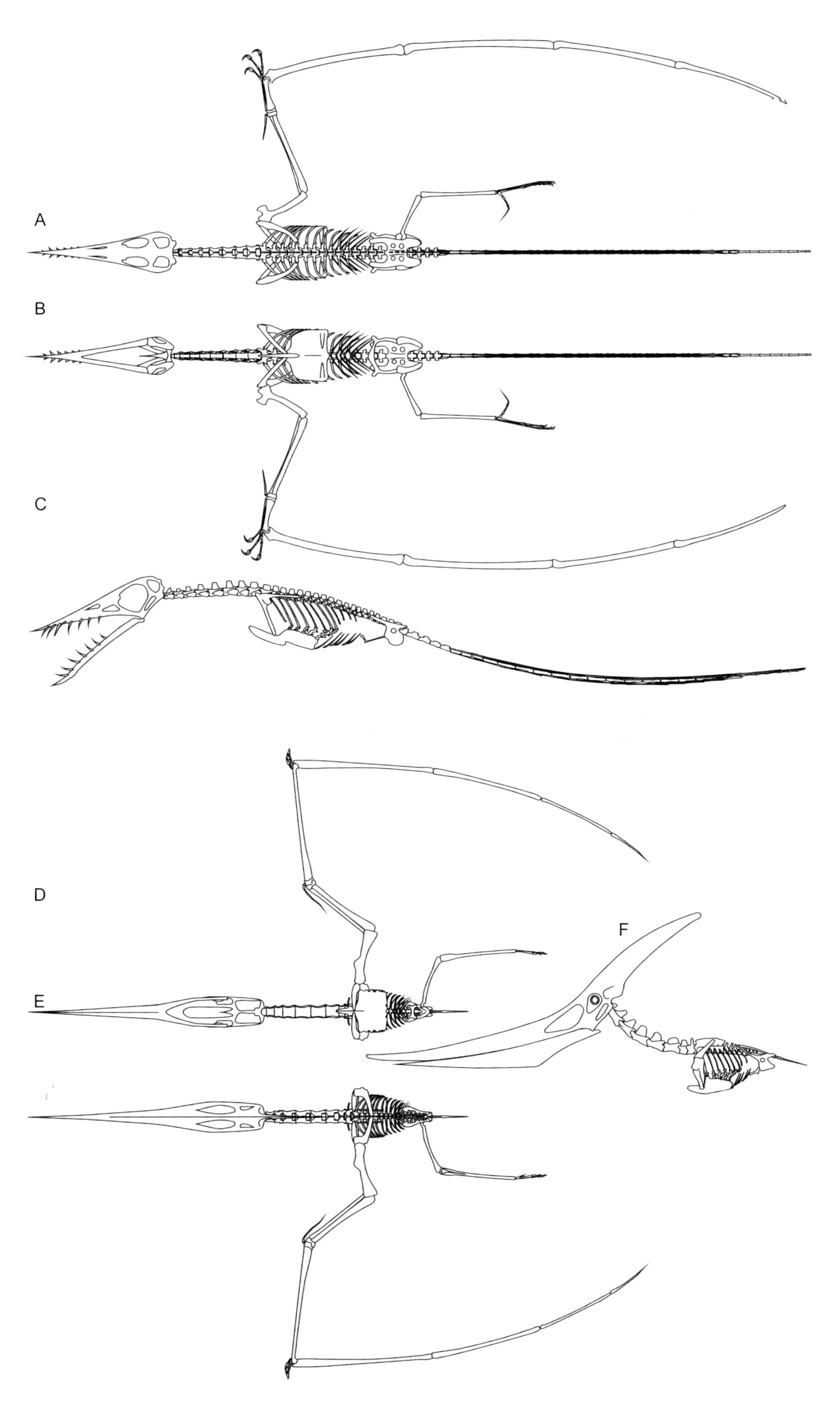

FIG. 4.2. Skeletal anatomy of two very well-known pterosaurs. A–C, the long tailed Jurassic pterosaur *Rhamphorhynchus muensteri* in dorsal, ventral, and left lateral view (respectively); D–F, the short-tailed Cretaceous species *Pteranodon longiceps* in the same views. A–C, based on Wellnhofer (1975) and Claessens et al. (2009); D–F, based on Bennett (2001) and Claessens et al. (2009). Images not to scale.

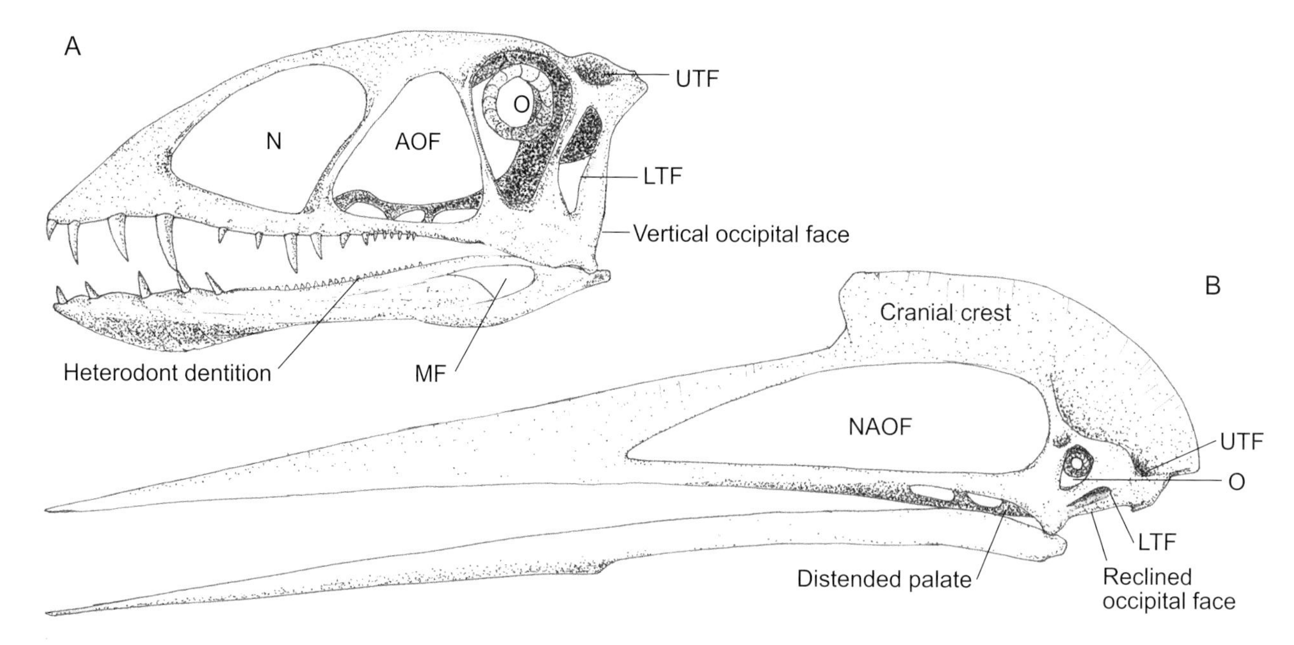

Fig. 4.3. Pterosaur skull diversity and osteology. A, the early pterosaur *Dimorphodon macronyx*, demonstrating a separate nasal and antorbital opening; B, the skull of the derived pterodactyloid *Quetzalcoatlus* sp., showing a confluent nasal and antorbital opening, and toothless jaws. AOF, antorbital opening; LTF, lower temporal fenestra; MF, mandibular fenestra; N, nasal opening; NAOF, nasoantorbital opening; O, orbit; UTF, upper temporal fenestra. A, based on specimens; B, based on Kellner and Langston (1996). Images not to scale.

angle in some forms (see fig. 4.3A–B). There is a fair amount of variation in the orientation of the occipital face in pterosaurs, which means the occipital condyle, a subspherical projection on the posterior face of the skull that articulates with the neck vertebrae, projects at many different angles in different species. In early forms, it projects directly behind the skull, but can face posteroventrally or almost entirely ventrally in more derived species. The upshot of this is that the neck does not articulate with the skull in the same place in all pterosaurs, which may betray some variation of habitual head posture in different forms. The jaws articulate with a hinge that is either in line with or somewhat ventrally displaced from the jawline. Typically, the jaw joint region is the widest point of the skull with the orbital and temporal regions marginally narrower. The hinge itself is formed by a roller joint that is often slightly asymmetrical, meaning that the lower jaw of some species bulged slightly outward when opened, possibly facilitating the ingestion of slightly larger prey items.

The palatal surface of pterosaur jaws is extremely variable. The biting, or occlusal, surfaces of most pterosaur jaws are flattened, but some species have scissorlike, ventrally tapered jaw tips. Some taxa have prominent ridges extending along the rostral palatal midline, which may continue posteriorly into a markedly distended palatal region that projects well beneath the lateral jaw margins. Other pterosaurs show contrasting conditions where the palatal surface is strongly concave, or may have a flattened, unremarkable oral roof. In all pterosaurs, holes in the posterior palatal region allow passage of jaw muscles and other soft tissues from the temporal region to the lower jaw.

MANDIBLE

Like the skull, the lower jaw, or mandible, is also comprised of several bones, but is a much simpler element than its dorsal counterpart. Each side of the mandible is comprised of both a mandibular ramus, a bony strut that extends from the anterior jaw region to the jaw joint, and a mandibular symphysis, a region formed from the fusion of the mandibular rami at their anterior ends. In some forms, this fusion is so short that the jaw forms a U or V shape, but other pterosaurs have up to 60 percent of the anterior mandibular fused together. Some pterosaurs develop keel-like crests beneath their mandibular symphyses, the size and extent of which are very variable between species.

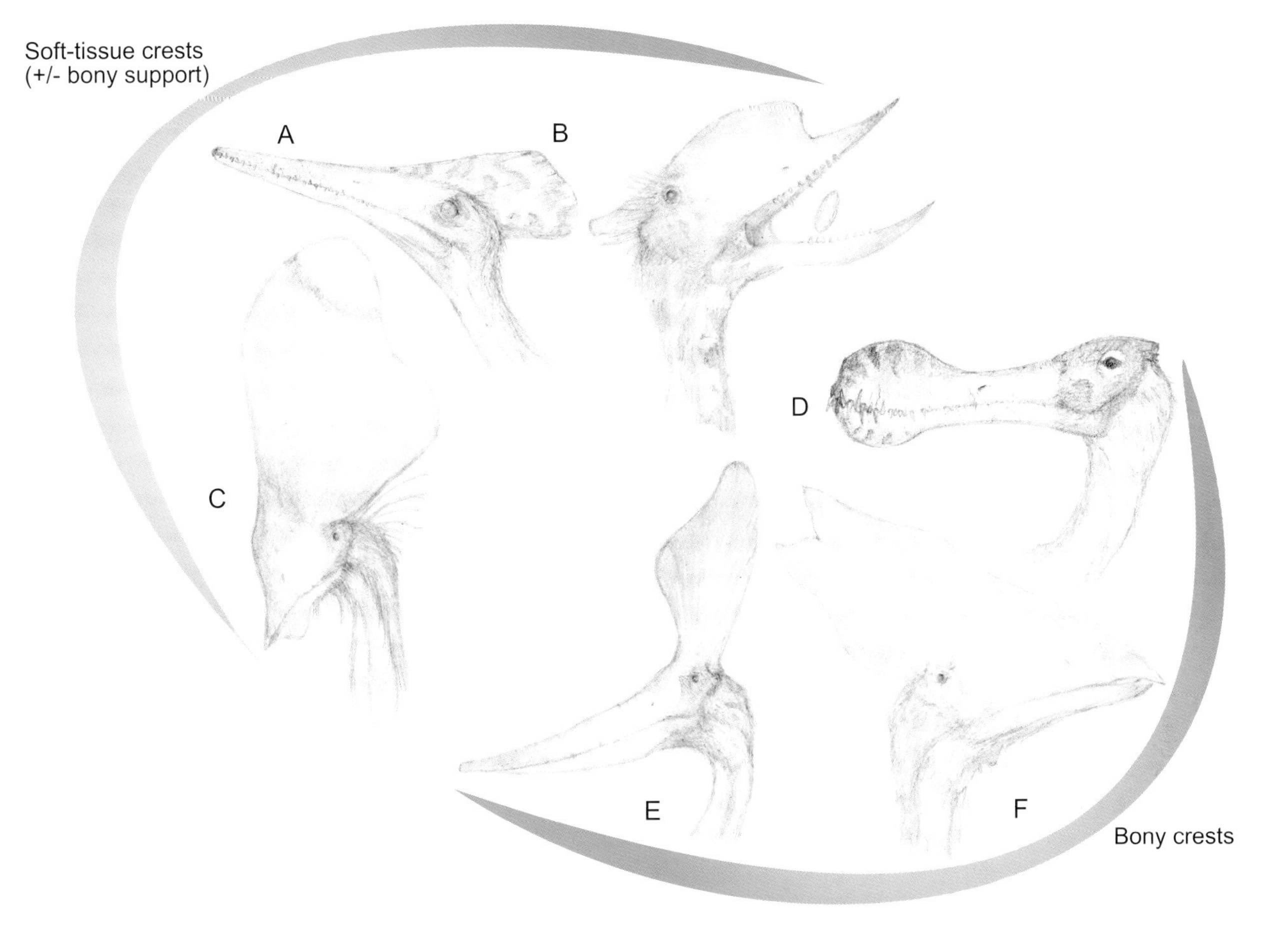

FIG. 4.4. Pterosaur headcrest diversity. A, *Pterodactylus antiquus*; B, *Dsungaripterus weii*; C, *Tupandactylus imperator*; D, *Ornithocheirus mesembrinus*; E, *Pteranodon sternbergi*; F, *Thalassodromeus sethi*. Sketches not drawn to scale.

The jaw joints articulate with the skull via depressions in the posterior region of the mandibular rami, behind which lies a posteriorly directed extension of bone known as the retroarticular process. The size of this process is quite variable, but they are relatively small in most pterosaurs. Some early pterosaurs possess an opening posteriorly in each of their mandibular rami, known as the mandibular fenestra. This feature was lost early in the evolutionary history of the group and most species possess no trace of this opening.

TEETH

Not all pterosaurs possess teeth, and instead posess birdlike, toothless beaks. Toothed pterosaurs, however, all possess dentition with enamel restricted to the tooth tip, and like all reptiles, replaced their teeth continuously throughout their lifetime. The mechanism developed for this is quite unusual, however, with replacement teeth emerging *behind* the older teeth, not directly *underneath* them as in other archosaurs. It is thought that this mechanism dampened the effects of tooth loss, because replacement teeth achieved 60 percent of their maximum size before the old tooth was shed (Fastnacht 2008). Thus, pterosaur jaws always maintained a more or less fully functioning array of teeth.

Pterosaur teeth are highly varied in shape and size to the extent that pterosaurologists often use them to distinguish different groups. Tooth size, and the distribution of differently sized teeth, can vary dramatically between species, but there is a general trend in tooth size reduction toward the back of the jaw. Some pterosaurs reverse this condition, or possess their largest teeth at the middle regions of their toothrows. Tooth counts differ considerably between pterosaur species, ranging from only a handful in some taxa to hundreds or even a thousand in the most extreme forms. Like other reptiles, tooth counts increase with age in most pterosaur species. Perhaps the commonest pterosaur tooth morphology is a laterally compressed cone, a shape ideally suited for holding small prey, but the teeth of other lineages were modified for different diets. Different species possessed large, curving tusks for spearing sizeable prey; highly compressed teeth with serrated or "razor" edges for slicing through soft tissues; multicusped teeth for chewing: or extremely fine, needlelike teeth that were tightly packed into the jaws for filter feeding. Perhaps the most unusual pterosaur dentition belongs to a group that deposited bone over their short, broad teeth to completely encase them within their jaw bones.

The Vertebral Column

As with all reptiles, pterosaurs have four types of vertebrae in their spinal column: cervical vertebrae in their neck (fig. 4.5), dorsal vertebrae in their torsos (fig. 4.6), sacral vertebrae in their hip region (fig. 4.6), and caudal vertebrae in their tails (fig. 4.7). Total vertebral counts range from around 34 to 70, with these differences primarily caused by the variable caudal counts between species. The vertebrae of each region share the same basic morphological features, but their development varies along the vertebral column. The main vertebral body, the centrum, is crowned with a neural arch. This holds an opening at each end to form the neural canal for the spinal chord. Atop the neural arch sits the neural spine, a tall, blade-like structure that extends along the vertebral midline. Transverse spines project laterally from the neural arch and centra, providing facets for articulating ribs. In all but the caudal vertebrae, the centrum extends posteriorly beyond the rest of the vertebra as a prominent condyle, which articulates with a cavity in the preceding vertebrae, the cotyle. Polished surfaces mark where these vertebrae articulate with one another, and can give a rough indication of how much movement was permitted at each vertebral joint. Pterosaur cotyles often possess anteriorly deflected ventral processes, the hypapophyses, which limit downward flexion of the vertebrae. In some pterosaurs, these insert between bulbous, posterolaterally projecting exapophyses of the corresponding condyle. Above the centra, zygapophyses project anteriorly (the prezygapophyses) and posteriorly (postzygapophyses), and articulate with their neighboring zygapophyses along smooth surfaces. Several openings and foramina are found across pterosaur vertebrae, but their positions, number, and size are highly variable.

CERVICAL VERTEBRAE

Pterosaur cervical vertebrae are generally the largest and most complex elements of the entire axial column. There are nine in total, with seven "free" cervicals, and two "dorsalized" cervicals incorporated into the dorsal column. The cervicals are often the longest vertebrae in the body and, in some long-necked pterosaurs, may be eight times longer than wide (e.g., fig. 4.5C–E). Many have prominent perforations on their lateral faces, and around the openings of the neural canal. These "pneumatic foramina" permit invasion of pneumatic tissues into the bone (see chapter 5) and their distribution can characterize different pterosaur groups. The transverse processes are generally small or entirely reduced (apart from those in the dorsalized cervicals 8 and 9), but neural spine morphology is very variable. Some are tall and blade-like, others long and low, and some practically absent. The cervicals of early pterosaurs retain facets for the articulation of cervical ribs (fig. 4.5A, B), but these are lost, along with the cervical ribs themselves, in more derived forms.

The articular faces between pterosaur cervicals are broad and, because of this, we assume that most had reasonably flexible necks, though the tight interlocking between some long-necked pterosaur cervicals may have limited their motion somewhat (e.g., Wellnhofer 1991a; Frey and Martill 1996). Paleontologists often struggle with ascertaining how mobile the necks of fossil animals were, and how they habitually held their necks when going about their daily business. This is partly because a crucial component in determining the range of motion between articulating bones, the cartilaginous caps that exist between joint surfaces, are frequently not preserved in fossils. These are essential components for determining the range of motion between joints, and without them paleontologists are left to glean what arthrological data they can from the bones alone. The position of cervical articular faces in pterosaur neck vertebrae suggest they were more capable of arching their necks skyward than looking at their bellies, but the tight fits between each vertebra appear to have limited their ability for rotation.

Articulated pterosaur neck fossils confirm that many species could arch their necks high without disarticulating their vertebrae (e.g., Wellnhofer 1970; Bennett 2001), though how often they assumed such postures is not clear. It seems that all nonmarine tetrapods generally hold their necks in a S shape when awake and alert, and not engaging in an activity requiring use of their necks (e.g., feeding, some forms of locomotion). In such postures, the neck base is angled up, the midlength is held more or less straight, and the anterior neck bones are flexed somewhat downward (Vidal et al. 1986; Graf et al. 1995; Taylor et al. 2009). This may seem counterintuitive when we consider the appearance of some of the pets and livestock that we are familiar with, but x-rays of these animals reveal that their neck soft tissues are obscuring the same basic configuration seen in all other nonmarine tetrapods (this also includes your neck, by the way). Given how ubiquitous this neck posture is across modern, distantly related tetrapods, it seems likely that it is a deeply rooted trend of tetrapod evolution and probably also applies to all extinct nonmarine tetrapods, including pterosaurs. If this is the case, we may even assume that pterosaur necks were probably elevated for much of the time, as the S-shaped postures of modern tetrapod necks are more pronounced in species with higher basic metabolic rates (Taylor et al. 2009), which were probably also found in pterosaurs.

DORSAL VERTEBRAE

It is often difficult to determine where pterosaur dorsal vertebrae stop and their sacral series begin, because their hip vertebrae can often "capture" posterior dorsal elements in older individuals. This leaves some

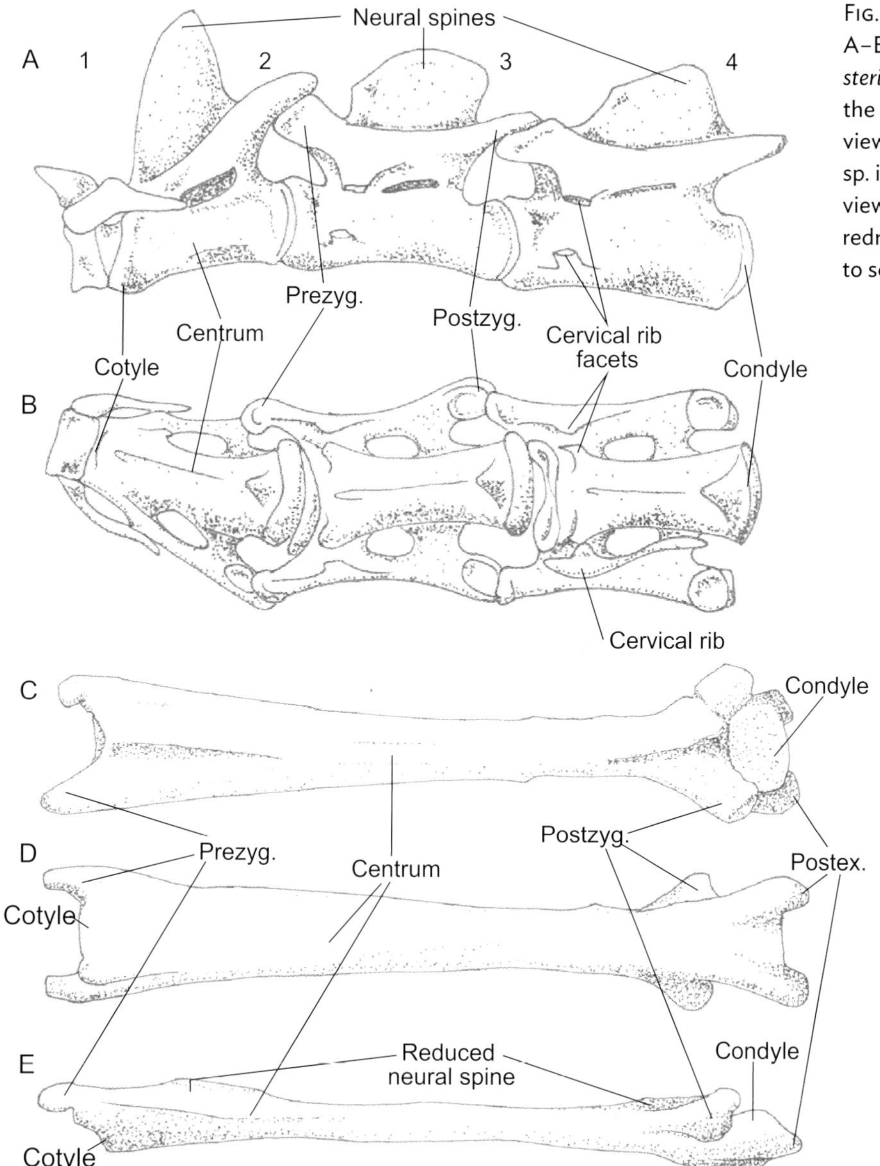

FIG. 4.5. Anatomy of pterosaur cervical vertebrae. A–B, anterior cervicals of *Rhamphorhynchus muensteri* (numbers denote position of each vertebra in the cervical series) in left lateral (A) and ventral (B) view; C–E, fifth cervical vertebra of *Quetzalcoatlus* sp. in dorsal (C), ventral (D), and left lateral (E) view. A–B, redrawn from Wellnhofer (1975); C–E, redrawn from Witton and Naish (2008). Images not to scale.

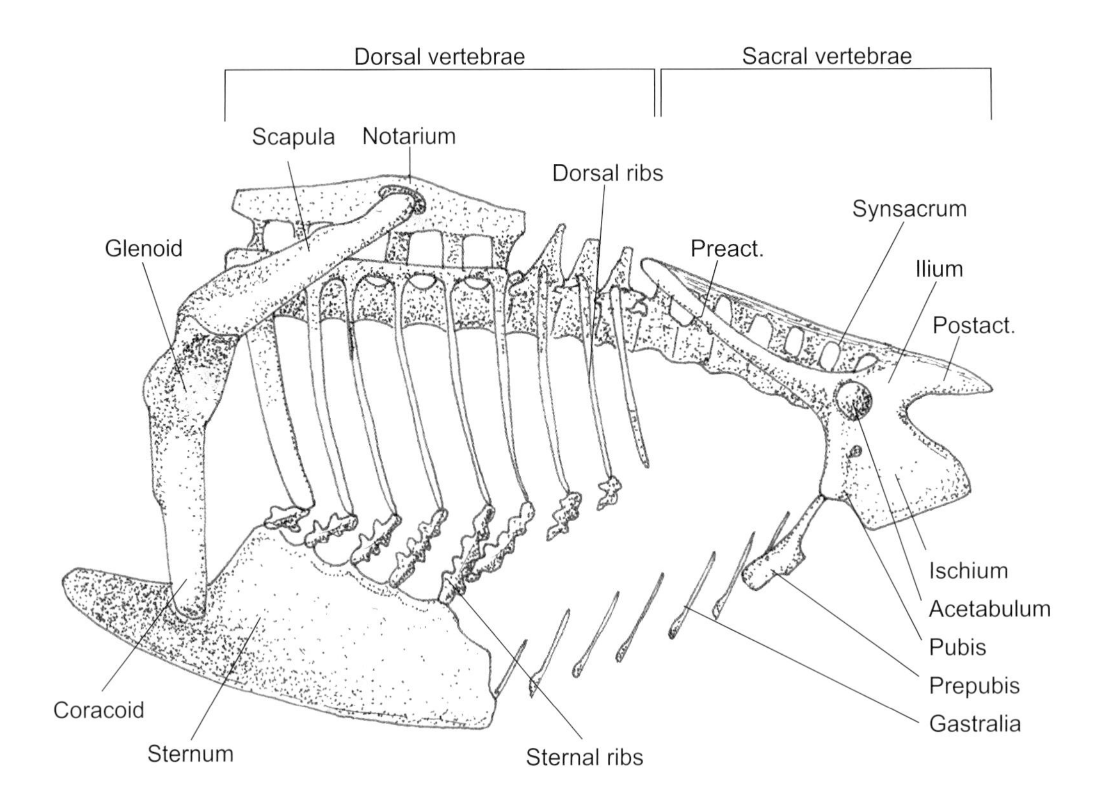

FIG. 4.6. Detailed torso, pectoral and pelvic anatomy of *Pteranodon* sp. Preact. denotes preacetabular process; Postact., postacetabular process. Based on Bennett (2001) and Claessens et al. (2009).

mystery as to how many dorsals originally existed in many species, but up to 18 have been counted in some. The dorsals themselves are generally quite uniform in size and shape with short centra, tall neural spines, and pronounced transverse spines that support large dorsal ribs, with some decrease in process size occurring posteriorly. In some large, fully grown pterosaurs, at least the anterior three dorsal centra fuse together to solidify the axial column over the shoulder region. This fusion may occur through ossification of tendons along the top of the neural spines, or through anteroposterior expansion of the neural spines until they grow into their neighbors. This fusion forms a structure known as the supraneural plate, a long slab of bone that extends across the neural spines and may articulate with the shoulder blades via small recesses on each lateral surface. Further fusion among the anterior dorsals occurs between the chunky anterior dorsal ribs and their corresponding centra. The resultant structure, a complex mass of fused vertebrae and ribs, is known as the notarium (see fig. 4.6), and in particularly old pterosaurs, it may encompass as many as seven anteriormost dorsal vertebrae.

SACRAL VERTEBRAE

The sacral verebrae sit between the bones of the pelvis, completely fused with each other and, via their transverse spines, the dorsal pelvic bones. The sacrum, the name for this collection of fused sacral vertebrae, is comprised of at least three or four sacrals in all pterosaurs, but some mature forms incorporate numerous posterior dorsals to raise the sacral count to nine. In pterosaurs with particularly well-developed notaria, a second supraneural plate is seen extending along the top of the sacral neural spines (Hyder et al., in press). This does not seem to contact the notarial supraneural plate in any pterosaurs, however, leaving about four dorsals free of neural fusion.

CAUDAL VERTEBRAE

The number of vertebrae in pterosaur tails ranges from as few as 10 in some species to 50 in others. They are quite diverse in size and shape, ranging from elongate tubes with incredibly long zygapophyses, which overlap several neighboring caudals, to very short, featureless cylinders. The former are found in many species of long-tailed pterosaurs and, together with elongate bones known as chevrons positioned underneath the caudal centra, form a tail of similar flexibility to a riding crop. Presumably, these long tails had some inherent rigidity but were capable of bending elastically when swished around or brought into contact with other objects. The centra of most pterosaur caudals do not have the well-formed condylar/cotyle articulation

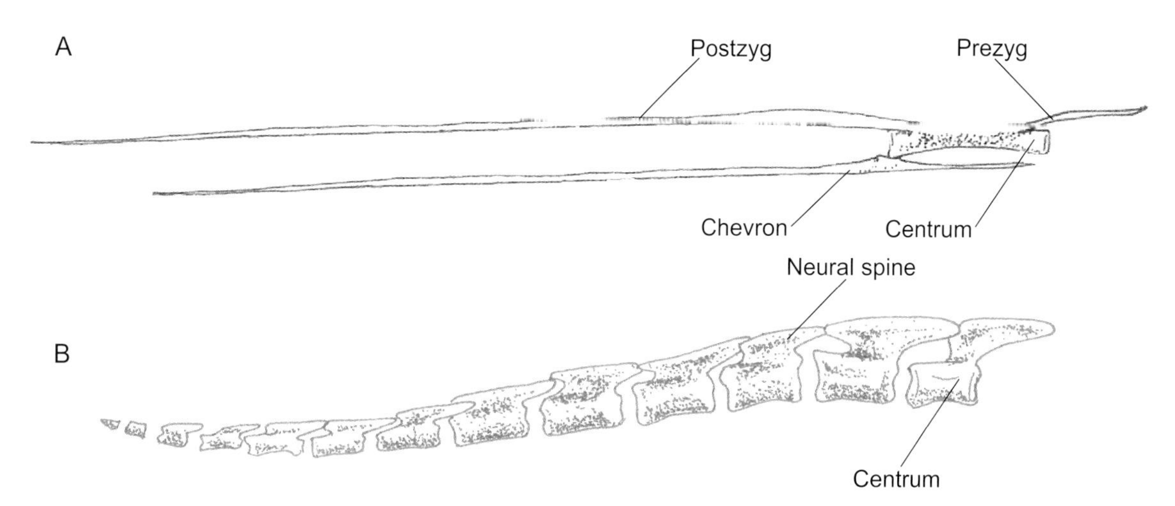

FIG. 4.7. Pterosaur caudal vertebrae. A, right lateral view of midseries caudal of *Rhamphorhynchus muensteri*; B, caudal series of *Pterodactylus* sp. Images not to scale. Redrawn from Wellnhofer (1991a).

of the other vertebrae, so could probably move rather freely in all planes.

RIBS

Only the cervical and dorsal vertebrae of pterosaurs bear ribs. The cervical ribs are always much shorter than the dorsal ribs and appear to be absent altogether in many species. Recent discoveries show that these "absent" ribs may be present in juvenile pterosaurs, but are extremely reduced and fuse to the centrum at an early age (Godfrey and Currie 2005).

The dorsal ribs, and particularly the anterior two pairs, are the largest ribs in the pterosaur body. The anterior ribs form broad lateral arcs from their double-headed articulations with the dorsal vertebrae and extend toward the breastbone, or sternum. They do not quite contact this element, however, instead articulating with short, flat bones with lobed margins known as sternal ribs. These small bones complete the bony arc between vertebral column and breastbone and make the anterior trunk skeleton rather rigid (fig. 4.6). Behind the sternum, the dorsal ribs and sternal ribs meet V-shaped belly ribs, or gastralia, that lay in the soft tissues lining the abdominal region. The ribs of the posteriormost dorsals are either rather tiny or may be absent altogether.

Pectoral Girdle

As is typical of flying animals, the pterosaur shoulder skeleton, or pectoral girdle, is large and robust (fig. 4.6). Each side of the girdle is formed from two stout bars, the dorsally situated scapula (shoulder blade) and coracoid. These elements are often fused around the shoulder joint, or glenoid, forming the continuous "scapulocoracoid." The pectoral girdle is morphologically complex, assuming a broad "V" or "U" shape in anterior or posterior view and, in most species, a lopsided boomerang shape when viewed laterally. The scapula, which is usually the longer of the two pectoral bars, mostly lies flat along the laterodorsal face of the dorsal ribs and, in some pterosaurs, articulates directly with the supraneural plate of the notarium. In most pterosaurs, the scapula extends anterolaterally from the vertebral column when viewed dorsally, but this was reconfigured in some pterosaurs to project at a right angle from the backbone.

The glenoid is situated entirely on the scapula in early pterosaurs, but migrates ventrally to span the junction between the scapula and coracoid in more derived forms. The glenoid is a rather complex, saddle-shaped joint that allows the forelimb considerable movement and rotational capacity. The forelimb can move through an arc of 90° when extended laterally (Padian 1983a; Wellnhofer 1991a; Bennett 2001) and, with the articular face of the glenoid wrapping onto the posterior face of the scapulocoracoid, it can be rotated toward the body to a 10° angle from the spinal column. In this position, the forelimb can also be swung anteroposteriorly beneath the shoulder in most pterosaurs (Bennett 1997b; Unwin 1997), though some species have abbreviated articular faces that restricted ventral movement of the forelimb (Wellnhofer 1970).

The coracoid articulates with the breastbone, or sternum, via laterally facing notches at the front of

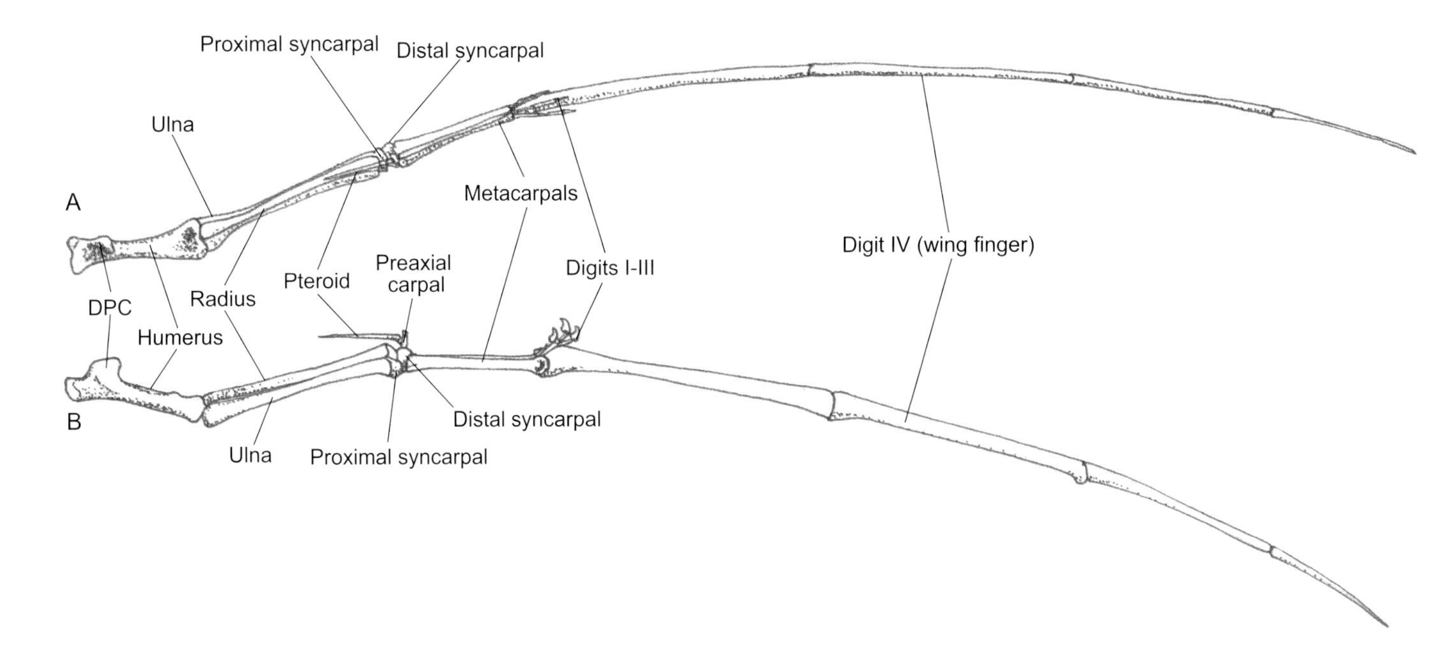

Fig. 4.8. Pterosaur forelimb osteology, as demonstrated by the forelimb of the ornithocheirid "*Santanadactylus pricei.*" DPC denotes the deltopectoral crest. A, anterior view; B, dorsal view. Redrawn and reversed from Wellnhofer (1991b).

the sternal plate. These are sometimes asymmetrically positioned, one being slightly in front of the other. Pterosaur sterna are necessarily large bones, occupying considerable proportions of the torso with large areas for anchoring substantial flight muscles. They are generally broad and dished with shallow keels along their ventral midlines. Each has a prominent blade of bone, the cristospine, projecting anteriorly from the coracoid articulation. The posterior sternal margins are variable in shape, being square, rounded, or triangular, and their lateral margins are sometimes lobed, with each promontory providing an articulatory facet for a sternal rib. Recent research on pterosaur chest construction suggests that the sternum is dorsally inclined when fully articulated with the coracoid and ribs, rather than being held horizontally within the body as traditionally assumed. This makes the posterior sternal region the deepest part of the pterosaur chest (Claessens et al. 2009).

Forelimb

The pterosaur forelimb contains the same bones that you have in your own arms, although they are so modified for flight that you could be forgiven for not recognizing them. At full extension (fig. 4.8), the forelimb bones project out from the pterosaur body to form the long leading edge of the wing, but the entire arm could be neatly folded to permit locomotion on the ground.

The stoutest, most proximal bone of the forelimb is the humerus, a relatively short element with a robust, saddle-shaped proximal articulation. The most obvious feature of this bone is the deltopectoral crest, a large, forward-projecting, and proximally located flange that anchored powerful flight musculature extending from the shoulders to the forelimb. This crest is variable in size and shape and is very characteristic for specific pterosaur groups. Many pterosaurs have pneumatic foramina located on their proximal humeri, and their location and number are useful criteria for distinguishing pterosaur groups. The distal end of the humerus has two prominent condyles that wrap around the anterodistal surface to form the elbow joint. It seems that the pterosaur elbow was a simple hinge joint, which allowed the forearm to be tightly folded against the upper arm, or opened widely (Bennett 2001; Wilkinson 2008). Articulating the elbow joints of well-preserved pterosaur specimens suggest they had an articulatory range of 30–150° (although some uncertainty exists over the former value—see Bennett 2001). When closed, the slightly skewed nature of the distal humeral condyles means that the radius and ulna, the subparallel bones of the forearm, were somewhat deflected medially.

The pterosaur radius and ulna are relatively long, straight tubes that lie alongside each other, with the ulna lying on the lateral side of the radius. In some derived pterosaurs, the ulna is relatively large compared to its neighbor. When the elbow is flexed, the radius slides forward to flex the wrist about 50°, along with instilling some minor medial motion in the wrist elements (Wilkinson 2008). The wrist itself, or carpus (figs. 4.9 and 4.10), is a complex series of four bones: the proximal and distal syncarpals (formed from smaller carpal elements that fuse in adults); the preaxial carpal; and a unique pterosaur bone, the pteroid. Several sesamoids, bones that lie within tendons, are also associated with the carpus. Pterosaur wrists have proved rather controversial topics among researchers, with debates raging over the range of motion in the wrist, and the origin, positioning, and orientation of the pteroid (see Unwin et al. 1996; Bennett 2001, 2007a; Wilkinson et al. 2006; and Wilkinson 2008 for overviews). The two syncarpals share a complex articulation with each other that was once thought to have restricted all movement (Hankin and Watson 1914; Bramwell and Whitfield 1974; Padian 1983a), but recent appraisals have revealed that these bones bore a sliding joint permitting at least 30° of rotation between them (Bennett 2001; Wilkinson 2008). The joint between the carpus and the metacarpals, the next bones in the pterosaur forelimb (see below),

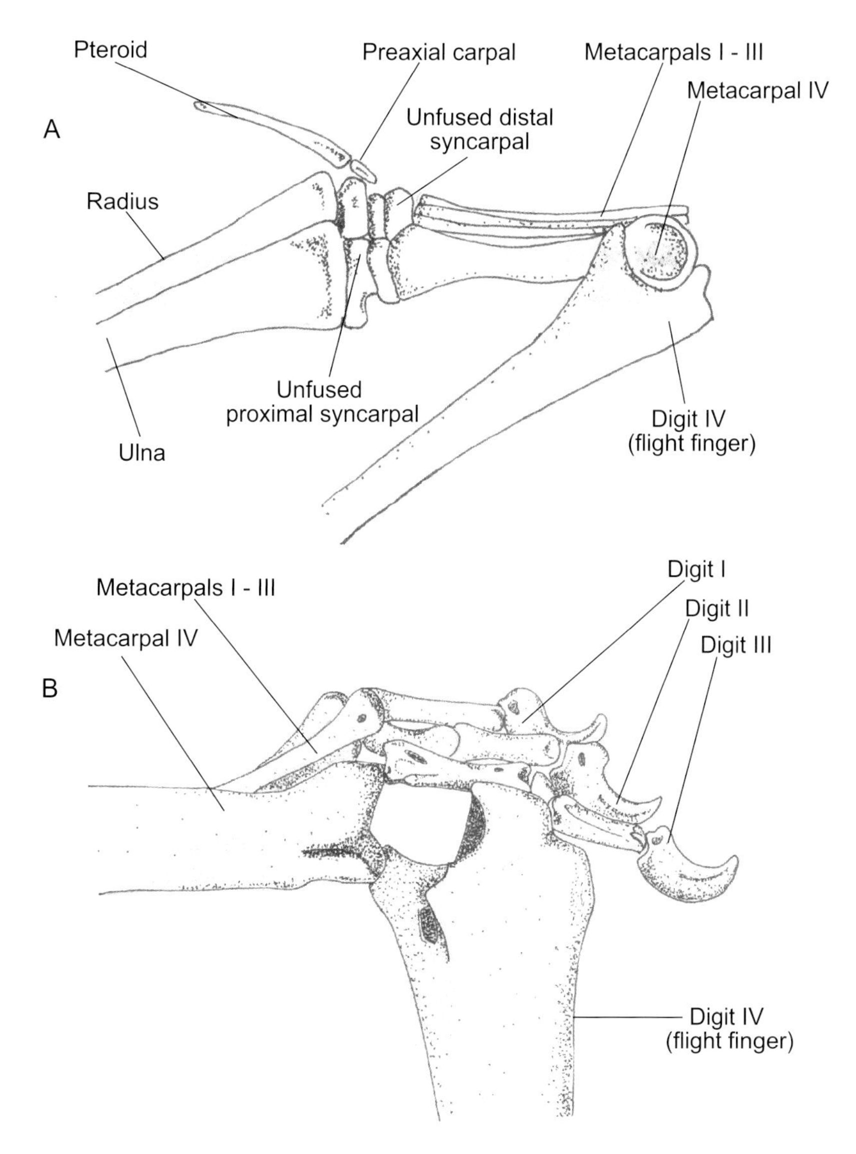

FIG. 4.9. Pterosaur wrist and hand osteology. A, detail of carpus and manus of an immature *Rhamphorhynchus muensteri* (digits I–III are missing, and note that the separate syncarpal elements would fuse later in life); B, distal metacarpal region of *Pteranodon* sp. Note the short metacarpals I–III. A, redrawn from Wellnhofer (1975); B, redrawn from Bennett (2001). Images not to scale.

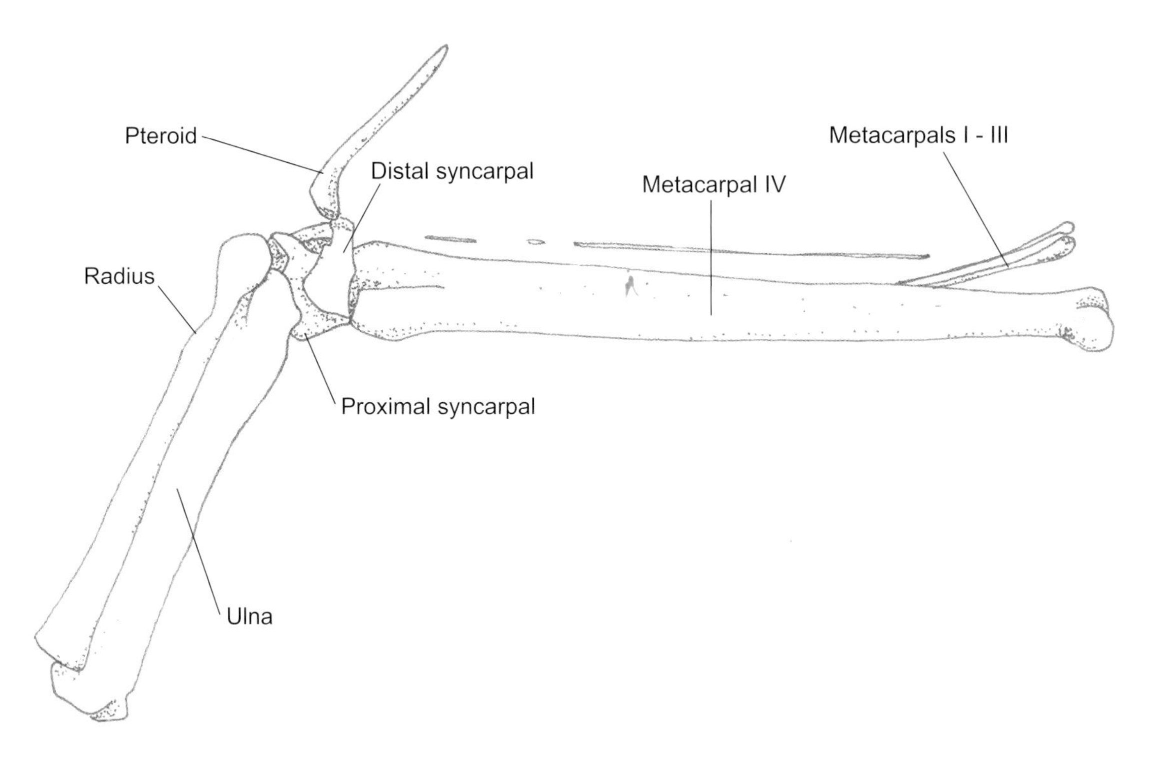

FIG. 4.10. Articulated specimen of *Pteranodon* sp. forearm showing possible maximum wrist flexion. Redrawn from Bennett (2001).

also seems capable of about 20° of movement. Thus, it seems that considerable motion was available at the pterosaur wrist, with their hands possibly flexing to near right angles to their forearms (fig. 4.10) or extending out to angles of 150–175° (Wellnhofer 1975; Wilkinson 2008). As we'll see in chapter 8, direct evidence that pterosaurs could flex their wrists has been discovered in specimens with arthritic wrist joints, caused by overuse and wear on their carpal joints!

The preaxial carpal bone articulates with the anterior face of the distal syncarpal, projecting forward from the wrist. This preaxial element articulates with the pteroid, a long, slender bone of varying shape that is completely unique to pterosaurs. There has been some dispute over the origins of this bone, with suggestions that it represents a modified thumb or an ossified tendon instead of a novel ossification within the wrist complex. Comparisons of pterosaur finger bone counts with other reptiles suggest it is unlikely to represent a modified first digit. Pterosaur fingers have, from the first digit to the fourth and last, two, three, four, and four bones in each digit respectively, which almost perfectly matches the reptilian configuration of two, three, four, and five (the discrepancy in the last digit is explained by the loss of the claw from the wing finger, as mentioned in chapter 3). This indicates that the first digit of pterosaurs is homologous with that of other reptiles, and was not co-opted for use as the pteroid (Unwin et al. 1996). Histological analyses suggest that pteroids were not ossified soft tissue either (Unwin et al. 1996), leading us to draw the conclusion that the pteroid must be "real" bone. Perhaps its origins will be revealed should some pterosaur ancestors, like those hypothesized in the previous chapter, be discovered.

The pteroid has traditionally been reconstructed as sitting in a small cradle at the end of the preaxial carpal, but recent work indicates that this space was actually occupied by a sesamoid, a small, rounded bone that develops within tendons where they pass over certain joints (Bennett 2007a). Instead, the pteroid probably articulated in a depression on the medial side of the preaxial carpal (Bennett 2007a), and may have also contacted the proximal syncarpal (Peters 2009). Some recently discovered pterosaurs with undisturbed wrists regions support this new interpretation (Wang et al. 2010). Such configurations would force the pteroid to project medially toward the body, which contradicts other proposals that the pteroid projected forward (e.g., Frey and Reiss 1981; Unwin 2005; Wilkinson et al. 2006; Wilkinson 2008). The idea of anterior-projecting pteroids has often been criticized for not only being unsupported by fossil evidence (e.g., Bennett 2007a) but also for being very unlikely on biomechanical grounds (we'll discuss the reasons for the latter in the next chapter).

The pterosaur hand, or manus, has wholly different proportions from those of all other tetrapods. Their palms are comprised of four elongate bones, the metacarpals, which articulate with the distal syncarpal. The first three are extremely slender, but the fourth is comparatively massive, bearing a large roller joint at its end, which supported and flexed the enormous pterosaur wing finger. The three smaller metacarpals are tightly appressed to the larger, or "wing," metacarpal, with the first metacarpal, which supports the first digit of the hand, takes the dorsal position (fig. 4.8). Some or all of the three smaller metacarpals may not reach the carpus in some species, instead tapering to fine points at their proximal ends. Like other reptiles, the first three digits of the pterosaur hand contain two, three, and four phalanges (digit bones, including the claws) in digits I–III, respectively. Generally speaking, the manual claws of pterosaurs are larger and more curved than those of the feet, and their movement was sometimes assisted mechanically by small bones, antungual sesamoids, situated immediately behind each claw.

The enormous pterosaur wing finger, which usually comprises the majority of the wing length, was relatively inflexible between its individual phalanges, with its only real point of motion situated at its joint with the fourth metacarpal. This allowed the wing finger to swing open as far as 165–175° from the metacarpal, or to tightly fold up against the palm. When folded in, the distal end of the wing finger was deflected slightly outward from the elbow, thanks to a slightly asymmetrical roller joint. Most pterosaurs possess four phalanges in their wing digits, and the proportions of these phalanges differ so markedly between groups that their relative lengths are used to identify different lineages.

Pelvic Girdle

At first glance (fig. 4.6), it seems that pterosaurs were at the back of the queue when hindquarters were being handed out, their legs and pelves appearing rather small compared to their enormous arms and heads. This is something of an illusion, however. For their body size, pterosaur hindlimbs and their girdle are quite proportionate and were powerfully built (Padian 1983a; Bennett 1997a). Although some variation in pterosaur pelvic morphology is now being documented (Hyder et al., in press), the structure of the pelvis itself is generally fairly consistent across pterosaurs groups. As with all tetrapods, pterosaur pelves are comprised of three bones on each side: the ilia, pubes, and ischia, which all fuse into a solid unit in adult pterosaurs without any clear suture lines. The ilium is a long, low bone extending along the top of the pelvis with two prominent projections, the preacetabular and postacetabular processes marking its fore and aft extremities. These extend beyond the anterior and posterior margins (respectively) of the lower pelvic bones. The ventral region of the pterosaur pelvis is often, but not always, formed into a shield-like ischiopubic plate. This is a fusion between the anteriorly lying pubis and the broad, posteriorly positioned ischium. The latter fuses with its mirroring element along their ventral margins, forming a closed pelvic canal. The pubis is revealed as a relatively narrow, ventrally projecting bone in individuals without complete ischiopubic plates, while the ischium is generally somewhat broader. All three of these bones contribute some bony material to the acetabulum (the hip joint), an imperforate, hemispherical depression in the lateral face of the pelvis that permitted great freedom of movement for the hindlimb (see chapter 7).

The anteroventral surface of the pubis bears a small articulation for another unique pterosaur bone, the prepubis. This fork- or spatula-shaped bone projects forward toward the sternum, joining with the gastralia and the mirroring prepubis to form a belly-spanning cradle. The prepubic bones and gastralia were capable of moving up and down with each breath taken by their owner, while the chest cavity remained relatively static (Claessens et al. 2009).

Hindlimb

As with the forelimb, the pterosaur hindlimb contains the same bones as our own (fig. 4.11). The femur is a long, slightly curved bone with a prominent, bulbous head projecting at an acute angle from the proximal end. Opposite this sits the greater trochanter, a site for muscle attachment used in hoisting the leg into the flight position. The distal end of the femur has two large articular condyles that provide the hinge-like knee joint a range of motion similar to our own.

Pterosaur shins are comprised of two bones, the long, robust tibiotarsus and the relatively slender fibula. In most cases, this latter bone never extends as far as the distal end of the tibia and generally fuses at both ends to its larger neighbor. In some pterosaurs, the fibula is extremely small and slender, occupying less than one third of the shin length. The tibiotarsus bears a large cnemial crest, a ridge of bone along the proximal anterior border, but becomes increasingly slender distally. The proximal ankle bones are

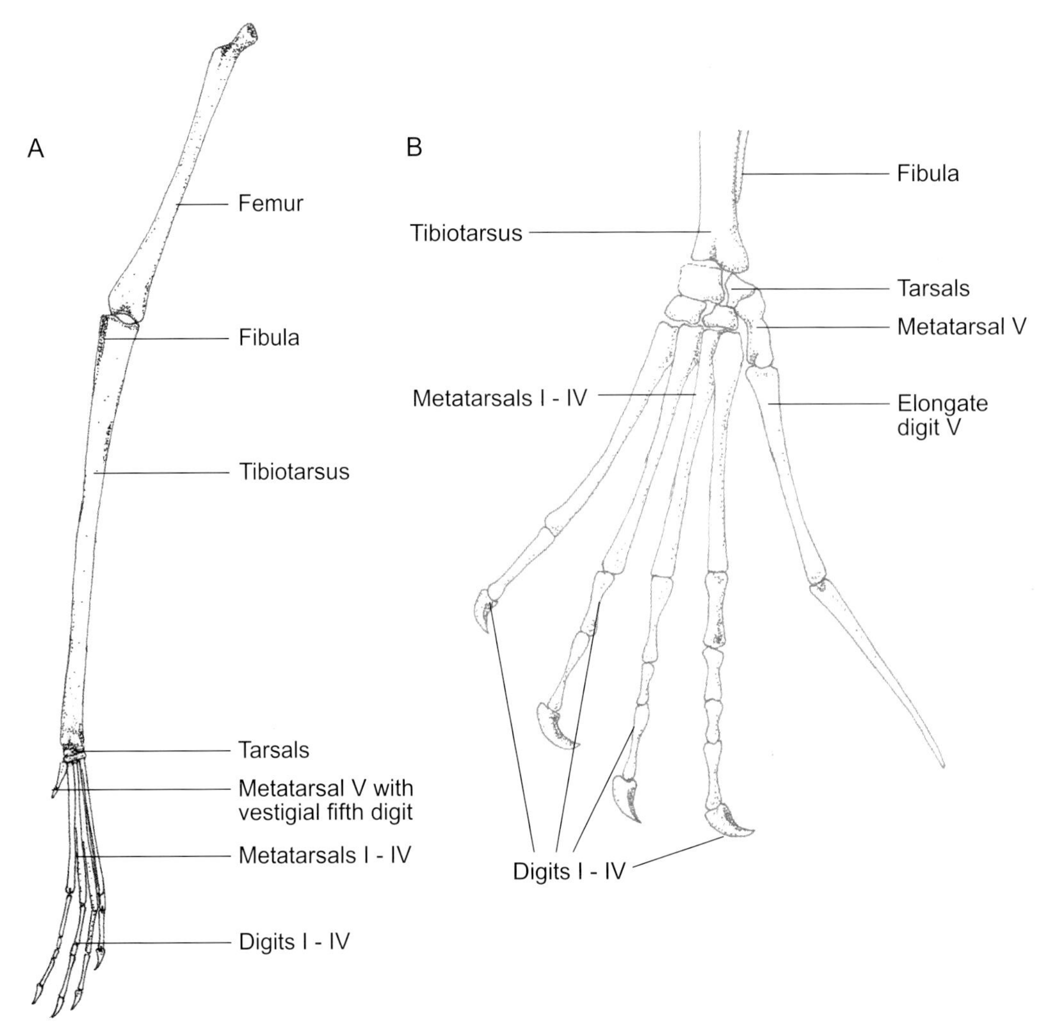

FIG. 4.11. Pterosaur hindlimb osteology. A, anterior view of the hindlimb of *Anhanguera santanae*; B, detail of pes of an immature *Rhamphorhynchus muensteri*. Note the differences in fifth digit anatomy, and the unfused tarsal elements of B. A, redrawn from Wellnhofer (1991b); B, redrawn from Wellnhofer (1975). Images not to scale.

fused with the end of the tibiotarsus in adult pterosaurs, forming a hinge-like ankle joint that allows the foot, or pes, to swing from 45° of the tibiotarsus to lie roughly in line with the same bone (Bennett 1997a).

Two other ankle bones, the distal tarsals, lie between the pterosaur tibiotarsus and the foot. These articulate with the metatarsals, the long, slender bones that make up the posterior portion of the pedal skeleton. The first four metatarsals of all pterosaurs are well-developed and generally spread somewhat distally, resulting in a splayed foot structure. In some species, however, they are aligned in a subparallel fashion to create a more compact pedal structure. The toes on these metatarsals are long and are comprised, from digit I to IV, of two, three, four, and then five phalanges, with the third or, in rarer cases, the fourth toe being the longest. The last phalanx of each toe forms a curved claw; these claws are generally smaller than those of the hands. The articular surfaces between the individual toe phalanges are generally quite flat, indicating limited movement took place between these elements (Clark et al. 1998).

The fifth metatarsal and digit of pterosaur feet are very different from their other toes. Early pterosaurs bear a short, chunky fifth metatarsal with complex articulations that probably allowed movement in several planes. Pterosaurs with such a metatarsal bore a long, clawless fifth toe comprised of two long phalanges that, thanks to the mobile metatarsal, could move in broad arcs around the lateral side of the foot. Later

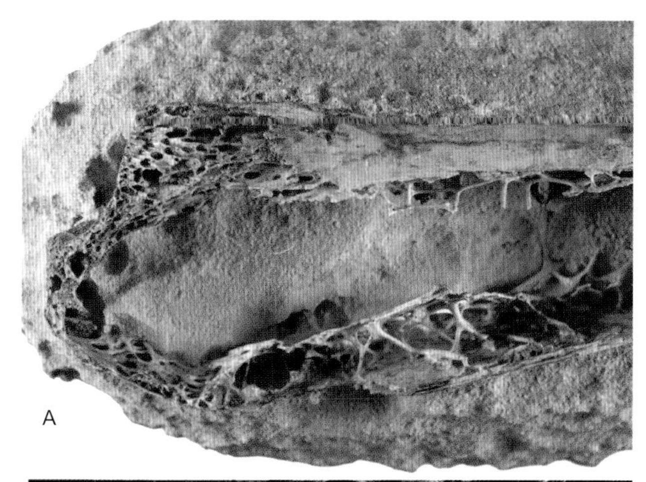

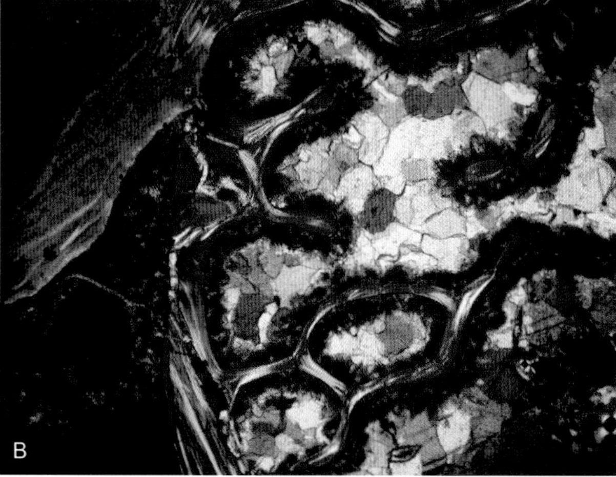

FIG. 4.12. Pterosaur bone histology. A, detail of pterosaur long bone from the Lower Cretaceous Santana Formation showing spongiose bone, bony trabeculae, and the vast, hollow void once occupied by pneumatic tissues; B, thin section of spongiose bone; C, thin section of fibrolamellar bone wall and trabeculae. Photographs courtesy of David Martill.

pterosaurs however, considerably reduced their fifth toe to only a single phalange, or lost it altogether. A small but stocky fifth metatarsal remained in the feet of these pterosaurs but it was probably not particularly mobile.

In the Bone

It seems that most—perhaps even all—pterosaurs bore at least some pneumatic bones, air-filled structures that are almost entirely hollow and supported by thin bony struts (trabeculae) spanning their internal chambers (fig. 4.12; Bonde and Christiansen 2003; Butler, Barrett, and Gower 2009; Claessens et al. 2009). Bones of this type are not unique to pterosaurs. Many animals, including yourself, have some pneumatic bones, but only one living group has developed their skeletal pneumaticity to a level comparable with pterosaurs: birds. As with their feathery archosaur relatives, it seems that pterosaur bones were lined with air-filled pneumatic tissues linked to a number of air sacs, including the lungs, via pneumatic foramina in the bone walls (including those mentioned on the cervical vertebrae and humerus, above). Pneumaticity was limited to the skull and portions of the vertebral column in early pterosaurs (Bonde and Christiansen 2003; Butler, Barrett, and Gower 2009), so their limb bones possess relatively thick bone walls not dissimilar to those seen in other reptiles. Later pterosaurs became some of the most pneumatized animals known, with the entire forelimb, pectoral girdle, more of the vertebral column, and even some elements of the hindlimbs being filled with air. During this process, pterosaur bone walls became unprecedentedly thin (Claessens et al. 2009). A similar trick has occurred in different parts of the skeletons of modern birds, with almost every bone pneumatized in some large species.

Outside of birds, pterosaurs, and some nonavian dinosaurs, extensive skeletal pneumaticity is virtually unknown. Its function is not well understood and multiple interpretations of its physiological role have been published (Witmer [1997] provides a very good overview of this topic). One common explanation for the extensive pneumatization of bird and pterosaur skeletons, that it makes them "lightweight" compared to grounded beasts, is likely untrue. Although this idea works well intuitively (both groups are generally flighted, and less weight to carry into the air equates to more economical flight, right?), it may not reflect reality. Bats demonstrate that fairly large wingspans can be reached without any pneumatized skeletal components, and at given body masses, bird skeletons are just as heavy as those of mammals (Prange et al. 1975). The latter point is especially important, as it suggests pneumaticity does not lower the ratio of skeletal mass to overall body mass and does not, therefore, provide a lightweight skeleton.[3] This creates an obvious paradox—how can a skeleton with hollow bones be as massive as one with marrow-filled, solid bones? The observation that pneumatized bones are often larger than their counterparts in nonpneumatized animals may offer a solution. Bones become pneumatized through "invasions" of pneumatic diverticula, offshoots from the soft-tissue air sacs that absorb the cores of bones as they grow into them (e.g., Witmer 1997; O'Connor 2004). The eroded bone is not simply "removed" from the skeleton however, but is redeposited elsewhere as part of the same bone. In effect, the bone is "inflated" until it reaches a mechanical limit (Witmer 1997). As extensively pneumatized animals like birds, pterosaurs, and sauropod dinosaurs demonstrate, this can lead to bones attaining huge linear dimensions with minimal additions of bony tissue, a consequence of the air sacs, or pulmonary system, aggressively pushing skeletal mechanics to its limits. Thus, a pneumatized bone may be thought of as an inflated balloon: it weighs the same as when it's deflated, but is stretched to a much larger size. The walls of pneumatized bones become thinner and thinner as the bone mass is stretched further, meaning that the bones expand without gaining mass. We may therefore interpret heavily pneumatized animals as small forms inflated to the dimensions of big ones, making them lightweight compared to nonpneumatized species of the same proportions but not, strictly speaking, making their skeletons lightweight.

Both birds and pterosaurs possess notably expanded and pneumatized shoulder girdles and wing bones, perhaps for obvious reasons. Expanding these components means larger attachment sites and better leverage for their flight muscles, and greater length of the wing bones increases their wing dimensions. Pneumaticity was arguably taken further in pterosaurs than birds to a certain extent. Pterosaur bones were expanded so far that their bone walls are mere millimeters thick even in the largest forms. This feature is often used to suggest pterosaur bones were very delicate and fragile in life, but this is another misconception about pneumatic bones. For a given mass, hollow bones like those of pterosaurs are more resistant to bending and twisting than solid tubes, though they are less resistant to impacts (Habib 2008; Witton and Habib 2010). Pterosaur bones also possess "fibrolamellar" bone tissues, where each thin bone lamina is deposited perpendicular to the preceding layer, creating a plywood-like bone material that is resistant to bending and torsion (Ricqlés et al. 2000; Steel 2008). The trabeculae within the bones are orientated to optimize their resistance against flight forces (Habib 2007), and spiraling structures (endosteal ridges) lining the inner surface of the bone walls provide even more reinforcement (Ricqlés et al. 2000; Steel 2008). A bony honeycomb structure, the so-called spongiose bone, is also present in all pterosaur bones, but is restricted to the ends of their long bones in order to strengthen their joints. With this reinforcement and deft natural engineering, pterosaur bones were far from being fragile, and even heavily pneumatized parts of the skeleton were well suited for withstanding the stresses and forces inflicted on them in everyday life. The fact that pterosaur bones appear strong enough to break off dinosaur teeth (see chapter 7) is further evidence that their skeletons were not fragile, easily broken structures as they are often perceived.

Alas, the same features that optimize the pterosaur skeleton for strength and size make them terrible material for fossilization. Even well-engineered bone flakes apart rapidly when decaying, so thin-walled pterosaur bones would rot quickly once their owners died. Thin bone walls also render their fossils easily eroded and fragile, so their preservation potential is generally low. A possible silver lining to this is that the thinness of pterosaur bone is highly characteristic among fossil reptiles, so even scrappy pieces can be recognized fairly easy. Thus, while pterosaurs remain rare fossils, at least we can be fairly certain when we have found them.

[3] There is some evidence, however, that at least some nonpneumatized bones are heavier than their pneumatized counterparts (Fajardo et al. 2007). It has yet to be proven that animals with pneumatized skeletons have lower ratios of skeletal mass to overall body mass than their nonpneumatized counterparts, however.

5
Soft Bits

The fossil record is primarily comprised of the hard parts—shells and bones—of organisms. Soft tissues are rarely preserved because they are scavenged and decayed by a multitude of organisms and quickly disappear once an animal dies. If preserved rapidly enough, however, soft tissues can be preserved as sediment impressions or thin mineral films lying alongside fossil bones. The latter are not always visible to the naked eye, but can be detected through the use of ultraviolet (UV) light. This process can transform unassuming specimens into truly fantastic fossils. Undeniably the leader in UV fossil revisualization is Helmut Tischlinger, and many images in this book are examples of his work (see Frey and Tischlinger 2000; and Frey, Tischlinger, et al. 2003 for more of his UV pterosaur photography). Specimens enhanced by this technique often reveal shocking truths about the anatomy of ancient animals that could not have been inferred from bones alone.

Fossils containing soft-tissue remnants are typically only found in fossil Lagerstätten, those sites with exceptionally high-quality fossil preservation that paleontologists are so fond of. This means that soft-tissue data is unknown for most fossil species. In these cases, we can look for osteological signatures of certain soft tissues and, using closely related animals as a guide, deduce something about the nature the soft parts even when they are not preserved. Both observations of exceptionally preserved pterosaur fossils and these educated inferences have allowed us to build up a fairly detailed picture of pterosaur soft-tissue anatomy, including many specifics of their external appearance (fig. 5.1). Piecing together this soft tissue data also provides an important insight into pterosaur physiology and behavior. Thus, as we go through the pterosaur body here, starting with what we know of their internal soft anatomy and working outward, we should begin to get a real flavor for what pterosaurs were like as living animals, and not just mere petrified skeletons.

Inside a Pterosaur

THE RESPIRATORY MECHANISM

We mentioned in chapter 4 that hollow pterosaur bones were filled with pneumatic tissues related to a system of soft-tissue air sacs, and that this configuration is mirrored in modern birds. This tells us a great deal about the pterosaur respiratory system even without any direct preservation of their lungs or other respiratory structures. The position of pneumatic foramina across ornithodiran skeletons correlates with specific parts of the soft-tissue pulmonary system (Wedel 2003; O'Connor and Claessens 2005; O'Connor 2006), and those seen in pterosaur skeletons suggests that they bore air sacs lining their necks (cervical air sacs) and at least two sets of abdominal air sacs, one fore and the other aft (fig. 5.2) (Claessens et al. 2009). The forelimbs of some pterosaurs were pneumatized with diverticula from the anterior thoracic sacs, and the presence of cavities and pneumatic foramina around their antorbital fenestrae suggests this structure and much of the skull was probably also filled with pneumatic tissue (Witmer 1997; Claessens et al. 2009). If pterosaurs possessed a very birdlike pulmonary system, they would also possess additional air sacs around the ventral region of the trunk (Claessens et al. 2009).

The existence of these air sacs suggests that pterosaurs almost certainly possessed avian-like lungs, which are rather different and far more efficient than those of other reptiles and mammals. Birds have a "flow through" lung system, where air sacs squeeze richly oxygenated air into solid lungs in a single direction only, with the oxygen-depleted air then being passed through other parts of the pulmonary system to be exhaled. In contrast, our bellow-like lungs mix old, deoxygenated air and fresh, oxygenated air with each breath, so our breathing efficiency is limited by the constant circulation of half-used air. We use a muscular, mobile rib cage that expands and contracts in order to inhale and exhale, but birds have fused anterior trunk skeletons that deny them this ability. Only their belly regions can flex readily, so movements of this region squeeze or inflate the posterior thoracic air sacs to pump air around the pulmonary system. The almost certain presence of air sacs, with a rigid chest skeleton but mobile belly regions, suggests pterosaurs also employed this respiratory mechanism, perhaps using muscles anchoring on the gastralia and prepubes to facilitate breathing motions (Unwin 2005; Claessens et al. 2009). Like birds, pterosaurs probably found a flow through pulmonary mechanism very ef-

FIG. 5.1. *Tupandactylus navigans* glowing in the sunset over the Cretaceous Crato hinterland. Preservation of pterosaur soft tissues demonstrate that much of their bodies were covered in dense, hairlike "pycnofibers" analogous to the simple feathers of some dinosaurs or the fur of mammals.

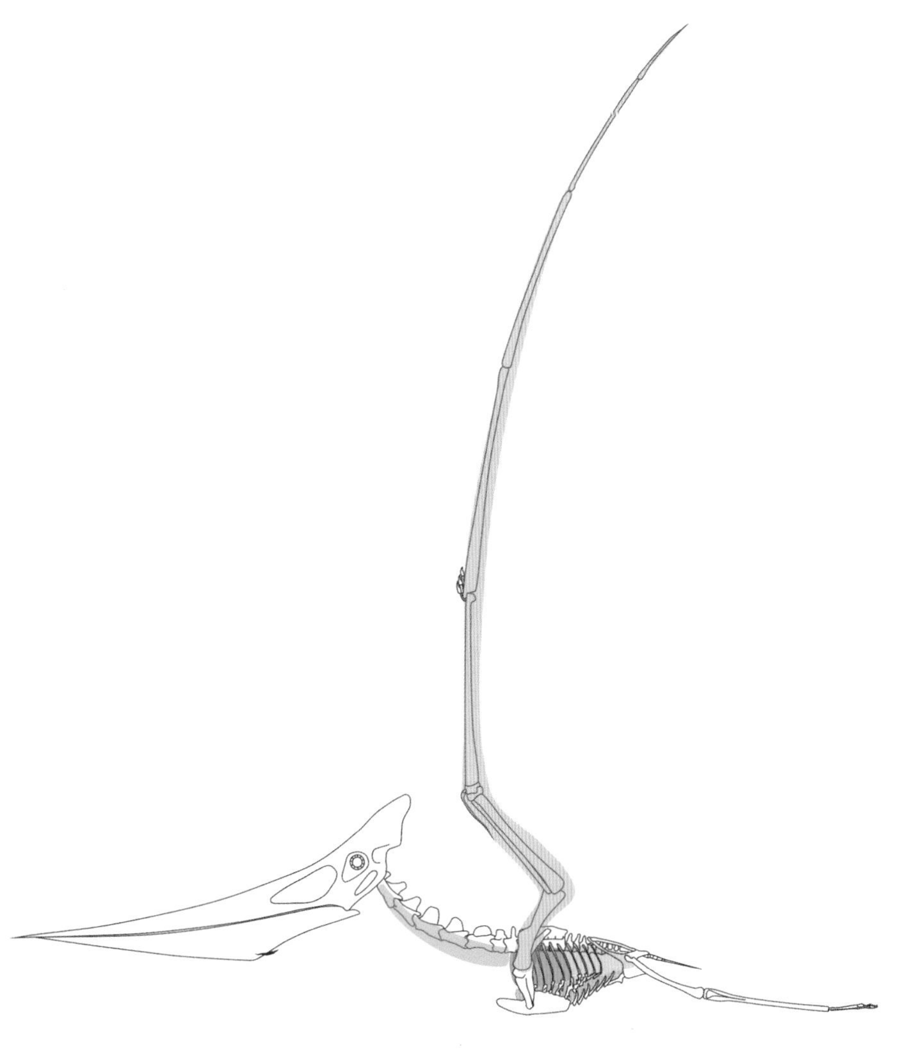

Fig. 5.2. Distribution of the pulmonary system and air sacs in the highly pneumatized *Pteranodon*. Green shading, cervical air sacs; red, lungs; blue, abdominal air sacs; purple, distribution of pneumatic diverticula throughout the wing skeleton; and gray, thoracic air sacs. Pulmonary system distribution based on Claessens et al. 2009.

ficient at harvesting oxygen (Claessens et al. 2009), which means they join birds and bats in having been able to respire very effectively, a necessary adaptation for sustaining powered flight (bats have enormous lungs, compensating for the relatively inefficient mammalian respiratory system).

THE PTEROSAUR GUT

Unlike the respiratory system, pterosaur intestines left no osteological signatures. Guts have very little preservation potential, being comprised of soft muscle, lined with bacteria, and partially filled with acid, so it is unsurprising that they have rotted away from every known pterosaur fossil. However, plenty of indirect evidence provides clues to the nature of the pterosaur digestive system, including the presence of an extremely elastic esophagus (the pipe carrying food from the mouth to the stomach). A small specimen of *Rhamphorhynchus* from the late Jurassic of Germany shows a completely swallowed fish within its bowels that occupies a tremendous *60 percent* of the torso length (Wellnhofer 1975). This fish has also been partially digested at its anterior end, so it may have been even bigger when eaten. Preserved outlines of pterosaur necks indicate they didn't have unusually flappy, jowly throats, so the only way such a relatively large meal could fit down the throat of a small pterosaur was through an esophagus that could distend well beyond its normal volume (fig. 5.3; also see fig. 8.10B for an image of the fossil under discussion). This, of course, indicates that their other ventral neck tissues were also stretchy and, because most pterosaurs do not appeared to have chewed their food (e.g., Ősi 2010), pterosaur necks must have bulged tremendously whenever they ate any large foodstuffs.

The pterosaur stomach seems to have been of a typically archosaurian nature in having two distinct com-

FIG. 5.3. Restoration of the Jurassic pterosaur *Rhamphorhynchus muensteri* swallowing an entire fish, based on fossil *Rhamphorhynchus* gut content. Pterosaur throats must have been capable of gross expansion to swallow such large prey. See chapter 8 and fig. 8.10 for more about the specimen behind this reconstruction.

partments. The front region, the proventriculus, acted similarly to our own stomach by chemically attacking food with enzymes, while the back, the ventriculus or gizzard, ground food mechanically through powerful muscular action. Some archosaurs swallow grit and stones to assist with this mechanical breakdown (once adopted for this function, said grit and stones are known as "gastroliths"), but others rely on the action of the gizzard alone. The existence of such a stomach in pterosaurs is not only predicted through their likely evolutionary relationships (Reily et al. 2001), but was verified recently with the first discovery of pterosaur gastroliths found within the guts of the South American filter feeder, *Pterodaustro* (Codorniú et al. 2009).

We also have little evidence of what happened to pterosaur food once it left the stomach. Like other archosaurs, pterosaur guano was probably a pasty mixture of urine and fecal matter, but intriguingly, there is some evidence that not all products of pterosaur digestion left via the back door. A fossil pellet (a form of fossil vomit) comprised of four juvenile bird skeletons from the Lower Cretaceous Los Hoyas deposits of Spain, may have been regurgitated by a pterosaur (fig. 5.4; Sanz et al. 2001). Regurgitating indigestible material is not uncommon among modern animals and, happily, particularly dedicated biologists have routed through enough animal vomit to identify pellet signatures distinctive to different groups. It seems that fish, amphibians, and mammals regurgitate bones in a loose, disarticulated fashion, but lizards, snakes, and crocodilians digest most bone and only regularly vomit the most resilient, proteinaceous parts of their food (claw sheaths, hair, feathers, and so on; Fisher 1981a, 1981b). Some birds also almost entirely digest bone, but most regurgitate bones that have been lightly etched with acid during the digestive process. The Las Hoyas pellet does not match any of these pellet signatures, suggesting that it was not made by any type of animal known in the modern world, and a small dinosaur or pterosaur is therefore its likely source (Sanz et al. 2001). More definitively identified pterosaur vomit is known in another pterosaur fossil, but it probably represents material regurgitated into the mouth during death throes rather than purposely rejected material (see chapter 8; Bennett 2001; Witton 2008b).

THE BRAIN

A number of fossilized pterosaur brain casts, or endocasts, have accrued over the last century (fig. 5.5; Newton 1988; Edinger 1941; Kellner 1996a; Lü et al. 1997; Bennett 2001; Witmer et al. 2003). Many authors have commented on how birdlike (rather than reptile-like) pterosaur brains were, presumably reflecting common neural requirements for processing and coordinating powered flight. Pterosaur brains were relatively large compared to those of other reptiles, though they were not quite as large as those of birds (Witmer et al. 2003). The regions of the pterosaur brain devoted to balance—the semicircular canals—are huge (even bigger than those of birds), suggesting they had an excellent sense of composure and spatial awareness in complex, three-dimensional environments. Similarly, the flocculus, the brain region associated with muscle coordination and keeping a steady, focused gaze with the eyes, is larger in pterosaurs than any other tetrapods. Coupled with the enlargement of the optic lobe and their generally large orbits, it appears that

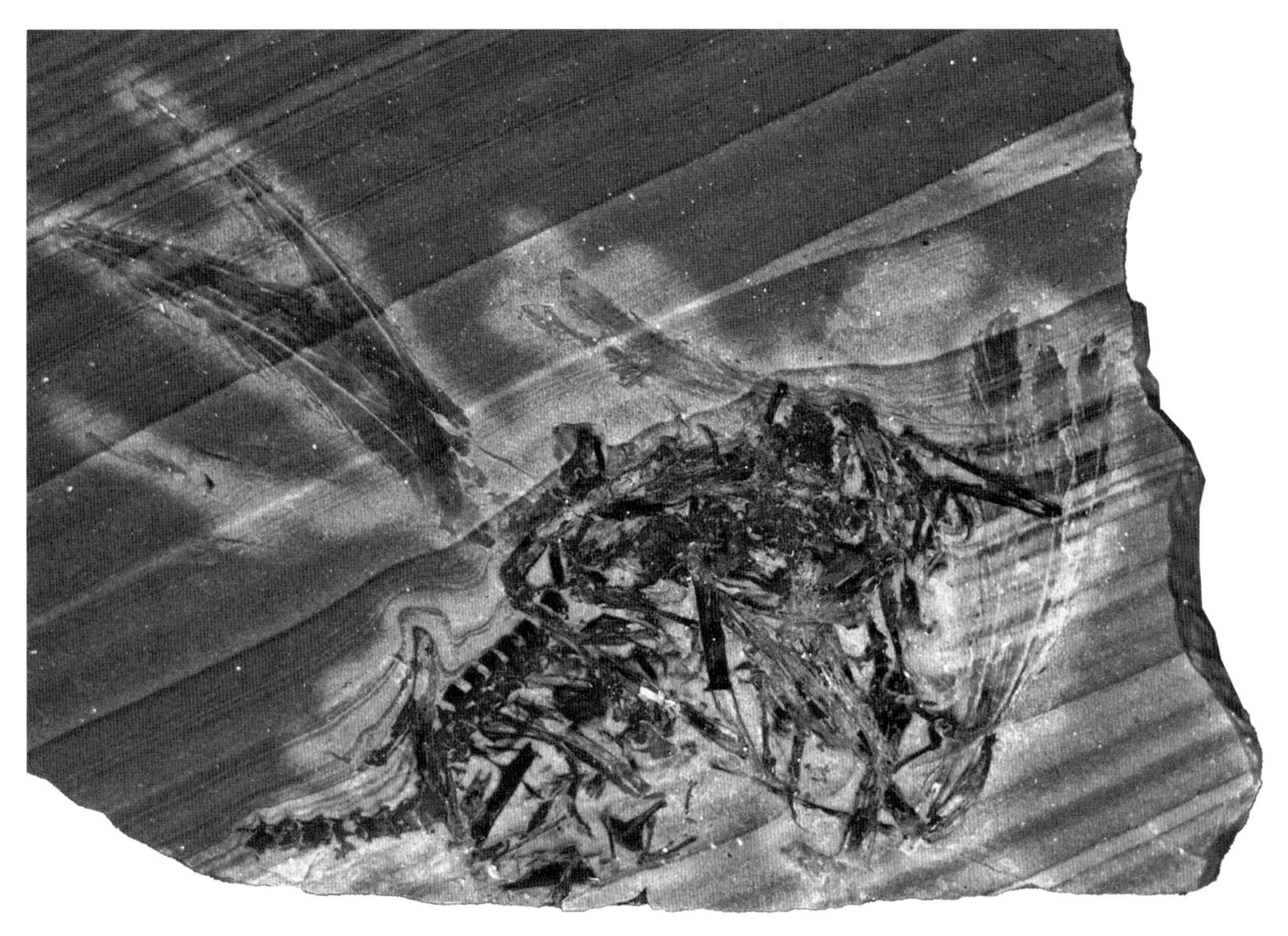

FIG. 5.4. Photograph taken under ultraviolet light of four bird skeletons preserved in a dinosaur or pterosaur gut regurgitate from Lower Cretaceous deposits of Las Hoyas, Spain. The feathers in this specimen are not discernible in normal light. Image courtesy of Jose Sanz and Luis Chiappe.

pterosaurs had excellent eyesight, though their tiny olfactory bulbs, which process information on smells and odors, suggest their sense of smell was not quite as impressive.

A further piece of behavioral information, pertaining to how the pterosaur head was generally held when awake, has also been linked to brain morphology. It has been suggested that the orientation of semicircular canals, a pair of tube-like structures on the side of the brain, reflects habitual head posture. Pterosaur semicircular canals have been interpreted as indicating a habitual horizontal head orientation in some forms, or slightly downturned in others (Witmer et al. 2003; Chatterjee and Templin 2004). This idea has not been without its critics, however. The relationship between semicircular canal orientation and head posture seems quite variable across different animals, and may vary even within a single species, suggesting it is a poor indicator of habitual head orientation (Taylor et al. 2009). Thus, while pterosaurs probably adopted a reasonably constant head posture while moving (Witmer et al. 2003 and references therein), ascertaining the preferred attitude of their heads may be more complex than previously thought.

Musculature

Pterosaur muscles are virtually unknown from the fossil record, though one specimen of the tiny pterosaur *Anurognathus* preserves elements of its arm and leg musculature (see fig. 11.3; Bennett 2007b). Thus, we are mainly reliant on reconstructing pterosaur myology through comparative anatomy, correlating signatures of muscle attachment from living animals to those on pterosaur skeletons. This has been done in

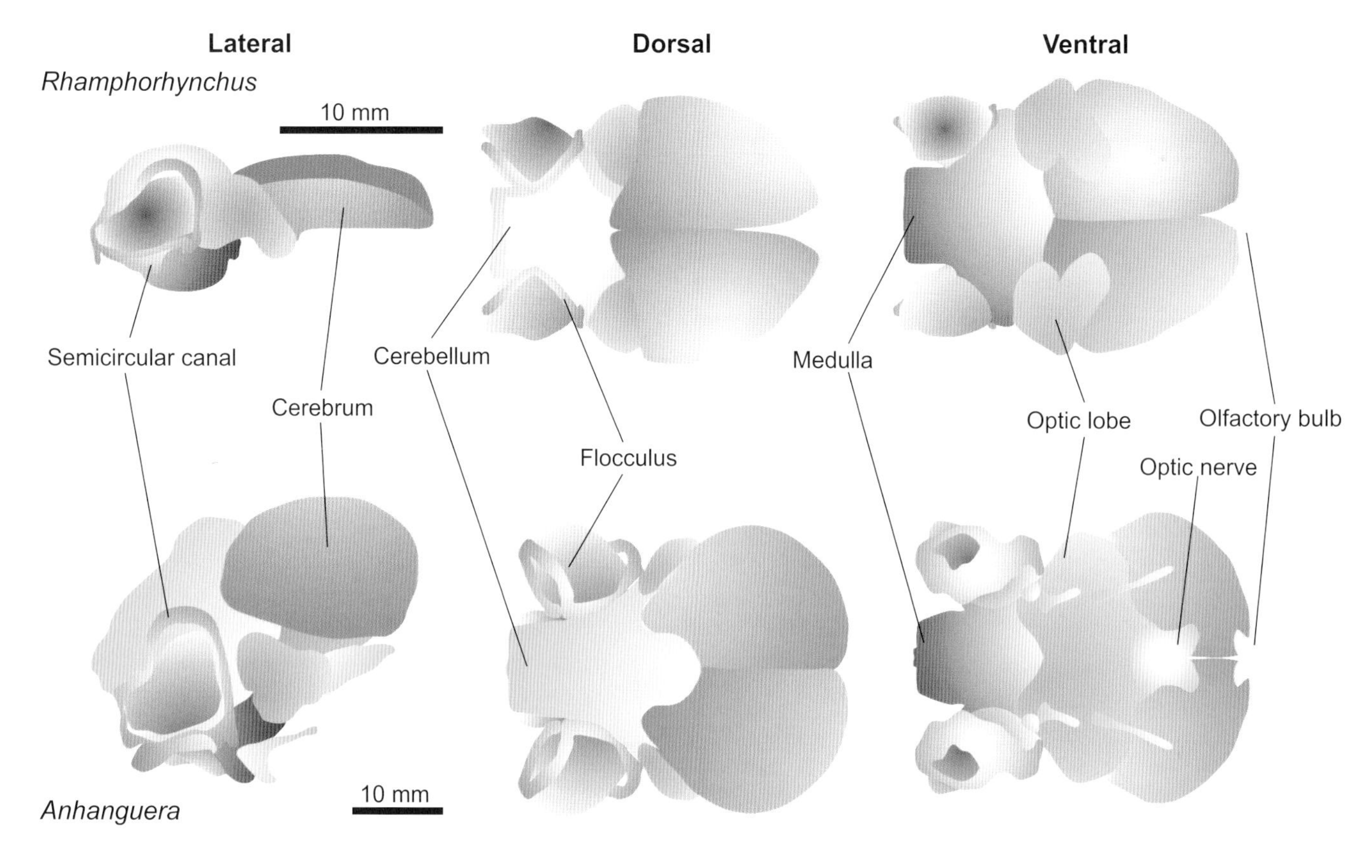

FIGURE 5.5. Pterosaur brain endocasts, colored to highlight neurological regions. All redrawn from Witmer et al. (2003).

detail for the pterosaur wing (Bennett 2003a, 2008), jaws (Fastnacht 2005a; Ősi 2010), and to a lesser extent, the hindlimbs (Fastnacht 2005b). Careful examination of pterosaur bone-muscle correlates suggests that they generally retained the "basic" tetrapod muscle plan rather than the unusual, derived myology common to birds (Bennett 2003a, 2008), and it seems probable that some parts of the pterosaur body—particularly their shoulders, proximal forelimbs, and perhaps some parts of their necks—were actually very muscular (Witton and Habib 2010). This idea contrasts with most pterosaur life reconstructions where their limbs and necks are stick thin, but it follows with with the massive muscles we see powering flight in birds and bats.

THE HEAD AND NECK

Like other reptiles, pterosaurs possessed three basic sets of jaw muscles but, in most forms, they were fairly small and probably provided relatively weak bites (fig. 5.6). The smallest jaw muscle, its size a consequence of its gravity-assisted job of merely opening the jaws, was *m. depressor mandibulae*. This attached to a small prong at the back of the skull and the tip of the retroarticular process. The other sets of jaw muscles were larger, since they were involved with the more taxing job of closing the jaws and processing food items. The external, or temporalis, adductor muscles were the largest jaw muscles in most pterosaurs, stretching from the posterior portion of the mandible into the temporal fenestrae, where these holes allowed the muscles to bulge when contracted. The final set, the internal adductors (a complex series of muscles including *m. pterygoideus* of fig. 5.6), extended along the underside of the antorbital region before descending toward the posterior mandible and wrapping around the retroarticular process. This muscle would be visible, as it is in birds and crocs, as a small bulge at the back of the jaw and is associated with delivering particularly powerful bites in some animals. The manner in which pterosaurs developed their external adductors over the internal is mirrored in modern birds and suggests that most were more interested in rapidly closing their jaws on food items rather than grappling with it. This interpretation agrees with their general lack of chew-

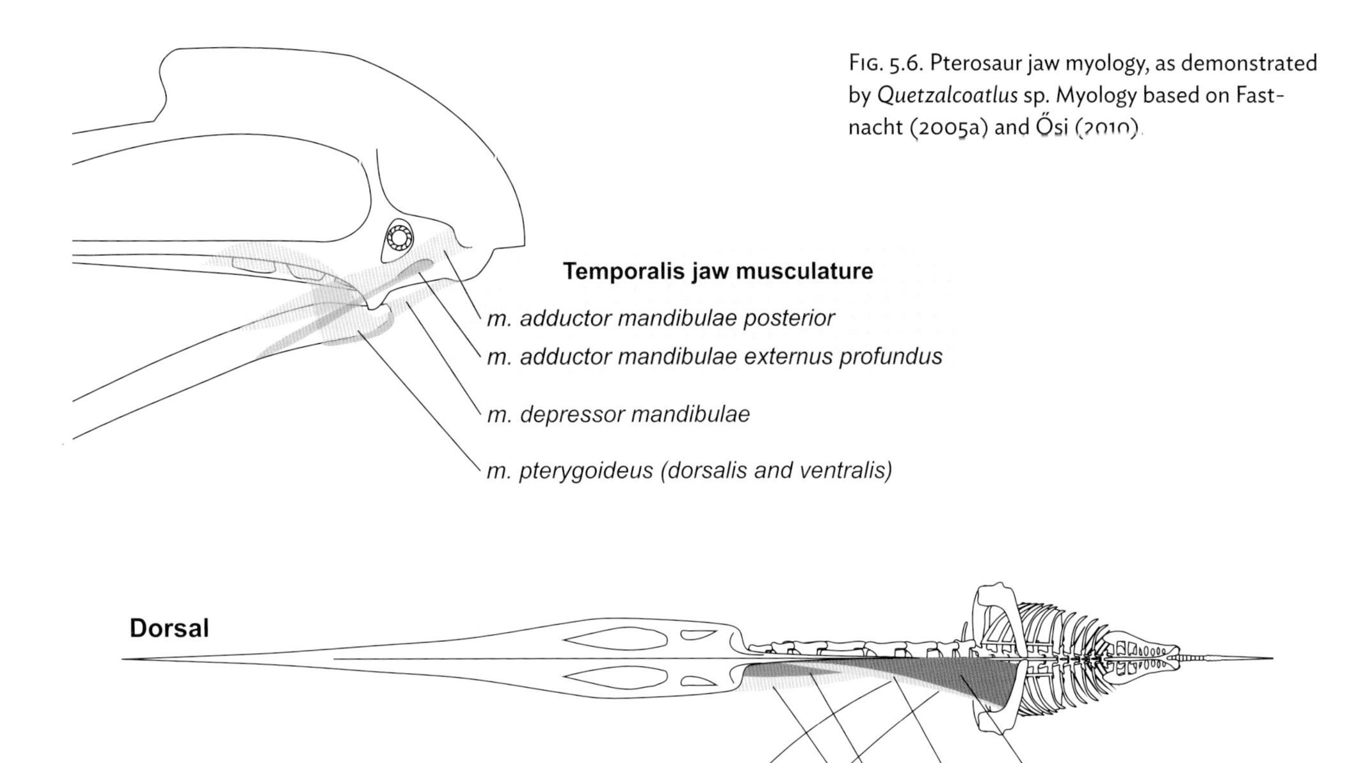

Fig. 5.6. Pterosaur jaw myology, as demonstrated by *Quetzalcoatlus* sp. Myology based on Fastnacht (2005a) and Ősi (2010).

Fig. 5.7. Pterosaur neck myology, as demonstrated by *Pteranodon longiceps*. Myology based in part on Bennett (2003a), but also Hildebrand (1995).

ing adaptations or other means of mechanically taking food apart.

Life restorations of pterosaurs typically present them with pencil-thin necks, perhaps because bird necks, once stripped of their feathers, are comparatively thin compared to their skulls and bodies. However, the processes and ridges across the occipital face of pterosaur skulls indicate that this surface anchored substantial muscles and ligaments responsible for elevating and turning the head, suggesting that pterosaur necks were fairly well endowed with soft tissues (fig. 5.7). Matching the muscle attachment scars of pterosaurs with those of modern reptiles suggests that these tissues extended to the top of the skulls or the base of the supraoccipital headcrests, and to the lateral extremes of the occipital face. Note that this face is one of the widest parts of the skull in most pterosaurs, but it is often reduced in birds, which explains their relatively thin necks. In most tetrapods—and probably pterosaurs—the muscles and ligaments attaching to the back of the head run from the skull right down to the shoulders and back, anchoring on the shoulder girdle, sternum, and dorsal vertebrae. Because the throat muscles were anchored between the mandibular rami, pterosaur necks were probably generally deeper than wide. Rarely preserved soft-tissue outlines of pterosaur necks corroborate these observations (e.g., Frey and Martill 1998). Thus, even the long, tubular neck vertebrae in some pterosaurs were likely surrounded by blocks of muscle, suggesting that artists need to put far more flesh on pterosaur necks than the scrawny, undermuscled depictions we typically see.

FORELIMBS

At one stage, it was thought that the flight muscles of pterosaurs were very birdlike, with the arm lifted by a muscle, *m. supracoracoideus,* anchoring on the sternum rather than the shoulders. In birds, this muscle arcs over the glenoid to attach on the dorsal surface of the humerus, elevating the wing with a pulley-like system (e.g., Kripp 1943; Padian 1983a; Wellnhofer 1991a). Detailed reconstruction of the proximal arm musculature of pterosaurs shows that this is not the case, however, and that the arm was more likely lifted by large muscles anchored on the scapula and back, and lowered by those attached to the sternum and coracoid (fig. 5.8; Bennett 2003a). Unlike birds, where two vastly expanded muscles are mainly used to power flight, it appears that pterosaurs used several muscle groups to form their flapping strokes. Most of these muscles anchored around the proximal humerus and particularly onto the deltopectoral crest. Like other flying tetrapods, pterosaurs appear to have concentrated their muscle mass proximally and bore relatively slender muscles on their distal forearms, though the muscles responsible for opening and closing the wing finger may have remained quite large. Like birds and bats, pterosaurs may have employed a system of ligaments to automatically open and close their wings when the elbow was moved (Prondvai and Hone 2009).

THE HINDLIMB AND TAIL

Pterosaur hindlimb musculature probably followed the same basic tetrapod muscle plan seen across the rest of their bodies (fig. 5.9; Fastnacht 2005b), though the development of a complete, shield-like ischiopubic plate suggests their hindlimb depressor muscles were unusually large. With the pterosaur wing membrane probably attaching to the leg (see below), Unwin (2005) suggested the extensive ventral hindlimb muscles may have allowed these limbs to power flight along with the forelimbs. However, the slender nature of the legs (and thus their propensity for bending compared to the robust forelimb) argues against this. To me, it seems more likely that the large pterosaur thigh muscles were used to stabilize the legs and wings in flight (primarily against lift forces attempting to elevate the posterior wing region), rather than power flight itself. Pterosaurs are sometimes restored with very large anterior thigh muscles because of their long preacetabular processes, but there are suggestions that this is erroneous (Hyder et al., in press). The variable length and orientation of these processes do not seem to correlate with enlarged muscle scars on the hindlimbs or other changes to hindlimb proportions. Moreover, unlike the broad, tall preacetabular processes of mammals and birds that anchor thick anterior thigh muscles, the preacetabular processes of pterosaurs are dorsoventrally flattened and slender. In some pterosaurs, this process also curves upward so much that it would offer very little purchase to any muscles attached to it (Fastnacht 2005b). Hence, it is likely that only the base of these processes anchored any substantial muscles, and the rest of the long preacetabular body was developed for another purpose, such as anchoring muscles and ligaments of the back (Hyder et al., in press).

Pterosaurs are unusual among archosaurs by possessing a very small *m. caudofemoralis,* a hindlimb retractor muscle anchored on the femur and caudal vertebrae. This muscle is tremendously long and powerful, and would clearly bulge from the side of the tail in most extinct archosaur species, as it does in modern crocodilians. Even pterosaurs with long tails have such slender caudals that this muscle could only be weakly developed at best, and the caudal vertebrae of short-tailed pterosaurs are so weakly developed that they appear unlikely to anchor much musculature at all (Persons 2010). Hence, unlike crocodiles or dinosaurs, pterosaur tails were not thickly muscled and their haunches were probably far more mammalian or birdlike in overall appearance, with most of their posterior hindlimb muscles anchoring onto their well-developed aft pelvic bones.

Integument

Pterosaur fossils reveal that their bodies were adorned with proteinaceous beaks, claw sheaths, scales, naked skin, callused footpads, and most surprisingly, soft, fuzzy "fur." We are almost entirely reliant on preservation of these tissues in fossils to have any idea of their existence, and when found, they often reveal startling anatomies that drastically alter our percep-

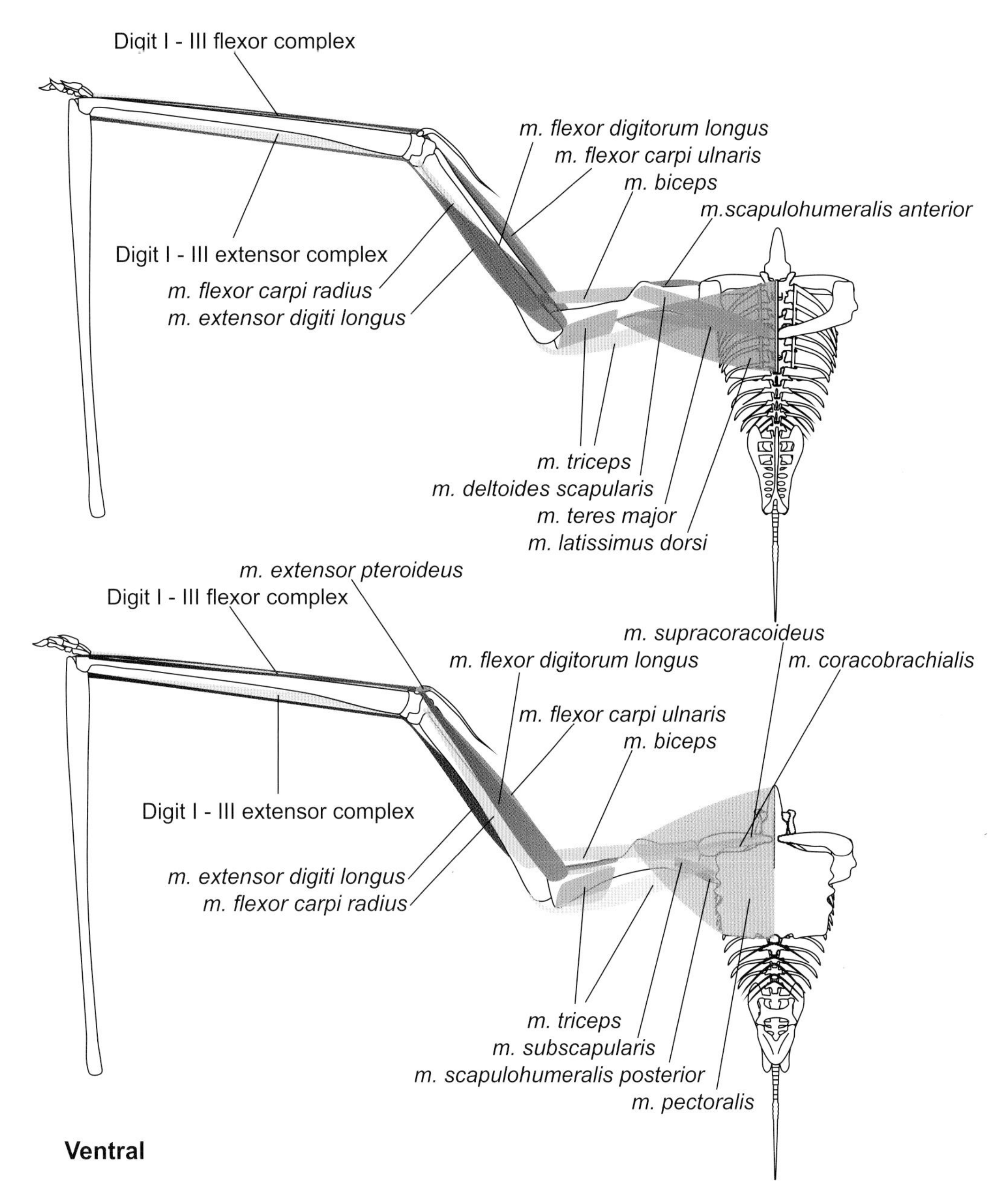

Fig. 5.8. Pterosaur forelimb myology, as demonstrated by *Pteranodon longiceps*. Myology based on Bennett (2003a, 2008).

tion of pterosaur appearance and biology (e.g., Czerkas and Ji 2002; Frey, Tischlinger, et al. 2003). We lack specific details of the composition and chemical makeup of these soft tissues in most instances, but it seems reasonable to assume that most were derived from keratin, a very versatile protein that, in modern animals, has been fashioned into all manner of scales, hair, feathers, beaks, and claws.

Being reptiles, you may expect that pterosaurs had scales somewhere on their bodies, and small, nonoverlapping circular scales can indeed be found on the soles of their feet (fig. 5.10; Frey, Tischlinger, et al. 2003). These scales were relatively large around the pronounced soft-tissue pads seen on their ankles and beneath the end of the metatarsals, which presumably cushioned pterosaur feet as they walked, ran, and

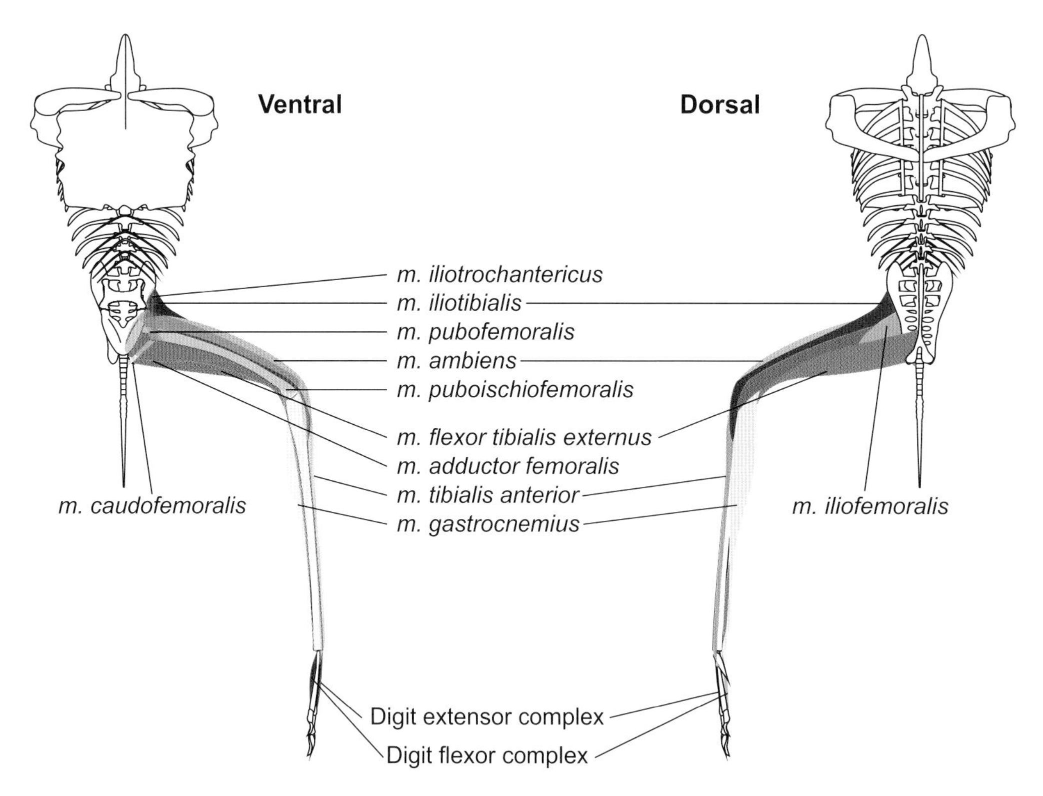

FIG. 5.9. Pterosaur hindlimb myology, as demonstrated by *Pteranodon longiceps*. Myology based in part on Fastnacht (2005b), but also Hildebrand (1995).

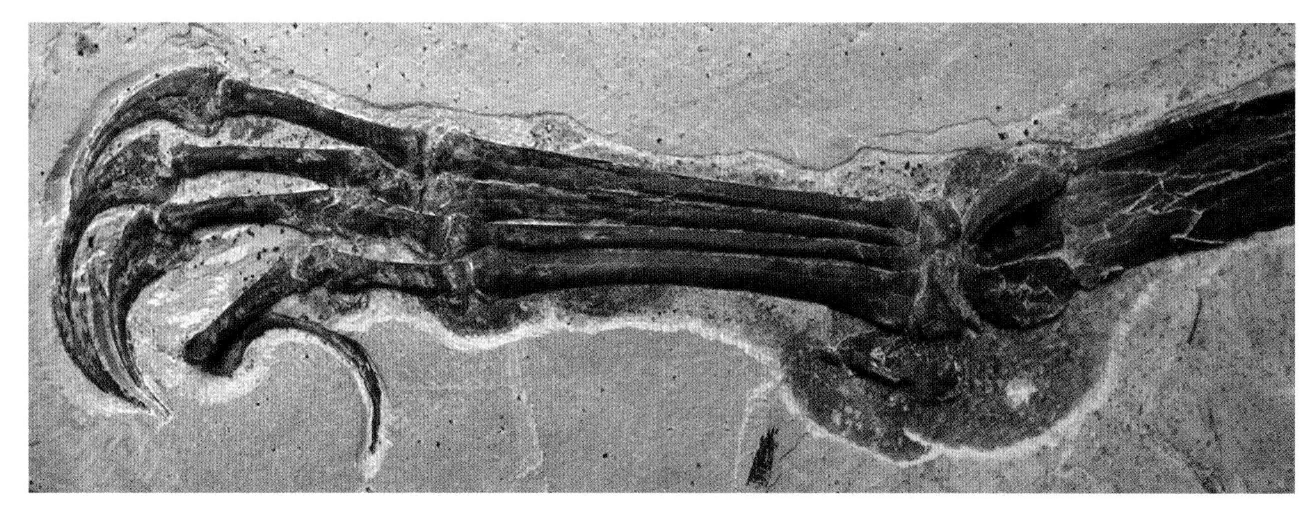

FIG. 5.10. Pterosaur foot pads and claw sheaths preserved in an indeterminate tapejarid pterosaur from the Lower Cretaceous Crato Formation, Brazil. Note the small, polygonal scales on the ankle pad. Image courtesy of Bob Loveridge.

landed. Interestingly, scales are not known from anywhere else on pterosaur bodies. The skin on their limbs appears to have a texture not unlike our own (Unwin 2005). Webbing between pterosaur toes and fingers is also known, with fine fibers running along their lengths (Frey, Tischlinger, et al. 2003). Both the foot and hand claws were adorned with long, curved, and finely tapered claw sheaths, and those of the hands were particularly large (Frey, Tischlinger, et al. 2003; Kellner et al. 2009). Impressions of these details of hand and foot

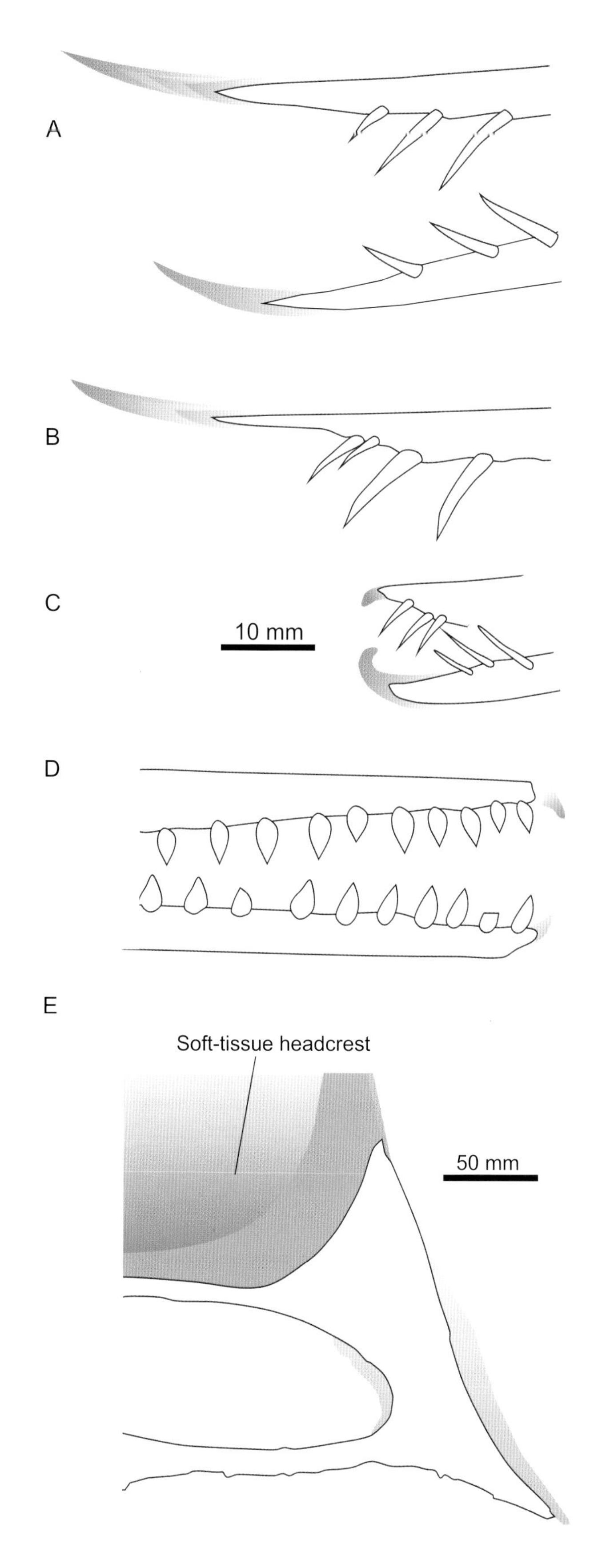

FIG. 5.11. Soft tissues of pterosaur crania. Rhamphothecae of *Rhamphorhynchus muensteri* (A–C), *Pterodactylus* sp. (D), and rhamphothecae, crest, and nasal septum of *Tupandactylus navigans* (E). Tan shading on E indicates extent of fibrous supportive bone for the soft-tissue headcrest. A–C, redrawn from Wellnhofer (1975); D–E redrawn from Frey, Tischlinger, et al. (2003).

anatomy have been found in well-preserved pterosaur footprints (e.g., Hwang et al. 2002).

The integument of pterosaur heads is also known in some detail. Many pterosaurs, even those that bore teeth, had hard, keratinous extensions of both jaw tips that we could consider "beaks" (figs. 5.11–5.12; Wellnhofer 1975; Frey, Tischlinger, et al. 2003). Exquisitely preserved fossils of toothless pterosaurs show that these rhamphothecae extend over much of the jaw (Frey, Martill, and Buchy 2003a), which agrees with observations that pterosaur jaw bones, like those of birds, often have deep blood vessel channels impressed into their surface by the existence of a tough, horny covering. Similar blood vessel imprints are also found on bony pterosaur crests , suggesting these may also have been covered in a similar horny material (e.g., Kellner and Campos 2002).

Many pterosaur headcrests, including the largest known, were almost entirely comprised of soft tissue (figs. 5.12–5.13). These were supported by a ridge of fibrous bone along the dorsal skull surface (e.g., Campos and Kellner 1997) or by small bone promontories at the anterior crest base (e.g., Czerkas and Ji 2002). A great number of pterosaurs possess these structures on their skulls, indicating that many more pterosaurs bore soft-tissue crests than we have direct evidence for in the fossil record. Some soft tissue crests were also supported by small, triangular soft-tissue structures projecting from the back of the skull, which superficially resemble the bony supraoccipital crests of other pterosaurs (Wellnhofer 1970; Frey, Tischlinger, et al. 2003). Astonishing preservation of color banding in one headcrest indicates that these structures could be strikingly colored, which has some significance for interpretations of pterosaur cranial crests as communication devices (see chapter 8) (Czerkas and Ji 2002). Recent finds also indicate that some pterosaurs bore soft-tissue headcrests without any clear osteological signatures at all (Frey, Tischlinger, et al. 2003; Tischlinger 2010), meaning that some doubt now exists over whether some pterosaurs were genuinely crestless, or have merely lost all trace of their headgear *en route* to fossilization.

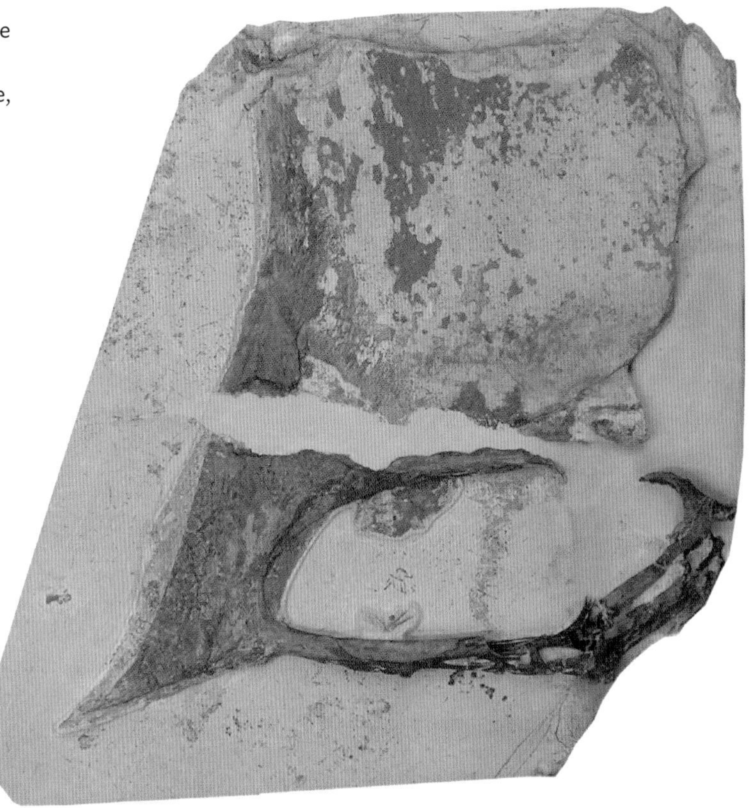

Fig. 5.12. The skull of *Tupandactylus navigans* from the Lower Cretaceous Crato Formation, Brazil, showing extensive soft-tissue preservation of rhamphothecae, crest, and nasal septum. Image courtesy of David Martill and Bob Loveridge.

Fig. 5.13. UV photograph of *Pterorhynchus wellnhoferi* from the Jurassic Tiaojishan Formation, China, showing a fantastic soft-tissue headcrest with color banding. Image courtesy of, and copyright to, Stephen Czerkas 2002.

Pterosaur bodies were adorned with pycnofibers, the recently coined name for pterosaur fuzz (Kellner et al. 2009). A hairlike integument was first reported for pterosaurs in 1831, and again in 1927 (Goldfuss 1831; Broili 1927), and despite some periodic controversies and initial resistance to the idea, it is now universally accepted that most or all pterosaurs were covered in some kind of pelage (see Wellnhofer 1991a; and Bakhurina and Unwin 1995a for reviews; and Kellner 1996b for a different view). Pycnofibers seem to have been fairly short (only 5–7 mm in some specimens), tapering, and flexible structures apparently lacking any internal detail aside from a central canal. They also may not have been anchored deeply into the skin, unlike the hair of mammals (Bakhurina and Unwin 1995a). Opinions differ over exactly how dense pterosaur pelts were. Unwin (2005) suggested that their fibers were no more concentrated than the hairs on a human arm, but others have compared the heaviness of their coats to those of furry mammals (e.g., Sharov 1971; Frey and Martill 1998; Czerkas and Ji 2002). The latter interpretation may be more accurate. Pterosaur pelts appear to be preserved in concentrated, dense mats of fibers similar to those surrounding fossil mammals, suggesting they were just as fuzzy as our fossil ancestors. Pycnofibers seem to have adorned the faces, but not the jaws, of some (perhaps most) pterosaurs, and they may have covered the entire heads of some species (Bakhurina and Unwin 1995a). The necks, bodies, and proximal limbs also bore pycnofibers, with particularly long fibers running along the back of the neck in some forms (Frey and Martill 1998). Some unusual pterosaurs possessed short pycnofibers on the distal trailing edges of their main wing membranes (Kellner et al. 2009), which may have provided predatory or aerodynamic advantages (see chapter 11).

Presently, it looks like pterosaurs developed their pelage independently of other animals, though it has been proposed that pycnofibers are early feathers homologous with those seen in theropod dinosaurs (Czerkas and Ji 2002). This intriguing idea would mean that all ornithodirans were ancestrally feathered, but close examination of pycnofibers suggests they are not structurally comparable to early dinosaur feathers (Unwin 2005; Kellner et al. 2009). Moreover, for the time being at least, incontrovertible feathers seem largely restricted to a distantly related group of dinosaurs, the coelurosaurian theropods. New dinosaur discoveries, however, reveal that feather-like or hairlike integuments may be more common to the group than previously appreciated (Zheng et al. 2009), so perhaps the idea of a completely fuzzy Ornithodira should not be ruled out just yet.

PTEROSAUR PYCNOFIBERS: INSULATING WARM BODIES?

Exactly why coats developed in pterosaurs is not fully understood, but the likeliest explanation may be that they provided insulation against external temperature changes. Body temperature is a rather fundamental component of animal physiology as our biochemistry is optimized at temperatures around 37°C, or perhaps a little higher in some species. Our proximity to this temperature determines how efficiently our muscles and organs work, and thus how generally active we can be. Body temperatures in reptiles rise and dip with that of the external environment (a condition known as heterothermy), but mammals and birds maintain a constant high body temperature close to our biochemical optimum by burning energy roughly 10 times faster than a comparably sized reptile (homeothermy). Burning energy this fast comes at a cost, and requires constant refueling with food to stoke metabolic fires. Thus, to conserve self-generated heat and keep energy requirements low, many homeotherms have developed insulating coverings such as feathers and fur. With heterotherms never developing similar structures for thermoregulation or any other reason, the presence of pycnofibers in pterosaurs is likely to indicate a homeothermic physiology more akin to that of mammals and birds than that of modern reptiles. Unwin (2005) cautioned against using pycnofibers as evidence of homeothermy, suggesting that they may have instead developed to hold air flows close to the body during flight to reduce drag. This is possible, but there are no modern gliding animals that have developed pelage for this purpose, and the absence of pycnofibers from pterosaur wings argues against their dedicated aerodynamic function. It seems far simpler to interpret pycnofibers as we do analogous integuments in fossil dinosaurs and mammals—as a form of insulation. In concert with features indicative of an active locomotory style and the capacity for rapid growth (see chapter 8), pycnofibers strongly indicate that pterosaurs were probably hot-blooded, active homeotherms.

Wing Membranes

Some of the most extensive soft tissue structures of the pterosaur body were their wing membranes. The presence of wing membranes was predicted early on for pterosaurs because of their elongate fourth fingers (Taquet and Padian 2004), and by the late 1800s, this was confirmed when remains of pterosaur membranes were

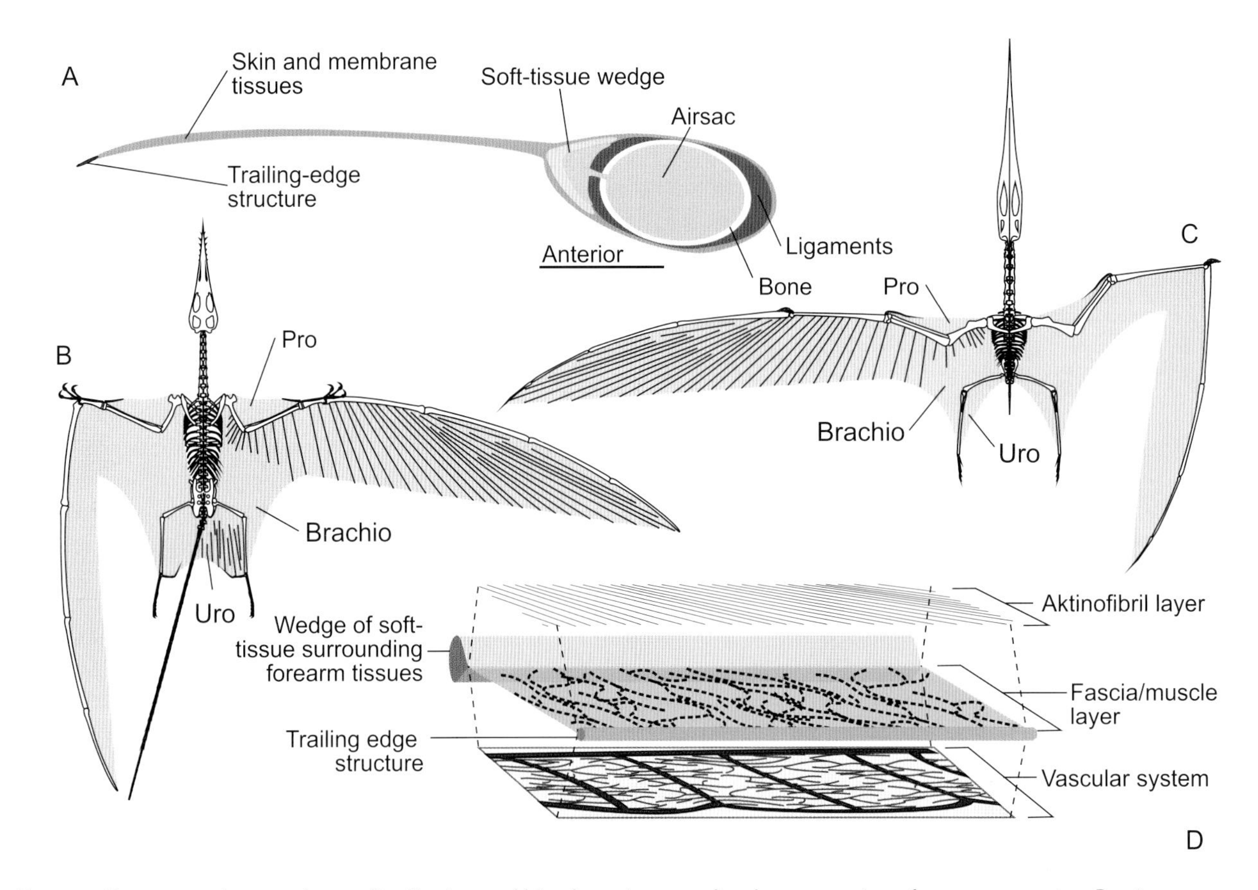

FIG. 5.14. Pterosaur wing membrane distribution and histology. A, generalized cross section of a pterosaur wing; B, wing membrane distribution in *Rhamphorhynchus muensteri* (note the extensive uropatagium supported by the long fifth toe); C, wing membrane distribution in *Pteranodon* (note the vestigial, split uropatagia); D, detailed membrane histology. Abbreviations denote wing membrane divisions. Pro, propatagium; Brachio, brachiopatagium; Uro, uropatagium. A, based in part on Tischlinger and Frey 2010; aktinofibril distribution of B and C largely based on Bennett 2000; D, based on Frey, Tischlinger, et al. (2003).

discovered in the fossil record (see Padian and Rayner 1993 for a review). Much research has been conducted into these structures because of their complex anatomy (fig. 5.14), functionality, and frequently controversial preservation states. It is now generally agreed that pterosaurs bore three sets of membranes: one between the wrist and shoulder (the propatagium); the main flight membrane between the arm, wing finger, body, and hindlimb (brachiopatagium); and another between their legs (uropatagium).[4] Agreement among many pterosaur workers on this issue ends there, however, with the shape, distribution, and histology of these membranes all hotly debated in recent years.

[4] Tail vanes, which could be considered a fourth type of membrane, are also known in some species of long-tailed pterosaur. It is generally assumed that all long-tailed forms had these structures, though direct evidence for them has only been found in one group, the Rhamphorhynchidae. We will discuss them more specifically in chapter 13.

MEMBRANE HISTOLOGY

The complexity of pterosaur wing membranes only became apparent recently as specimens preserving detailed membrane tissues, as well as the technology to view them, have only become available within the last few decades. At one time, the best insight into wing membrane histology was a controversial piece of possible wing tissue from Brazil, thought to reveal a thin epidermis, a mysterious spongy layer, stiffening fibers (see below), and a layer of muscle compressed into a millimeter or so of membrane tissue (Martill and Unwin 1989; Martill et al. 1990; Frey, Tischlinger, et al. 2003). Recent studies of the same specimen have argued that it does not contain tissues seen in other pterosaur wing fossils and may be better interpreted as a piece of skin from the body (Kellner 1996b; Bennett 2000; Hing 2011). Thankfully, other incontrovertible and exquisitely preserved wings have since provided

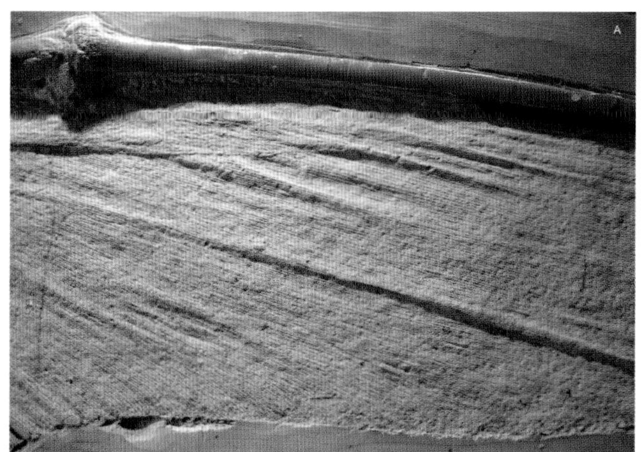

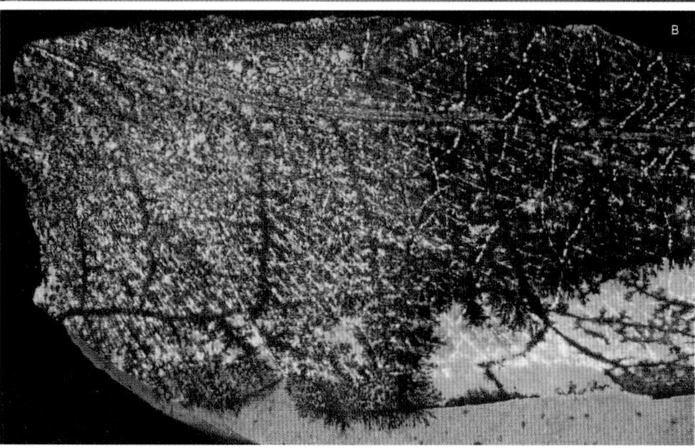

FIG. 5.15. Exquisitely preserved *Rhamphorhynchus* wing specimens from the Jurassic Solnhofen Limestone of Germany. A, the famous "Zittel Wing" (housed in the Bavarian State Collection of Palaeontology and Geology, Munich; used with permission); B, detail of the counter slab of the specimen known as the "Darkwing *Rhamphorhynchus*." Note the stiffening fibers, or aktinofibrils, in both images and the looping, branching blood vessels in the Darkwing specimen. Photograph B courtesy of Helmut Tischlinger.

independent evidence that the wings contained an extensive network of blood vessels and muscle fibers (Frey, Tischlinger, et al. 2003), and thereby shifted research focus away from the more ambiguous Brazilian material (fig. 5.15).

Though thin, pterosaur wing membranes seem to have been very sophisticated structures—far more so than the wing membranes of bats—that were segregated into several different layers and regions (Fig. 5.14D; Frey, Tischlinger, et al. 2003). The blood vessels seem to form the lowermost known layer, and occur beneath a layer of muscles and fascia, or connective tissues. The third and uppermost tier comprises a network of structural wing fibers, unique pterosaur inventions used to strengthen and stabilize the membrane. Proximal to the wrist, these fibers were short and highly flexible, apparently allowing the wing to be highly mobile and elastic, but those of the distal wing were long and relatively stiff (Unwin and Bakhurina 1994; Lü 2002). The longer of these fibers (or aktinofibrils, as they are sometimes called) radiate out from the wing bones toward the distal and trailing edges of the membrane and seem to have made the distal wing far tougher, and thus more likely to preserve than the proximal region. Even in complete pterosaur specimens where the distal wing membranes are excellently preserved, the proximal regions are often missing (Unwin and Bakhurina 1994). Accordingly, the structure of the distal wing is much better known than the proximal, and has been documented in detail (Wellnhofer 1975, 1987b; Padian and Rayner 1993; Bennett 2000; Frey, Tischlinger, et al. 2003; Kellner et al. 2009; Lü 2009a). Aktinofibrils may bifurcate as they approach the wing edge, and because each is only 0.2 mm across, can be up to 2000 times longer than wide. They appear to have been stiff enough to largely resist bending when the wing was folded, instead spreading or contracting in a fan-like fashion as the wing was opened and closed (Bennett 2000). In all likelihood, this would mean that the distal membrane would never totally disappear from sight, like those of bats, when the wing was folded. Recent discoveries have revealed that the aktinofibrils were anchored to the wing finger within a wedge of soft tissue (fig. 5.14A; Tischlinger and Frey 2010), a structure that may have served to not only strengthen the aktinofibril attachment but also to streamline the leading portion of the wing.

Exactly what role aktinofibrils had in the wing is controversial. Some argue that they were primarily involved in expanding the wing and creating tension while in flight (Bennett 2000; Unwin 2005), while others suggest they allowed the wing to camber and transfer flight loads to the forelimb bones (Padian and Rayner 1993). The ability of aktinofibrils to bend with mere folding of the wing elements would render them poorly suited to forming a neat camber, however (Bennett 2000), and it seems most likely that they acted in concert with a mysterious ligament-like structure identified along the trailing edge of the wing (the so-called tailing edge structure; Wild 1994; Frey, Tischlinger, et al. 2003) to hold the wing taut and prevent membrane flutter during flight.

SIZE AND SHAPE OF THE PROPATAGIUM

Although only small (fig. 5.14B–C), the propatagium is an important agent in generating lift in modern birds (Brown and Cogley 1996), and was probably of similar importance to pterosaurs (Wilkinson et al. 2006). The pteroid, the enigmatic bone attached to the pterosaur wrist, has long been recognized as a supporting agent of the propatagium, but its orientation from the wrist and the resultant shape of the propatagium has been hotly contested. While most have assumed that the pteroid pointed medially, several authors (e.g., Frey and Reiss 1981; Unwin 2005; Wilkinson et al. 2006; Wilkinson 2008) predict that the pteroid was directed straight forward, granting much greater lift by expanding the size of the propatagium considerably. This orientation would be impossible if the pteroid attached to the medial side of the preaxial carpal and anterior surface of the syncarpal, however, which it almost certainly did (Bennett 2007a; also see chapter 4). Moreover, biomechanical modeling of the pteroid suggests that even modestly sized, forward-pointing pteroids would generate so much lift during flight that their slender shafts would probably snap under the strain (Palmer and Dyke 2010). These observations provide good reason to assume that pteroid bones were pointed medially, making the propatagium a relatively small, triangular membrane between the wrist and shoulder, comparable in shape to those of birds and bats. Given that the "small" propatagia of these forms generate more than enough lift for flight, along with the likelihood that pterosaurs were probably no heavier than these animals (see chapter 6), there is no reason to assume that the propatagia created by medially pointing pteroids would be insufficient lift generators for pterosaur flight.

SIZE AND SHAPE OF THE BRACHIOPATAGIUM

The extent of the primary pterosaur wing membrane is one of the longest running controversies in pterosaur research (see Unwin 1999; and Elgin et al. 2011 for reviews). A wealth of fossil evidence demonstrates that pterosaur wings were narrow distally, but the shape and attachment site of the proximal wing has been strongly contested. Classically, pterosaurs were thought to have "bat-like" brachiopatagia that extended to their ankles or lower legs, but later proposals argued for hip (the "birdlike" model), thigh, or knee attachments. These arguments took into account all sorts of observations on membrane structure, terrestrial abilities and flight mechanics, but crucially, there was little in the way of fossil evidence for any particular model prior to the 1990s. Exceptions included a single *Pterodactylus* specimen (informally dubbed the "Vienna Specimen" after its housing in the Viennese Naturhistorisches Museum; see fig. 19.12) showing an ambiguous thigh attachment (Wellnhofer 1987b), and a few other specimens showing ankle attachments (Sharov 1971). The number of pterosaur fossils with preserved proximal wing regions has grown to 11, and to greater and lesser extent they all support the ankle attachment model (Naish 2010; Elgin et al. 2011). Because these 11 samples cover almost the entire spectrum of known pterosaur diversity, it is safest to assume that *all* pterosaurs bore ankle-attached brachiopatagia until proven otherwise (fig. 5.14B–C). Crucially, evidence for a hip or thigh attachment is still wanting and, because we now appreciate how flexible and elastic the proximal membrane region seemed to be, the controversial thigh attachment of the Vienna specimen has been reinterpreted to represent shriveling of the membrane after death (Elgin et al. 2011).

This has not quite stopped all arguments over brachiopatagial shapes, however. Instead, attention has turned to the curvature of this structure as it approached the body. Did it turn in slowly from the narrower distal wing, forming a broad proximal region, or sharply, continuing the reduced wing profile close toward the body and only extending to the distal leg as a narrow band of tissue? Narrower wings with sharply turning proximal borders are favored by those studying pterosaur flight, but we continue to wait for fossil evidence to verify their conclusions.

SIZE AND SHAPE OF THE UROPATAGIUM

The membrane spanning the hindlimbs of pterosaurs has been variably interpreted, and has been modeled as spanning the entire area between the legs; existing as two separate membranes trailing along the back of each limb; attaching at the hip or at the tip of the tail; or in some cases, not existing at all (e.g., Padian 1983a; Wellnhofer 1987b; Unwin and Bakhurina 1994; Bennett 2001). Once again, evidence in the shape of exceptionally preserved specimens suggests that two uropatagial configurations existed among pterosaurs. Pterosaurs with long fifth toes bore membranes occupying the entire space between the legs, with the long fifth pedal digit supporting its posterior edge (fig. 5.14B) (Unwin and Bakhurina 1994; Wang et al. 2002; Bakhurina and Unwin 2003). The rearward edges of these membranes were not straight, with the region between the supporting toes showing a V-shaped notch. Pterosaurs lacking large fifth toes, by contrast, appear to have had a rather reduced uropatagium comprised of two shallow membranes extending along the back of each leg from the ankle to the top of the thigh (fig. 5.14C) (Wellnhofer 1970; Unwin 2005). The splitting of the uropatagium doesn't actually reduce relative wing area that much (Witton 2008a), but these split-winged pterosaurs may have been more unstable when flying and, in the process, more acrobatic.

6 Flying Reptiles

The ability of pterosaurs to fly has almost certainly played a role in propelling them to paleontological stardom while other, equally interesting fossil species are left in the shadows (fig. 6.1). Our understanding of pterosaur flight capabilities, however, is still in its infancy and often rather controversial. Fundamental factors of pterosaur flight mechanics are being debated as our understanding of pterosaur anatomy—and animal flight in general—is continually developing. Consequently, there is still a lot to be learned about this defining pterosaur characteristic.

Why Fly?

We take it for granted that pterosaurs were volant, but why did they take to the air in the first place? Pterosaurs did not evolve flight simply because the Triassic world had a vacancy sign over flying vertebrate niches. Flight had to offer pterosaur ancestors clear advantages to drive its evolution. Pursuit of flying insect prey is perhaps the most widely reported catalyst for the development of pterosaur flight (e.g., Unwin 2005), but this idea is not without some problems. Insects may have taken wing as early as the Devonian period (416–359 Ma) and, for certain, were flying by the late Carboniferous (320–299 Ma; Engel and Grimaldi 2004). This gives them a head start over vertebrate fliers of at least 75 million years, allowing them to become agile, adept aeronauts while tetrapods were still learning to walk. Protopterosaurs, essentially flying with their training wheels on, would have been underpowered, sluggish, unsophisticated fliers by comparison (see chapter 3) and would have had as much chance of catching a flying insect as the first tetrapods would have of catching an Olympic sprinter. Modern birds and bats demonstrate that only the most agile and maneuverable vertebrate fliers are capable of routinely catching insects on the wing (e.g., Norberg and Rayner 1987; Rayner 1988; Paul 1991), and early pterosaurs were probably a long way from such aerial feats.

If not food, then what else might have driven the evolution of pterosaur flight? Perhaps pterosaur ancestors simply enjoyed the many advantages that volant locomotion offers, with the chief boon being the terrific efficiency of locomotion achieved by flying animals. Although energetically demanding, the amount of distance covered while flying far exceeds that of walking or running for the same energetic cost. It also presents the ability to pass over geographical barriers such as valleys, highlands, bodies of water, or simply the gulf between one elevated perch and the next without the arduous and possibly dangerous task of traversing them directly. Flying animals can also reach environments that many nonvolant animals would struggle to (such as islands or treetops), which makes it far easier to find safe places to reproduce or rest. Predator evasion is also much easier for a flying animal, as simply taking to the skies at the sign of trouble is a near perfect escape strategy. None of these reasons apply to pterosaurs more than any other flighted animals, and it's likely that the same factors promoted the development of flight in insects, birds, and bats.

One Hundred Years of "Ultralight Air Beings"

Quantitative research into pterosaur flight began at the turn of the twentieth century (Hankin and Watson 1914) and has continued with gathering momentum to the present day. A key assumption in much of this research brings many of its conclusions into question for modern pterosaurologists, however: our predications of pterosaur body masses may have been much too low in the majority of these studies. Body mass is one of the three critical parameters required for even rudimentary flight studies because it influences virtually all aspects of flight (wingspan and wing area are the two other essential values). Therefore, it is critical that we estimate it accurately if we hold any hope of understanding of pterosaur flight kineamtics. For much of the twentieth century, pterosaurs were thought to have extremely low masses for their body size. Pterosaurs with a wingspan of 7 m were widely cited to mass only 16 kg, and some predications of an individual with a 10–11 m wingspan are as low as 50 kg (see Bramwell and Whitfield 1974; and Unwin 2005 for examples; and Witton 2008a for a review). The assumption that pterosaurs were "ultralight airbeings"

Fig. 6.1. Azhdarchid pterosaurs soar over a Cretaceous swamp, while sauropod dinosaurs bathe in the waters below.

(as termed by Gregory S. Paul in 1991) developed from early researchers concluding that their highly pneumatized skeletons were indicative of "weight-saving [being] carried to the extreme" (Hankin and Watson 1914). This idea became entrenched in subsequent generations of pterosaur researchers to the point where it was almost an accepted "fact" of pterosaur paleobiology, so few questioned the remarkable dichotomy between the low estimated masses and pterosaur body size (e.g., Brown 1943; Bramwell and Whitfield 1974; Brower 1983; Brower and Veinus 1981; Rayner 1988; Hazlehurst and Rayner 1992; Wellnhofer 1991a; Chatterjee and Templin 2004; Unwin 2005). As we discussed in chapter 4, pneumatic skeletons do not necessarily reflect proportionally lower skeletal masses, so inferring that pterosaurs were extremely lightweight because of their pneumatic anatomy is probably erroneous. The low masses predicted for pterosaurs necessitated that they were also undermuscled relative to their skeletal volume, leading to suggestions that they were underpowered fliers that principally soared and glided, and could only flap sporadically. Takeoff could not occur without assistance from slopes, cliff edges or headwinds and, according to some researchers, they could only fly in relatively calm conditions. Too much wind or rain would buffet pterosaurs around the sky like oversize kites before dashing their weak bodies against rocks and breaking their delicate skeletons to matchwood. With prospects like that, it's a small wonder that they went extinct.

Under close scrutiny, however, the notion of ultralight pterosaurs fares poorly. For one thing, such mass estimates don't just make pterosaurs lightweight compared to nonflying animals; they also make them two or three times lighter than comparably sized birds and bats (fig. 6.2). While modern animals cannot directly tell us whether the ultralight masses are correct, they do demonstrate that animals of comparable size to smaller pterosaur species can load their wings with much higher masses and be flight worthy (Witton 2008a). Moreover, the overall density of an ultralight pterosaur body (that is, mass divided across the total body volume) is unrealistically low. Modern flying birds have densities ranging from 0.6–0.9 g/cm^3, but Paul (2002) calculated that a 16 kg *Pteranodon* with a 7 m wingspan would have a body density of 0.4 g/cm^3. A 50 kg, 10 m span pterosaur may have a density as low as 0.1 g/cm^3, or put another way, would be comprised of *90 percent air* (Witton 2008a). The explanation for these startling figures is that many pterosaurs were very big animals, comparable in stature with giraffes at their largest extreme. Stretching ultralight masses over animals of pterosaurian proportions barely provides enough meat to deck out the shoulder girdle in flight muscles, let alone anything else (Paul 2002). In short, it seems unlikely that ultralight pterosaurs would have had enough mass to *exist*, let alone fly.

The idea that pterosaurs may have not been ridiculously lightweight is catching on, with many authors now considering the idea that they were of comparable mass to modern fliers (e.g., Witton 2008a; Habib 2008; Palmer and Dyke 2010; Sato et al. 2009; Henderson 2010). Hence, we appear to be moving toward a world where a 7 m span pterosaur weighed something more like 35 kg, and a healthy 10 m span giant would hit 250 kg (Witton 2008a). The implications of this are interesting, because when armed with such figures, pterosaur flight starts to make a lot more sense. With greater masses overall, proportionally more bulk can be allocated to muscle tissues. This makes takeoff easier than it was at lower masses, because the greater energy output compensates for the greater overall weight. Likewise, pterosaur flight would be faster, more powerful, and capable of withstanding more adverse conditions than previously appreciated. In short, this fresh look at pterosaur flight is suggesting that not only were pterosaurs capable of flight, but that they actually excelled at it.

Four Legs Good?

Our understanding of another fundamental component of pterosaur flight—their takeoff—has also recently received an intellectual shot in the arm. Exactly how pterosaurs took off has long been a problem, and particularly so if they are considered to have the higher masses discussed above (e.g., Chatterjee and Templin 2004). Until recently, it has been assumed that pterosaurs launched like birds, using the hindlimbs to leap into flight. Under this take off regime, large pterosaurs would need running starts, slopes, headwinds, sheer drops, and possibly even different gravitational or atmospheric conditions to become airborne (e.g., Sato et al. 2009). Perhaps pterosaurs living in coastal environments may have been able to find enough cliffs or wind to regularly employ them in takeoff, but the same cannot be said for the many pterosaurs occurring in continental deposits, sometimes hundreds of miles away from the nearest coastline. Of particular note here is that the largest and heaviest pterosaurs, the azhdarchids, were most abundant in terrestrial settings (see chapter 25). Azhdarchids would need the longest runways, strongest headwinds, and steepest slopes to take off using avian launch mechanisms, and yet they seem to have frequented habitats with the least reliable wind and topographic conditions available. Adding more questions to this quandary are

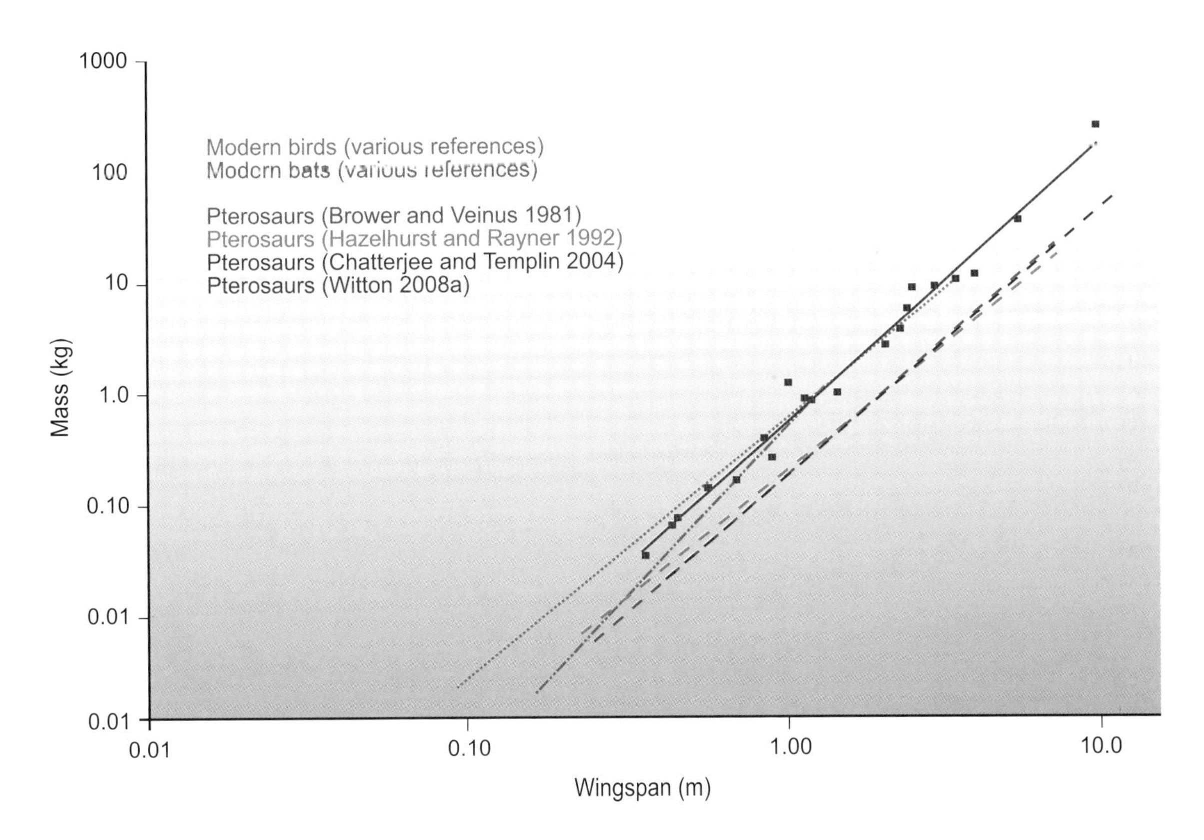

FIG. 6.2. Predicted pterosaur masses compared to those of birds and bats. Colored lines denote best fit lines of various pterosaur mass data sets; dotted line denotes bird masses; dot-dash line denotes bat masses. Note how most estimations of pterosaur mass have presumed far lower masses than those of modern fliers.

recent calculations suggesting pterosaurs weighing above 70 kg, or even 41 kg, were unable to become airborne *at all*, regardless of environmental assistance (Chatterjee and Templin 2004; Sato et al. 2009). Others have argued that these calculations cannot be correct because all pterosaur fossils, even the most gigantic, show characters indicative of flight (Buffetaut et al. 2002; Witton and Habib 2010).

It turns out that one of the core assumptions of many calculations of pterosaur launch, that pterosaurs took off in a birdlike fashion, may be the cause of these confused interpretations. The idea of birdlike, bipedal launching is not without merit as the strength of pterosaur hindlimbs has been noted on several occasions (Bennett 1997a; Padian 1983a), but biomechanicists Jim Cunningham and Mike Habib pointed out that assumptions of pterosaurian bipedal launch are countered by several lines of compelling evidence (e.g., Habib 2008). Firstly, birds and bats launch with the same gait with which they walk and run. Birds both walk and take off bipedally, while bats walk and take off as quadrupeds (see below). With good evidence that pterosaurs were quadrupedal (see chapter 7), it is likely that they also launched from a four-limbed start. Secondly, it turns out that hindlimb launching seriously pumps up the leg skeleton of its practitioners. Bird leg bones increase in size much faster with respect to body mass than other parts of their bodies—even their wings—to meet the demands of heaving themselves into the air. Pterosaurs hindlimbs, by contrast, do not increase in size as quickly as those of birds, but the opposite is true for their strong arms, which are particularly well-developed around the shoulders (Habib 2008). Under mechanical analysis, these bones are more than strong enough to catapult twice the weight of a given pterosaur into the air from a standing start, whereas their hindlimbs would fail at much lower stresses. Greater muscle power is also available to animals employing their forelimbs in launch, as their takeoff uses power from their extensive flight muscles rather than the relatively diminutive leg muscles alone. This allows quadrupedal launchers to heft much greater masses into the air than bipedal launchers, and is therefore consistent with pterosaurs producing much larger and heavier fliers than any bird (Habib 2008).

It seems highly probable, therefore, that pterosaurs took off quadrupedally (fig. 6.3). The launch cycle, as

FIG. 6.3. How to launch a pterosaur, modeled by *Pteranodon longiceps*. A, quadrupedal launch from a standing start; B, quadrupedal launch from the water surface, requiring multiple "hops" to become airborne (see chapter 18 for more on water launching).

predicted by Mike Habib and Jim Cunningham, runs as follows. At first, the pterosaur would crouch before shoving up and forward with its hindlimbs to vault over its own arms. Note that the subsequent push off from this crouch is more powerful than the energy output achieved in a running launch, so standing starts are more effective than moving ones. After the initial push with the legs, the muscular forelimbs push upward, changing the trajectory of the pterosaur body from moving primarily forward to moving forward *and* skyward. While this is occurring, the wing finger extends to open the distal wing. As the animal leaves the ground, a partial upstroke is achieved by the arms being swept above the shoulders. The wing is fully open by the end of the upstroke and full flap cycles can begin, which the pterosaur would continue as it climbs. All this would take place in a very rapid, highly synchronized fashion, and even the largest pterosaurs would be clear of the ground in a second or so.

Some verification that this launch system can work stems from a number of bats (most famously vampires) that launch into the air in a very similar fashion. Because they employ their powerful flight muscles for takeoff, they explode into the air in a particularly powerful and efficient fashion (Schutt et al. 1997). In fact, little vampire bats almost spring into the air vertically, a trick that would not be seen in larger pterosaurs. Climb out angles decrease with size and mass, meaning the largest pterosaurs would require relatively open environments to take off. It would seem odd for very large, gangly, flying animals to frequent dense woodlands or forests anyway, so this may not have been much of an issue. And if this evidence for quadrupedal launching is not convincing enough, there are rumors of a trackway (a series of fossilized footprints) that may even show a pterosaur taking off in just such a fashion (M. B. Habib, pers. comm. 2010).

Now We're Up

Understanding pterosaur flight kinematics is a complex task, but there are some aspects of their flight that we can assume with a fair degree of certainty. Firstly, it seems that no pterosaur yet discovered had become secondarily flightless in the way that many modern birds have (Witton and Habib 2010). Secondly, all pterosaurs appear equipped with large enough shoulder and forelimb anatomy to house muscles substantial enough to power flapping, and their predicted high metabolisms were probably sufficient to sustain this for long periods. Increased body size heightens the metabolic cost of flapping in all flying animals, so bigger pterosaurs may have struggled to flap continuously for as long as their smaller counterparts. Hence, larger pterosaurs likely employed flap-gliding (bouts of flapping interspersed with periods of soaring) or used external sources of lift (e.g., thermals, ridge lifts, and so on) to prolong their flight times (e.g., Marden 1994; Pennycuick 1990; Sato et al. 2010). Some readers may think this assumption contradicts observations that the heaviest distance fliers of modern times—some bustards and swans—are continuous flappers, powering their flight with flapping wings from the moment they take off to the time when they land. Swans and bustards are renowned for stretching the limits of animal flight however, using wings and flight muscles that are small and heavily loaded for their body size (Rayner 1988; Marden 1994; Pennycuick 1998). To date, there are no pterosaurs known that show such characteristics. Most reconstructions of pterosaur wings suggest they generally had relatively large, ex-

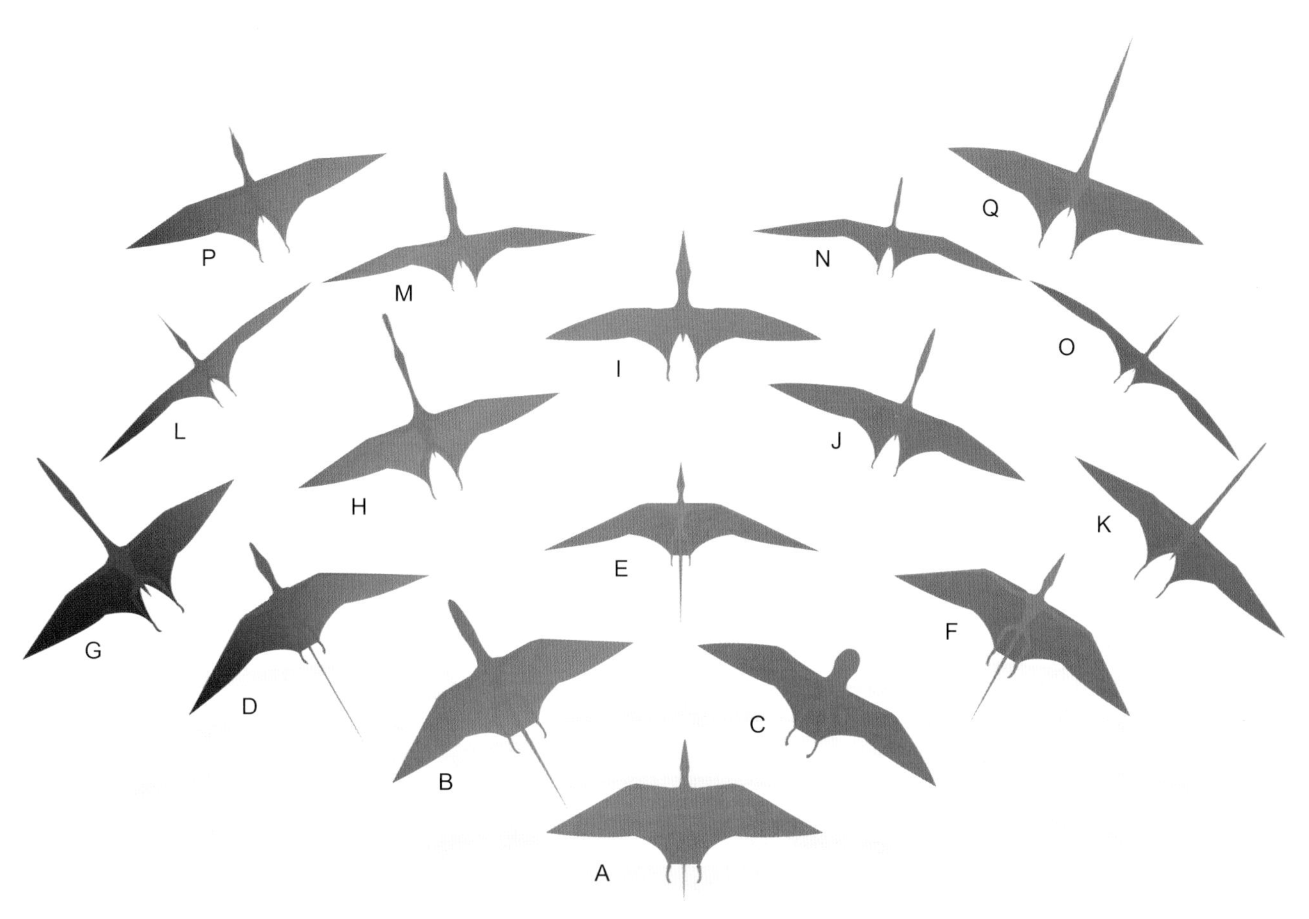

Fig. 6.4. Diversity of pterosaur wing shape, as modeled in Witton (2008a). Predicting the exact shape of pterosaur wings is fraught with difficulty, but these schematics may provide a rough indication of wing shapes in major groups. A, *Preondactylus*; B, *Dimorphodon*; C, *Anurognathus*; D, *Eudimorphodon*; E, *Rhamphorhynchus*; F, *Sordes*; G, *Huanhepterus*; H, *Ctenochasma*; I, *Dsungaripterus*; J, *Pterodactylus*; K, *Pterodaustro*; L, *Pteranodon*; M, *Istiodactylus*; N, *Anhanguera*; O, *Nyctosaurus*; P, *Sinopterus*; Q, *Quetzalcoatlus*.

pansive wing areas and proportionally large flight muscle masses, and were thus constructed rather differently to these heavy, power-flapping birds. Hence, it seems that pterosaur bauplans were better suited to flap-gliding at larger sizes than continuous flapping.

We may also assume that the anatomical and proportional differences between different pterosaurs are indicative of a variety of flight styles (fig. 6.4). Two values of the animal flight apparatus provide useful insights into this: wing loading (a measure of weight supported by a given area of wing) and aspect ratio (wing area divided by wing span). Generally speaking, longer and narrower (high aspect) wings are most efficient for steady flight, but broader (low aspect) wings permit more maneuverable flight. Wing loading directly influences flight speed and thus the ease of takeoff. Spreading the weight of a flying animal over large, broad wings (low wing loading) permits flight at lower velocities and subsequently a slower, less energetic takeoff, but also reduces overall flight speed and glide efficiency (Pennycuick 1971). We can gain some insight into the nature of pterosaur flight by comparing their estimated wing loading and aspect ratios with those of birds and bats. When standardized for size, we find that these modern fliers have convergently developed the same wing loading and aspect ratios for similar flight styles (Norberg and Rayner 1987; Rayner 1988; Hazlehurst and Rayner 1992). For example, natural selection has honed the wings of both birds and bats that chase aerial insects into similar wing shapes and proportions, despite not sharing any evolutionary history for hundreds of millions of years, providing them with the most effective wing shape to catch their prey. Because pterosaur wing shapes and proportions were similarly optimized by natural selection, comparisons of their likely wing loadings and aspect ratios with modern fliers may allow insight into their abilities to hawk insects, soar

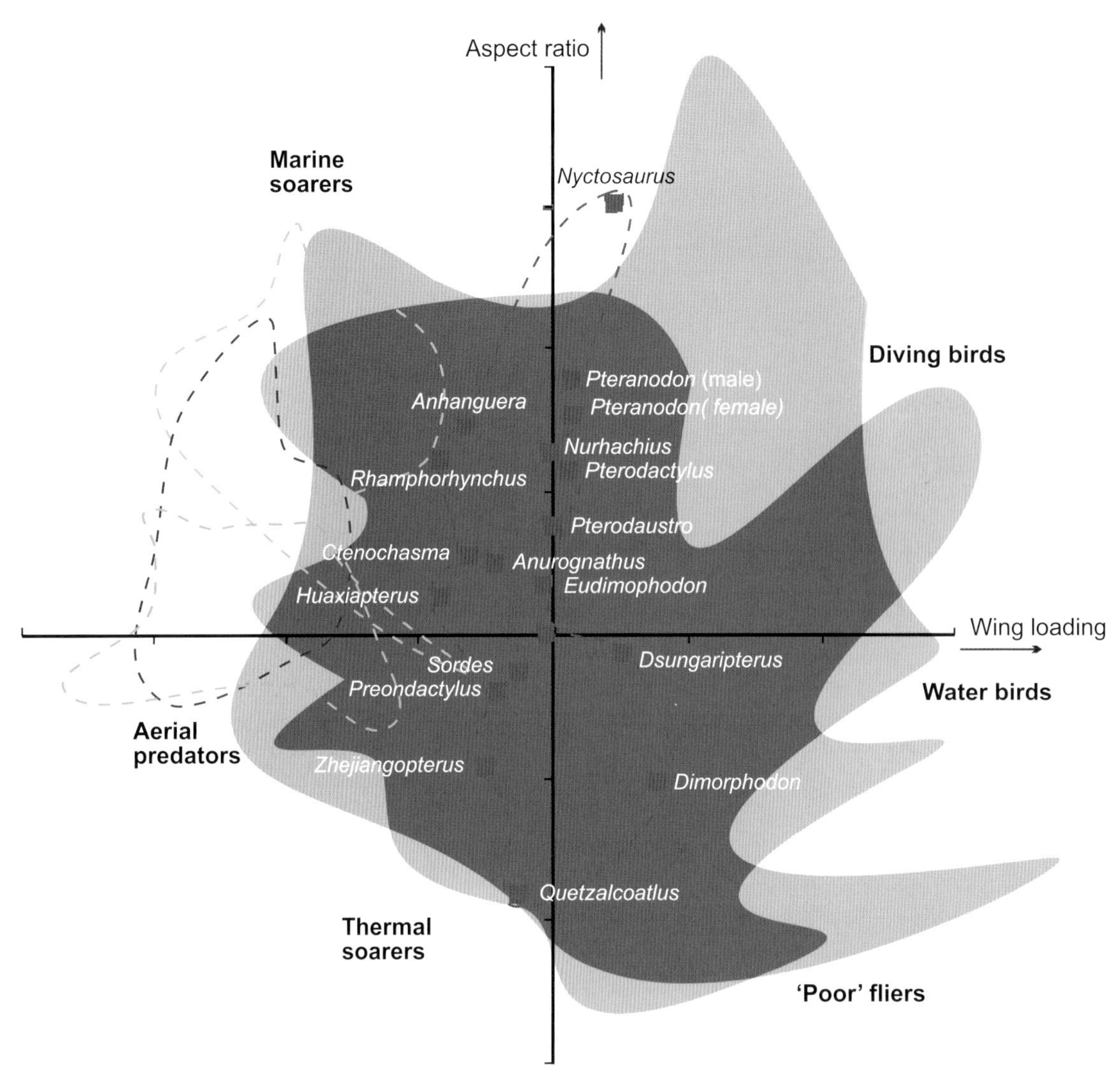

FIG. 6.5. Comparisons of flying vertebrate aspect ratio and wing loading, standardized for size. Gray shading indicates the wing ecomorphospace occupied by modern bats; blue shading reflects modern avian ecomorphospace. Dotted outlines reflect pterosaur data sets with the same color key as figure 6.2. Note how "heavier" masses bring pterosaurs more into line with modern fliers compared to those of previous "lightweight" estimates. Details for the methodology behind this graph can be found in Norberg and Rayner (1987); Rayner (1988); Hazlehurst and Rayner (1992); and Witton (2008a).

on thermals, glide over oceans, and so forth (fig. 6.5; Hazlehurst and Rayner 1992; Witton 2008a). When combined with biomechanical analyses of pterosaur skeletons, detailed insights into their flight ability can be obtained.

Unfortunately, much of this science is still in its early stages and the flight of specific pterosaurs has not been investigated in great detail. Most in-depth pterosaur flight studies have focused on *Pteranodon* (e.g., Hankin and Watson 1914; Bramwell and Whitfield 1974; Stein 1975; Sato et al. 2009), with a couple on the giant *Quetzalcoatlus* (Sato et al. 2009; Witton and Habib 2010), and a handful looking at the flight of a broad suite of pterosaurs (Brower and Veinus 1981; Hazlehurst and Rayner 1992; Chatterjee and Templin 2004; Witton 2008a). The majority of these studies use hyperlight mass estimates and, being performed at a time when the shape of pterosaur wings was less constrained than they are today, typically gave their pterosaurs higher-aspect wings than is supported by recently found fossil evidence. These factors contributed to most of these analyses drawing the common

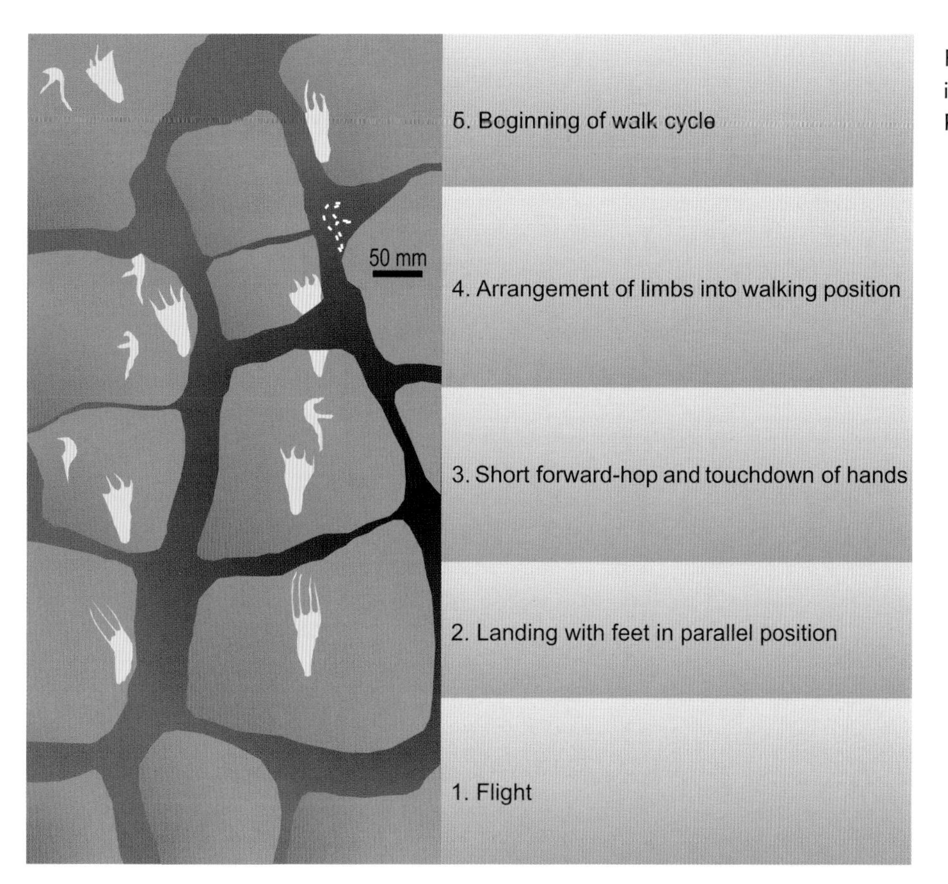

FIG. 6.6. Trackway thought to be made by a landing pterosaur, with interpreted stages of landing. Redrawn from Mazin et al. (2009).

conclusion that most pterosaurs were ancient seabird analogues, gull-like soarers that flew at slower, more maneuverable speeds than any modern fliers. Applying higher mass estimates more in keeping with those of modern animals to pterosaur flight calculations has challenged this idea (Witton and Habib 2010) and, in considering a broader range of wing shapes, opened up the scope of pterosaur flight considerably. There is some evidence that, rather than only using seabird and shorebird flight styles, pterosaurs were also flap-gliders, continental soarers, aerial predators, and sometimes even heavyset, short-burst fliers (Witton 2008a). Recognizing this, we will leave our discussion of pterosaur flight in favor of more specific discussions about the flight dynamics of each pterosaur group in our later chapters.

Grounding

Of course, this talk of flight has left one obvious question: How did pterosaurs land? A recently discovered landing trackway (fig. 6.6) indicates that landing was no trouble for pterosaurs (Mazin et al. 2009). They appear to have lost most of their flight speed before landing, apparently dropping on both feet simultaneously, rather than employing makeshift runways to "run out" their flight speed. Their neat landing strategy would have seen the entire body pitching up, swinging the legs into a landing position and orientating the wings somewhat vertically to act as an airbreak. The pteroid bone would be depressed to provide a deep wing camber, thus maintaining some airflow and lift (remember that the propatagium would generate a massive amount of lift on its own), and the animal would quickly slow to a point where it stalled. High in the air, this would be disastrous because the animal would plummet from the sky, but when close to the ground, the pterosaur would simply drop a short distance to the floor (Chatterjee and Templin 2004; Wilkinson 2008). From this bipedal pose, the animal would then drop forward onto its hands and within a few moments, it was ready to scamper away (Mazin et al. 2009). This neatly brings us to the topic of our next chapter: How did pterosaurs move when they were not flying?

7
Down from the Skies

Pterosaurs may also be known as "flying reptiles," but they were not airborne indefinitely. Eventually, they would have to land in order to feed, rest, or reproduce, and this presented novel locomotory challenges for traversing both land and water. Exactly how pterosaurs did this has been the subject of much debate, and opinions on how well pterosaurs could walk have swung from one extreme to the other in the last 200 years. Until quite recently, most thought that walking pterosaurs had as much grace as a man tangled in a crumpled parachute, scrambling about on weak, flailing limbs and hoping that nothing large and hungry took notice (e.g., Bramwell and Whitfield 1974). Happily, real breakthroughs were made in unraveling pterosaur terrestrial capabilities in the 1990s, so their terrestrial competence is now much clearer (fig. 7.1). It is quite possible that these discoveries would not have been made if it were not for some of the unusual, and sometimes just plain wacky, discussions held in the previous decades about grounded pterosaurs. The history of this debate is just as important, therefore, as its conclusions.

A Brief Stroll through Research into Grounded Pterosaurs

Early pterosaurologists considered their subjects to be both competent terrestrial animals and adept fliers, though exactly how they imagined pterosaurs walking varied. Cuvier (1801) imagined pterosaurs as agile bipeds; Soemmerring (1812, 1817) considered them quadrupeds with a bat-like posture; while Seeley (1870, 1901), also favoring the quadrupedal gait, argued that they walked with erect limbs (fig. 2.4). Stieler (1922) agreed that pterosaurs would be strong terrestrial locomotors, suggesting that they may have used a phase of lizard-like bipedal running to enhance their takeoff. This optimistic view was not to last, however. Convinced that pterosaurs were adapted for flight to the detriment of virtually everything else, most paleontologists of the early to mid-twentieth century considered pterosaurs to be hapless grounded animals. Hankin and Watson (1914), Abel (1925), and Bramwell and Whitfield (1974) thought pterosaurs could barely stand, and instead used their legs to push themselves around on their bellies (fig. 7.2B). So useless were these grounded pterosaurs that they could only become airborne by flopping over a cliff edge to fall into flight (Bramwell and Whitfield 1974)! Similarly, Kripp (1943) suggested that pterosaurs could only rest by hanging on cliffs. Toward the end of the century, Wellnhofer (1988, 1991a) and Unwin (1988b) granted pterosaurs some grace by suggesting they could stand and walk quadrupedally, but they maintained that they would be cumbersome, inelegant forms when grounded (fig. 7.2D).

These interpretations were based solely on studies of pterosaur anatomy, as the first 170 years of pterosaur research had not provided any direct evidence of how pterosaurs walked or ran. Fossil pterosaur footprints or trackways, which could provide vital details of the stances, gaits, and speeds of pterosaurs to these early researchers, had not been identified. Thankfully for pterosaurologists, an unusual fossil trackway was discovered in the Sundance Formation of Wyoming in 1952 that seemed to be of pterosaur origin (fig. 7.3; Stokes 1957). It recorded the rapid movement of a quadrupedal animal with dramatically different manus and pes prints. The hand prints were asymmetrical, three-pronged structures with the digits facing laterally or posterolaterally, while those of the feet were relatively symmetrical and triangular, with an obvious "heel" impression, and four long, laterally curved toes. William Lee Stokes, the discoverer of these tracks, linked their unusual digital configuration to pterosaurs and named them *Pteraichnus,* meaning"wing trace," in honor of their likely pterosaurian origin. (This is not an unusual practice; paleontologists have created a taxonomy and nomenclature for distinctive trackways and other trace fossils in the same mold as that of body fossils). Similar tracks were soon discovered (Logue 1977; Stokes and Madsen 1979), all of which provided undeniable proof that pterosaurs were both capable of walking effectively and were also fully quadrupedal. Finally, a real breakthrough had been made.

What happened next, in retrospect, is baffling. The discovered tracks were ignored for decades in most

FIG. 7.1. A group of azhdarchid pterosaurs mosey down a Cretaceous riverbed, leaving trackways that we will later name *Haenamichnus.* Scenes like this are relatively new for pterosaur artists as the likely nature of pterosaur terrestrial locomotion was only grasped by pterosaur workers in the late 1990s, following centuries of uncertainty, and, at times, heated debate.

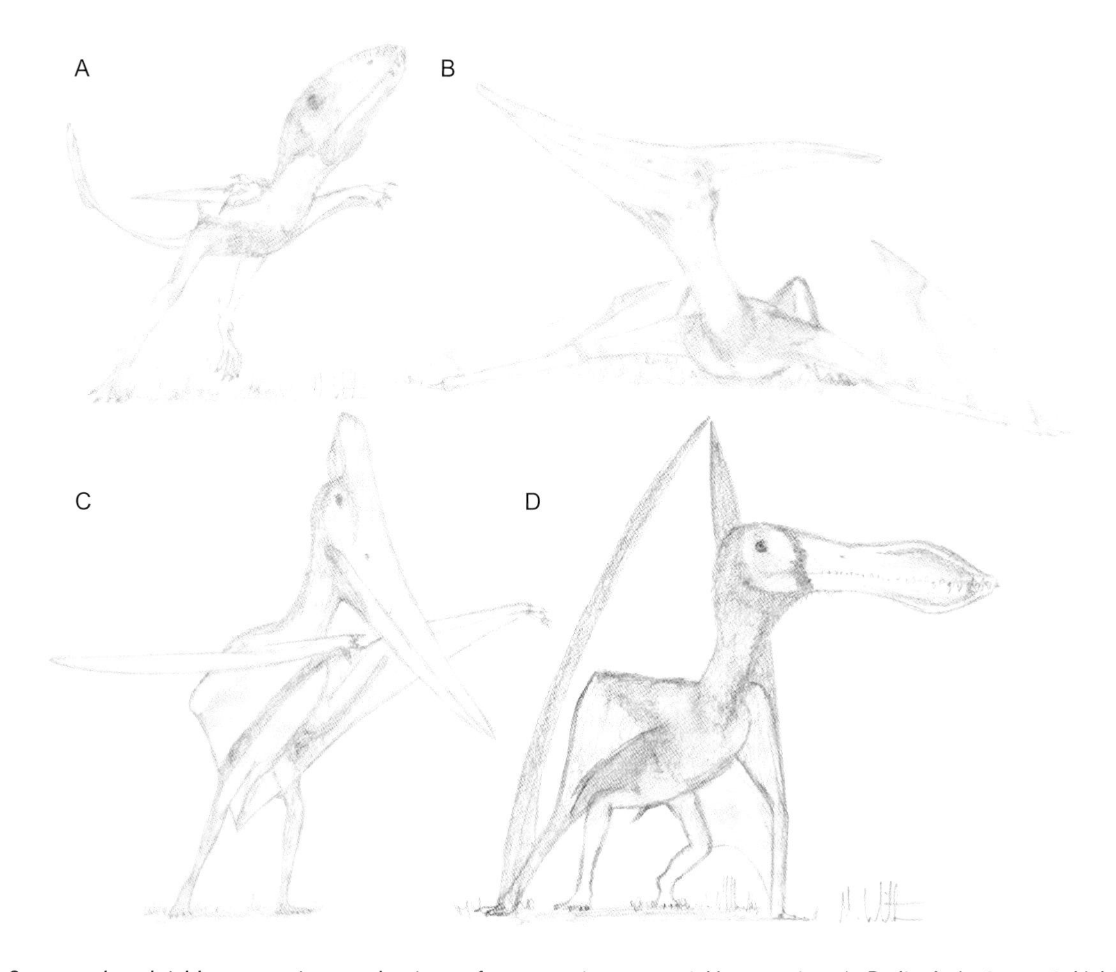

Fig. 7.2. Suggested, and richly contrasting, mechanisms of pterosaurian terrestrial locomotion. A, Padian's (1983a, 1983b) bipedal dinosaur-wanabee *Dimorphodon*; B, Bramwell and Whitfield's (1974) *Pteranodon*, an animal so helpless it must shove itself around on its belly with its legs alone; C, Bennett's (1990, 2001) teetering, fully erect bipedal *Pteranodon*; and D, Wellnhofer's (1988) sprawling, quadrupedal *Anhanguera*.

discussions of pterosaur terrestrial locomotion, and were dismissed decades later as not originating from pterosaurs at all (Padian and Olsen 1984; Unwin 1989; Wellnhofer 1991a). Padian and Olsen were particularly scathing of the idea that *Pteraichnus* was pterosaurian, favorably comparing it with the tracks of a modern caiman to suggest they were made by a crocodile-like animal instead. Unwin (1989) was also unimpressed with the pterosaur identification of *Pteraichnus*, interpreting the skeletal anatomy of pterosaurs to suggest they bore rather different stances and print morphology to those recorded in the tenuously identified trackway. *Pteraichnus* was well and truly sidelined by the time pterosaur terrestrial locomotion became a hot topic once more. By the end of the 1980s, several rather contrasting ideas had emerged after much discussion (fig. 7.2). Perhaps the most famous of these is Kevin Padian's notion that some pterosaurs were bipedal (Padian 1983a, 1983b). This idea stemmed from research into the hindlimb of the Jurassic, long tailed pterosaur *Dimorphodon* and the assumption that the brachiopatagium attached at the hips (see chapter 4). Padian concluded that the pelvic structure of pterosaurs would permit their hindlimbs to stand in a manner identical to those of birds, with erect limbs and digitigrade feet. Crucially, the glenoid fossa was interpreted as having bony stops that prevented the arm from being used in terrestrial locomotion (Padian 1983a, 1983b), so the forelimb was neatly folded away in a birdlike fashion. Chris Bennett (1990) later suggested that short-tailed pterosaurs were also bipedal, albeit in a far more erect stance to compensate for their lack of a balancing tail (fig. 7.2C). This stance was possible, said Bennett, because the pterosaur acetabu-

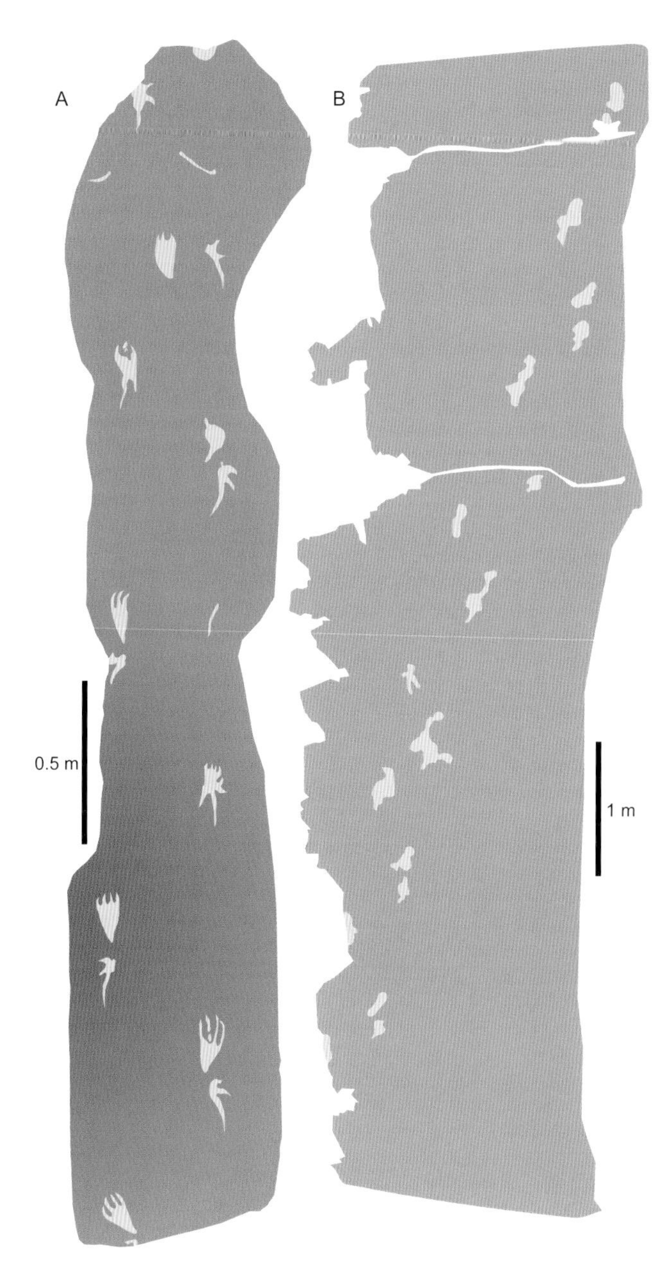

FIG. 7.3. Pterosaur trackways, the once-ignored Holy Grail of research into pterosaur terrestrial locomotion. A, the Late Jurassic *Pteraichnus saltwashensis*, from the Morrison Formation of Utah; and B, the Late Cretaceous *Haenamichnus* sp. from the upper Cretaceous Uhangri Formation of South Korea. Note the narrow gauge of the trackways in both images, and the distinct manus and pes tracks. A, redrawn from Stokes (1957); B, redrawn from Hwang et al. (2002).

lum (hip joint) has a well-developed anterior "shelf" that permitted an upright pose without dislocation of the femur from the hip joint. In addition, the elongate forelimbs of giant pterosaurs were suggested to inhibit anything but a bipedal stance, forcing all large pterosaurs into a bipedal posture (Bennett 2001).

Other authors disagreed with this, suggesting that pterosaurs were quadrupedal animals (a conclusion reached independently of *Pteraichnus*). Using a three-dimensionally preserved *Anhanguera* pelvis from the then newly opened Santana Formation, Wellnhofer (1988, 1991b) stated that the pterosaur femur could not swing below the body in a parasagittal manner (that is, with the limbs held more or less straight beneath the body) because the acetabulum faced dorsolaterally, and the pterosaur femoral head was not sufficiently angled to compensate for this. Wellnhofer criticized Padian's bipedal reconstructions of pterosaurs, noting that the pterosaur foot does not have metatarsals of subequal length and therefore would not permit their owners to assume, as birds do, a digitigrade posture. He further observed that the wings of pterosaurs do not fold up as neatly as Padian suggested and would be hard to stow away when moving bipedally. Wellnhofer accordingly suggested that pterosaurs had a lizard-like semi-sprawling stance, their humeri projecting laterally or posterolaterally and their femora projecting anteroventrally when standing. A further hypothesis, that many pterosaurs were climbing animals that spent most of their downtime scampering through trees or over cliffs, was proposed by Unwin (1987, 1988b). Unwin's idea explained the sprawling stance permitted by the pterosaur hindlimb, their large claws, and the phalangeal configurations of many pterosaur taxa. Paul (1987) disagreed with the notion that pterosaurs had bowed hindlimbs and considered most pterosaurs to be elegant, erect quadrupeds with limbs held almost entirely beneath the body. According to Paul, some earlier forms were also occasional bipeds and skilled climbers.

Back to the Start

Clearly, with so many conflicting interpretations and hypotheses existing at the start of the 1990s, there was a real need for a fresh look at the entire problem. A new approach began in the mid-1990s when pterosaur workers returned to the controversial *Pteraichnus* tracks to reassess their previous dismissal. With fresh eyes, it was thought that *Pteraichnus* had been ig-

FIG. 7.4. *Germanodactylus cristatus* walks this way, showing the narrow gait predicted for pterodactyloid pterosaurs by their tracks.

nored through some pretty shaky arguments. The caiman tracks used to support a crocodilian affinity for *Pteraichnus* (Padian and Olsen 1984) were criticized as being rather poor impressions of pterosaur tracks for a number of reasons. The foot and hand prints are quite unlike those of *Pteraichnus*; the prints were placed in rather different positions; and a pronounced tail drag runs along the middle of the caiman tracks but was missing from purported pterosaur trackways (Lockley et al. 1995). Unwin's (1989) hypothesized pterosaur trackway, once dismissed as being nothing like *Pteraichnus*, was appreciated as being far closer to *Pteraichnus* than previously realized, even if the placement of the individual tracks in the model was a bit off (Lockley et al. 1995). Indeed, it seemed that pterosaur hand and foot anatomy fit the distinctive *Pteraichnus* prints so well that it could not have been manufactured by any fossil reptile but a pterosaur. After decades of neglect, *Pteraichnus* was welcomed back to the debate on pterosaur terrestrial locomtion with open arms (well, mostly: see Padian 2003).

Following this reappraisal, it became clear that pterosaur tracks had been documented from around the world in the decades since Stokes' original *Pteraichnus* discovery (Lockley et al. 1995). The marks of pterosaurs were identified in the United States, Britain, Spain, Mexico, Korea and a host of other places (e.g., Wright et al. 1997; Calvo and Lockley 2001; Hwang et al. 2002; Rodriguez-de la Rosa 2003; Zhang et al. 2006). The crown jewel of pterosaur track sites was found in Crayssac, of southwestern France (Mazin et al. 1995; Mazin et al. 2003; Mazin et al. 2009), where their tracks are both abundant and behaviorally diverse. Sets of giant pterosaur footprints in England and Korea were named *Purbeckopus* (fig. 8.11B) and *Haenamichnus* (fig. 7.3B), respectively (Delair 1963; Wright et al. 1997; Hwang et al. 2002; but also see Billon-Bruyat and Mazin 2003). Recent work has also noted that *Agadirichnus*, a mysterious reptile track identified from Late Cretaceous deposits of Morocco in 1954, may also have pterosaurian affinities (Billon-Bruyat and Mazin 2003). After years of debate, these tracks finally gave pterosaur workers direct insight into the posture, gaits, and locomotory methods of grounded pterosaurs.

How It Worked

Pteraichnus and other pterosaur tracks demonstrated that much of what had been hypothesized about pterosaur gaits and postures from bones alone was largely incorrect. Notions of pterosaur bipedality were the first to be dismissed. All known pterosaur trackways indicate that they habitually walked and ran on four limbs. Their feet were also entirely plantigrade, as shown by the clearly defined ankle prints in each pes track. Clark et al. (1998) provided further details on this plantigrade foot posture using a well-preserved pterosaur foot skeleton from Mexico, which suggested that their toes were incapable of the motion necessary for a digitigrade stance. Thus, the birdlike posture proposed by Padian (1983a, 1983b) was found to be flawed, as was the the upright bipedal posture proposed by Bennett (1990, 2001). The latter was also proved doubtful by suggestions that the muscle layout of pterosaur hips impeded their ability to walk or

Fig. 7.5. An Early Cretaceous azhdarchid makes off with half a dinosaur carcass, pursued by a gang of dsungaripteroids. Pterosaur trackways demonstrate that, in addition to being efficient walking animals, pterosaurs could also run quadrupedally.

run in an upright fashion (Fastnacht 2005b). Bennett's (2001) argument that the elongate forelimbs of giant pterosaurs would inhibit their use in quadrupedal locomotion has also been doubted, because trackways of giant pterosaurs are incontrovertibly quadrupedal (the giant trackmakers in question have estimated shoulder heights of 2.5 m, which is about as large as pterosaurs are known to get) (Hwang et al. 2002).

The carriage of pterosaur limbs was another surprise. Wellnhofer (1988) and Unwin (1989) argued that the disproportionately long pterosaur forelimb dictated that their manus prints should be much further away from the trackway midline than the pes prints. Most pterosaur tracks show a different story, however, with the manus prints almost directly in line with the pes prints, or only slight lateral displacement at most (fig. 7.4; e.g., Stokes 1957; Lockley et al. 1995; Hwang et al. 2002 Mazin et al. 2003). It seems that by swinging their humeri behind and below their shoulder joints, the forelimb was able to move fore and aft directly beneath the body (Bennett 1997a; Unwin 1997). Reassessment of pterosaur pelvic structure reveals that their acetabula were not, as Wellnhofer (1988) argued, dorsolaterally orientated but posterolaterally, allowing the legs to move not only beneath the body but to actually cross before they would dislocate (Bennett 1997a). This gives pterosaurs a relatively efficient stance with their weight carried on nearly vertical limbs, rather than the relatively energy-sapping sprawling postures of living reptiles and amphibians. However, trackways at Crayssac show that pterosaurs did occasionally move with laterally splayed forelimbs, a feature that correlates well with increased stride length (Mazin et al. 2003). Indeed, the stride length of these tracks are so great that, no matter what contemporaneous pterosaur taxon is plugged into them, the animal had to be running (fig. 7.5; Unwin 2005). This suggests that pterosaurs splayed their forelimbs when moving at high speed, a consequence of rotating the body forward to give extra reach to the shoulders.

Pterosaur tracks have also revealed several insights into their walking limb mechanics. We now appreciate that only the first three digits of their hands regularly touched the ground when walking because impressions of the fourth digit, the flight finger, are very rare. Thus, it seems this digit was tightly folded and stowed alongside the body (fig. 7.6), while the other three fingers hyperextended to splay laterally and posterolaterally when placed on the ground (Bennett 1997b; Mazin et al. 1995; Mazin et al. 2003). Of further interest is that pterosaur pes prints are consistently found *in front* of their hand prints, indicating that their hands had to be clear of the ground before the hindlimb could be placed in front of it. This means that pterosaurs had a walk cycle akin to that of long-limbed mammals like camels and giraffes, where the limbs on one side of the body are moved before those of the other side (i.e., the back left limb is moved first, then the front left, then right back, right front and repeat). This contrasts with walk cycles of other quadrupedal reptiles, where the moving limb at any point in the step cycle is diagonally opposite the limb that moved previously (Bennett 1997b; Padian 2003). Pterosaur hand prints also are comparatively deeper than those of their feet, a consequence of their front-heavy bodies. This may explain why some pterosaur tracksites mysteriously only preserve manus prints (e.g., Parker and Balsley

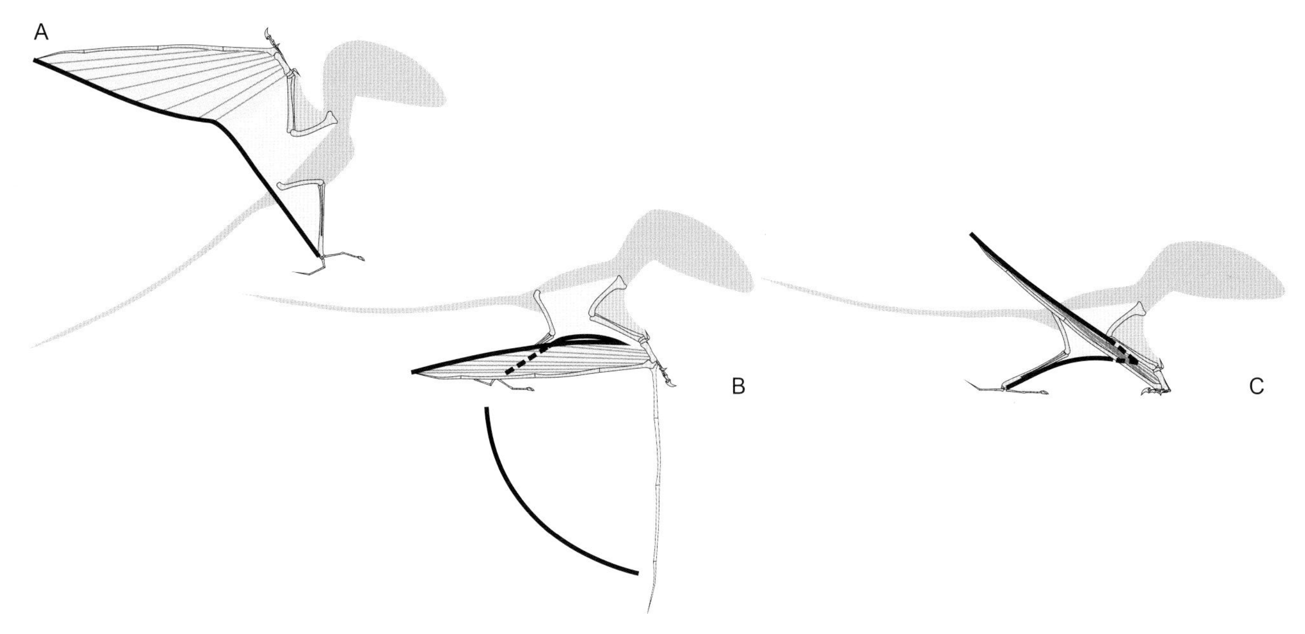

Fig. 7.6. How to fold a pterosaur wing, with the stiffening fibers shown aiding wing collapse as predicted by Bennett (2000). A, wing fully extended during landing; B, touchdown with the hindlimbs, and the body falls forward. Simultaneously, the wing finger rotates and collapses the wing; C, quadrupedal terrestrial pose, with the wing fully stowed. Note how the reinforced distal wing membrane stays *above* the wing finger once collapsed. Gray shading represents the proximal, unstiffened portion of the brachiopatagia; green represents the stiffened region; red lines represent stiffening fibers; thick black line represents distal margin of the brachiopatagia.

1989); some sediments may resist their comparatively light footfalls but buckle beneath the greater pressure of their hands (Mazin et al. 2003).

But Wait: Is This the Case for *All* Pterosaurs?

Although pterosaur tracks are found throughout the world, it is noteworthy that they only occur in rocks dating from the Middle Jurassic onward, despite the pterosaur body fossil record extending well beyond this time, into the upper Triassic (e.g., Wild 1984b). Where, then, are the Triassic and Early Jurassic pterosaur tracks? It has been suggested that the earliest pterosaurs, with their legs bound together by extensive uropatagia, and the assumption that they had a very low, sprawling gait, were simply less adept at terrestrial locomotion than later forms (fig. 7.7; Unwin 1988b, 2005; Henderson and Unwin 1999; Unwin and Henderson 1999). Accordingly, these early species may have simply not produced as many footprints as their descendants, explaining their lack of trackways in the fossil record. It is thought that the tracks of these pterosaurs should be easy to identify due to the presence of their long fifth toes but, with one possible exception (Lockley and Wright 2003), none have been identified.

This idea may require some reevaluation, however. Firstly, the functional anatomy of early forms does not really support the notion that they were inept terrestrial locomotors. Several workers have pointed out that *all* pterosaur hindlimbs and pelves, including those of early forms, appear suited to an upright, erect stance rather than a sprawled one (Padian 1983a, 1983b, 2008a; Bennett 1997a; Hyder et al., in press). The suggestion that the legs are "bound" by the uropatagium ignores the probable elasticity of this membrane (it has to be elastic to enable it to adopt different postures in flight, after all) and also assumes that early pterosaurs had to walk or run with each leg moving independently of the other. As rabbits, frogs, and many other animals can attest, bounding along with alternating movements of the fore- and hindlimbs is a perfectly effective and potentially rapid form of travel. We do not, of course, have any footprint data to con-

Fig. 7.7. Sprawling *Eudimorphodon*. Compared to pterodactyloids, little research has been done into the terrestrial ability of early pterosaurs, but they have been suggested to have adopted the sprawling, lizard-like posture shown here. Others argue that at least their hindlimbs were capable of an erect gait, and some species may have also been able to rotate their forelimbs beneath their body (see chapter 10).

firm that early pterosaurs moved in this way, but there is not any anatomical reason to suggest it was not possible. Furthermore, recent investigation of the glenoid joint of one early pterosaur, *Dimorphodon*, suggests that it was very open behind and beneath the shoulder (Witton 2011), allowing its forelimbs to move as freely as later, terrestrially competent forms (though see Padian 2008a for a contradictory view). Some other early pterosaurs, by contrast, do seem to have more restricted shoulder joints and may have possessed only semi-erect forelimbs. Regardless, with vertical hindlimbs, both of these forelimb posture conditions will elevate the carriage of their owners far enough above the ground to allow their limbs plenty of room for efficient walking, so there seems little reason to assume that early pterosaurs were low-lying, sprawling crawlers. Inefficient limb mechanics may, therefore, not explain the lack of early pterosaur tracks.

A second reason to be cautious about reading too much into the absence of early pterosaur footprints is more pragmatic. How can we tell the difference between the footprints not existing, and our failure to have found them? The latter is a very real possibility given how patchy the fossil record can be. Plenty of preservational and sampling biases affect our chances of finding fossils, including those that may ultimately prove common. Moreover, even if their absence is genuine, it does not mean that early pterosaurs were poor terrestrial locomotors per se; there are many other factors that may influence the generation, or not, of early pterosaur footprints. Perhaps early pterosaurs were simply rather rare, or did not frequent environments conducive to footprint preservation. The latter seems most likely to me, assuming that the absence of tracks is genuine. The feeding anatomy of later pterosaurs suggests they exploited far more niches than we would predict of their ancestors, which perhaps brought them into realms more conducive to leaving footprints. (This could be particularly true among the wading-adapted ctenochasmatoid pterosaurs; shallow bodies of water can be fantastic sedimentary environments to preserve fossil trackways. See chapter 19 for more on these animals). In short, because there are many ways we can interpret the deficit of early pterosaur footprints, we must be careful not to favor one hypothesis over others until a more complete picture of early pterosaur evolution is obtained.

Up High

On a related note, some pterosaurs show evidence of frequenting environments with no chance whatsoever of generating a track record; namely, trees and rock faces. In early pterosaurs, keen climbers are identified by their deepened, highly recurved manual claws with comparatively large flexor tubercles and antungual sesamoids, features that allow the digit ligaments more leverage when extending the claw bones (fig. 7.8; Unwin 1988b). Some particularly skilled climbing pterosaurs also bear these features on their feet, suggesting they could use all four limbs equally to power their way across vertical faces like the climbing equivalents of 4 × 4 trucks.

The climbing characteristics seen in later pterosaurs are quite different from those of earlier species (Wang, Kellner, et al. 2008). The chief difference is their limb lengths, with those of later forms being much longer and consequently holding their centers of mass away from the climbing surface. This likely lessened their

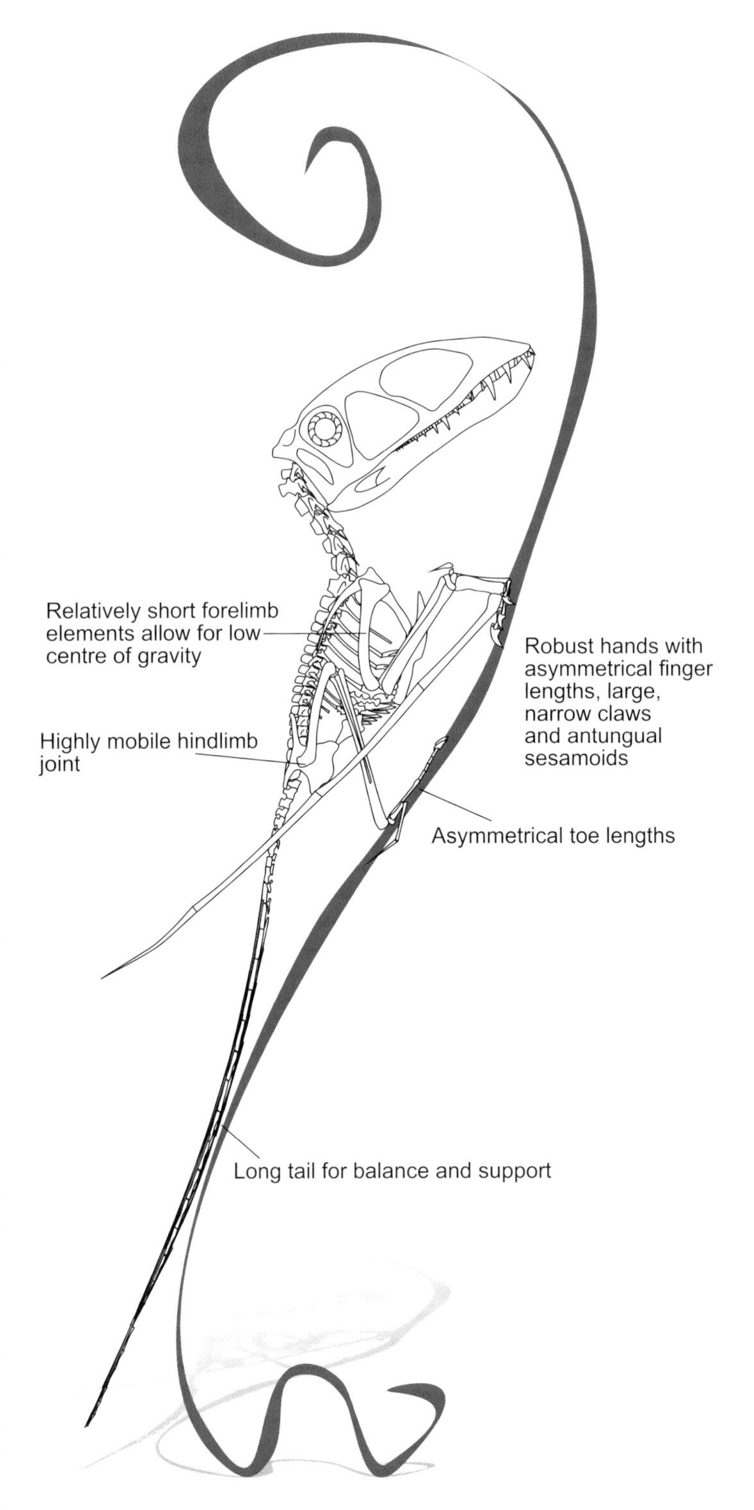

FIG. 7.8. Climbing adaptations of *Dimorphodon*, one of the most scansorially adapted of all pterosaurs (see chapter 10).

stability on vertical surfaces (see Hildebrand 1995 for a review of climbing adaptations in vertebrates). These pterosaurs may, therefore, have been more at home hanging from branches like sloths than scaling vertical surfaces like cats or squirrels. Pterosaurs are sometimes depicted as hanging from cliffs or trees by their toes in a bat-like fashion (fig. 2.6; see Wellnhofer 1991a, and Bramwell and Whitfield 1974 for examples and discussion), but their feet seem ill suited for this task. Most pterosaurs bear small feet with slender metatarsals and toes, and phalanges of limited mobility, making them unlikely grappling devices. As such, while the idea of pterosaurs hanging from trees and cliffs is a fine idea, the notion of bat-like suspension is poorly founded.

Dunk-a-dactyl

Our discussion in this chapter so far has focused almost entirely on grounded pterosaurs, but what happened to them when they entered water? Little dedicated research has been conducted into swimming pterosaurs, but digital modeling of likely postures is underway (Hone et al. 2011a). In the interim, we can at least be content that good evidence exists to confirm that pterosaurs *could* swim. The discovery of tracks left by swimming pterosaurs, amounting to partial pes prints and a plethora of scratch marks made when paddling their way through shallow lake margins, demonstrates that they were both happy to get their feet wet and able to propel themselves through water (fig. 7.9; Lockley and Wright 2003). The lack of manus prints with these tracks suggests that swimming pterosaurs held their hands higher in the water column than their feet. Perhaps floating pterosaurs held their arms in front of their bodies, resting them on the water surface to operate as pontoons and prevent their front-heavy bodies from capsizing forward, while their webbed feet paddled them around (fig. 7.10). Other possible records of waterlogged pterosaurs include the aforementioned manus-only trackways, which Lockley et al. (1995) suggested may record pterosaurs wading in shallow water with buoyed-up hindquarters.

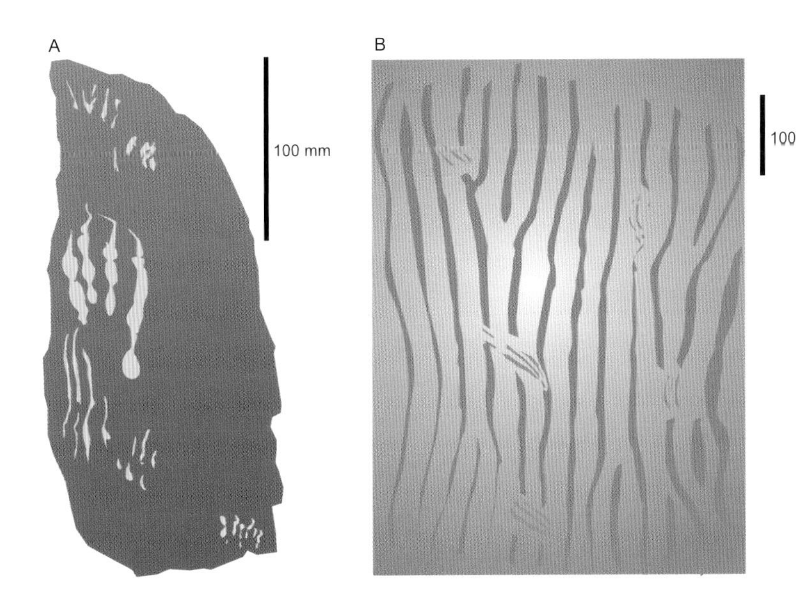

Fɪɢ. 7.9. Swimming pterosaur traces, showing toes and claws raking the sediment while propelling buoyant pterosaurs over or through water. A, Late Jurassic tracks from the Summerville Formation, Utah; B, swimming tracks made over ripple marks in the Sundance Formation, Wyoming. Redrawn from Lockley et al. (2003).

Fɪɢ. 7.10. Two *Ornithocheirus mesembrinus* punt their way around a shallow Cretaceous seaway, their arms stretched out across the water surface to prevent them from faceplanting the sea.

8
The Private Lives of Pterosaurs

As with our opinion of their terrestrial capabilities, our perception of pterosaur paleoecology has changed dramatically in the last few decades. Typically, pterosaurs were assumed to be very birdlike in all aspects of their lives, nesting and growing in a very avian fashion before assuming a seabird-like ecological destiny. More recently, a rather different picture has emerged. New discoveries show that pterosaur reproduction and growth was not comparable to that of modern avians after all, and it seems that many—perhaps most—pterosaurs were not the seabird analogues they were long held to be. Glimpses into diseases and injuries, social structure, sex lives, and even how some pterosaur individuals may have met their ends (fig. 8.1) have enriched our perception of pterosaur lifestyles. We can still only paint the broadest picture of their day-to-day existence, but these discoveries provide at least a flavor of what life as a pterosaur may have been like.

In the Beginning

Proving just how finicky the fossil record can be, it took over 200 years for pterosaur workers to discover any pterosaur eggs, and then three turned up in the same year (Wang and Zhou 2004; Chiappe et al. 2004; Ji et al. 2004). Each of these eggs is remarkable for the preservation of their developing embryos in detail (fig. 8.2), and with all three we now have a record of pterosaur embryology from its early phases to a well-developed near-hatchling (fig. 8.3; Chiappe et al. 2004; Unwin 2005; Unwin and Deeming 2008). A recently found fourth egg is an exception to this, preserving no trace of the embryo within. This egg is still remarkable, however, for being preserved alongside its deceased mother (fig. 2.8; Lü, Unwin, et al. 2011). Presumably, this association indictates that egg was at an extremely early stage of development, and unlikely to contain any preservable remains of the embryo.

The shells of all four known pterosaur eggs are fairly similar. Though the eggshell histology varies a little, it is consistently very thin at only 0.03–0.25 mm across, and contains only one layer of calcitic, shelly material. This would give the eggs a rather soft, parchment-like texture, more like those of most modern reptiles than the hard-shelled eggs of birds (Ji et al. 2004). Confirmation of this comes from the lack of cracking in any of the egg specimens. Hard-shelled fossil dinosaur or bird eggs are typically rather cracked by the time paleontologists find them, but there's not a break to be seen on any pterosaur egg specimens.

Such a thin shell has interesting implications for pterosaur reproductive strategies. Thin-shelled eggs are able to take up water from their surrounding environment, which means they do not have to be laid with a store of water on board. The latter is necessary in thicker-shelled, hard eggs however, because the eggshell is a total barrier to absorption of external moisture (Unwin and Deeming 2008). Thin shells were therefore good news for pterosaur mothers, because they would not have had to provide vast water stores for their developing offspring, though the eggs could risk desiccation if laid in dry settings. Moreover, the eggs could have been more easily squashed if the parents attempted contact-incubation (i.e., sitting on them) (Grellet-Tinner et al. 2007). Both of these problems could be alleviated if the eggs were buried in a humid nest or underground chamber. Such practices are seen in many modern lizards, crocodiles, and megapode birds,[5] and it was likely pterosaurs did the same. The incubation times for buried eggs are also remarkably consistent among modern animals—about two to three months—suggesting that pterosaur eggs probably enjoyed a similar, leisurely incubation time (Unwin and Deeming 2008).

The embryos of all egg buriers share a further trait. They are all very precocial, or well developed, before hatching, and are independent of their parents (to greater and lesser extents) from the moment they leave the egg. There is some evidence that pterosaur hatchlings were similarly precocial. Pterosaur embryo

[5] Megapodes are a group of very unusual birds with extremely precocial offspring. Hatchling megapodes dig their way out of their nests and look after themselves without any parental care, a condition which starkly contrasts with other birds. Most bird species look after their chicks until they are nearly full-grown, which is reflected in the altricial state of the chicks when they hatch.

FIG. 8.1. When spinosaurs attack: a helpless ornithocheirid is ripped from the water in Cretaceous Brazil by the spinosaurid *Irritator*. Fossil evidence indicates that the latter certainly dined on the former, though whether *Irritator* hunted and killed pterosaurs, or merely scavenged them, remains to be clarified.

FIG. 8.2. The remarkably preserved egg of the Cretaceous pterosaur *Beipiaopterus chenianus* from the Cretaceous Yixian Formation, China. Image courtesy of David Unwin.

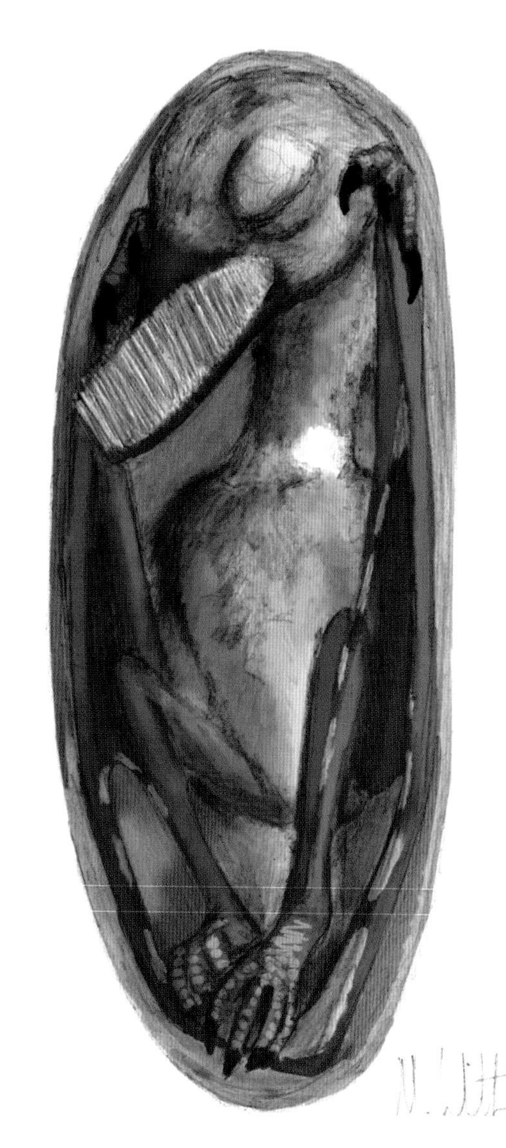

FIG. 8.3. A late-stage *Pterodaustro guinazui* embryo, fit for hatching. Note the adult-like limb proportions and fully developed wing membranes, features that suggest neonatal pterosaurs were capable of flight soon after hatching.

bones are well ossified (even their toe phalanges are completely formed at birth), bear similar proportions to their parents, and possess a full complement of wing membranes (Chiappe et al. 2004; Wang and Zhou 2004). Indeed, the proportions of embryonic pterosaurs are so similar to those of adults that we can assign all of their currently known embryos to specific pterosaur groups, and in one case, even to a particular species. Such advanced development indicates that hatchling pterosaurs may have been capable of looking after themselves—maybe even flying—very soon after hatching. This sets pterosaurs apart from virtually all other flying vertebrates (only megapodes can also boast similar flight ready, superprecocial offspring) and creates the possibility that the embryos could incubate and hatch with complete independence of their parents (e.g., Unwin 2005; Unwin and Deeming 2008), in the manner of most reptiles. Others argue that the association of one pterosaur egg with a large number of adult pterosaur remains indicates the opposite, that pterosaur chicks were be looked after by their parents (Chiappe et al. 2004). It is hard to know either way with the limited data currently available, but I wonder if a compromise between these ideas is most likely. All modern archosaurs—birds and crocodiles—generally guard and maintain their nests, even if the latter do

FIG. 8.4. Azhdarchid pterosaurs enter the world. A lone parent has opened the top of their nest to assist their hatching, but it may well have little else to do with its offspring now that they have emerged.

only exhibit limited parental care once their hatchlings have emerged. Nest guarding may, therefore, be deeply ingrained by archosaur evolutionary history. If so, pterosaurs could have demonstrated the same behavior, even if they did not worry too much about their offspring once they hatched (fig. 8.4).

Growing Up

The first stages of avian and mammalian lives revolve around two activities: eating and growing. We spend large amounts of time eating high-quality, nutritious food that is rapidly transformed into the tissues we need to become self-sufficient. Often, our parents go to great lengths to ensure that we are also suitably warm, so that we do not waste potential growing energy keeping our own metabolic fires stoked. This is not the case for precocial offspring. Along with growing, they have to invest considerable energy into ensuring their own survivial including regulating their body temperatures, finding their own food, avoiding predation, and so forth. As such, even precocial animals with high metabolisms—and therefore potentially fast growth rates—cannot grow as quickly as their mollycoddled cousins.

There is good evidence that pterosaurs faced these arduous childhoods. Although their bones possess fibrolamellar textures indicative of rapid bone growth, the lines of arrested growth (LAGs), or growth rings, suggest that pterosaurs may have taken quite a few years to hit their maximum adult size. Counting bone

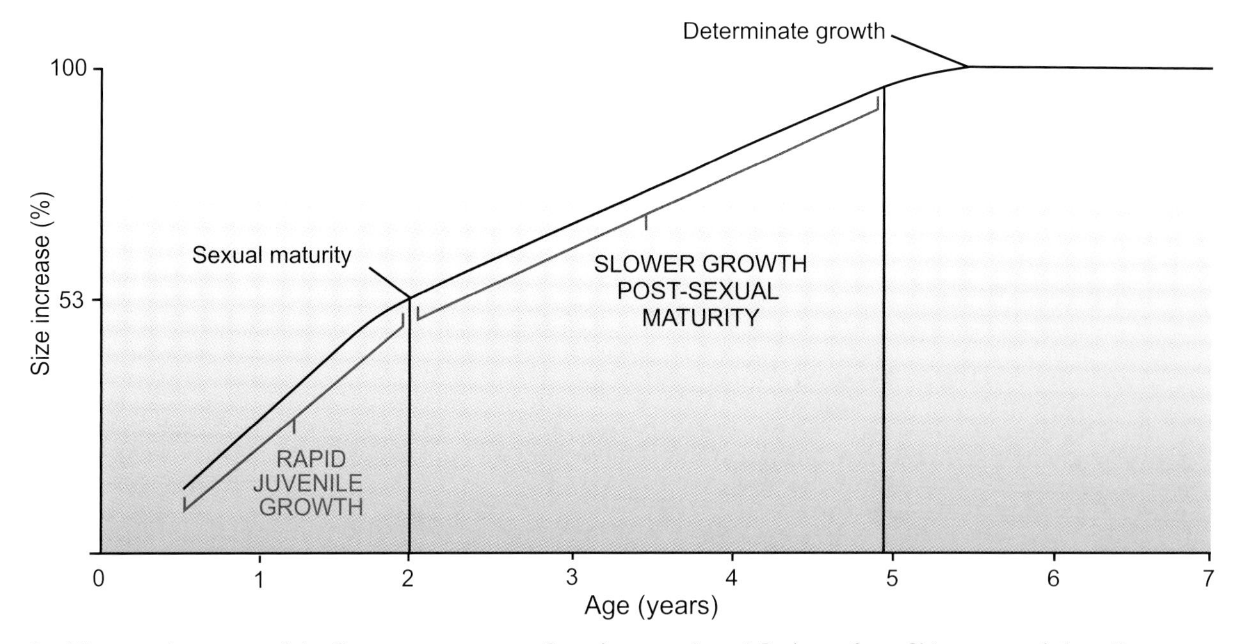

FIG. 8.5. The growth strategy of the Cretaceous pterosaur *Pterodaustro guinazui*. Redrawn from Chinsamy et al. (2008).

LAGs allows us to accurately age their owners in much the same way that tree rings can be used to age tree trunks. Analysis of LAGs in *Pterodaustro*, a Cretaceous pterosaur with a 3 m wingspan, suggests it took seven years to reach full size (see fig. 8.5) (Chinsamy et al. 2008). Further evidence for slow growth rates comes from the preservation of distinct size classes in several pterosaur species that correspond to different stages of growth. This is thought to represent regular annual mortalities of immature individuals (Bennett 1995, 1996b). The number of size classes in a single pterosaur species is high (up to 5 in the case of the Jurassic, 2 m wingspan form *Rhamphorhynchus*), which corroborates the idea that pterosaurs did not race to their adult proportions as quickly as pampered mammals and birds.

As with other reptiles (including nonavian dinosaurs), pterosaurs seem to have grown relatively rapidly early in life before slowing down when they achieved approximately half their maximum size (Bennett 1995; Chinsamy et al. 2008; Chinsamy et al. 2009). The onset of this slower growth has been correlated with the onset of sexual maturity. The type of bone fibers deposited by growing reptiles, nonavian dinosaurs, some early birds and pterosaurs change once their reproductive cycles begin, and this allows us to detect the beginning of sexual maturity from bones alone (Chinsamy et al. 2008; Chinsamy et al. 2009). The additional drain on resources brought on by reproductive behaviors and processes may explain why the growth rate drops a little once pterosaurs reached this phase of their lives.

Interestingly, while this form of reproduction and growth is classically "reptilian," pterosaurs are more like modern mammals and birds in having finite body size. Many reptiles continue to grow for their entire lives, albeit extremely slowly in adult forms (so-called indeterminate growth). Pterosaurs, by contrast, demonstrate "determinate" growth, a state that can be readily detected in their skeletons through changes in bone texture and fusion. Growing pterosaurs possess richly vascularized bones and only partially fused sutures in their compound bony elements (e.g., skulls, shoulder girdles, pelves, etc.), but those of a certain size and age developed entirely ossified skeletons and laid down final, avascular layers around their bones (Bennett 1993; De Ricqlés et al. 2000; Chinsamy et al. 2008; Chinsamy et al. 2009). This signifies the end of the growing stage and "sets" the skeletal morphology of the animal, denying them further last-minute changes to their skeletons (see below) or additional increases in size.

Like virtually all animals, some parts of pterosaur bodies grew at different rates than others. Such growth, referred to as "allometric" (as opposed to "isometric," which refers to steady growth without disparate proportional changes), means that adult pterosaurs look rather different than their offspring

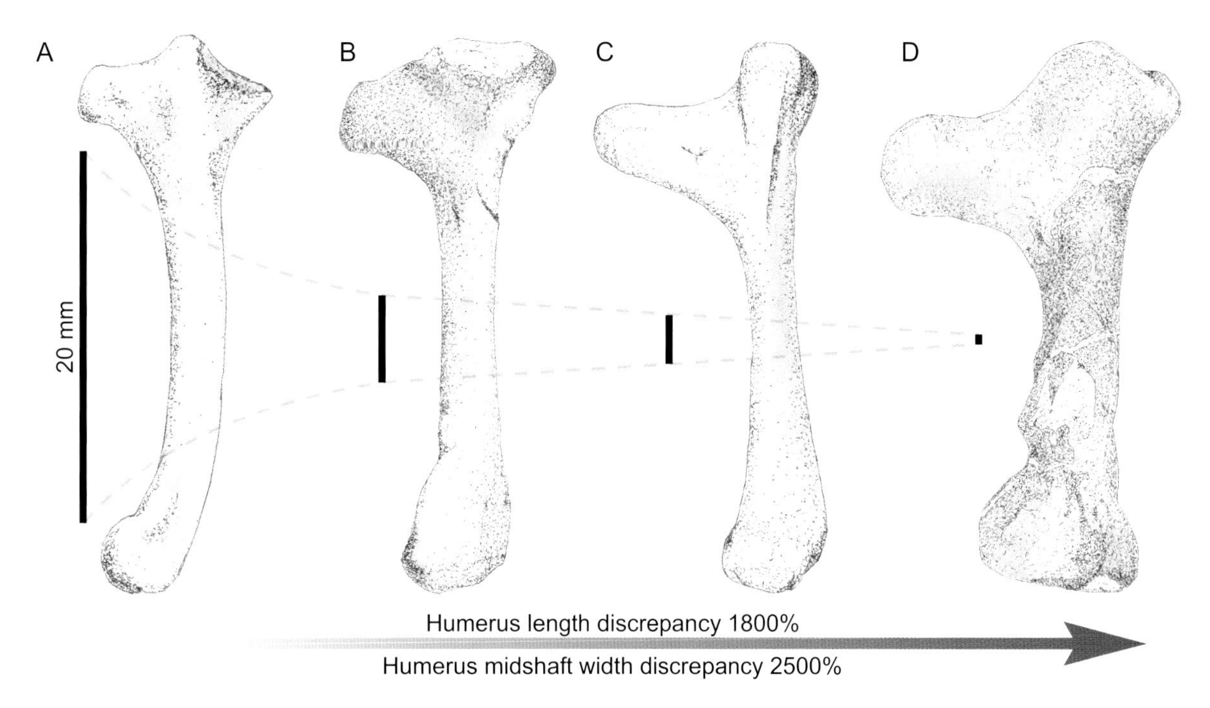

FIG. 8.6. The mechanical impact on pterosaur size, demonstrated by differently sized lophocratian pterosaur humeri. Although the humeri shown here are from different species, the same scaling effects can be seen in the growth regimes of individual species. A, *Pterodactylus antiquus*; B, *Lonchodectes* sp.; C, *Tupuxuara leonardii*; D, *Quetzalcoatlus northropi*. Note the increases in the size of the bone articulations, deltopectoral crests, and humeral shaft thickness (increasing by 2500 % from A–D) relative to the increase in overall length (1800 % from A–D).

(see Wellnhofer 1970; Brower and Veinus 1981; Bennett 1995, 1996b; *and* Codorniú and Chiappe 2004 for pterosaur scaling regimes). Their skeletons generally became more robust with age, and particularly so around the wing joints (fig. 8.6). The proximal arm region tends to shorten slightly, while the wing metacarpals and legs increase in length. Their necks seem to have increased in length dramatically compared to other bodily components, as did their jaws (fig. 8.7). As these lengthen, most toothed pterosaurs added more teeth to their dental array in a typically reptilian fashion, although some pterosaurs bucked this trend (Bennett 1995, 1996b).

The most dramatic allometry of pterosaur growth, however, involves their cranial crests (fig. 8.7B). Crests are absent in all known juveniles of crested species but began to sprout in pterosaur teenagers, growing quickly so that fully formed headgear was present by the time they achieved maximal size (Bennett 1992; Martill and Naish 2006). That said, not *all* individuals of crested pterosaur species received a hastily grown crown. Some only produced low, unspectacular crests, or may have remained entirely unornamented. This relatively recent discovery is one of the most intriguing findings of modern pterosaur research, not only providing details of social strategies and sex lives but also of the probable function of their otherwise enigmatic crests.

Crests, Sex, and the Pterosaur Social Scene

Explaining pterosaur headcrests has caused many headaches for pterosaurologists. They have classically been interpreted as strictly mechanical devices, such as aerodynamic rudders and airbrakes (Bramwell and Whitfield 1974; Stein 1975); jaw stabilizers for dip-feeding taxa (Wellnhofer 1987a, 1991a; Veldmeijer et al. 2006); jaw muscle anchors (Eaton 1910; Mateer 1975); or devices for thermoregulation (e.g., Kellner 1989; Kellner and Campos 2002). These explanations have been hard to swallow as the tremendous disparity of pterosaur crests has become increasingly apparent (see figs. 1.1 and 4.4 for examples) and they ignore a very obvious question: If crests were critical for flight, feeding, or temperature control, how did crestless individuals survive without them?

Fig. 8.7. Life restorations of juvenile and adult individuals of two pterosaur species. A, the Jurassic pterosaur *Rhamphorhynchus muensteri*; B, the Cretaceous form *Tupuxuara deliradamus*. Each juvenile and adult portrait is to scale, but the species are not to scale with each other.

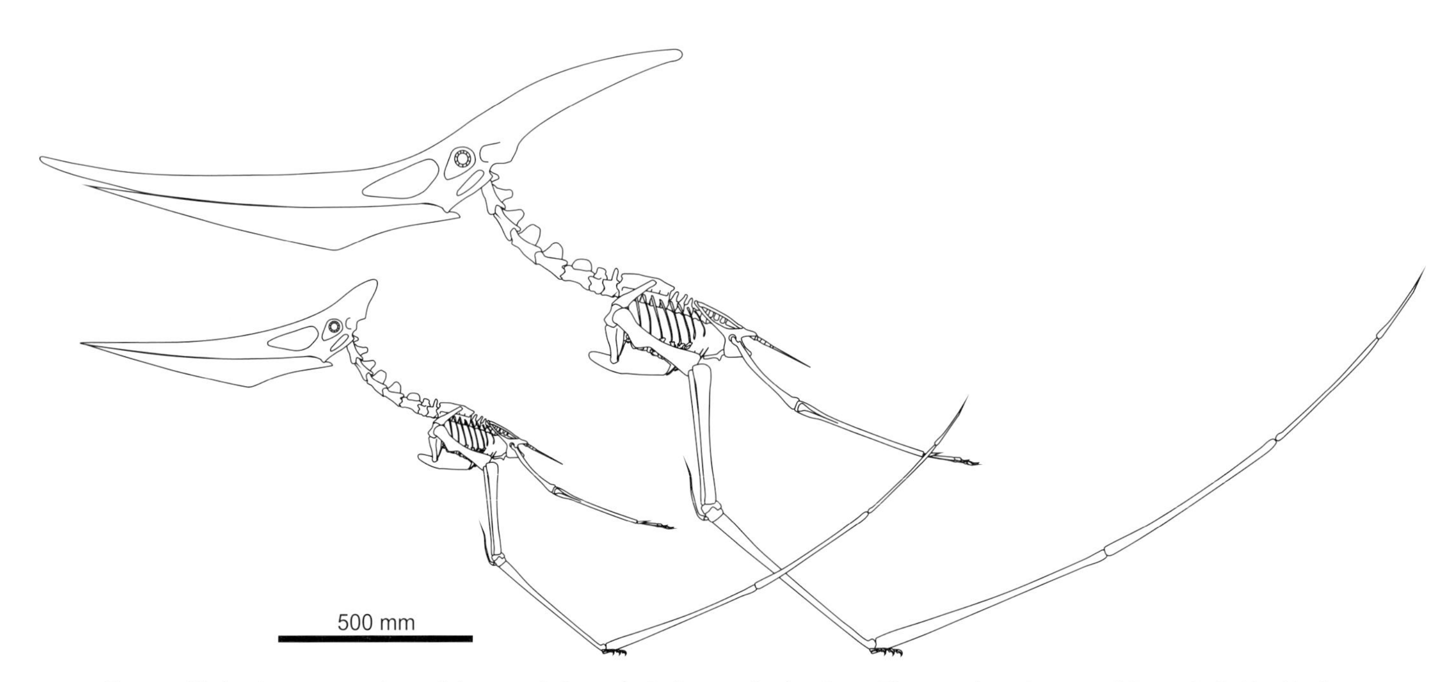

Fig. 8.8. Skeletal reconstructions of the sexual dimorphs in *Pteranodon longiceps.* The prominently crested, larger individual is the presumed male. Based on Bennett (1992, 2001).

Chris Bennett's (1992) study into the headcrests of the quintessential crested pterosaur *Pteranodon* was perfectly timed with the discovery of many new crest morphologies and growing discomfort with traditional interpretations of pterosaur crest use (fig. 8.8). *Pteranodon* is famous for bearing a large, bladelike crest projecting posterodorsally from its skull, but only a minority of *Pteranodon* specimens shows such a crest. Most bear rather stunted, rounded crests, and because these small-crested individuals show clear osteological signs of adulthood, this cannot be explained by their immaturity. The anatomies of these two *Pteranodon* morphs are otherwise nearly identical, save for two more distinctions: the larger-crested morphs are roughly 50 percent bigger than the smaller-crested variants, and their pelvic canals are narrower. In modern animals, such differences often reflect sexual distinctions within a species. Males have no need to pass eggs or offspring through their pelvic canals, and subsequently tend to have smaller pelves than those of females. The males of promiscuous species that regularly compete for reproductive access to multiple females are often larger because size is an advantage in fighting off reproductive competitors. Bennett (1992) noted these comparisons in his *Pteranodon* sample: the larger forms with narrow pelves (representing about one third of the *Pteranodon* specimens known) were probably males of a sexually dimorphic species, with their large crests being sexually selected structures that embellished their appearance. This is rather exciting, because the degree of dimorphism and showmanship seen in these pterosaurs implies that they may have used "lek" mating strategies, where males intensely compete with one another for harems of females (chapter 18; Bennett 1992). Further studies into the *Pteranodon* headcrest corroborate their likely sexual function. The rapid growth of their crests vastly exceeds that necessary for thermoregulatory requirements alone (Tomkins et al. 2010), and even the largest *Pteranodon* crests have, at best, modest aerodynamic consequences (Elgin et al. 2008).

Evidence for headcrests as primarily communicative devices is not only found in *Pteranodon*. Cranial crests only occur in the adult forms of at least two other pterosaur species: the antler-crested *Nyctosaurus* and the sail-crested *Thalassodromeus* (Bennett 2003b; Martill and Naish 2006; but also see Witton 2008b). This suggests that headcrests were also only of importance to mature individuals of these species and, as with *Pteranodon*, the thinly constructed, antler crest of *Nyctosaurus* has also been shown to perform poorly as a rudder and air brake (Xing et al. 2009; also see chapter 18). Another pterosaur, *Darwinopterus*, has been found to show the same trends in pelvic and crest morphology as *Pteranodon*, with presumed females bearing crestless skulls and broader pelvic canals than the crested males (fig. 8.9; Lü, Unwin, et al. 2011). The striking color pattern of a pterosaur crest has also been preserved in one fossil (fig. 5.13), which is conducive to the idea of their use as display structures (Czerkas and Ji 2002). Accordingly, while we cannot definitively say that all pterosaurs were dimorphic or that all crested individuals were males, there is growing evidence for their crests being sexual display devices above all else, and any mechanical impacts they had were likely side effects of their structure. Showy headcrests may have served to demonstrate the health of male suitors to females or to settle disputes between rivals without the need for physical engagement. Of course, this could also work the opposite way; the general "showiness" of many fossil reptiles may stem from *both* sexes selecting potential mates by the impressiveness of cranial ornamentation, and this may apply to some species of pterosaur in which both males and females bear headcrests (Hone, Naish, et al. 2011).

A marked downside of this insight into crest development, however, is that the use of cranial crests in pterosaur taxonomy is now rather questionable. While basic crest morphology—their general position on the skull, bony composition, and so forth—remains a useful characteristic for distinguishing broad pterosaur groups, it is not certain that subtle differences in crest shape, or even crest possession, are taxonomically significant among closely related species. The possession, size, and shapes of sexually selected structures in modern animals often vary considerably within a species, which means we must be cautious about interpreting minor differences in pterosaur crest morphology as being taxonomically significant. Alas, a number of pterosaur species have been diagnosed by such features, which raises some questions over their taxonomic validity. This problem is theoretically negated through extensive sampling of pterosaur remains that could reveal variation in crest morphology within a single species, as per *Pteranodon*. However, the rarity of pterosaur fossils means that this is unlikely to happen for the majority of species, and pterosaurologists simply have to tread carefully when viewing these structures for taxonomic purposes.

The idea that pterosaur crests were prominent, visually communicative structures relates to suggestions that the animals spent at least some of their time in large groups or flocks. Hard evidence for this is rare in fossil animals with patchy fossil records such as pterosaurs, but the occurrence of thousands of pterosaur bone fragments in a single Cretaceous flash flood deposit of Chile may represent a large number of cohabiting pterosaurs being killed simultaneously (Bell and

FIG. 8.9. Sexual dimorphism in the Jurassic pterosaur *Darwinopterus modularis*. Unlike *Pteranodon*, the sexes of this species appear to be generally similar in size, but only the presumed males possess headcrests.

Padian 1995). In other instances, complete pterosaur skeletons found in close association have been interpreted as the remnants of small social groups (Witton and Naish 2008), and pterosaur fossils are simply so numerous in some deposits that they probably record the existence of high pterosaur populations in certain areas. (They are the most abundant tetrapod in several fossil Lagerstätten for example, including the Santana, Crato, and Solnhofen Formations.) It is not clear if pterosaurs formed tight, socially complex groups or merely aggregated in the same locations, but it seems probable that large numbers of pterosaurs at least cohabited in certain settings, even if they were not paying specific attention to each other.

Fueling Leathery Wings

Much remains unknown about pterosaurian feeding habits, and it's only within recent years that pterosaurologists have really started to address this issue (e.g., Humphries et al. 2007; Witton and Naish 2008; Ősi 2010). Direct evidence of pterosaur dietary preferences and feeding behavior is rare, but is available from a small array of fossils. Fossil gut content is the best indicator of these habits. Remains of half-digested fish have been found in the gut regions of *Rhamphorhynchus* and *Eudimorphodon* (Wellnhofer 1975; Wild 1978; Frey and Tischlinger 2012), and from the throat region of *Pteranodon*. The latter probably represents perimortem vomiting (Brown 1943) (fig. 8.10). Fishy gut content has also been reported for *Pterodactylus* (Broili 1938), but the specimen purported to show this has been lost and the claim cannot be validated. A particularly unusual piece of possibly ingested food—a large, sharp plant frond—is associated with the skull of the likely fish-eating pterosaur *Ludodactylus* after, so the story goes, the pterosaur mistakenly stabbed the leaf into its gular pouch when feeding (fig. 16.4D; Frey, Martill, and Buchy 2003b). Vomited pterosaur gut content may also be known from the mysterious Las Hoyas pellet discussed in chapter 5 (fig. 5.4).

Good evidence of pterosaurian foraging activities is known from several pterosaur track sites (fig. 8.11;

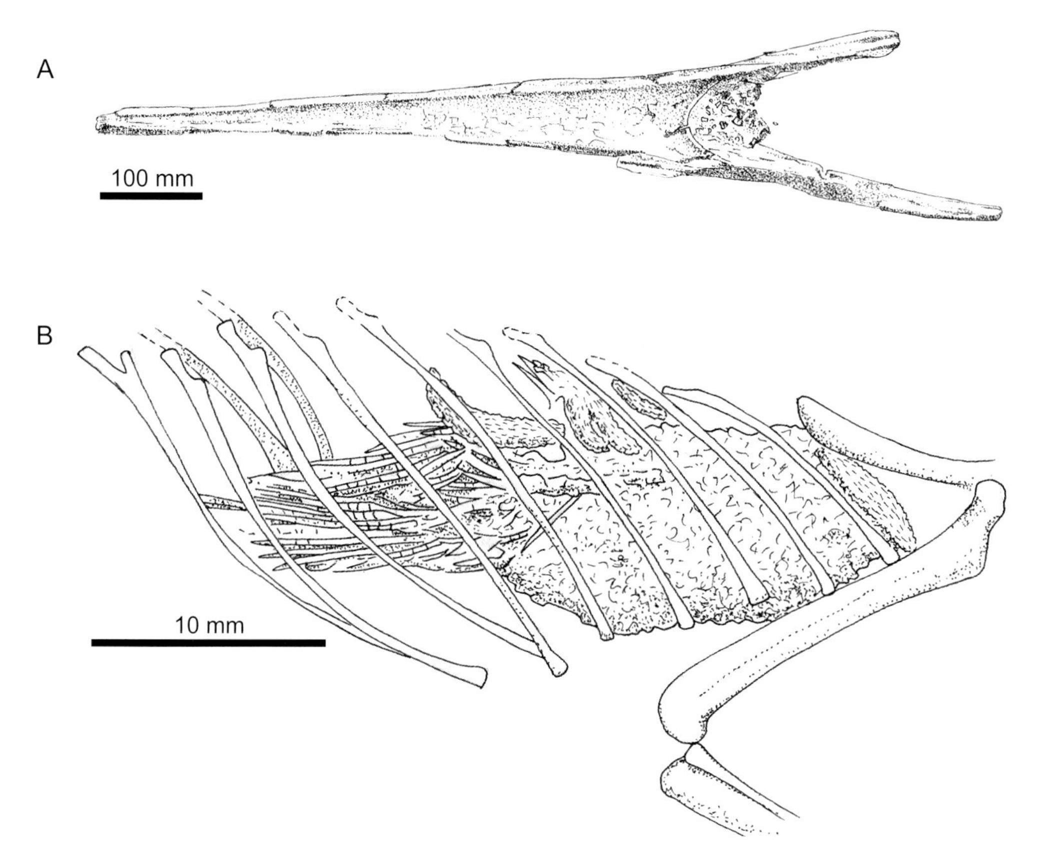

Fig. 8.10. Pterosaur gut content. A, *Pteranodon* mandible with partially digested, regurgitated fish vertebrae situated between its mandibular rami; B, *Rhamphorhynchus muensteri* torso with the remains of a partially digested fish that was swallowed whole. See figure 5.3 for a depiction of the ingestion of this fish. A, drawn from photographs kindly supplied by Carl Mehling; B, redrawn from Wellnhofer (1975).

Parker and Balsley 1989; Wright et al. 1997; Lockley and Wright 2003). Some pterosaurs appear to have patrolled mudflats and watercourses with their jaws held close to the sediment surface, pecking at possible foodstuffs and leaving characteristically paired peck marks in the substrate as they went (fig. 8.12). These marks were almost certainly caused by pterosaur jaws because, in addition to being intimately associated with pterosaur tracks, they are dead ringers for the paired beak traces left by modern foraging birds. Bird and pterosaur beaks are not so different that their pecking traces should be dissimilar, so their conformity is useful corroboration of this identification. Viewed from above, pterosaur peck marks superficially resemble U-shaped burrows, common types of trace fossils left by small burrowing invertebrates, but this similarity is purely superficial because the peck marks only penetrate the uppermost layers of sediment (Lockley and Wright 2003). It should be stressed that the pterosaurs leaving these traces were not *probing* the sediment in the manner of some modern waders, but grabbing either food from the sediment surface or prey suspended in the water column immediately above the substrate. The walking traces left by the feeding pterosaurs are often quite chaotically arranged, suggesting the animal or animals responsible were not feeding in a particularly systematic way, but were merely wandering about, snapping at food as they saw it. Repeated occurrences of such tracks in multiple sedimentary horizons at the same locality indicate that pterosaurs returned to the same feeding sites again and again (Unwin 2005).

Sweeping gouges made by pterosaur jaws held close to the substrate are also known (fig. 8.11A), but their identity is not widely acknowledged (Lockley and Wright 2003). Indeed, these sweeping traces, though sometimes associated with prod marks, are often thought to be tail drags made by the wandering track makers (Parker and Balsley 1989; Rodriguez-de la Rosa 2003). As the track makers in these instances were almost certainly short-tailed pterosaurs (and in one case, probably a form with particularly long legs),

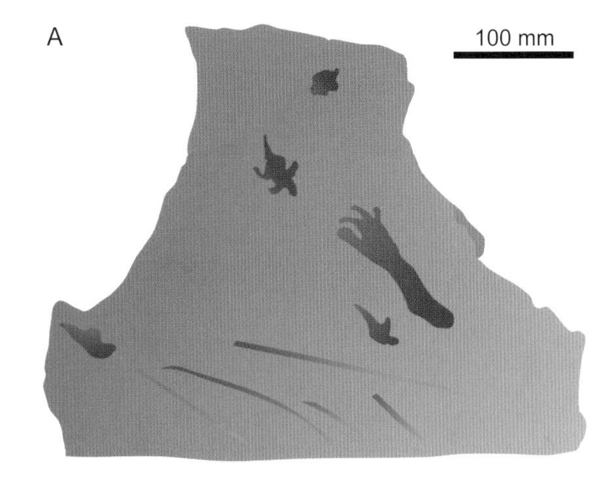

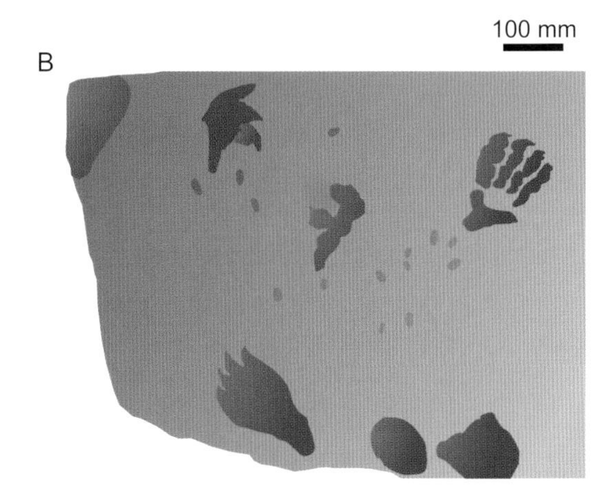

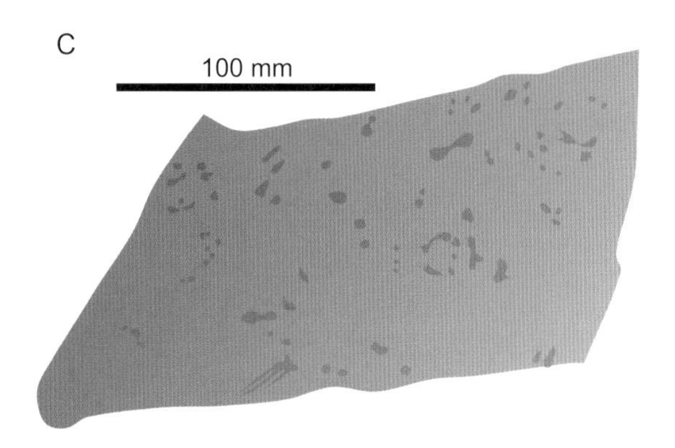

FIG. 8.11. Possible pterosaur feeding traces, beak scrapes, and peck marks (purple shading). A, "*Pteraichnus*" sp. tracks from the Late Cretaceous Cerro del Pueblo Formation, northern Mexico with "sweeping" traces; B, *Purbeckopus pentadactylus* tracks from the Early Cretaceous Purbeck Limestone Formation, England; C, Late Jurassic pterosaur tracks from the Del Monte Mines, Utah. A, redrawn from Rodriguez-de la Rosa (2003); B, redrawn from Wright et al. (1997); C, redrawn from Lockley and Wright (2003).

this identification seems very unlikely unless, perhaps, these tracks record a pterosaurian party where the track makers crouched right down to shake their tail feathers (or pycnofibers) (Witton 2008b). Pterosaur jaws can reach the ground with much greater ease, and beak sweeping is a commonly observed behavior in modern, foraging waders. Sweeping beaks, therefore, is probably a more parsimonious interpretation of these marks than tail dragging.

Interestingly, pterosaur feeding traces first appear in the Upper Jurassic and then appear periodically in the pterosaur record until almost the very end of the Cretaceous, suggesting that pterosaurs were combing shallow waterways and their adjacent mudflats for food throughout the latter half of their evolutionary history. Unfortunately, the identities of the pterosaurs responsible for these feeding marks are unknown in most cases, and in one instance—the feeding traces associated with the large pterosaur track *Purbeckopus* (fig. 8.11B)—even the pterosaurian affinity of the track maker has been questioned (Billon-Bruyat and Mazin 2003). The lack of tooth impressions with the prod marks, an expected and unavoidable result from the oversize, procumbent dentition found in many pterosaurs, suggests that the foraging animals at least possessed edentulous jaw tips, narrowing the likely candidates down to a handful of groups. Nevertheless, their actual identity remains mysterious.

A Mesozoic Whodunit: Did Pterosaurs Attack the Solnhofen Insects?

Evidence of pterosaurs attacking other animals is extremely rare, and probably restricted to two fossil insects that are tentatively considered to have been victims of pterosaur assaults (fig. 8.13; Tischlinger 2001). The specimens, from the Jurassic Solnhofen Limestone of Germany, are of a fossil lacewing (*Archegetes neuropterum*) and a dragonfly (*Cymatophlebia longialata*). Each had large chunks of their wing tips torn off prior to fossilization, the damage to both far exceeding that expected from typical wear and tear. (Butterfly wings, for example, will lose no more than 10 percent of their wing area through successive nicks

Fig. 8.12. How to generate beak feeding traces. Disturbed substrates left by foraging pterosaurs do not suggest they inserted their beaks deep into sediments in search of food, but instead grabbed it from the substrate surface or the water column immediately above it.

Fig. 8.13. Insects from the Solnhofen Limestone with wing pathologies attributed to pterosaur attacks. A, the dragonfly *Cymatophlebia longialata*; B, the lacewing *Archegetes neuropterum*. Photographs courtesy of Helmut Tischlinger.

to their wing margins during their lives; Hendenström and Rosén 2001.) The dragonfly was given a particularly hard time with two portions of its right wings removed, with the proximal margin of the forewing damage continuous with that on the hindwing.

While there is no direct evidence to link pterosaurs to these assaults, their involvement can be reasoned through logical deduction (Tischlinger 2001). The insects are intact save for damage to the wings, suggesting that a marine aggressor can be ruled out. It seems unlikely that an aquatic predator would make a partially successful attack on a passing flying insect without eating the prize once it hit the water. Likewise, the insects show no signs of transportation damage or desiccation, suggesting they were not washed into the lagoon from the Solnhofen hinterland. The localization of the damage at the wing edges is also consistent with bite marks left on insect wings after failed attacks by modern aerial insectivores (Wourms and Wasserman 1985). Pterosaurs may not be the only flying animals known from Solnhofen deposits because the first birds, specifically *Archaeopteryx*, also existed in this setting and could have been capable of at least limited flight. It is very unlikely that *Archaeopteryx*-grade birds had the flight ability necessary to catch insects like dragonflies, however (Paul 1991). Just as we discussed regarding early pterosaurs in chapter 6, the flight capabilities of early birds were probably far inferior to those of dragonflies or lacewings, and their chances of catching such prey were probably low. A process of elimination, then, suggests that pterosaurs may be the only likely assailants of these insects. Alas, the bite traces on their wings do not record any specific features supporting this hypothesis, but it is noteworthy that one species of Solnhofen pterosaur, *Anurognathus*, seems well adapted to pursuing and eating insects on the wing (Bennett 2007b; see chapter 11). Perhaps this, or another pterosaur, made flying insects quake in their exoskeletons when they took to the skies.

Bumps, Scrapes, Knocks and . . . Throat Parasites?

Despite possessing the abilities to terrorize small animals like the aforementioned insects, pterosaurs were far from omnipotent; they were just as vulnerable to attack, personal injury, and disease as the ani-

Fig. 8.14. Distribution of known pterosaur pathologies, mapped onto the skeleton of *Pteranodon*. Compiled from various sources.

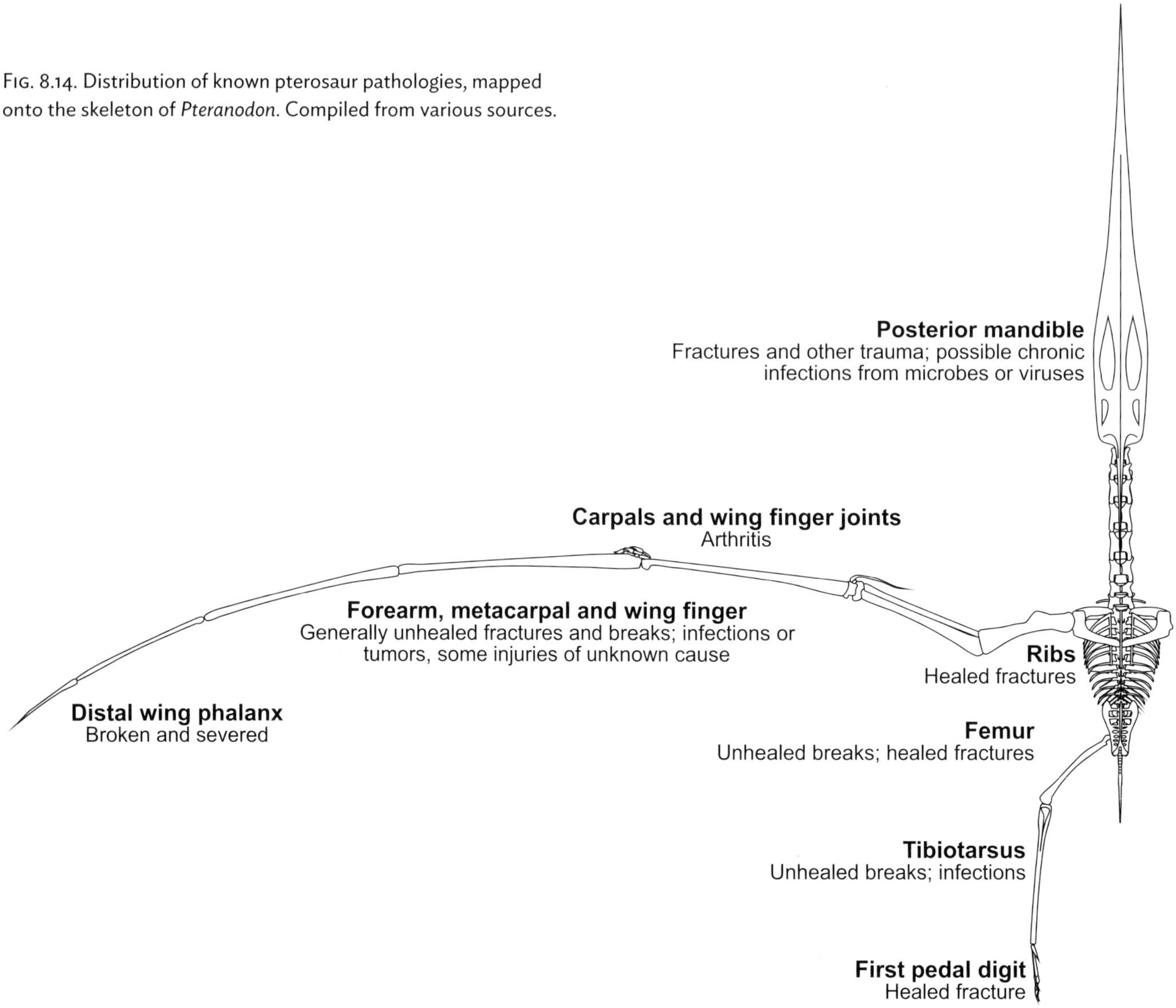

mals on which they preyed. Many pterosaur skeletons are found with pathological—diseased or otherwise damaged—bones, suggesting their lives were far from easy (fig. 8.14; see Bennett 2003c for a review). Necroses, caused by soft-tissue infections close to the bones that resulted in skeletal deformities, are relatively common in pterosaur fossils and particularly so in places where little soft tissue covered the bone, such as the distal limbs, skull, and posterior mandible. The cause of these injuries and their infections is often unknown, but clumsiness, fights with other animals, or perhaps parasitic infections and diseases may be responsible. Some pterosaur skull and mandible pathologies are particularly interesting in this respect. While some are associated with bones broken through compressive trauma (Kellner and Tomida 2000), others are represented by gnarled bony overgrowths, pitting, or depressions of various sizes (Bennett 2003c). These could have been caused by traumatic wounds and tissue inflammation, or they could be the result of diseases and infections. Depressions, bone remodeling, and pitting of the mandibular rami similar to that of some pterosaur specimens are seen in modern birds suffering from infections by the protozoan *Trichomonas,* the nematode *Capillaria,* or a number of other agents. It is quite conceivable that some pathological pterosaur skulls and mandibles reflect very similar afflictions (E.D.S. Wolff, pers. comm. 2011), an idea that finds some indirect support from the identification of *Trichomonas*-like infections in nonavian dinosaurs (Wolff et al. 2009).

Other pterosaur pathologies include fractures and breaks to distal limb elements, the most extreme example being the severing of a portion of the terminal wing phalanx in a *Pteranodon* individual (Bennett 2003c). Many such injuries show signs of completely healing, however, and suggest that the animals lived on long after their calamities. Complete breaks to limb bones have been rarely found (for exceptions, see Wang et al. 2009, and Lü, Xu, et al. 2011) and are generally thought to have been fatal to pterosaurs. A broken limb bone would compromise use of their wings and limit their options for terrestrial movement. Even if the unlucky animal didn't starve, it would be at high risk of becoming lunch for a predator.

A final type of pterosaur affliction is one many of us are all too familiar with: arthritis. The wrists and metacarpal/wing finger joints of several large pterosaur specimens were so heavily used in life that the cartilage pads between each bone wore away, leaving the bones to gouge deep grooves into each other (Bennett 2003c). The development of this condition is unusually extreme in pterosaurs, and has been likened to the arthritic limbs developed by heavyset cattle or draught horses (Bennett 2003c). Arthritic pterosaurs must have been repeatedly sustaining heavy loads on these joints throughout their lives, perhaps by long, sustained bouts of flapping (Bennett 2003c), or possibly by repeated launch activities. The wrist and wing finger were sure to be intensely strained during quadrupedal launching (probably even more than experienced during flapping), so their use over a lifetime may have taken a toll on particularly old individuals.

When Animals Attack

It was not just disease, clumsiness, and old age that made life difficult for pterosaurs. Other animals, great and small in size, were also after their blood. Perhaps the most intriguing group of assailants, though also the hardest to clearly link with pterosaurs, are the Nakridletia, a group of Jurassic insect species that have been suggested as specialist pterosaur parasites (fig. 8.15; reviewed by Vršanský et al. 2010). Other flea-like creatures, such as the Early Cretaceous forms *Saurophthyrus* (Ponomarenko 1976) and *Tarwinia* (Jell and Duncan 1986), have also been labeled as pterosaur parasites. None of these insects exceed 10 mm in length but all were armed with bloodsucking mouthparts and long, savagely hooked appendages. These allowed them to cling to pterosaur pycnofibers and membranes while draining them of their blood. While the adaptations of these insects are compelling evidence for their parasitic lifestyle, some have questioned why they should be considered specialist pterosaur parasites since they may well have been equally comfortable on the hides of feathered dinosaurs or mammals (Grimaldi and Engel 2005). Other flight membranes were also on the menu in the Mesozoic; gliding mammals and reptiles may have proven equally tempting hosts for Nakridletia and their kin if they were, as supposed by some, "membrane specialists."

More convincing evidence exists to show that large Mesozoic animals certainly dined on pterosaurs, and in a much more destructive manner than stealing their blood. Indeed, it seems that pterosaurs were eaten by other animals throughout their evolutionary history. The oldest known case of predation on pterosaurs—a fish regurgitate containing nothing but loosely associated small pterosaur bones—was found in Italian rocks of Late Triassic age (Dalla Vecchia et al. 1989). The identity of the pterosaur is uncertain, but the local paleofauna and composition of the vomited remains (see chapter 5) are consistent with the idea that

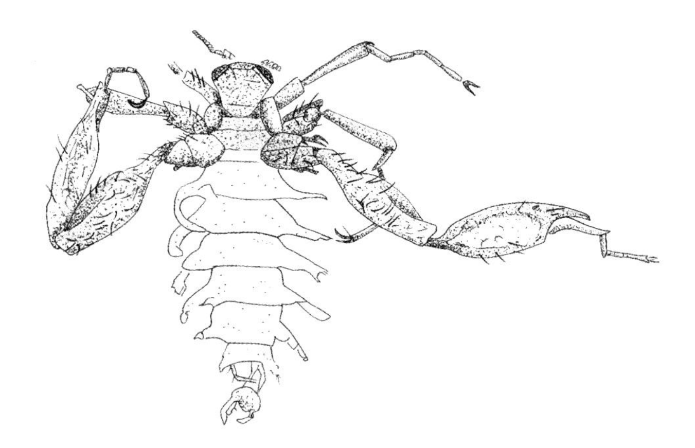

FIG. 8.15. The supposed pterosaur parasite *Strashila incredibilis* from Jurassic deposits of Russia. Redrawn from Vršanský et al. (2010).

a large fish—perhaps the meter-long *Saurichthys*—ate and regurgitated the pterosaur (Dalla Vecchia 2003a). Fishy predation of pterosaurs is recorded again in Jurassic rocks from Germany with another gastric pellet comprising partially digested bones of the pterosaur *Rhamphorhynchus* (Schweigert et al. 2001). This pellet was originally attributed to one of the large marine crocodiles found in the same deposits, but if the stomach acids of ancient crocodiles were like those of their modern relatives, they would dissolve thin pterosaur bones rapidly. A large fish assailant again seems most likely. More evidence of fishy pterosaur predation was reported from German sediments by Frey and Tischlinger (2012), who described multiple associations between a large predatory fish (*Aspidorhynchus*) and *Rhamphorhynchus*, suggesting that they frequently died together after blundered predation attempts by the former.

Sharks also had their fill of pterosaur meat over time. The Cretaceous shark *Squalicorax*, a form noted for its likely scavenging activities, seems to have been particularly fond of pterosaur flesh (see Mike Everhart's [2005] excellent overview of Western Interior Seaway paleontology). *Squalicorax* dwelled in the Western Interior Seaway, a shallow sea that once divided North America, and seems to have dined on the remains of virtually all vertebrates occurring in this ancient sea. Their bite marks can be recognized through the serrated tooth gouges left on many fossil bones, including those of *Pteranodon* (S. C. Bennett and M. J. Everhart, pers. comm. 2008). One does have to wonder if *Squalicorax* was a strict scavenger in this case. Typically under 3 m in length, a full-grown *Squalicorax* would weigh somewhere between 100 to 250 kg, which is much more than even a generously massed, large *Pteranodon* (about 35 kg if the rationale in chapter 6 is accurate). With such a weight ratio, it seems possible that *Squalicorax* could have seized *Pteranodon* in the manner that modern sharks apprehend seabirds alighted on the water surface. A second type of tooth mark, without serrations, also occurs on *Pteranodon* remains and indicates that another denizen of the Western Interior Seaway took interest in *Pteranodon* meat. The identity of this second animal remains mysterious, however.

The alleged occurrence of pterosaur bones in the gut cavities of plesiosaurs and marine crocodiles have been cited as evidence that these marine reptiles also ate pterosaurs (Brown 1943; Martill 1986), but doubt has been expressed over the validity of both claims (Forrest 2003; Witton 2008b). More definitive records of reptile-pterosaur predation stem from three specimens incontrovertibly demonstrating that Cretaceous dinosaurs ate pterosaurs. These predators include the large Brazilian spinosaurid *Irritator*, and the rather smaller dromaeosaurs *Saurornitholestes* from Canada, and *Velociraptor* from Mongolia (figs. 8.1 and 8.16; Currie and Jacobsen 1995; Buffetaut et al. 2004; Hone, Tsuhiji, et al. 2012). The size disparity between the consumers and consumed contrasts strongly in the former two cases. The spinosaur, recorded through a broken tooth embedded in a pterosaur vertebra, dwarfs its meal, but the pterosaur eaten by the *Sauronitholestes* (who lost a tooth gnawing the pterosaur's shin) had a wingspan three times that of the dinosaur's length. While the latter seems to be a good indication that the pterosaur was scavenged, the size relationship between the former individuals provides a plausible predatory scenario as well as one of scavenging. Modern crocodiles and cats are perfectly capable of catching and killing even relatively large birds, so there's little reason to assume that a savvy dinosaur predator could not do the same with pterosaurs.

FIG. 8.16. A group of *Saurornitholestes langstoni* pilfer pieces of flesh from a deceased, giant azhdarchid alongside a Late Cretaceous River in Canada, while a plucky azhdarchid (lower left) attempts to dominate the carcass and startle the local avian population.

9
The Diversity of Pterosaurs

My PhD supervisor, pterosaur guru David Martill, once told me of pterosaurologists who consider all pterosaurs to basically represent the same animal, albeit with a different head and neck bolted onto each species. This really could not be further from the truth, and we're going to spend most of the rest of this book investigating why. The proportions and anatomies of different pterosaurs are not only distinct enough that we can identify different lineages from fragmentary remains, but we can also detect a broad array of ecological strategies and locomotory methods across many groups and species (fig. 9.1).

To appreciate the scope of pterosaur diversity, the next 15 chapters will focus on the specific groups of pterosaurs known at the time of writing, covering their history of discovery, anatomy, and paleoecology. To assist us in this endeavor, we will need a taxonomic scheme to follow. Fortunately, pterosaur workers have dedicated considerable time to unraveling pterosaur interrelationships and something of a handle on their evolutionary pathways has been obtained. Early studies on this topic were entirely subjective, with deductions of shared ancestry made by considering the morphological similarity of pterosaur species and their relative position in time (e.g., Young 1964; Wellnhofer 1978). Since the 1980s, some of this subjectivity has been removed by applying computerized phylogenetic and cladistic analyses to pterosaur species and genera (see Unwin 2003 for a history of these studies). These analyses split pterosaur taxa into groups, or clades, based only on shared anatomical characteristics, and attempt to present the most parsimonious, or simplest, evolutionary pathways between them. Nowadays, there are two broadly accepted schools of thought on pterosaur phylogeny. One is derived from a data set compiled by David Unwin (discussed at greatest depth in Unwin 2003, but presented in its most modern guises in Lü, Unwin, et al. 2006; Lü, Unwin, et al. 2008; Lü, Unwin, et al. 2010; Lü, Unwin, et al. 2011; and Lü, Unwin, et al. 2012), and the other is based on the data of Alexander Kellner (best detailed in Kellner 2003; but featured in revised and updated form in Wang, Kellner, Zhou, et al. 2008; Wang et al. 2009; and Wang et al. 2012. The evolutionary trees generated by these datasets agree with one another in many respects, but also differ in some rather fundamental ways. Other phylogenetic work (e.g., Martill and Naish 2006; Andres and Ji 2008; Dalla Vecchia 2009a; Andres et al. 2010) is more or less derived from either the "Unwin" or "Kellner" data set, so we're starting to see something of a split in opinions over pterosaur interrelationships as workers favor one scheme over another.

To progress any further, we have to make something of a choice over which of these schemes to follow. With testing, the "Unwin" school of pterosaur interrelationships has just about emerged as the more reliable of the two schemes (Andres 2007; Unwin and Lü 2010), and the results of my own phylogenetic studies have generally found more congruence with this school of pterosaur taxonomy than any other. Thus, although there are good reasons to prefer some taxonomic arrangements within the "Kellner" phylogenies, we will more or less follow the "Unwin" model in the subsequent chapters. Their content broadly follows a recent and detailed incarnation of the "Unwin" pterosaur phylogeny (fig. 9.2; Lü, Unwin, et al. 2010), and while this choice will not please everyone, significant disagreements over the placement of controversial species or groups, and disputes over the name or composition of a group, will also be discussed.

Tooling Up

Before we get started, it may be pertinent to quickly arm ourselves with a few basics concerning the content of the following pages. Rather than merely referring pterosaurs to the major time periods of the Mesozoic, we'll be placing them in specific ages, the smaller stretches of time that make up our more familiar geological periods (see fig. 9.3 for reference). Hopefully, this will give us a greater appreciation of pterosaur distribution in time than if we used the broader Early/Middle/Late epochs of the Mesozoic periods.

It will be useful to know that pterosaur taxonomists recognize two incontrovertible groups of fly-

Fig. 9.1. A flock of istiodactylids soar over what will become China in the Early Cretaceous. Istiodactylids were probably excellent at soaring flight and may have used it to search for carrion. For more on istiodactylids, see chapter 15.

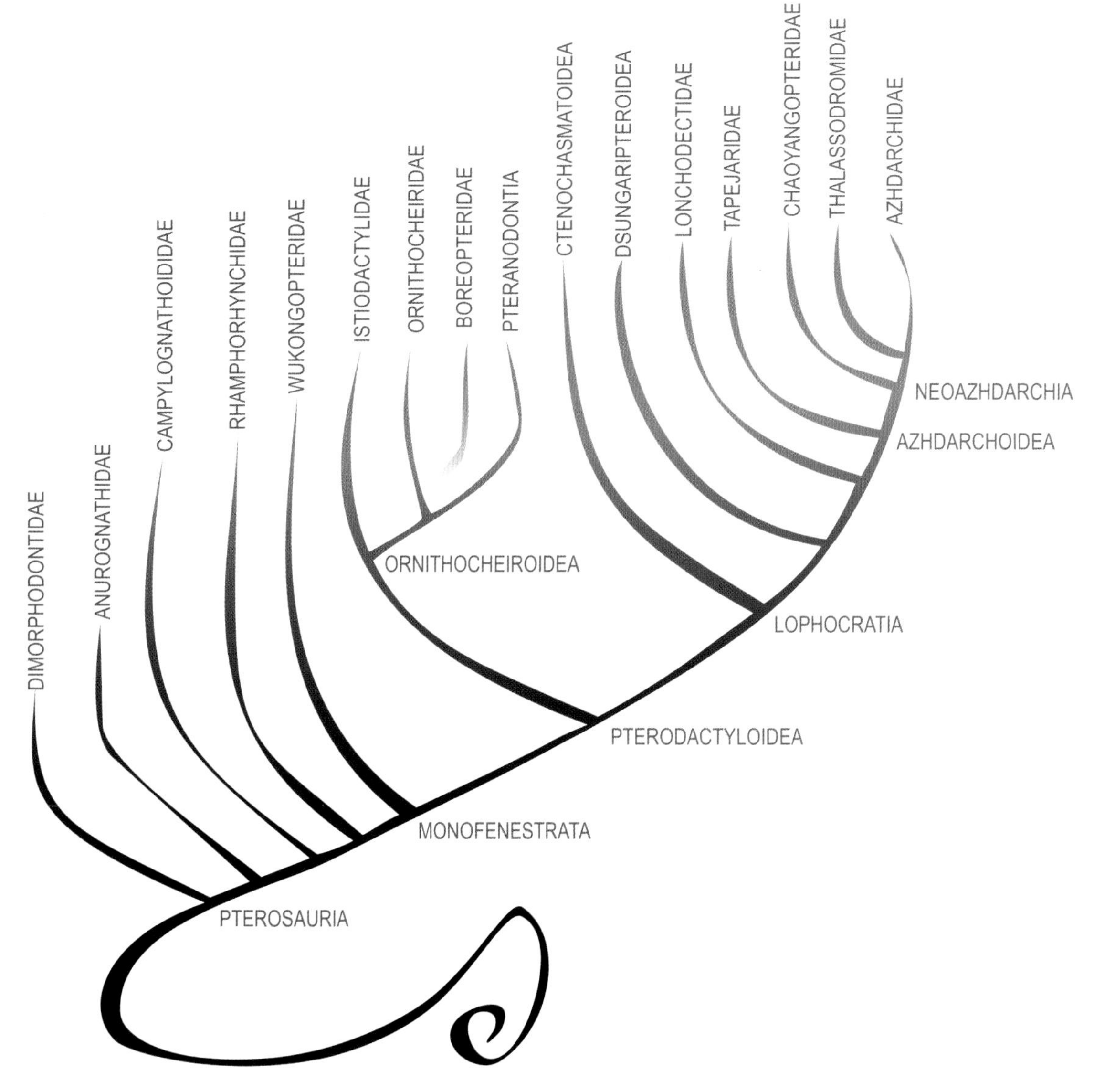

Fig. 9.2. The pterosaur phylogeny of Lü, Unwin, et al. (2010), and the taxonomic scheme the following chapters are based on.

ing reptiles within Pterosauria. The first is a ragtag assemblage of species that successively split from the pterosaur tree in the Triassic and Jurassic, and are generally identified by their long tails and lengthy fifth toes. The second is the large group Pterodactyloidea, a Jurassic/Cretaceous radiation of pterosaurs with short tails, relatively long metacarpals, very thin bone walls, and short or absent fifth toes. Veterans of pterosaur studies may know the first group by the moniker "rhamphorhynchoids," a term reflecting the pre-1990s view that pterosaur evolution was split early on into two groups: the long-tailed "rhamphorhynchoids," and the short-tailed pterodactyloids. Nowadays, it is pretty much universally agreed that only Pterodactyloidea represents a natural group, and the concept of "Rhamphorhynchoidea" is long abandoned. We know of many more pterodactyloid species than we do non-pterodactyloids. It seems that the former reached an acme of diversity and disparity in the lower Cretaceous, and to such a degree that this seems to have been something of a pterosaur heyday; pterosaurs as a group were never more diverse than they were in the Early Cretaceous (fig. 9.3). Pterodactyloids were also longer-lived and more geographically

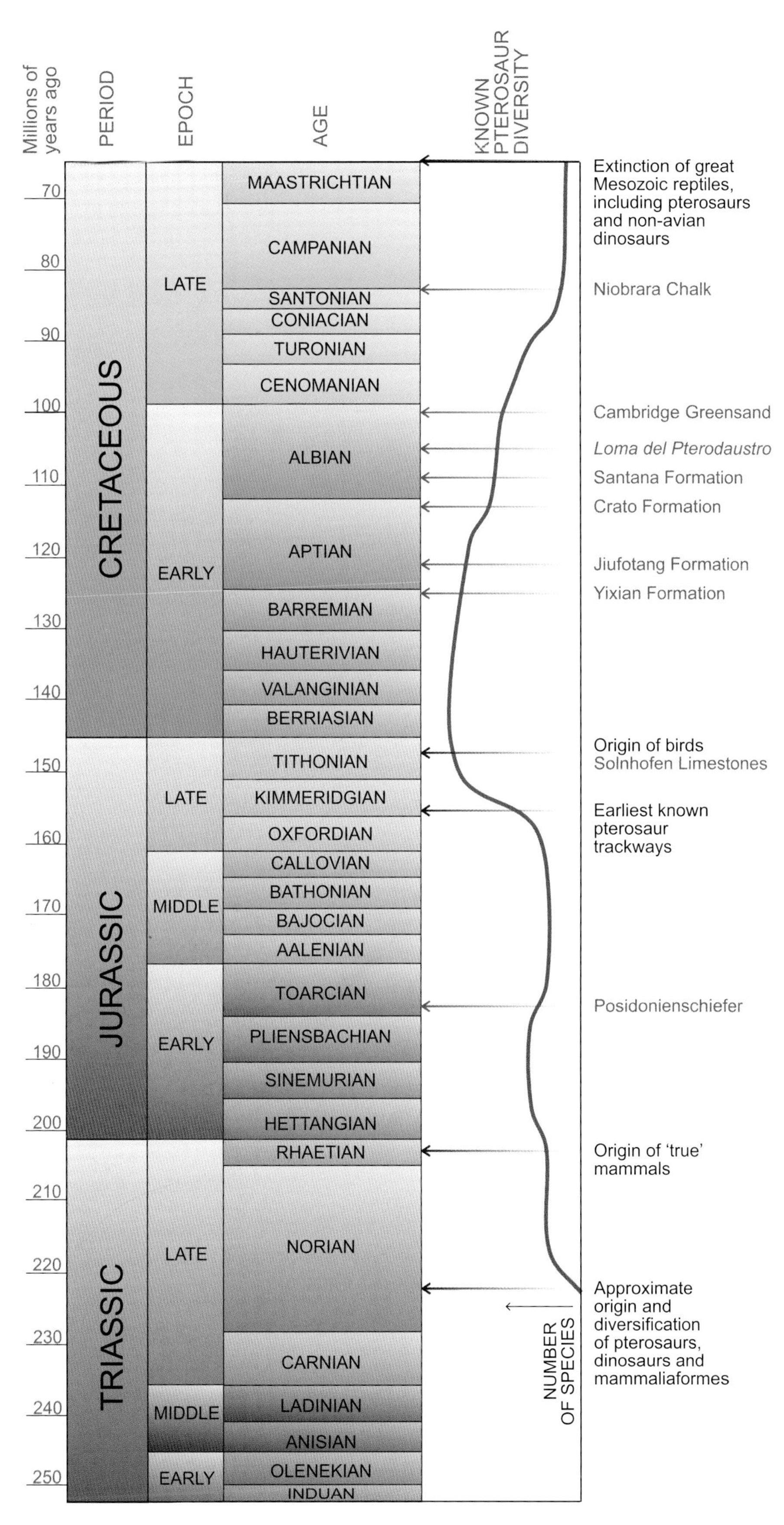

Fig. 9.3. Pterosaur diversity through geological time. Note how this diversity is enormously inflated through the Tithonian-Aptian, precisely the ages of the most productive pterosaur-bearing Lagerstätten, the stratigraphic positions of which are labeled at right in blue. See chapter 26 for more on pterosaur diversity through time. Diversity curve from Butler, Barrett, Nowbath, and Upchurch (2009).

widespread than their non-pterodactyloid ancestors. How much we should read into these facts, however, is another topic, and one that we'll address in the final chapter.

One final thought to consider as you read on is that the following chapters contain a far more diverse array of species than they would have if this book had been written ten years ago. At the time of writing, about 130 to 150 species of pterosaur are recognized (though as we'll see, this is an overly conservative estimate in some eyes). Should anyone write another book of this type in another decade, the number will have undoubtedly increased even more, and perhaps the arrangements of the chapters themselves would be very different. The following pages are only a snapshot of pterosaur diversity as we know it in 2012–13, and it will almost certainly look very different in years to come.

10
Early Pterosaurs and Dimorphodontidae

PTEROSAURIA > *PREONDACTYLUS* + DIMORPHODONTIDAE

Frustratingly little is known about the earliest phase of pterosaur evolution (fig. 10.1). Scrappy remains of the earliest pterosaur lineages are distributed across the globe (fig. 10.2; Barrett et al. 2008), with their best fossils found in Europe, North America, and more recently, Brazil. The latter occurrence is potentially rather important, as it may represent the oldest pterosaur yet found, stemming from the Carnian/Norian boundary (216 Ma) rather than, as with most Triassic pterosaurs, the upper Norian (ca. 210 Ma). Named *Faxinalipterus minima* (Bonaparte et al. 2010), the fragmentary remains representing this animal bear some pterosaurian features, such as hollow bones and a somewhat pterosaur-like humerus and coracoid. Interestingly, some attributes of its anatomy appear rather primitive compared to most flying reptiles, which may be predicted for an animal several million years older than the other representatives of its group. Its fibula, for example, is as long as the tibia, a configuration almost unheard of in any other pterosaurs but found in purported pterosaur ancestors (chapter 3). Some pterosaur workers are doubtful that *Faxinalipterus* is a pterosaur however, because plenty of other Triassic animals bear hollow bones, and its remains are so fragmentary that clear-cut pterosaur features cannot be seen. *Faxinalipterus* may, therefore, represent one of several groups of small Triassic reptiles, and is not necessarily pterosaurian.

This is not the first time a purported, and potentially important, ancient pterosaur genus has been doubted. In 1866, the famous American paleontologist Edward Drinker Cope reported a cluster of small bones thought to represent a pterosaur from Triassic deposits of Pennsylvania. These bones were not only the first pterosaur fossils to be reported from the United States, but also the first from Triassic strata anywhere in the world. Cope eventually named them *Rhabdopelix*. Over a century later, reappraisal of these important remains suggested that they were insufficient to denote a pterosaur identity, and as with *Faxinalipterus*, could belong to a number of Triassic reptiles (Dalla Vecchia 2003b; see also Witton 2010 for more on the history of this discovery). We will return to Cope's alleged Triassic pterosaur in chapter 18.

For undoubted remains of early pterosaurs, we have to turn to middle or late Norian deposits of northern Italy. Some of the pterosaur fossils from this region are generally considered to represent the earliest offshoots from the pterosaur lineage, including *Preondactylus bufarinii*, an animal known from a virtually complete skeleton (fig. 10.3; Wild 1984b; Dalla Vecchia 1998). Unfortunately, *Preondactylus* is not known as well as it could be, thanks to an error worthy of inclusion in an episode of *The Simpsons*. The collector of the only known skeleton accidentally washed most of its fossil remains down a sink when cleaning the fossil, leaving only sediment molds of its bones to work from. Thankfully, enough data can be gleaned from these to interpret *Preondactylus* as a very basal, if not the most basal, member of Pterosauria (Dalla Vecchia 2009a; Andres et al. 2010; Lü, Unwin, et al. 2010; Lü, Unwin, et al. 2011; Lü, Unwin, et al. 2012; though see Wang et al. 2009 for a differing opinion).

Contemporaneous rocks elsewhere in northern Italy have yielded a close relative of *Preondactylus*, the dimorphodontid *Peteinosaurus zambellii* (Wild 1978). This animal is known from two (possibly three) incomplete specimens representing limb bones, limb girdles, and a mandible (Dalla Vecchia 2003b). This provides enough information to deduce that *Peteinosaurus* was probably closely related to *Dimorphodon macronyx*, a relatively large and historically important early pterosaur from Jurassic rocks (Sinemurian; 197–190 Ma) of southern England (Buckland 1829). *Dimorphodon* is something of a pterosaurian A-lister, as it was the first pterosaur fossil identified from Britain, the first pterosaur identified outside of the Solnhofen Limestone quarries of Germany, and was associated

FIG. 10.1. *Dimorphodon macronyx* bounds through a Sinemurian forest in pursuit of an unfortunate early mammal, propelling itself through the trees with anatomy honed for climbing.

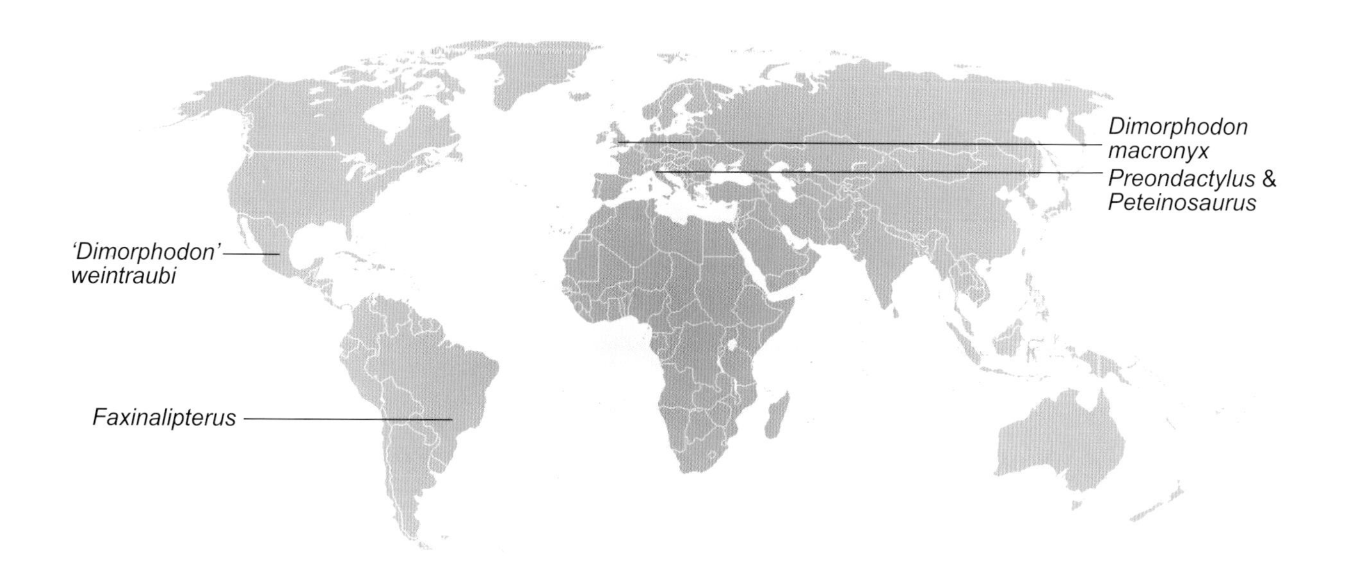

FIG. 10.2. Distribution of dimorphodontid and other early pterosaur taxa.

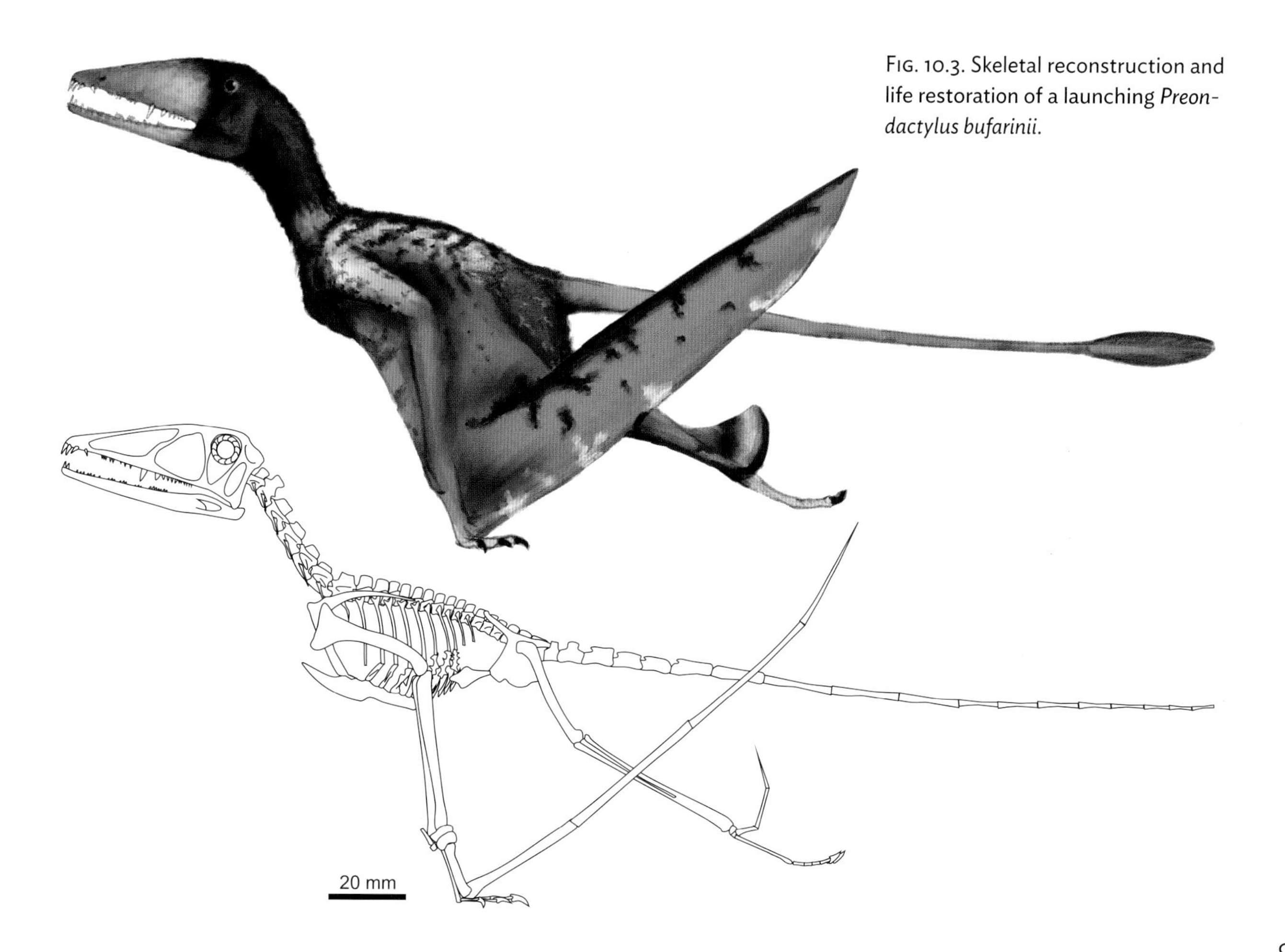

FIG. 10.3. Skeletal reconstruction and life restoration of a launching *Preondactylus bufarinii.*

FIG. 10.4. A nearly complete specimen of *Dimorphodon macronyx from the* Sinemurian Lower Lias of Britain and described in 1870 by the Victorian paleontologist heavyweight, Richard Owen. Photograph copyright Natural History Museum, London.

with the Victorian heavyweights of British paleontology. Mary Anning, perhaps the first professional fossil collector, pulled the first *Dimorphodon* specimens from the Lower Lias of Britain's famous "Jurassic Coast" in 1827. These shallow marine deposits also yield the gigantic marine reptiles that Anning is probably better known for discovering. She passed the pterosaur remains to Reverend William Buckland, the early paleontologist and geologist best known for providing the first account of a dinosaur fossil in 1822. Buckland named the pterosaur *Pterodactylus macronyx*,[6] thinking it was rather similar to the pterosaur specimens already named from Germany (see chapters 2 and 19). A more complete specimen of *Dimorphodon* was later found (fig. 10.4) and passed to Richard Owen, the preeminent natural historian of his day who not only coined the term "dinosaur" but also spearheaded the construction of London's fantastic Natural History Museum. Owen was a prominent figure in the early history of pterosaur research, naming numerous species from Jurassic and Cretaceous deposits of Britain and documenting their anatomy with detail unrivaled by his nineteenth-century peers. He considered the second specimen of *Pterodactylus micronyx* so distinctive from the German *Pterodactylus* that it warranted its own genus (Owen 1870); the name *Dimorphodon* was the result. Since then, a steady trickle of *Dimorphodon* material has been discovered in Liassic deposits and a possible second species, the slightly larger *D. weintraubi*, has been recovered from Early to Middle Jurassic deposits of Mexico (Clark et al. 1998). (Whether the latter is referable to *Dimorphodon* has been questioned in recent years however, and its remains await full description and taxonomic appraisal.) Despite its familiarity, *Dimorphodon* continues to be a prominent creature in pterosaur research by providing insights into pterosaur posture (e.g., Padian 1983b; Unwin 1988b; Bennett 1997a; Clark et al. 1998) and the relationships of pterosaurs to other reptiles (e.g., Nesbitt and Hone 2010). Excitingly, another *Dimorphodon*-like form was recently found in Pliensbachian (190–183 Ma) deposits of southern Britain, and may represent a new genus or species of dimorphodontid (Unwin 2011).

In the scheme used here, *Dimorphodon, Peteinosaurus,* and the new British form all comprise Dimorphodontidae, an early radiation of large-headed pterosaurs that may have been well adapted for climbing. They are united by a suite of features seen in their

[6] The practice of naming pterosaurs as species of *Pterodactylus* or "*Ornithocephalus*" was common in the nineteenth century (see chapter 19 for more on the origins of these names). As we will see in subsequent chapters, most pterosaurs discovered during this time were placed in these genera. The application of novel generic names to pterosaur fossils became more commonplace towards the end of the nineteenth century however, and at present, most pterosaur genera only contain one species.

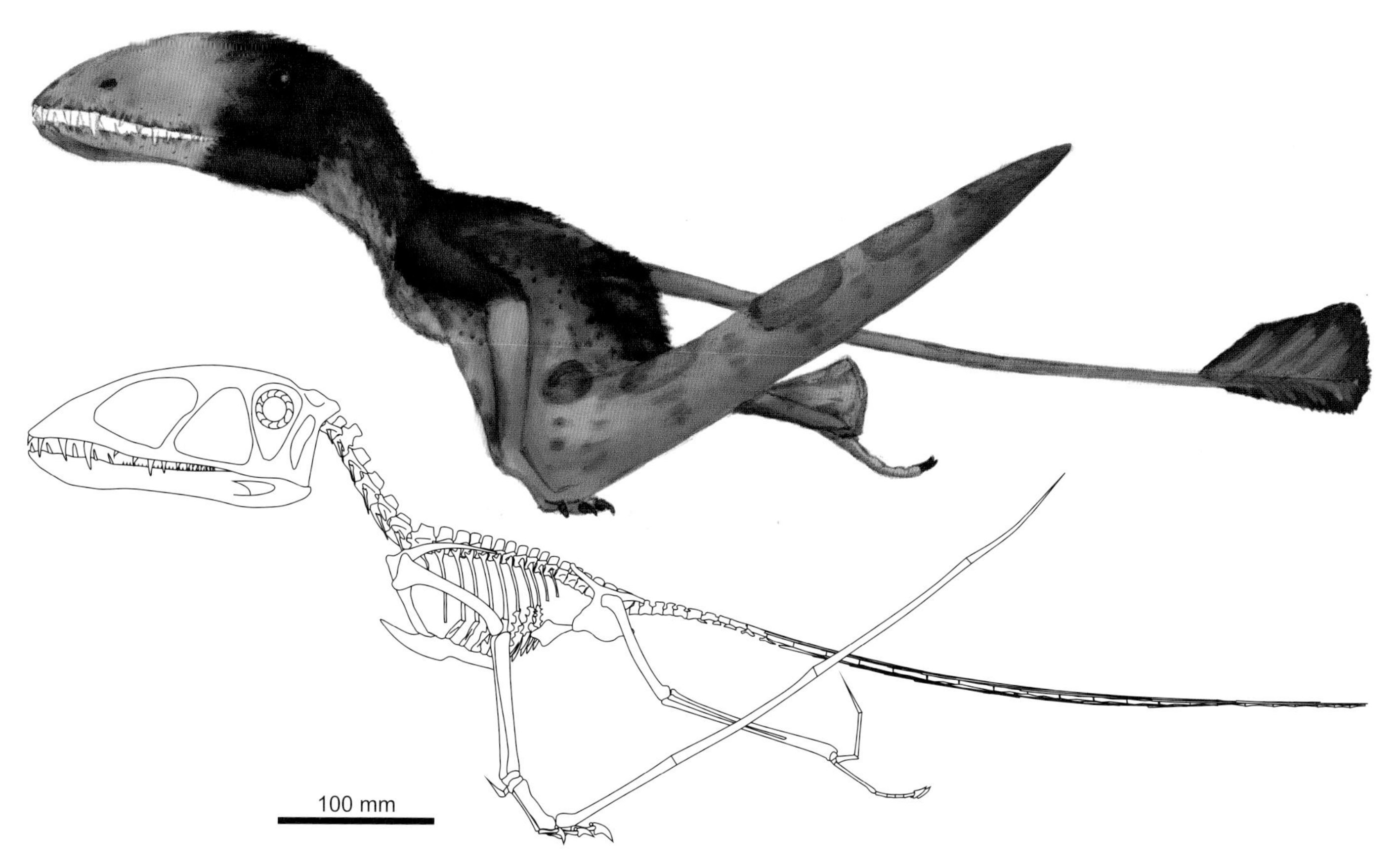

Fig. 10.5. Skeletal reconstruction and life restoration of a launching *Dimorphodon macronyx*.

lower jaws and forelimbs, including the robustness of their proximal wing phalanges and the shapes of their deltopectoral crests (fig. 10.5). While this group is not recognized in other phylogenies (Dalla Vecchia 2009a; Wang et al. 2009; Andres et al. 2010), a close relationship between *Peteinosaurus* and *Dimorphodon* at the base of the pterosaur tree is generally accepted, even if they are not always thought to be each other's closest relatives.

Anatomy

So far as can be seen, the first pterosaurs share a suite of features that are documented, along with the rest of their anatomy, by Owen (1870), Wild (1978, 1984b), Padian (1983b), Unwin (1988b), and Dalla Vecchia (1998). They are generally small, with wingspans of 0.5 m in *Preondactylus*, 0.6 m in *Peteinosaurus*, and 1.3 m *Dimorphodon*. They do not have the outlandish proportions seen in other pterosaurs with heads, necks, and forelimbs that are large, but not so much to make their bodies and legs look tiny. Their skulls are rather deep and perforated with large fenestrae so that, despite their size (the skull of *Dimorphodon* is about 20 cm long), they were probably quite lightweight. The only known skull of *Preondactylus* is somewhat crushed and its exact shape cannot be determined, but it is often reconstructed with a rather shallow snout (e.g., Wellnhofer 1991a; Dalla Vecchia 1998). This may not be accurate however, as the length of the bony strut between its antorbital fenestra and nasal opening dictates that it must have had a deeper snout than is typically reconstructed (fig. 10.3; Unwin 2003). The skulls of these pterosaurs are generally unornamented, but *Dimorphodon* does bear a slight keel along the front part of its lower jaw. *Dimorphodon* also possesses an external mandibular fenestra, an archosauriform trait retained from pterosaur ancestry that was lost in later pterosaurs (Nesbitt and Hone 2010). Sadly, the remains of other early pterosaurs are too poorly preserved determine if they bore the same feature.

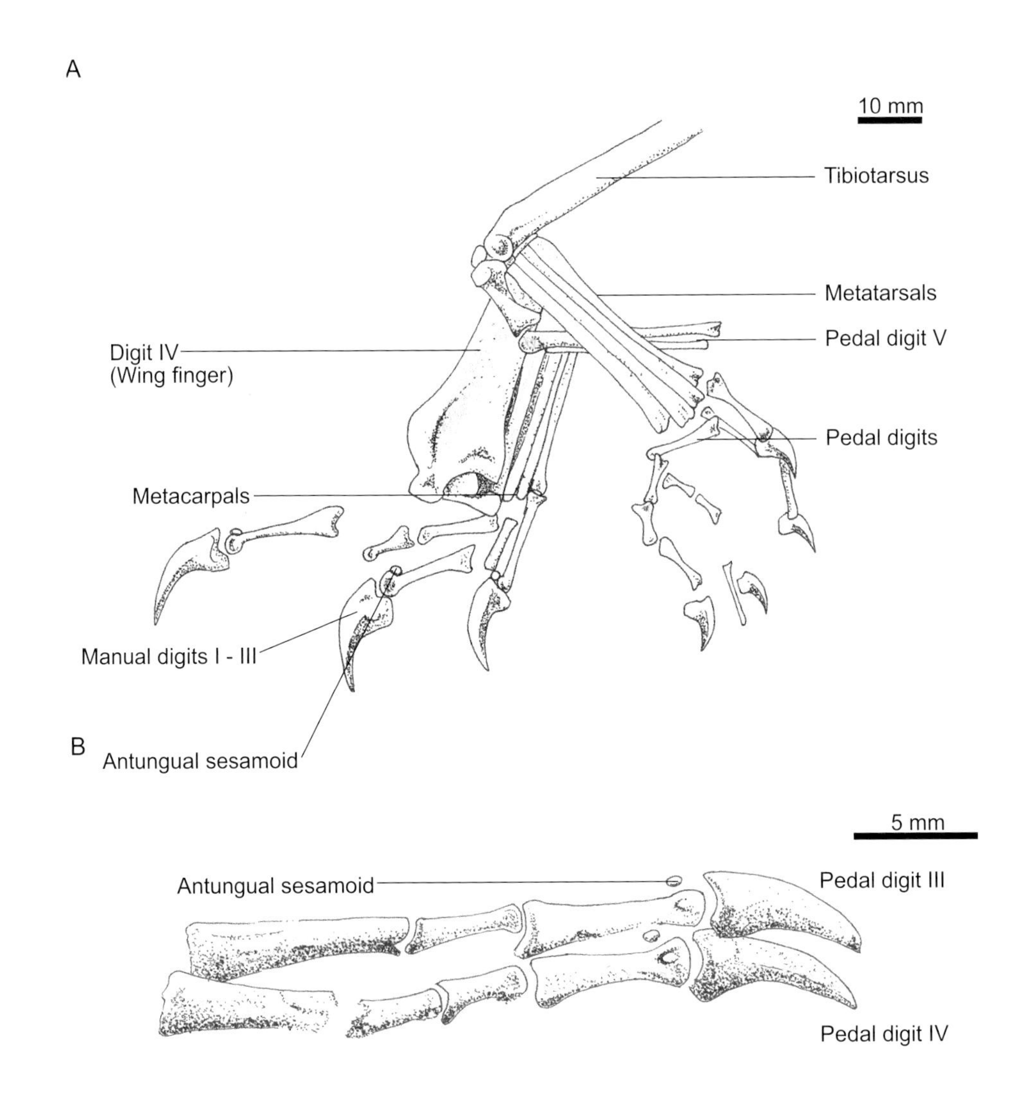

FIG. 10.6. The robust hands and feet of *Dimorphodon macronyx*. B, redrawn from Unwin (1988b).

These early pterosaurs all possess teeth that vary markedly in size throughout the jaw. All possess large, laterally compressed, fang-like teeth at the jaw tips and smaller teeth toward the rear. Some of these teeth (such as the anterior teeth of *Dimorphodon*, and the large teeth under the antorbital fenestra in *Preondactylus*) bear very fine serrations. In *Dimorphodon*, the upper jaws contain much larger, well-spaced teeth than the lower, where they are triangular in profile, tightly packed, and only millimeters tall. The teeth of *Preondactylus* are also generally small, save for the aforementioned teeth at the jaw tips and two rather large teeth at the midpoint of the upper jaw (a trait also seen in some other early pterosaurs; see chapter 12).

The forelimb/hindlimb ratios (ignoring the length of the wing finger) of *Preondactylus* and *Dimorphodon* are among the lowest known for any pterosaurs, and their wing fingers are particularly short. These are generally considered to be "primitive" characteristics, and may indicate that these early forms had not had the time to evolve the very disproportionate, and perhaps more wing-like, limb proportions of more derived pterosaurs. Thus, their limbs are of rather equal length and their wingspans are quite small for their body size. This is particularly so in *Dimorphodon*, which has such a large body, head, and legs that it may have weighed as much as 1.3 kg, twice as much as other pterosaurs of similar wingspan (Brower and Veinus 1981 Witton 2008a). These small wingspans may be related to the limited pneumatization of the *Dimorphodon* skeleton, which seems confined to the skull and cervical series (Butler, Barrett, and Gower 2009).

Despite their antiquity, the flight anatomy of these early forms is fully developed, with robust shoulder girdles, blocky carpals, pteroids and wing fingers all present and leaving little doubt that these early pterosaurs were capable of flight. Though their legs are long and robust, their hips are unusually small and bear particularly short preacetabular processes. Their hands and feet are strongly developed and especially so in *Dimorphodon macronyx*, which has proportionally large claws and antungual sesamoids on all claw-bearing digits (fig. 10.6; Unwin 1988b). These sesamoids seem to be absent in *D. weintraubi*, however. The possession of sesamoids on the pedal digits is not seen in any other pterosaur and may indicate that *D. macronyx* was rather specialized for a particular type of terrestrial locomotion. As with most long-tailed pterosaurs, the caudal vertebrae of dimorphodontids are lined with elongate zygapophyses for much of their length, but the tail of *Preondactylus* lacks these features. The exact number of tail vertebrae is unknown in this pterosaur (suggested as 20 by Wild 1984b), but is relatively high at 30 in *Dimorphodon*.

Locomotion

FLIGHT

Little research has been performed on the flight of the earliest pterosaurs, which is somewhat surprising given how much it may tell us about the early evolution of volancy in vertebrates. The breadth of the *Dimorphodon* flapping stroke has been determined, with its glenoid morphology permitting the wing to move through a full a 90° arc (Padian 1983b). At least some early pterosaurs, therefore, already possessed some anatomical hallmarks of powerful flapping, though the lack of sternal remains means we cannot presently estimate how large their flight muscles were. The comparatively short forelimbs and long hindlimbs of these early pterosaurs are predicted to create relatively broad, low-aspect wings (Witton 2008a). When compared with the wing loading and aspect ratios of modern fliers, *Preondactylus* plots among birds and bats that are rather "generalized" in their flight strategy (fig. 4.5; Witton 2008a). These animals are powerful, skilled aeronauts, but not particularly specialized to any particular flight style (Rayner 1988), and are represented in modern birds by widespread and adaptable groups, such as crows, parrots and pigeons. Interestingly, these birds are adapted for flight in cluttered terrestrial settings, which corroborates the idea that early pterosaurs may have lived and evolved in multi-tiered, terrestrial environments (see chapter 3).

The flight of dimorphodontids appears to be a different case altogether. The high mass and relatively short, broad wings of *Dimorphodon* indicate it was a relatively highly loaded flier akin to modern birds that take to the wing infrequently, and often only for short bursts of time (e.g., rails, grouse, woodpeckers, tinamous) (Witton 2008a). Takeoff and flight for such animals is rather energetically demanding because their glide angles are comparatively steep; should they stop flapping, their flight paths resemble a graceful plummet rather than efficient gliding. Equally, because of their weight, they are not particularly maneuverable fliers. Predicted wing attributes of *Dimorphodon* suggest it may have flown in a similar manner, requiring a hefty leap to become airborne and then rapid flapping to maintain level flight. *Dimorphodon* may have used undulating flight, an energy-saving rhythm of flapping bursts punctuated with short periods of gliding (Rayner et al. 2001) in order to maximize its flight efficiency. This would have given their flight a distinctively "swooping" pattern like that of heavily loaded woodpeckers, a bird group found to be close flight analogues of *Dimorphodon* (Witton 2008a). These predicted wing attributes may also indicate that dimorphodontids were infrequent fliers, perhaps only becoming airborne in an emergency or when needing to travel long distances (fig. 10.7). This is an interesting prediction, as pterosaurs probably more "primitive" than *Dimorphodon*—such as *Preondactylus*—appear comparatively better adapted to flight. This implies that the comparably limited flight abilities of dimorphodontids evolved *after* pterosaurs learned to fly competently, rather than representing a crude, early effigy of pterosaur flight. Considering that this development would have taken place as early as the lower Jurassic, it seems that pterosaurs wasted no time in experimenting with their locomotory mechanics.

ON THE GROUND

If *Dimorphodon* was a reluctant flier, it would need to be well adapted to moving around in some other fashion. As with *Preondactylus* and *Peteinosaurus*, *Dimorphodon* seems well suited to scampering around complex terrestrial environments. All three of these species have robust, probably well-muscled limbs and extremities, suggesting they would have had no trouble walking or bounding along when grounded. Their proportionate limbs indicate that the stride lengths and power of all four limbs could be fully utilized when walking or running, unlike many other pterosaurs in which the short, relatively weak hindlimbs

Fig. 10.7. How to make *Dimorphodon* fly. As a heavy pterosaur for its size, *Dimorphodon* may have found flight quite strenuous. Perhaps, like comparably loaded birds, it only took to the air for short hops between trees or when under duress (duress here being a hungry-looking theropod dinosaur).

compromised the stride potential of the forelimbs (see chapters 15–18 for the most extreme examples). In addition, while very few pterosaur workers still accept Kevin Padian's (1983b) idea that *Dimorphodon* and its kin were bipedal (chapter 7), most of his observations about the cursorial adaptations of their hindlimbs remain sound; they were clearly powerful legs adapted for high-octane terrestrial locomotion (also see Bennett 1997a).

It is likely that much of this terrestrial locomotion would not have taken place on the ground itself, however. The highly developed and uniquely adapted hands and feet of *Dimorphodon* cast it as a particularly sprightly clamberer, and perhaps among the best of all pterosaur climbers. It bears exaggerated versions of the pterosaurian climbing adaptations we discussed in chapter 7, such as particularly large, hooked, and narrow claws that are well suited to finding purchase on tree trunks or rocky crags (Unwin 1988b). Its powerful hindlimbs would have been of obvious utility in climbing, as would its long, balancing tail. *Dimorphodon*'s relatively conservative body proportions are also potentially significant, as these would have kept the center of mass close to the climbing surface, thereby reducing the pull of gravity when climbing a vertical face. Accordingly, we may imagine *Dimor-*

phodon moving through treetops or rock faces like a leathery-winged squirrel. *Peteinosaurus* and *Preondactylus* also bear some of these attributes and were probably also capable climbers (Wild 1984b), but neither appears quite as adapted for life in the higher levels like *Dimorphodon*.

Paleoecology

Early pterosaurs are generally thought to have eaten either fish or insects (Wellnhofer 1991a; Unwin 2005; Ősi 2010), with the deep-snouted *Dimorphodon* often compared to modern puffins in artistic reconstructions (see, for example, images in Bakker 1986). There may not be any particular reason to assume that these animals fished, however. To my eyes, their jaws and necks are not long or mobile enough for seizing food from the water in flight, nor do their predicted flight styles lend themselves to the steady, level flight needed to strike at swimming prey from above. Birds utilizing this form of dip-feeding tend to be strong, stable fliers, and good gliders (Rayner 1988), none of which has been considered likely for these early forms. Without any obvious swimming adaptations either, there seems little reason to assume these animals were adapted for pursuing fish or other swimming organisms more than any other type of food.

The notion that early pterosaurs were insectivorous has recently been corroborated by a detailed study of their tooth and jaw mechanics (Ősi 2010). The absence of wear on the teeth of early pterosaurs demonstrates that their finely pointed teeth interlocked when the mouth was closed, a feature excellently adapted for holding small prey. The dagger-like nature of their larger, microserrated teeth at the jaw tips seem well suited for penetrating hard insect exoskeletons, and the presence of much smaller teeth on the posterior lower jaw is consistent with the tooth distributions seen in modern insect-eating reptiles.

Exactly how early pterosaurs caught their probable insect prey is less clear, however. Unwin (2005) considered that early pterosaurs hawked their meals in midair, but as we discussed in chapter 3 (and will touch on again in chapter 11), early pterosaurs almost certainly lacked the aerial agility necessary to catch flying insects. If modern insect hawkers are anything to go by, early pterosaurs may have been too heavy to effectively chase insects through the air. The largest modern aerial insectivores (the potoos, a close relatives of owls and nightjars) have rather small bodies, long wings, and relatively low masses (about 0.5 kg for the largest species, which has a 1 m wingspan). These birds, however, are still limited to catching fairly large, slow insects in flight (M. B. Habib, pers. comm. 2012). This suggests that there is a fairly low size limit for effective aerial insectivory, and the large, heavyset *Dimorphodon* (which may have been more than twice as heavy as the largest potoo, despite being almost the same size) was probably too big, heavy, and clumsy in flight to catch nimble airborne prey. Early pterosaurs also lack the very wide, short mouths found in modern insect hawkers, which maximize their chances of bagging insects in midflight. Thus, the case for early pterosaurs grabbing insects in flight, in my view, looks very poor. Although less dramatic, I wonder if these early pterosaurs simply grubbed around in leaf litter and the tree canopy for beetles, worms, and other small prey, using their robust limbs to walk or hop from one foraging site to the next or even to run down their prey should it try to escape (fig. 10.1).

Special mention should go to the functionality of the *Dimorphodon* skull. Although not as disproportionately large as other pterosaur skulls, it is still much bigger, relatively speaking, than those of other early forms. Bakker (1986) suggested that *Dimorphodon* bore a powerful bite, assuming that its very large antorbital fenestra was filled with the anterior extreme of its internal jaw musculature. This notion was firmly rejected by Larry Witmer (1997), who noted that the slender bones surrounding this fenestra would not permit such a bite. In his words: "Such a musculature system probably would not have had the opportunity to contract more than once!" Perhaps a more plausible interpretation of the large *Dimorphodon* head is that it was adapted for tackling prey of larger size than other early pterosaurs. The depth of its snout would have provided greater resistance to bending and twisting than those of other early pterosaurs (Fastnacht 2005a), suggesting particularly large insects, and perhaps small vertebrates, may have been on its menu. However, with jaw muscles better adapted to rapid jaw closure than providing a firm grip (Fastnacht 2005a), claws better suited to climbing than holding prey (Wild 1984b; Unwin 1988b), and no obvious means of processing large carcasses, *Dimorphodon* seems unlikely to have subdued particularly sizeable animals.

11
Anurognathidae

PTEROSAURIA > ANUROGNATHIDAE

Anurognathids, small, Muppet-faced, insect-catching pterosaurs (fig. 11.1), are one of the most intriguing, but in many ways one of the most mysterious, pterosaur groups. They are only known from a few sites around the world (fig. 11.2) and are represented by rare fossils that are often difficult to interpret. Subsequently, pterosaurologists have only recently developed a handle on their basic anatomy. It seems that anurognathids turned the bizarreness of pterosaur form up to 11, possessing a mosaic of "derived" and "basal" characteristics that has caused outright disagreements over where they fit into pterosaur phylogeny. Some authors view anurognathids as the most basal pterosaurs yet known (e.g., Kellner 2003; Bennett 2007b; Wang et al. 2009), while others set them as an evolutionary notch above dimorphodontids (Unwin 2003; Lü, Unwin, et al. 2010; Lü, Unwin, et al. 2011; Lü, Unwin, et al. 2012), nestling alongside pterodactyloids (Dalla Vecchia 2009a; Andres et al. 2010) or, most radically, as members of Pterodactyloidea (Young 1964). The latter idea has not gained acceptance among pterosaur workers (the long fifth toes and short metacarpals of anurognathids are not pterodactyloid features), but each of the other suggested phylogenetic placements are not easily dismissed. Perhaps these disputes on anurognathid origins will remain unresolved until more substantial data on their early evolution is discovered.

Only a handful of anurognathid fossils are known, a consequence of their delicate skeletons and an apparent preference for terrestrial depositional settings. Such environments are not generally conducive to preserving even robust fossils, let alone the fragile bodies of animals like anurognathids. The first anurognathid specimen was reported from the late Jurassic (Tithonian; 151–145 Ma) Solnhofen Limestone of Bavaria in 1923 by Ludwig Döderlein, but it was so poorly preserved that interpreting the specimen was very challenging; its nickname of "the road kill specimen" is not an ironic one. Enough was clear to Döderlein for him to realize that the animal represented a new type of pterosaur, which he christened *Anurognathus ammoni*. However, with little more than an impression of an incomplete skeleton to work from, he could not ascertain many specifics of its form. Döderlein noted several unusual features: a short, fused tail akin to the pygostyles (fused distal caudal vertebrae) of modern birds; particularly long wings; and four phalanges in the fifth toes. He also thought much of the skull was missing, assuming it was a long-snouted pterosaur with a rounded posterior skull region. This assumption was not unreasonable given the knowledge of pterosaur diversity at the time, but later workers disagreed and reconstructed the skull with a much shorter snout (e.g., Young 1964; Wellnhofer 1975). These later reconstructions were closer to the reality of anurognathid skulls than Döderlein's interpretation, but the true morphology of their cranial regions remained mysterious for many more decades. Other specifics of Döderlein's interpretation were also contested (see Bennett 2007b for a review), but his interpretations of long wings and "pygostyles" were mentioned in overviews of anurognathid form for many years to come (e.g., Wellnhofer 1991a).

The discovery of a second anurognathid specimen told us little else about their anatomy. *Batrachognathus volans*, stemming from Upper Jurassic (Oxfordian/Kimmeridgian; 161–151 Ma) strata of Kazakhstan (Riabinin 1948), is only known from an incomplete skull and small amounts of postcranial material. This revealed that the anurognathid skull was, unusually among pterosaurs, wider than long, but too little else of the animal was preserved for further insights into their skeletal form. With pterosaur research also moving through its mid-twentieth-century "dark age" (chapter 2) at this time and the specimen being held behind the Iron Curtain, it's hardly surprising that little attention was paid to *Batrachognathus* for many years (Bakhurina and Unwin 1995b). Happily, additional specimens of this pterosaur, and perhaps a new contemporaneous anurognathid species, have since been found and await description (Bakhurina and Unwin 1995b, Unwin and Bakhurina 2000).

More fragmentary remains of anurognathids were identified in the late 1980s, stemming from the Middle Jurassic Bakhar Formation of Mongolia (Bakhurina

FIG. 11.1. Why chiroptophobes may not have found much solace in the Mesozoic: hundreds of the small insect-hawking pterosaurs *Anurognathus ammoni* leave their daily hiding places in pursuit of an evening meal in the Jurassic of Germany.

FIG. 11.2. Distribution of anurognathid taxa.

1986), but it was almost the end of the century before the first complete anurognathid skeleton was found. This frog-faced Holy Grail, which finally revealed anurognathid anatomy in full, was discovered in the Lower Cretaceous Yixian Formation of China and christened *Dendrorhynchoides curvidentatus* (Ji and Ji 1998; note the precise age of the Yixian beds is controversial, though a Barremian age, 130–125 Ma, seems likely). *Dendrorhynchoides* scoops further accolades for being one of the first pterosaurs found in the fossil sites of China's Liaoning province, a region now famous for its enormous quantities of feathered dinosaurs and other spectacularly preserved fossils. This new specimen gave pterosaurologists their first opportunity to assess the quirky anatomy of anurognathids, and it was soon apparent that some oft-reported aspects of their morphology were incorrect. The wings of anurognathids were actually rather short, not particularly long, and like all other non-pterodactyloid pterosaurs, they only had two bones in their fifth toes, not four. An unexpected feature was the relatively long tail sported by *Dendrorhynchoides*. The latter suggested to Ji and Ji (1998) that *Dendrorhynchoides* was a member of Rhamphorhynchidae (chapter 13), a group of largely seagull-like, long-tailed pterosaurs that are morphologically very distinct from anurognathids. However, later assessments noted that its long tail was crafted from dinosaur bones added to the tail region by fossil dealers seeking to make the specimen more marketable. A short tail and many other characteristics of the skeleton suggested *Dendrorhynchoides* was better placed within the short-tailed Anurognathidae than Rhamphorhynchidae (Unwin, Lü, and Bakhurina 2000). A second *Dendrorhynchoides* specimen has given a recent twist to this story however, revealing that this taxon did indeed bear the longer tail like that found on the original fossil, and that the forgery had only filled in missing bones between the fossil's real tail tip and base. Thus, it seems that *Dendrorhynchoides* was an unusually long-tailed anurognathid, though its tail remains shorter than those of other early pterosaurs (Hone and Lü 2010).

Even better anurognathid specimens were unearthed at the turn of the new millennium. The first of these was the middle Jurassic *Jeholopterus ningchenensis* from Oxfordian/Kimmeridgian Daohugou beds of northeast China (Wang et al. 2002). Although the skull is rather messily preserved, this fossil is preserved in such completeness and detail that even individual pycnofibers and structural wing fibers can be seen

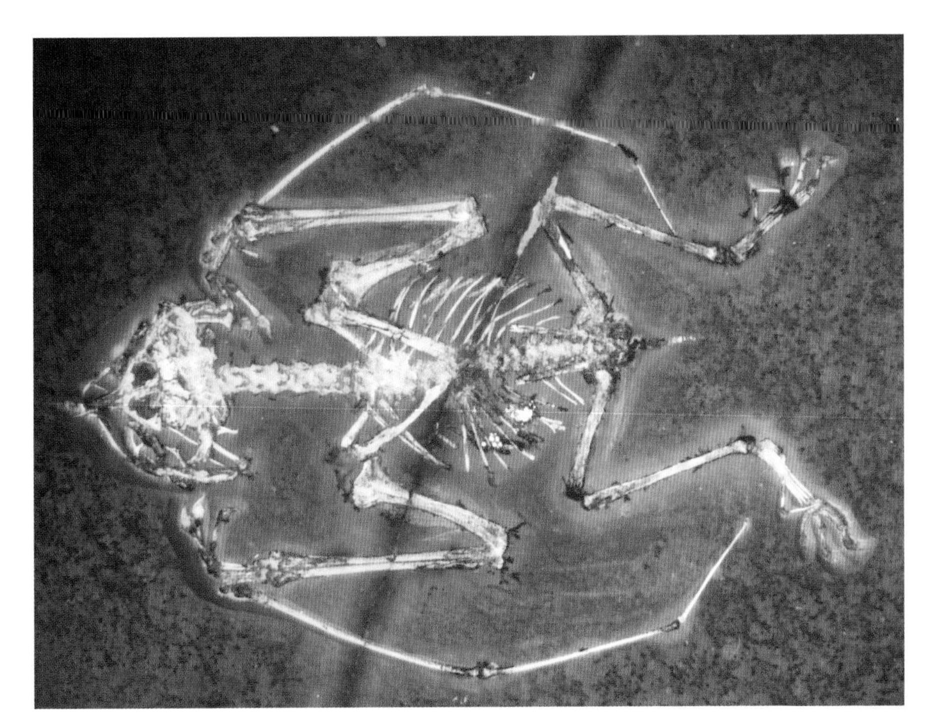

FIG. 11.3. UV photograph of the wonderfully preserved juvenile *Anurognathus ammoni* from the Tithonian Solnhofen Limestone of Germany, recently described by Chris Bennett (2007b). Note the ghostly impression of muscle tissues around the humeri and femora. Photograph courtesy of Helmut Tischlinger.

(Kellner et al. 2009). Another specimen of *Jeholopterus* was reported in 2002, and though the skeleton is not quite as well preserved as the first, the soft-tissue preservation is equally spiffing (Ji and Yuan 2002). Thanks to these specimens, we now have a very good idea of the external appearance of anurognathids, and an equally good view of their internal anatomy was also recently uncovered. The best preserved anurognathid skeleton yet, a Solnhofen specimen of a juvenile *Anurognathus*, finally provided a well-preserved skull in addition to a skeleton preserved in crystal clarity. Even more excitingly, it also possesses traces of muscle tissues and wing membranes that glow under ultraviolet light (fig. 11.3; Bennett 2007b). Muscle tissues are virtually unknown in pterosaur fossils and pterosaurologists have wasted no time in using this new data for biomechanical studies into anurognathid anatomy (Habib 2011).

The next chapters of anurognathid research are already being written. Within the last few years an almost complete anurognathid skeleton was discovered in early Cretaceous strata of North Korea (Gao et al. 2009). This specimen seems to differ from other anurognathids in a number of ways, and it is currently awaiting description. Elsewhere in the world, work on the poorly known Upper Jurassic pterosaur "*Mesadactylus ornithosphyos*" from the Morrison Formation, Colorado (Kimmeridgian-Tithonian; 156–145 Ma) suggests it may partially represent the first American anurognathid. "Partially" is an appropriate term here as "*Mesadactylus*" is represented by fragmentary material that may be a chimera of several different pterosaur anatomies, including an anurognathid pelvis (Bennett 2007b) (see chapter 19 for more on this problematic taxon; see also Jensen and Padian 1989, and Smith et al. 2004 for other interpretations). With work to follow on these fossils, not to mention the many novel and fascinating discoveries being made by biomechanicists studying anurognathid anatomy, these are exciting times to be interested in these pterosaurs.

Anatomy

OSTEOLOGY

Anurognathids have some of the most distinctive anatomy of any pterosaur group, having tweaked most aspects of pterosaur osteology to create a very unique bauplan (fig. 11.4). They seem to have been a very conservative group, demonstrating no major changes to their general anatomy in 40 million years of evolution (Unwin, Lü, and Bakhurina 2000). All anurognathids are compact animals, with *Dendrorhynchoides* the smallest at 40 cm across the wings, *Anurognathus* spanning 50 cm, *Batrachognathus* spanning 75

Fig. 11.4. Skeletal reconstruction and life restoration of a launching *Anurognathus ammoni*.

cm, and *Jeholopterus* the largest them of all with a 90 cm wingspan (Wang et al. 2002). Undeniably, the best description of anurognathid osteology to date is that of Bennett (2007b), while their soft tissues have been documented most extensively by Kellner et al. (2009).

Anurognathid skulls are about 25 percent wider than long, a trait not seen in any other pterosaurs. The broad, U-shaped profile of their muzzles when viewed in dorsal or ventral aspect is another unique anurognathid characteristic. Their skulls were shortened by a drastic reduction of the snout, considerably diminishing the nasal and antorbital openings to narrow, forward-facing slits (Bennett 2007b). Indeed, anurognathid rostra are so tightly constructed that identifying their nasal and antorbital openings is difficult in their crushed fossils, and some have concluded that they were combined into one opening, as they are in wukongopterids (chapter 14) and pterodactyloids (chapters 15–25) (Andres et al. 2010). The orbit, by contrast, is enormous and occupies over half of the skull length, and the temporal fenestrae are also relatively large. Most of the skull bones are reduced to very slender rods, and their delicacy probably explains why anurognathid skulls are so rarely preserved intact. The roof of the mouth is also constructed with thin, scaffold-like bones instead of the broad palates of other pterosaurs. The mandible is, again, very gracile and, like the skull, assumes a broad, U-shaped arc in dorsal or ventral view. Their retroarticular processes, unlike those of other pterosaurs, flare laterally so as not to impede opening of mouth when stretched to a 90 degree maximum gape. Anurognathid jaws are occupied by low numbers of small, pointed and widely spaced teeth, which are slightly recurved and show no significant variation in shape or form throughout the jaw length.

Anurognathid necks and bodies appear relatively long compared to their shortened skulls, but are actually of pretty standard proportions for early pterosaurs. Their neck vertebrae lack cervical ribs and of their 13 dorsal vertebrae, only the anterior eight bear ribs. Four to five vertebrae comprise the sacrum, and beyond this, we find the characteristically short anurognathid tail. These are of variable length between species, being very short in *Anurognathus* but almost as long as the thigh in *Dendrorhynchoides* (Hone and Lü 2010). Short tails are typically regarded as a pterodactyloid trait, but they may well have developed independently in anurognathids. Subtle differences in the shape of the caudal vertebrae in these forms support this idea (Bennett 2007b); the caudals of pterodactyloids are longer than wide, whereas those of anurognathids are the opposite. It should also be noted that anurognathid tails are not fused into a "pygostyle" as originally supposed, although their flat articular surfaces probably permitted little flexion nevertheless.

The forelimb bones of anurognathids are generally rather long and robust, but the metacarpal is unusually short and adds very little to the wing length. Excluding the wing finger, the bones of the forelimbs are almost fifty per cent longer than those of the leg. Their pteroids are extremely reduced and may be the smallest, proportionally speaking, of any pterosaurs. The first three manual digits of anurognathids are typical of other early pterosaurs, being relatively large and bearing strong, moderately hooked claws, but the wing finger is more distinctive. In addition to being comprised of progressively shortening bones (an unusual trait for a non-pterodactyloid wing finger), anurognathid wing phalanges appear to have had flexible joints, which is unheard of in any other pterosaur group. This allowed the wings to be somewhat furled toward the body and, indeed, complete anurognathid specimens are often preserved in this posture (e.g., fig. 11.3; also see Ji and Ji 1998; Wang et al. 2002). For some additional quirkiness, *Anurognathus* has also lost its fourth wing phalange, leaving it with only three (Bennett 2007b).

The hindlimbs of anurognathids are short for their body size, though their feet remain large with robust, hooked claws on the first four digits. Their fifth toes are comprised of two long and straight phalanges. Another unusual feature common to all anurognathids is their especially long preacetabular processes, projections that are markedly more elongate than those of most other early pterosaurs (Unwin 2003). The rest of their pelvic regions are poorly known, however, so it cannot be seen if this unusual iliac morphology translates to further unusualness in other parts of the pelvis.

SOFT TISSUES

Although we only have a very small number of anurognathid specimens, a heck of a lot is known about their soft tissues. Some of these tissue types—wing membranes and claw sheaths—are also well known in other pterosaurs, but those of anurognathids are particularly well preserved and reveal details of their delicate medial membranes and propatagia (Wang et al. 2002; Ji and Yuan 2002; Bennett 2007b; Kellner et al. 2009). Other soft tissues preserved in anurognathids are virtually unknown from other pterosaur groups, including the remnants of their limb musculature.

Muscle preservation is extremely uncommon in fossil tetrapods, but can be seen along the front and back of the humerus, forearm, and thighs of the juvenile *Anurognathus* described by Bennett (2007b; also see fig. 11.3). These give a good indication of the minimum size and extent of the muscles in these regions, but unfortunately are not quite well enough preserved to determine exactly which muscles they represent (Bennett 2007b; see chapter 5 for more on pterosaur limb musculature).

Several anurognathid fossils provide a good insight into how fuzzy pterosaurs were (Kellner et al. 2009). The pycnofibers of *Jeholopterus* are preserved in a shapeless mat similar to those left by the hair of many fossil mammals, suggesting they were covered, from nose to elbows and knees, in dense pelts comparable to those we see in hairy mammals. The manner in which their pycnofibers cover their snouts and jaws seems to be unique among pterosaurs (presumably because most pterosaurs thought it just wasn't cool to look like the Cookie Monster from *Sesame Street*). The slightly rugose bone texture around the jaws of *Anurognathus* has been suggested to mark where a series of thick bristles once anchored (Bennett 2007b), but I'm not too sure about this interpretation. Hairs or feathers rarely leave such obvious marks on their owners' skulls and, moreover, no anurognathid fossil has shown these features yet despite excellent preservation of their facial pycnofibers. What is known for certain, however, is that some anurognathids bore a small tuft of short pycnofibers on the trailing edges of their wing tips, a feature unrecorded from any other pterosaur (Kellner et al. 2009).

Locomotion

FLIGHT

Anurognathids have been considered to be agile, nimble aeronauts since their discovery, a necessary requirement for their probable diets of flying insects (e.g., Döderlein 1923; Wellnhofer 1975, 1991a; Unwin 2005; Bennett 2007b; see also the discussion below for more on anurognathid paleoecology). Bakhurina and Unwin (1995b) interpreted the robust nature of the *Batrachognathus* humerus as an indicator of a powerfully flying animal, while Bennett (2007b) considered that their combination of short, broad wings, abbreviated tails, and small size would grant them slow, maneuverable flight. Among flying animals, such controlled flight styles are very advanced and require low wing loading, minimal turning inertia, and extremely sharp reflexes. These findings were echoed in contrasts between anurognathid wing attributes and those of modern birds; they favorably compared to those of swifts and falcons, birds capable of highly maneuverable, controlled flight (Witton 2008a).

The most interesting insights into anurognathid flight may be yet to come, however. Biomechanicist Michael Habib's analyses of anurognathid flight anatomy are revealing that astonishing power and strength once existed in these little pterosaurs. The proximal wing bones of *Anurognathus* seem capable of loading *22 times* their own body weight before breaking, and their wing spars were much stronger than those, on average, of birds (Habib 2011). The distribution of the wing area and the leading edge profile of *Jeholopterus* suggest that their wings were adapted for producing substantial lift, which would combine with their powerful forelimbs to result in explosive, high-angle launches. At the same time, the possibility that anurognathids could flex their wing fingers may have allowed for unusually fine control over their wing shapes, and this may have equated to greater agility in the air. The pycnofiber tufts at their distal wings may have also played an aerodynamic role, helping to maintain a streamlined airflow along the trailing edge of the wing and preventing stalling at higher angles of attack. This feature could have also dampened sound in the same manner that the comb-like fringes of owl wings minimize the noise of their flapping (M. B. Habib, pers. comm. 2011). The picture we're building, then, is that anurognathids were absolutely dynamite fliers, adapted for powerful flapping and dynamic twists and turns while maintaining a stealthy silence. They may not have the size or elaborate crests of other groups to make them stand out as highlights of the pterosaur menagerie, but the flight of these fuzz balls may make them a peak achievement of pterosaur evolutionary engineering.

ON THE GROUND

Little has been said about the grounded capabilities of anurognathids, but there is not much reason to assume that they were poor terrestrial locomotors. Bennett (2007b) pointed out that their claws would do a commendable job of supporting them when climbing, so we may have found anurognathids among the treetops should we walk through a Jurassic forest. It is thought that anurognathids may have been fairly inconspicuous animals because complete specimens are consistently preserved with their limbs folded close to their bodies, possibly betraying a common,

FIG. 11.5. The standard anurognathid resting posture? The small size and inconspicuous nature of anurognathid anatomy, combined with frequently compact preservational postures in their fossils, indicates that anurognathids may have had a typical, inconspicuous resting posture, allowing them to remain undetected by potential prey or dangerous predators.

compact resting posture suited to resting in nooks and crevices (fig. 11.5; see fig. 11.3 for a fossil example of this common posture) (Bennett 2007b). Other pterosaur fossils, by and large, don't demonstrate any real frequency of posture in preservation and in many cases, the animals in question may have been too large or awkwardly shaped to hide in cracks and holes. Anurognathids also lack cranial ornamentation, making them a rarity among pterosaurs, but this would be ideal should they have wanted to remain inconspicuous (though this could also reflect their optimization for agile flight; D.W.E. Hone, pers. comm. 2011). With these points in mind, the possibility that anurognathids were cryptically colored to help blend into their surroundings has also been raised (Bennett 2007b).

Paleoecology

The enormous orbits of anurognathid skulls are far larger, proportionally speaking, than those of any other pterosaur. Presumably, this reflects the presence of very large eyes, which may have afforded high visual acuity in dim light conditions. This presents the possibility that anurognathids were regularly active in dark environments, such as dense forests, or at darker intervals of the day (Bennett 2007b). Anurognathids may have used this darkness, possibly along with their silenced wings, to stealthily hunt their most likely prey items, aerial insects. Insects have long been recognized as the probable food source for anurognathids (e.g., Döderlein 1923; Wellnhofer 1975; Bennett 2007b) because of their widely spaced, conical teeth and jaw musculature adapted for very rapid closure (Ősi 2010). Indeed, they are unlikely to have subdued much larger prey because of the extremely delicate nature of their skulls. It is generally agreed that anurognathids caught most of their insects by hawking them in midair like modern swifts and nightjars. Like these birds, anurognathids have wide jaws and adaptations to vastly increase their gapes, attributes that granted them the best their chance of bagging insects as they flew toward them (Bennett 2007b). These birds also have particularly large, forward-facing eye sockets housing enormous eyeballs used for detecting small flying insects and judging striking distances; the same use of large eyes is suspected in anurognathids.

The awesome flight capability of anurognathids was probably critical to this foraging method. Catching flying insects, as discussed in chapter 6, is no mean feat. Many insects fly much slower, in absolute terms, than vertebrate predators but are extremely nimble, allowing them to easily outmaneuver cumbersome, backboned aeronauts in midair chases. Anurognathids likely countered this agility by being lightly loaded and rather slow fliers, ensuring that they would not shoot past their quarry should it suddenly alter its course. In such an event, it seems anurognathids also had the power and strength to perform tight turns and accelerate quickly, allowing them to give chase to their prey. With such strong evidence for anurognathids being aerial insect hawkers, it is very easy to imagine Mesozoic sunsets full of anurognathids wheeling and dancing above woodlands and lake margins, snatching up mouthfuls of insects in the same way that modern sunsets are associated with hawking swifts, nightjars, and bats.

The possibility that anurognathids procured some food on the ground has also been raised. Bennett (2007b) noted that some modern insect hawkers (such as frogmouths and potoos) are also able to pursue terrestrial prey, snatching up even relatively large vertebrates in some cases. These birds are excellently camouflaged with patterns and resting postures that match tree branches perfectly, so potential prey items saunter toward them ignorant of danger. Terrestrial foraging like this, or some other type, does not seem beyond anurognathids as they also may have been well adapted for remaining inconspicuous (see above). However, their delicate skulls would probably have been challenged by large prey, and they may have let such items walk by without risking an assault. Moreover, their adaptations for aerial pursuit indicate that terrestrial foraging was very much a secondary feeding strategy, and that they would have been most at home in the air.

12

"Campylognathoidids"

PTEROSAURIA > "CAMPYLOGNATHOIDIDS"

Plenty of debate surrounds the taxonomy of various pterosaur groups, but the "campylognathoidids," are by far the most taxonomically contentious of them all. There is virtually no agreement between modern pterosaur workers on how the species under discussion in this chapter are related to one another or other pterosaurs (see Unwin 2003; Kellner 2003; Dalla Vecchia 2003a, 2003b, 2009a, 2009b; Wang et al. 2009; Andres et al. 2010; Lü, Unwin, et al. 2010; and Lü, Unwin, et al. 2012 for different arrangements of these forms around the base of the pterosaur tree). Accordingly, I've had to make something of a subjective call over which animals should be discussed in this chapter. In keeping with the other pages in this tome, we will principally following the scheme of Lü, Unwin, et al. (2010), but we will also have a dash of input from Fabio Dalla Vecchia's (2009a) recent analysis of these animals. Ergo, "campylognathoidids" are regarded here as non-pterodactyloid pterosaurs with generally complex dentition; downturned mandibular tips; large temporal openings; oversize, posteriorly flared sterna; and proportionally elongate wing fingers (fig. 12.1). Although these shared features may suggest shared ancestry between these forms, only the vernacular term "campylognathoidid" will be used here to stress the contentious nature of this grouping.

The evolution of "campylognathoidids" appears to have been one of the first major pterosaur radiations, their earliest fossils occurring in Triassic European rocks of mid to late Norian age (210–204 Ma). They enjoyed a relatively long evolutionary history, lasting for around 40 million years until the Jurassic Toarcian stage (183–176 Ma) before disappearing (Dalla Vecchia 2003b; Barrett et al. 2008). "Campylognathoidid" evolution experimented with all sorts of novel anatomies that render them not only distinct from other pterosaurs but also among each other. Undoubtedly, their evolutionary innovations play some part in our poor understanding of their position within Pterosauria.

Fossils of "campylognathoidids" are so far mostly constrained to Europe (fig. 12.2). The first discovery of their kind is now the best known "campylognathoidid" of all—the Toarcian age *Campylognathoides* from the Posidonienschiefer of southern Germany (183–176 Ma; see Padian 2008b for a historical overview of this taxon). The Posidonienschiefer is famous for possessing enormous quantities of exceptionally preserved marine reptile, fish, and marine invertebrate fossils, along with the rhamphorhynchid pterosaur *Dorygnathus* (chapter 13). Two species of *Campylognathoides* are recognized: *C. liasicus* (initially christened *Pterodactylus liasicus*, but the first remains of this taxon, found in 1858, were pretty scrappy and misidentified) and the larger, rarer *C. zitteli* (found in 1894 and called "*Campylognathus*" until it was realized that this name was already taken by a stinkbug, of all things. It was renamed "*Campylognathoides*" in 1928). *Campylognathoides liasicus* is known from seven specimens in total, including a number of complete skeletons (fig. 12.3A), while only two fossils of *C. zitteli* have been discovered. The two species are only separated by minor anatomical differences and may be part of a growth series (Padian 2008b), but a current lack of specimens spanning the size gap between them prohibits confirming this. Jain (1974) proposed that a third *Campylognathoides* species occurred in the Early Jurassic of India (*C. indicus*), but its remains are rather scrappy and its referral to *Campylognathoides* or even, for that matter, Pterosauria, has been doubted (Padian 2008b).

More "campylognathoidid" species were not uncovered until the 1970s when another was unearthed—quite literally—with a bang. In 1965, a landslide of Triassic limestone (late Norian, ca. 205 Ma) deposits of Cene, Italy, which was large enough to bury nearby quarrying machinery, opened up a treasure trove of Late Triassic fossils. These included the articulated remains of the possible dimorphodontid *Peteinosaurus* (chapter 10) as well as *Eudimorphodon ranzii*, an early "campylognathoidid" (Zambelli 1973; see Paganoni 2003 for more on the discovery of this animal). Sadly, much of the *E. ranzii* specimen was destroyed in the landslide and the legs, tail, and wing fingers are mostly

FIG. 12.1. The stilt legged *Caviramus filisurensis*, one of the coolest-looking pterosaurs known, explores a coastal cave during a storm. The large soft-tissue crest shown here is not entirely known from fossils, but inferred to exist based on the bone texture of the bony crest components.

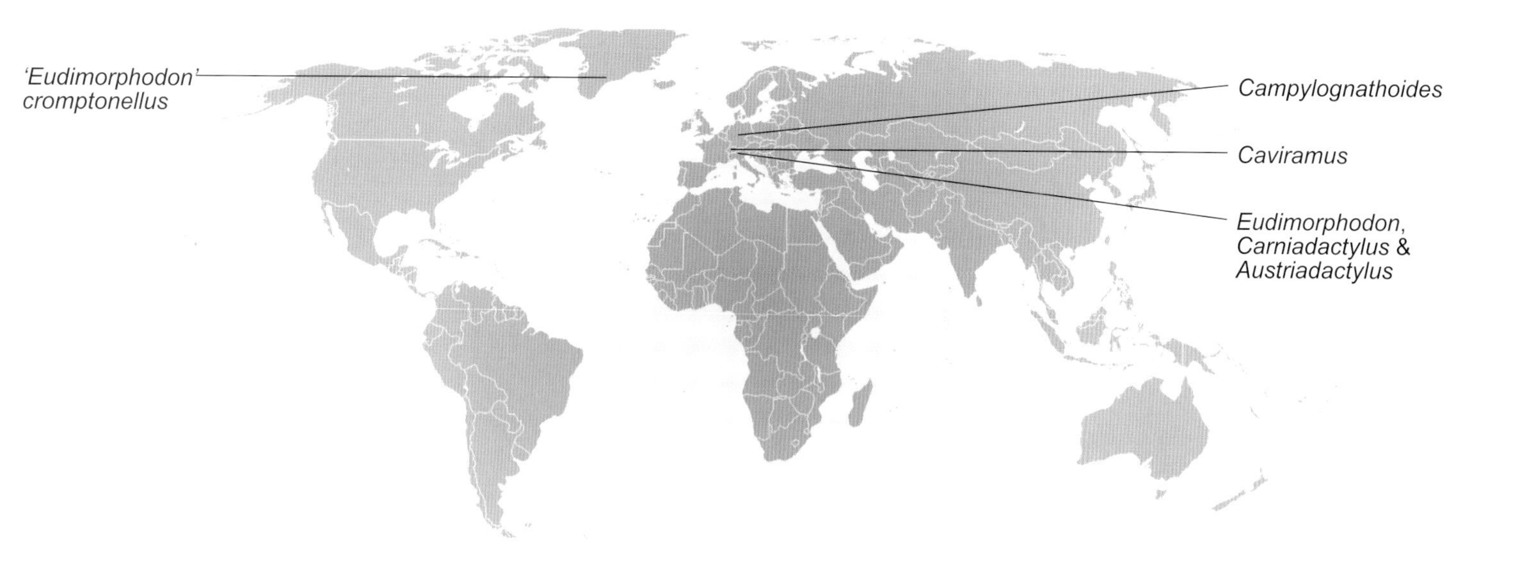

Fig. 12.2. Distribution of "campylognathoidid" taxa.

Fig. 12.3. Complete "campylognathoidid" fossils. A, *Campylognathoides liasicus* from the Toarcian Posidonienschiefer of Germany; B, *Eudimorphodon ranzii* from the Norian of Cene, Italy. Photograph A courtesy of Ross Elgin; B, courtesy of Attila Ősi.

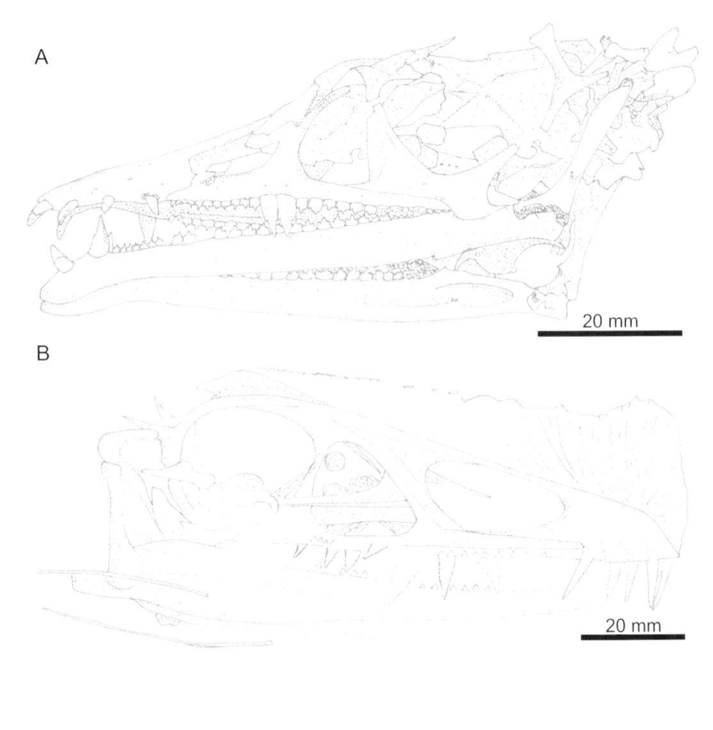

FIG. 12.4. "Campylognathoidid" skulls. A, *Eudimorphodon ranzii*; B, *Austriadactylus cristatus*. A, after Wild (1978); B, after Dalla Vecchia et al. (2002).

missing from an otherwise excellently preserved specimen (figs. 12.3B and 12.4A), but the find was still of great significance. It represented the first discovery of a genuine Triassic pterosaur (see chapter 10) and revealed that some pterosaurs possessed mouthfuls of closely packed, multicusped teeth, the like of which had never been seen before. Over the following decades, many Triassic pterosaur specimens were referred to this genus (e.g., Wild 1994; Wellnhofer 2003), and two new *Eudimorphodon* species were named: *E. rosenfieldi* (also from Triassic rocks of Italy; Dalla Vecchia 1995) and *E. cromptonellus* (Norian-Rhaetian; 228–202 Ma) of Greenland (Jenkins et al. 2001). Neither of these are considered species of *Eudimorphodon* any more however, with Dalla Vecchia (2009a) erecting a new genus name, *Carniadactylus*, for *rosenfieldi* and suggesting that the same be done for "*E.*" *cromptonellus*. *Carniadactylus* was long thought to be a juvenile specimen of *Eudimorphodon* due to its small size, but Dalla Vecchia (2009a) found it to bear many of all the hallmarks of maturity discussed in chapter 8 and argued that it represented a particularly small bodied "campylognathoidid" species instead. The upshot of such revisions is that the once abundant and widespread *Eudimorphodon* is now only represented by the specimen collected from the Cene landslide in 1965.

The year 2002 saw the next major "campylognathoidid" discovery, *Austriadactylus cristatus* (Dalla Vecchia et al. 2002). This middle Norian (ca. 210 Ma) animal is now known from relatively complete skeletons found in both Austria and Italy (Dalla Vecchia 2009b) and is particularly noteworthy for being the first Triassic pterosaur found with a cranial crest (fig. 12.4B). It was soon joined by another species, *Caviramus schesaplanensis*, from Norian-Rhaetian (228–201 Ma) limestones of Switzerland (Fröbisch and Fröbisch 2006; Stecher 2008). *Caviramus schesaplanensis* was at first solely known from a distinctive but incomplete mandible that bore large attachment sites for jaw musculature and was clearly once lined with numerous teeth. Another *Caviramus* species (*C. filisurensis*) was reported from contemporaneous Swiss deposits by Rico Stecher shortly after, in 2008 (fig. 12.5). *Caviramus filisurensis* was initially given its own genus, "*Raeticodactylus*," but comparisons with *C. schesaplanensis* have suggested to other authors (Dalla Vecchia 2009a, Ősi 2010) that the two are at least congeneric, if not entirely synonymous. Happily, *C. filisurensis* is far more complete than *C. schesaplanensis* and it remains one of the best preserved of all "campylognathoidids."

Anatomy

The anatomy of "campylognathoidids" is well recorded in a number of lengthy scientific papers. Rupert Wild's 1978 monograph on *Eudimorphodon* is perhaps the best known and also includes plenty of detail on *Carniadactylus,* as does Dalla Vecchia's (2009a) *Carniadactylus* redescription. The anatomy of *Caviramus filisurensis* was documented in great detail by Stecher (2008), and Kevin Padian recently redescribed every *Campylognathoides* specimen known (2008b). Such descriptions reveal that "campylognathoidids" are generally quite small animals, as is common to most early pterosaurs. *Campylognathoides zitteli* is the largest of all, with a wingspan up to 1.8 m, while *Caviramus* and *Austriadactylus* span 1.35 and 1.2 m, respectively. At the lower end of the size spectrum, *C. liasicus* and *Eudimorphodon* only span 1 m and *Carniadactylus,* the smallest, spans a mere 70 cm across the wings.

"Campylognathoidid" skulls have fairly low profiles, large cranial regions, big orbits, and tapering snouts. *Austriadactylus* and *Caviramus* sport prominent

FIG. 12.5. The Norian-Rhaetian remains of *Caviramus* (= *Raeticodactylus*) *filisurensis*. Photograph courtesy of Rico Stecher.

crests on their rostra that, in the former, extend right along the skull's dorsal surface to give it a rather rectangular profile. *Caviramus*, by contrast, has a large triangular crest that prominently projects anterodorsally from the snout tip. Both of these animals have bony crest tissues with fibrous textures and margins, creating the possibility that their crests supported soft-tissue extensions in life. The upper temporal fenestra is characteristically large in "campylognathoidids" and, aside from the orbit, may be the largest opening in their skulls (Unwin 2003). Their mandibles bear bulbous, slightly downturned tips, and are generally slender, but the lower jaw of *Caviramus* is particularly deep and bears a long, swollen retroarticular process and depressed jaw joint. *Caviramus* also has a prominent, triangular keel along the underside of its mandibular symphysis and a series of deep pits at the mandibular jaw tip. These may represent attachment points of a soft-tissue beak (Stecher 2008), or channels for blood vessels and sensory tissues. "Campylognathoidids" also bear elevated coronoid processes, dorsal projections of the posterior mandibular rami that anchored their jaw muscles. These processes range in size from rather small (*Eudimorphodon*) to huge (*Carniadactylus*). At least some "campylognathoidids" also retain external mandibular fenestrae (Nesbitt and Hone 2010).

"Campylognathoidid" jaws are pterosaur equivalents of Swiss Army knives, brimming with teeth of all sizes, shapes, and orientations (fig. 12.6; see Stecher [2008] for a detailed overview of the dental arrays of "campylognathoidids"). Unusually for pterosaurs, some "campylognathoidid" teeth are worn to the extent that their crowns have been stunted (Ősi 2010), suggesting extensive oral processing of their food. The arrangement and size of their teeth is extremely variable, although all "campylognathoidid" fossils possess several of their largest, most procumbent teeth at the jaw tips. Four of these typically well-spaced teeth are found on each side of the upper jaw, and two or three occur on the lower. In some forms, these fangs may be serrated (*Austriadactylus*), or bear ridged

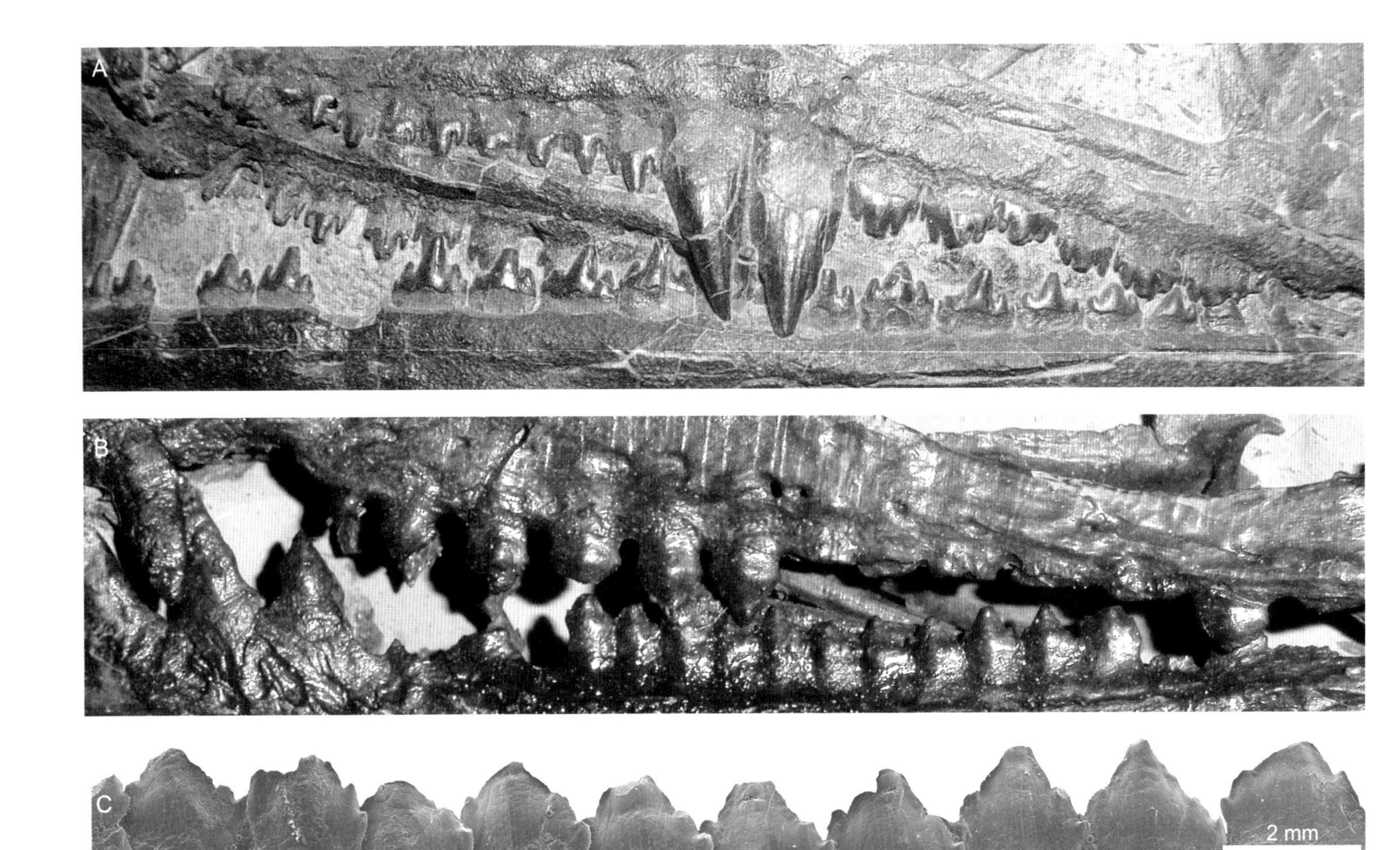

FIG. 12.6. The fearsome dentition of "campylognathoidids." A, *Eudimorphodon ranzii*; B, *Caviramus filisurensis*; C, Scanning electron microscope images of *C. filisurensis* mandibular teeth. Note the worn cusps and scratched surfaces of numerous teeth. Photographs courtesy of Attila Ősi.

(*Eudimorphodon*) or wrinkled enamel (*Caviramus*). Unlike their other teeth, it seems that their anterior fangs did not directly occlude when the jaws were closed.

Beyond the fearsome looking jaw tips, most "campylognathoidids" jaws are stuffed with small, often multicusped teeth. The shape and size of these teeth differs both within the jaws of individuals and among species, but those on the upper jaw are generally larger and more variable than those on the lower. Their multicusped teeth can bear three, four, or five cusps on a single tooth, and are so tightly packed into the jaw that they create a continuous cutting surface. The teeth of *Caviramus* are actually squeezed into the jaw so tightly that they have to offset and overlap one another to fit. Some genera (*Eudimorphodon*, *Austriadactylus*) possess particularly large teeth halfway along their upper jaws, creating a dental profile superficially like that of *Preondactylus* (chapter 10). *Austriadactylus* and *Campylognathoides* are notable for possessing nothing but single-cusped dentition, although the teeth of the former do bear coarse serrations. By contrast, *Campylognathoides* has entirely smooth enamel, no serrations, and unlike the rest of the group, low numbers of widely spaced teeth.

"Campylognathoidid" postcranial anatomy is just as varied as their skull morphology. Their vertebral column is fairly typical of early pterosaurs, with short, complex, rib-bearing cervicals, and a series of elongate caudals. Vertebral counts are not well known in most taxa, due to poor or incomplete preservation, but *Campylognathoides* is reported to have eight cervicals (a finding somewhat at odds with the interpretation that all pterosaurs bore nine cervicals), around 14 dorsals, four or five sacrals, and up to 38 caudals (Wellnhofer 1974). *Campylognathoides* and *Carniadactylus* are known to possess stiffening structures on their caudal vertebrae, but these are absent in *Austriadacty-*

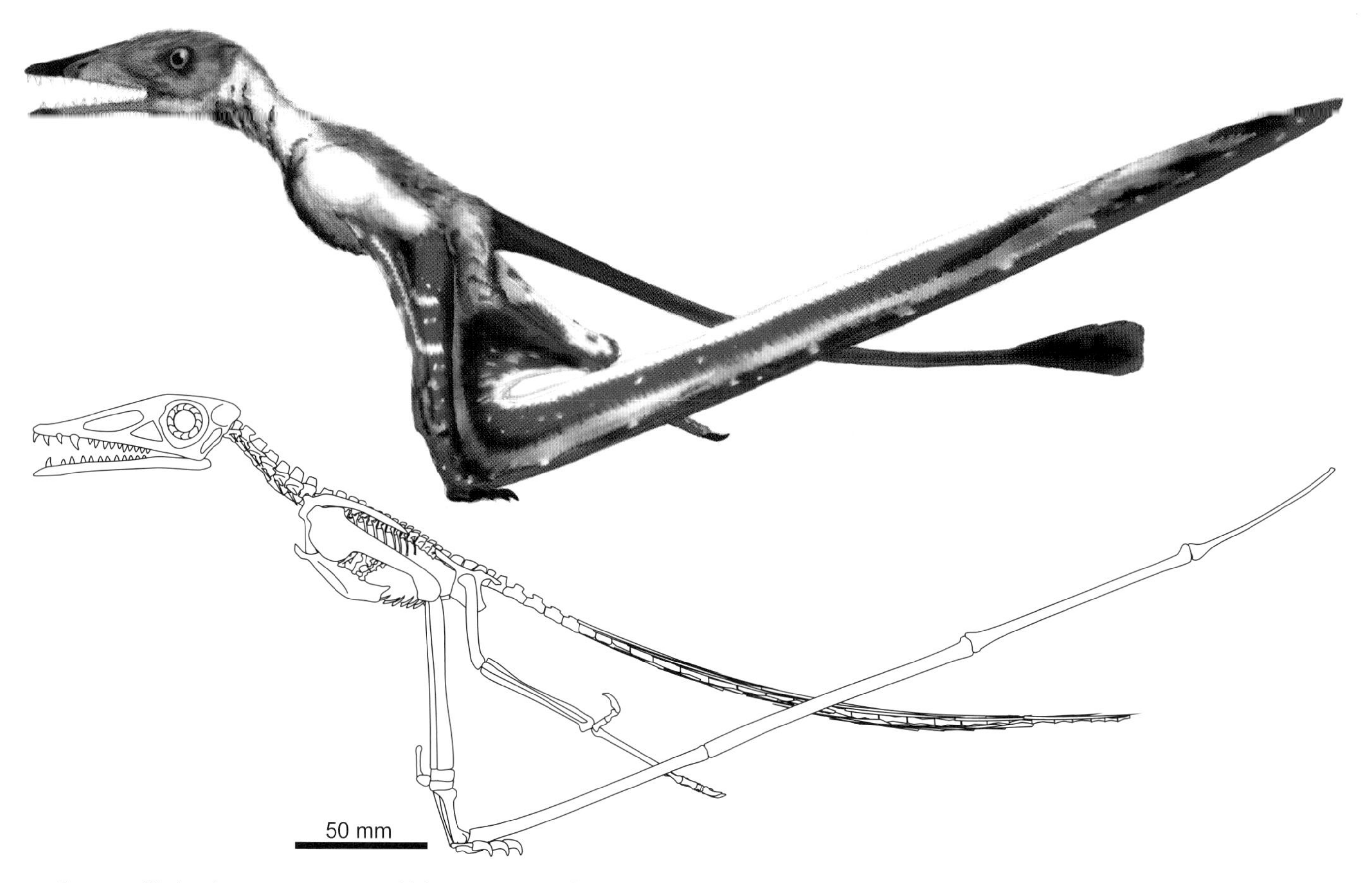

FIG. 12.7. Skeletal reconstruction and life restoration of a launching *Campylognathoides liasicus*.

lus and "*Eudimorphodon*" *cromptonellus* (see Wild 1994, and Dalla Vecchia 2009a). At least some of the cervicals and dorsals were pneumatized in *Caviramus* and *Campylognathoides*, but most "campylognathoidids" are too poorly preserved to determine the distribution of their pneumatic features (Bonde and Christiansen 2003).

Apart from their skulls, "campylognathoidid" forelimbs are their most distinctive features, and they come in two rather opposing flavors. The most typical forelimb condition shows forearm bones, pteroids, and metacarpals of similar proportions to those of *Preondactylus* and dimorphodontids, but they are much more massive and robust. The humeri are especially well developed, with shovel-shaped proximal ends, expansive deltopectoral crests, and shafts much thicker than expected for pterosaurs of their size. These are connected to large shoulder girdles, very long, strap-like scapulae, and enormous sterna. These breastbones are characteristically square shaped with laterally flared posterior regions, and although their cristospines are short, it is clear that the muscles around the shoulders of most "campylognathoidids" were something to be reckoned with. *Campylognathoides*, in particular, has shoulder anatomy so developed that it looks like a little pterosaurian gorilla (fig. 12.7). A rather different approach to forelimb construction was taken by *Austriadactylus* and *Caviramus,* the humeri of which are extremely slender. The humerus of the latter is nearly 20 times longer than its midshaft width (fig. 12.8), an unprecedented metric among pterosaurs. Both the humerus and forearm of *Caviramus* are proportionally elongate and slender, granting it one of the longest forelimbs of all non-pterodactyloid pterosaurs.

The hands of most "campylognathoidids" are poorly known. Those of *Eudimorphodon* and *Carniadactylus* possess antungual sesamoids and relatively well-developed claws on each of their first three digits, but it is not clear if this is the "standard" configuration for this group. Enormous wing fingers are common to these pterosaurs, however, and these comprise a staggering 67 to 79 percent of the wing length (Unwin

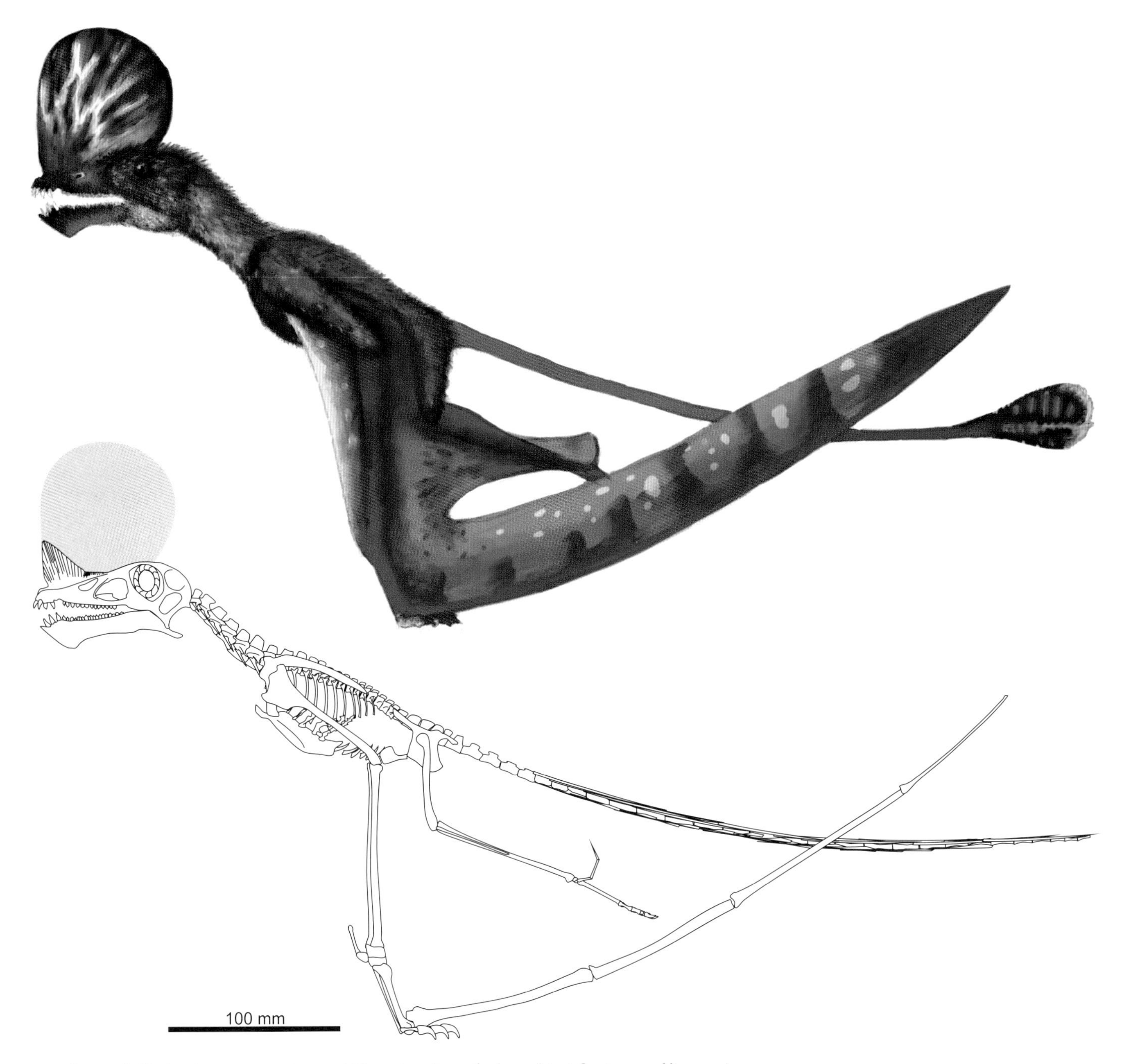

FIG. 12.8. Skeletal reconstruction and life restoration of a launching *Caviramus filisurensis.*

2003). The wing finger of *Campylognathoides* is particularly large, being almost as long as its entire body. Even slender-boned *Caviramus* bore a very long wing finger, though its exact length cannot be ascertained at present. No other pterosaurs have wing fingers of such proportions, and these features grant "campylognathoidids" some of the longest wings, proportionally speaking, of any pterosaur group.

Frustratingly, most "campylognathoidid" fossils are missing the majority of their hindlimbs. Consequently their leg anatomy is not well understood. What is known of their hindquarters suggests they have similar hindlimb proportions and morphology to *Preondactylus* and *Dimorphodon*, though the hindlimbs are much shorter than the forelimbs (even excluding the oversize wing finger). The legs of *Caviramus* are, like its forelimbs, atypically long and slender so, compared to other early pterosaurs, *Caviramus* would look like it stands on stilts. Their femora seem to have femoral heads angled perpendicularly to the main shaft, a contrasting condition to the obliquely angled femoral heads seen in other pterosaurs (Stecher 2008). The fibula of *Campylognathoides* is notable in that it almost touches the ankle, a primitive condition not seen in other "campylognathoidids" (Wild 1994; Dalla Vecchia 2009a). "Campylognathoidid" fifth toes are comprised of long, straight phalanges except for *Campylognathoides*, which has an unusually stunted fifth pedal digit. Presumably, this toe supported a broad uropatagium as it did in other early pterosaurs, but whether the shape of this membrane was modified in accordance to their stunted toes remains to be seen. We can at least be confident that "campylognathoidids" possessed this membrane however, as excellent preservation of a specimen once referred to *Eudimorphodon ranzii* (and currently in need of a new name) clearly reveals the attachment of the uropatagium onto the fifth toe (Bakhurina and Unwin 2003). The same specimen also shows that the brachiopatagium attached at the ankle and possessed a "trailing-edge structure" supporting its posterior margin (Wild 1994).

Locomotion

FLIGHT

Despite their unusual wing proportions, not a great deal has been said about "campylognathoidid" flight. Chatterjee and Templin (2004) modeled the flight of *Carniadactylus* (or "*Eudimorphodon*," as it was considered at the time) to include hovering for at least a few seconds. Hazlehurst and Rayner (1992) found "campylognathoidids" occupied a flight niche not reflected among modern fliers, an "extreme" flight style, perhaps like that of a frigate bird. I must admit some skepticism of these findings because both utilized the very low pterosaur masses discussed in chapter 6. Modeling the wing shape of *Eudimorphodon* with masses comparable to birds and bats, by contrast, resulted in predictions of a "generalized" flight style (Witton 2008*a*), the same sort of adaptable, unspecialized flight that we discussed for *Preondactylus* in chapter 10.

These studies have only scratched the surface of "campylognathoidid" flight analysis, however. The abnormally long wing fingers and massive shoulder regions of some "campylognathoidid" species (taken to its extreme in *Campylognathoides*) have tremendous implications for their flapping kinematics, aerial agility, power output, and all manner of other flight attributes, and clearly are worthy of detailed study. In lieu of such analysis, we may provisionally interpret the enlarged sites for pectoral muscle attachment (see chapter 5) as indicative of a particularly strong downstroke, perhaps allowing sustained, powerful flapping and rapid flight. Interestingly, their long, narrow distal wing—indicated by the hypertrophied wing finger—would aid this further. In flapping animals, the distal wing is principally responsible for generating thrust, and a narrow profile reduces drag. (The medial wing, by contrast, provides lift. See, for example, Hildebrand [1995] for generalities of wing function.) A wing structure comparable to that of some "campylognathoidids," with short proximal skeletal components, elongate distal regions, and a high flapping capacity, exists in modern birds with "high speed wings." These include fast, agile fliers such as falcons and mastiff bats. If "campylognathoidid" wings were as comparable to these fliers as they seem, they may be similarly suited for powerful, rapid flight.

Such a wing construction is not consistent across "campylognathoidids," however; the wimpy-winged *Austriadactylus* and *Caviramus* were clearly using their wings in a different fashion than their powerfully armed relatives. Their slender wing bones and deltopectoral crests argue against power flapping and high-speed flying, and may hint at a flight style with substantially more gliding. As many soaring birds show, gliding is much kinder to long, slender wing bones. Interestingly, the humerus of *Caviramus* is proportionally more slender than its femur, a very aberrant trait for a pterosaur. This brings the ability of *Caviramus* to use the quadrupedal launching strategy discussed in chapter 5 into question. It may not rule it out in entirety because pterosaur humeri are generally overengineered, particularly at small body sizes (Habib 2008; Witton and Habib 2010), so the slender humerus of *Caviramus* may have been capable of launching its owner without failing, but evidence for this launch strategy is less obvious in this pterosaur than in others.

ON THE GROUND

As one of the few early pterosaurs known with fairly well-preserved pelvic remains, *Campylognathoides* was once deeply involved in the controversy over the terrestrial ability of pterosaurs. The crushed pelvic remains of this taxon were contradictorily interpreted to suggest both ventrally open (Wellnhofer 1974; Wellnhofer and Vahldiek 1986) and ventrally fused pelvic constructions (Padian 1983a), with the latter eventually supported by other pterosaur remains (see

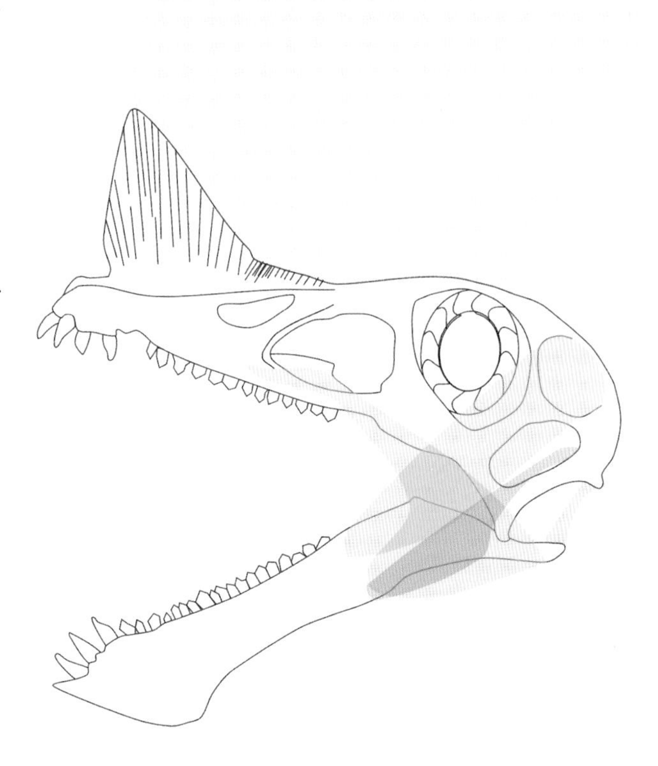

Fig. 12.9. Tentative reconstruction of the jaw muscles of *Caviramus filisurensis*. Note the complex, sculpted regions around the back of the jaw where the muscles would anchor.

chapter 7). This allowed *Campylognathoides*, like other pterosaurs, to have erect hindlimbs when walking. The glenoid of *Eudimorphodon* indicates that the forelimb could not rotate under the body for use in walking, so its arms were probably sprawled somewhat when grounded. Sadly, most other "campylognathoidid" glenoids are poorly preserved and crushed, so it is not clear if this posture was common to them all. Antungual sesamoids on the hands of *Eudimorphodon* and *Carniadactylus* suggest they were probably capable of climbing. The particularly low femur:tibiotarsus ratio and generally elongate, gracile limbs of *Caviramus* are features seen in particularly agile modern animals (Coombs 1978), possibly indicating it was a more sprightly grounded animal than its relatives.

Paleoecology

Some "campylognathoidids" developed a unique trick among pterosaurs: the ability to crudely chew their food. The extensive wear facets on their teeth (discussed in detail in Ősi 2010) is compelling evidence for this, revealing that *Caviramus* and *Eudimorphodon* wore their dentition away by tooth-on-tooth contact, and that the latter also experienced tooth-on-food abrasion. The depression of the jaw joint in some "campylognathoidids" would allow the entire toothrow to occlude at once, a characteristic of animals that chew their food efficiently. The tooth wear patterns of *Eudimorphodon* suggest that their jaws could move somewhat laterally when chewing, perhaps resulting from an unfused mandibular symphysis that allowed their jaws to flex as they worked (Ősi 2010). The unusually large upper temporal fenestrae and coronoid processes of these pterosaurs indicate that powerful muscles operated this chewing motion (fig. 12.9), and as may be expected, their tooth wear is most marked at the back of the jaw where the highest bite forces were generated. That these animals processed their food in this way suggests they were ingesting tough prey items, a notion supported by the digested ganoid fish scales found in the gut of *Eudimorphodon* (Wild 1978). By contrast, the lack of tooth wear in other "campylognathoidids" indicates that they avoided hard foods and performed little, if any, oral processing.

It is assumed that the multifunctional dentition of "campylognathoidids" enabled them to enjoy a broad diet of insects, soft invertebrates, tastier plant matter, small vertebrates, and carrion (Stecher 2008; Padian 2008b; Ősi 2010). If their wings were, as postulated above, adapted for agility and high speeds, it seems possible that some "campylognathoidids" could procure some food items in flight, opportunistically seizing fish or other relatively slow-moving prey as they rocketed past them. The particularly large wings of *Campylognathoides* may have provided additional lift to assist the transport of large prey, but like all "campylognathoidids," its anatomy is not obviously adapted for one particular foraging method over any other. They certainly do not have any skim-feeding adaptations (con. Stecher 2008; see Humphries et al. 2007; Witton 2008b; Ősi 2010; also chapter 24), but perhaps their downturned, often bulbous mandibular tips reflect frequent grubbing about in loose substrates with their jaws when foraging. These ideas are only very provisional however: as with their flight mechanics, the unusual jaws of these pterosaurs are also overdue for a dedicated functional analysis.

13
Rhamphorhynchidae

PTEROSAURIA > RHAMPHORHYNCHIDAE

Whereas the "campylognathoidids" seemed hell-bent on tweaking their anatomy at every opportunity, the rhamphorhynchid pterosaurs were a more conservative bunch that quickly organized themselves into two distinct bauplans and stuck with them for millions of years. Rhamphorhynchinae (fig. 1.1), a seagull-like lineage, seem to have been the more successful of these groups and were some of the most abundant pterosaurs of the Jurassic, enjoying 40 million years of evolution. The second lineage—the crow-like members of Scaphognathinae (fig. 13.1)—were just as widespread as their sister group, though their fossils are generally rarer and restricted to the latter 15–20 million years of the Jurassic. Happily, although this split in rhamphorhynchid relationships is not recovered in all analyses of pterosaur relationships, there is general agreement about the close affinities of purported rhamphorhynchid taxa (Dalla Vecchia 2009a; Andres et al. 2010; Lü, Unwin, et al. 2010; Lü, Unwin, et al. 2012; but also see Wang et al. 2009).

Rhamphorhynchids seem to be the first pterosaurs to achieve a relatively global distribution, with fossils known from the Americas, Asia, and Europe (fig. 13.2; Barrett et al. 2008). Unlike many early pterosaurs, their remains are not confined to fossil Lagerstätten, but when they do occur in such deposits they are often the most common members of the pterosaur fauna (Padian 2008a; Wellnhofer 1991a). The relative abundance of good quality rhamphorhynchid fossils means that we have an excellent understanding of their anatomy and that we can perform statistical testing on various attributes of some species. The idea of such number crunching and computing statistics may sound dull, but it reveals exciting insights to growth, sexual dimorphism, and even behavior that single fossils—even the most fantastically preserved ones—simply cannot provide. Similar testing can only be applied to a handful of other pterosaur species, most of which are pterodactyloids of younger geological age. Thus, rhamphorhynchids are an extremely important window into the paleobiology of early pterosaurs.

The Long, Long Research History of Rhamphorhynchidae

The research history of rhamphorhynchids stretches back to the very first pterosaur discoveries, with their fossils being documented in 1825. As is so often the case with nineteenth-century paleontology, the naming of the first rhamphorhynchids was a complicated affair with numerous names applied to the same specimens (see Wellnhofer 1975 for a review). With so much to cover, we can only skim the surface of the group's research history here. The first rhamphorhynchid fossil discovered was a delicately built skull from the Tithonian (151–145 Ma) Solnhofen Limestone of Germany, which was initially thought to represent the first known fossil bird. These remains were studied by Professor Samuel Thomas van Soemmerring, a key figure in the early history of pterosaur research and a proponent of the idea that pterosaurs were of mammalian origin (Wellnhofer 1991a). Soemmerring noted that the skull particularly resembled that of *Larus*, the modern gull, but he then sought the opinion of another professor once he realized his "bird" had a set of large, procumbent teeth. This opinion came from Georg August Goldfuss (1831) who suggested the skull actually belonged to a pterosaur, which he named *Ornithocephalus muensteri* after its discoverer Georg Graf zu Münster. In the same publication naming *O. muensteri*, Goldfuss described another new Solnhofen pterosaur fossil represented by a partial skeleton that was missing its legs, tail, and wing tips. He named this *Pterodactylus crassirostris*; the species name ("wide snout") reflecting the robust rostrum of the animal. More complete remains of both animals were eventually found, and after considerable taxonomic vacillation, it was suggested that each should be given novel generic names. Forms with a gull-shaped skull (fig. 13.3) were put into the genus *Rhamphorhynchus* (Meyer 1847), while the more robustly skulled forms (fig. 13.4) were christened *Scaphognathus* (Wagner 1861).

One hundred and eighty years since their discovery, *Scaphognathus* remains extremely rare and is still only

FIG. 13.1. *Sordes pilosus* investigates a Jurassic land snail in a Late Jurassic forest of Kazakhstan. In a few moments, the snail may be involuntarily investigating the beginnings of the *Sordes* alimentary tract.

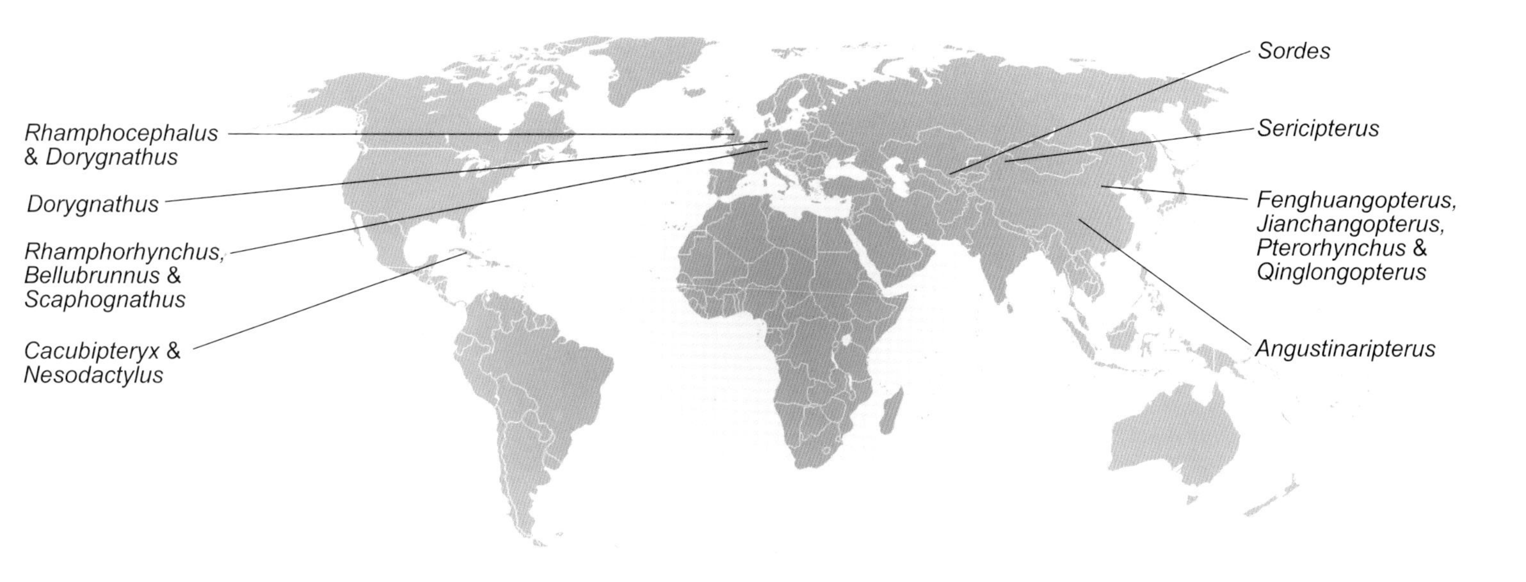

FIG. 13.2. Distribution of rhamphorhynchid taxa.

known from three specimens. *Rhamphorhynchus*, by contrast, is represented by over one hundred examples and is one of the most comprehensively understood pterosaurs of all. Many *Rhamphorhynchus* specimens have their soft tissues preserved in fantastic detail, allowing this species to contribute more to our understanding of pterosaur wing tissues than any other (see chapter 5). Up to 14 different species of *Rhamphorhynchus* have been named, but subsequent revisions by Wellnhofer (1975) and Bennett (1995) suggested these "species" actually represented the growth series of a single taxon, *R. muensteri*.

Other Rhamphorhynchines . . .

Rhamphorhynchus and *Scaphognathus* were not the only rhamphorhynchids to be discovered in the nineteenth century. Fragmentary fossils of rhamphorhynchine pterosaurs were recovered throughout the 1800s from the Jurassic (Toarcian; 183–176 Ma) Posidonienschiefer of Germany. Following a standard nineteenth-century nomenclatural fumbling, these were eventually christened *Dorygnathus banthensis* (see Padian and Wild 1992; and Padian 2008a for details of said fumble). Over time, more specimens, including complete skeletons, of this early, relatively "primitive" rhamphorhynchine were unearthed (e.g., fig. 13.5) so that, by the late 1900s, we had a good knowledge of its anatomy and had extended its range into northern France. In the 1970s, a second, larger *Dorygnathus* species was named: *D. mistelgauensis* (Wild 1971). However, new work on this taxon suggests it may merely represent a particularly large *D. banthensis*, and not a distinct species (Padian 2008a).

Dorygnathus may have also flown over Toarcian seas of Britain. An isolated, incomplete skull from the Alum Shale of Yorkshire was reported by E. T. Newton (1888) with particular excitement because, unlike most pterosaur remains known at that time, it was not squashed to paper-thin proportions. It also provided the first ever glimpse of a pterosaur brain, though only the dorsal region of the endocast could be examined. Newton considered these remains to represent a new species of *Scaphognathus* and christened it *S. purdoni*, but it was given the new generic name *Parapsicephalus* by Gustav von Arthaber in 1919. Further taxonomic revision has suggested that *Parapsicephalus* is congeneric with *Dorygnathus*, although it is still likely that the British skull represents a species separate from its German counterparts (Unwin 2003, 2005).

Another British rhamphorhynchine, *Rhamphocephalus*, is known from slightly more substantive remains than is *D. purdoni*, including a lower jaw and

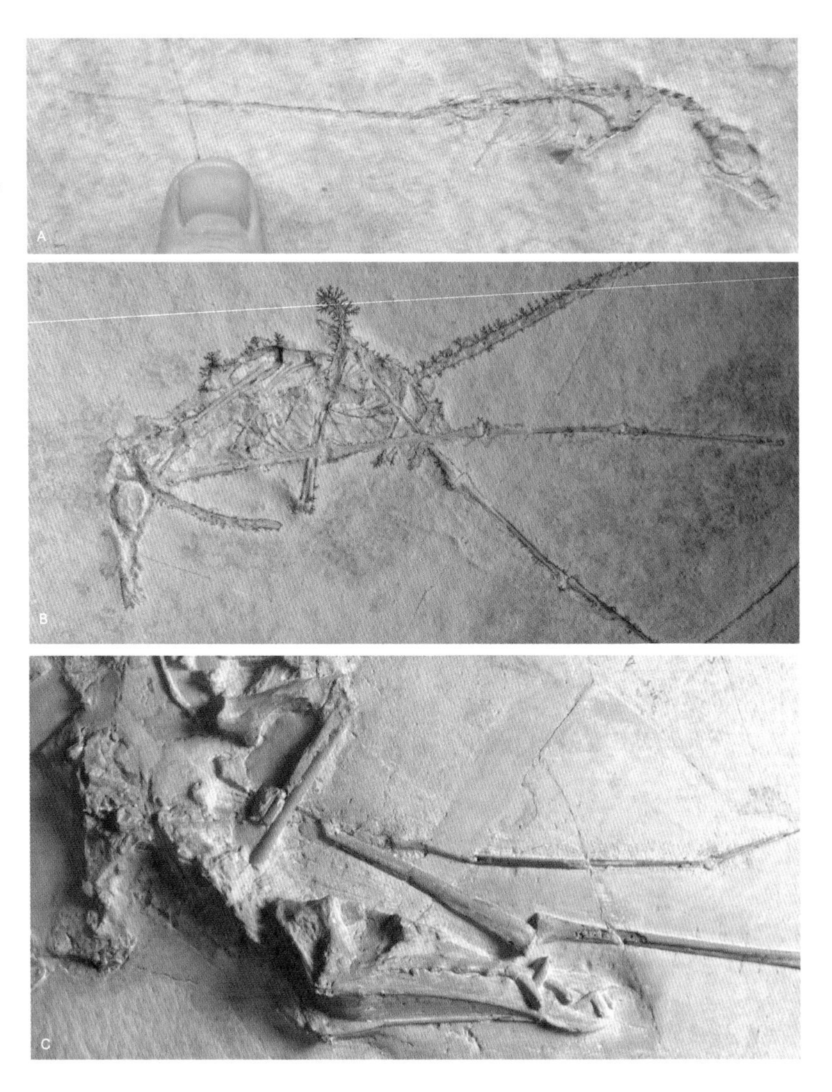

Fig. 13.3. *Rhamphorhynchus muensteri* fossils from the Tithonian Solnhofen Limestone, showing different growth stages. A, juvenile stage (note thumb for scale); B, subadult stage; C, large adult stage. The last stage is particularly rare, being represented by only two specimens. A and C, copyright Natural History Museum, London; B, courtesy of Helmut Tischlinger.

Fig. 13.4. Juvenile specimen of the rare scaphognathine *Scaphognathus crassirostris* from the Tithonian Solnhofen Limestone. Photography courtesy of David Martill.

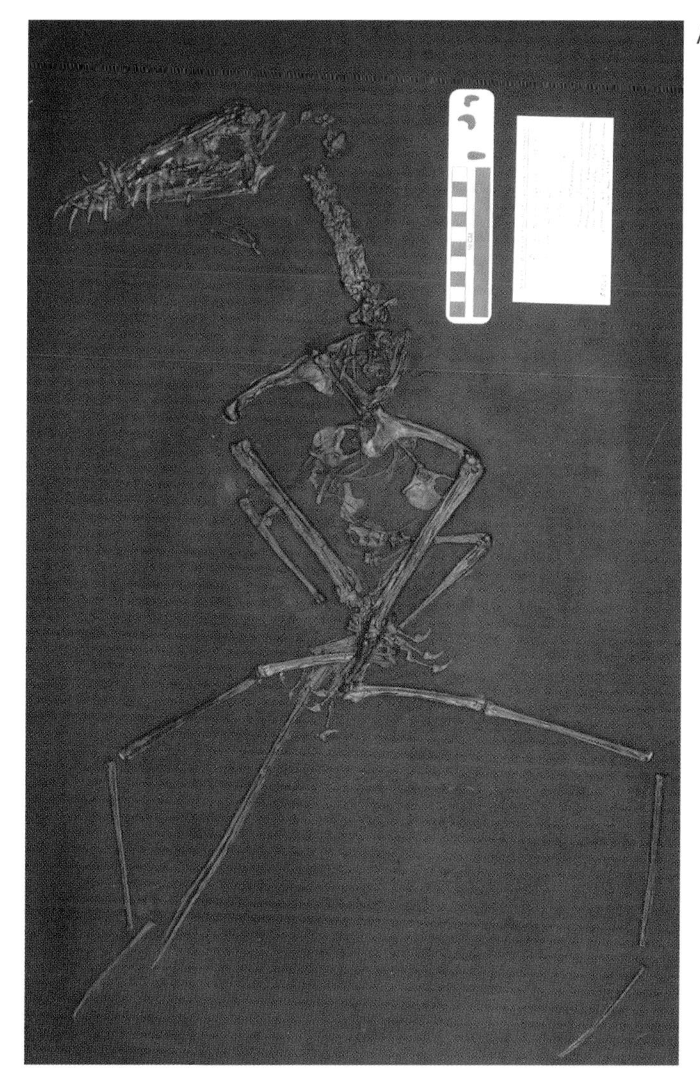

A

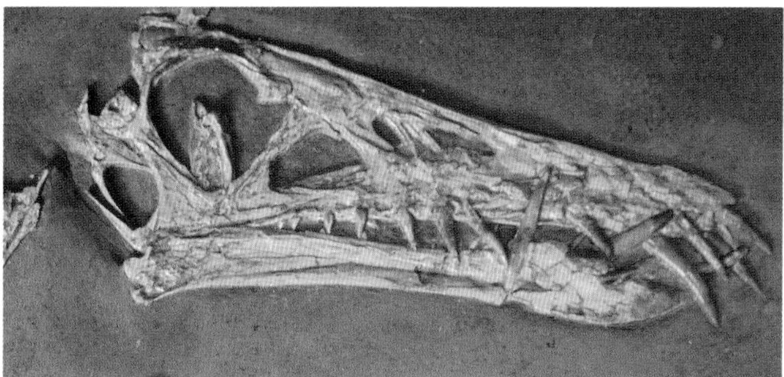

B

FIG. 13.5. Fossilized remains of *Dorygnathus banthensis* from the Toarcian of Germany. A, almost complete skeleton; B, complete skull and mandible. Note the tiny teeth in the posterior region of the lower jaw. Photographs courtesy of Ross Elgin.

numerous fragmentary fossils from the Middle Jurassic Stonesfield Slate of Oxfordshire (Bathonian; ca. 168–165 Ma). Alas, exactly how much of this material reflects *Rhamphocephalus* is not clear. At one time, every Stonesfield pterosaur fossil was referred to *Rhamphocephalus*, but the recent discoveries of possible wukongopterids (Andres et al. 2011; chapter 14) and ctenochasmatoids (Buffetaut and Jeffery 2012; chapter 19) in the same unit suggests some caution must be applied to this interpretation. Happily, a review of this material is now underway. *Rhamphocephalus* was another victim of convoluted Victorian naming practices that are still causing headaches today (we may, indeed, not be using the name "*Rhamphocephalus*" to refer to the Stonesfield pterosaur remains for much longer; M. O'Sullivan, pers. comm. 2012). By the end of the nineteenth century, three species were thought to have existed, only two of which (*R. depressirostris* and *R. bucklandi*) are presently considered valid (Unwin 1996). Other Middle Jurassic deposits across Britain have also yielded material referable to Rhamphorhynchidae, but much of it is so fragmentary that it has little, if any, taxonomic importance (Unwin 1996).

Other rhamphorhynchines come from far more exotic climes than rainy old England. The discovery of these pterosaurs in upper Jurassic rocks of Cuba was aided by the famous American fossil hunter Barnum Brown, who is perhaps best known for his role in discovering *Tyrannosaurus*. Brown collected rock specimens from western Cuba for the American Museum of Natural History between 1911 and 1919. These specimens spent much of the next few decades sitting in AMNH storage, an occurrence which may sound outrageous, but is not uncommon given how long it can take to process the discoveries of fossil-hunting expeditions. When the specimens were eventually assessed, one was found to hold a disarticulated and headless rhamphorhynchine skeleton that was named *Nesodactylus hesperius* (Colbert 1969). Conversely, only cranial remains were found with China's first Jurassic pterosaur, the Bathonian rhamphorhynchine *Angustinaripterus longicephalus* (He et al. 1983). China would prove to be a rich hunting ground for rhamphorhynchines, with upper Jurassic (Oxfordian; 161–156 Ma) deposits revealing the large, crested species *Sericipterus wucaiwanensis* (Andres et al. 2010) and the slightly older (Callovian-Oxfordian; 165–156 Ma) *Qinglongopterus guoi* (Lü, Unwin, et al. 2012). The latter appears to be a close relative of *Rhamphorhynchus* but is only represented by an extremely immature individual that limits detailed comparisons with other rhamphorhynchines.

Most recently, another new rhamphorhynchine was named from the Solnhofen-like Rhytmic Plattenkalk of Brunn, Germany, which dates to the late Kimmeridgian (ca. 152 Ma). Like *Qinglongopterus*, this animal is only represented by a very young individual and has been named *Bellubrunnus rothgaengeri* (Hone, Tischlinger, et al. 2012). Easily the most interesting aspect of this species is that the distal wing phalanges may have be deflected *anteriorly*. The preservation of the wings in the only specimen of this pterosaur is somewhat unusual however, so pterosaurologists are hoping to find another specimen to confirm its unusual anatomy (Hone, Tischlinger, et al. 2012).

. . . And More Scaphognathines

In contrast with rhamphorhynchines, most scaphognathine taxa are relatively new discoveries. Indeed, 140 years passed between the discovery of *Scaphognathus* and the naming of the next scaphognathine, *Sordes pilosus*, from the Upper Jurassic (Oxfordian-Kimmeridgian; 161–151 Ma) Karabastau Formation of Kazakhstan (Sharov 1971). It was well worth the wait. This second scaphognathine was preserved with a full complement of wing membranes with clear attachment sites, including the presence of a broad uropatagium supported by elongate fifth toes and brachiopatagia that extended to its ankles (see chapter 5; Sharov 1971; Bakhurina and Unwin 1992). *Sordes* also put debates about the fuzzy nature of pterosaur hides to rest (Sharov 1971), with two of the seven known *Sordes* specimens clearly showing pycnofibers around their bodies, heads, and necks. Close examination reveals surprising details about the density, length, and structure of the integument (Sharov 1971; Bakhurina and Unwin 1995; Unwin 2005), and the remarkable preservation of these fibers was the inspiration for its Latin name—"hairy devil." Alas, the Latin behind this title was somewhat mistranslated because *Sordes pilosus* actually means "filthy hair" (Bakhurina and Unwin 1995). Oops.

Following several more decades of quiet in the scaphognathine discovery ledger, a number of species were found in quick succession after the turn of the millennium. The first was the Chinese *Pterorhynchus wellnhoferi* from the Middle Jurassic (possibly Callovian-Oxfordian; 165–156 Ma) Tiaojishan Formation (Czerkas and Ji 2002). This animal is represented by a single complete skeleton with an entirely preserved soft-tissue headcrest, including color banding (fig. 5.13) and a covering of pycnofibers. The placement of this animal has typically been among scaphognathines (Unwin 2005; Barrett et al. 2008; Lü, Unwin, et al. 2010), but an alternative home among rhamphorhynchines has been recently proposed (Lü, Unwin, et al. 2012).

Documentation of the large scaphognathine *Harpactognathus gentryii* from the Upper Jurassic (Kimmeridgian-Tithonian; 156–145 Ma) Morrison Formation of Colorado followed shortly after reports of *Pterorhynchus* (Carpenter et al. 2003). *Harpactognathus* is known through a fragmentary rostrum that, as in *Pterorhynchus*, appears to have borne a headcrest, but its structure appears a little different from its Chinese counterpart. Another scaphognathine *Cacibupteryx caribensis* was reported in 2004 from the same Cuban deposits that yielded *Nesodactylus* (Gasparini et al. 2004). Alas, this animal is only known from a skull and a few postcranial bones. However, the most recently discovered scaphognathines, again from the Tiaojishan Formation, *Fenghuangopterus lii* and *Jianchangopterus zhaoianus*, are represented by virtually complete skeletons (Lü, Fucha, and Chen 2010; Lü and Bo 2011). The latter could well represent a juvenile wukongopterid, however, and not a scaphognathine.

Anatomy

OSTEOLOGY

Though looking rather different from each other, rhamphorhynchines and scaphognathines are united as a group through a number of features (Unwin 2003). Most notably, their teeth are very simple in structure and low in number, with no more than 11 tooth pairs found in their upper jaws. Their deltopectoral crests are typically pinched at the bases but have expanded terminations, with this condition taken to its most extreme in derived rhamphorhynchines. They also bear bent phalanges in their fifth toes, which are either gently curved or sharply boomerang-shaped. In other ways, however, rhamphorhynchines and scaphognathines are different enough that it will be wise to describe their skeletal anatomy independently.

RHAMPHORHYNCHINAE

The lithe and streamlined anatomy of rhamphorhynchines (fig. 13.6) is very well documented, and especially so for *Rhamphorhynchus* and *Dorygnathus* (Wellnhofer 1975; Padian 2008a). These chaps were among the largest of Jurassic pterosaurs, with *Dorygnathus*, *Rhamphorhynchus*, *Rhamphocephalus,* and *Sericipterus* attaining wingspans of 2 m or more. Rham-

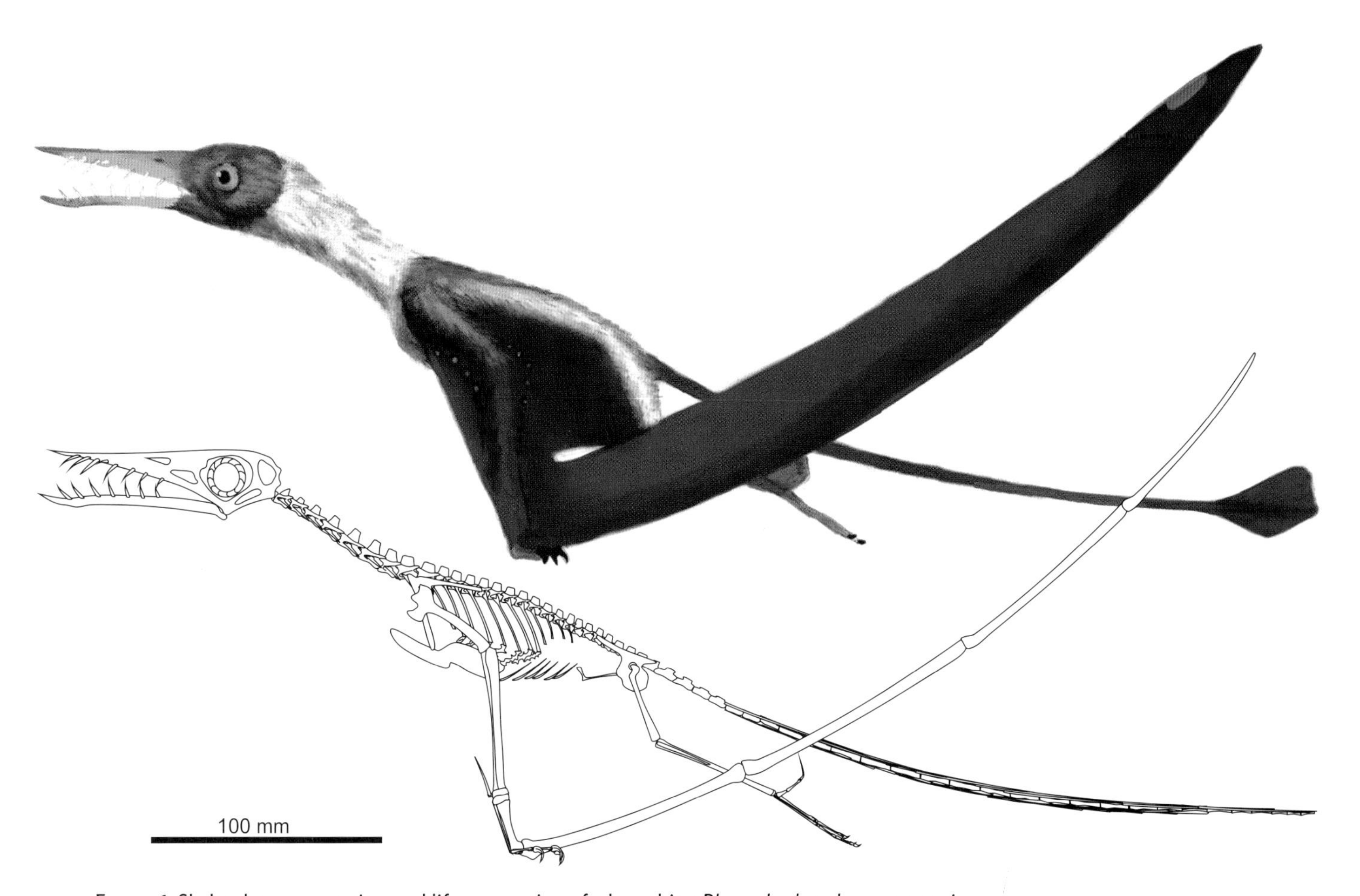

Fig. 13.6. Skeletal reconstruction and life restoration of a launching *Rhamphorhynchus muensteri.*

phorhynchine skulls are low and slender, with small, slit-like nasal openings and antorbital fenestrae. Their jaw tips are prominent, bony prows that extend well beyond the anterior teeth, and in the lower jaw these represent extensions of a short mandibular symphysis. The shape of the mandibular prow is quite variable among species and also changes with age, but typically ends up being relatively hooked and robust in adult forms (figs. 8.7A, 13.7; Bennett 1995). Their teeth are generally procumbent and are particularly so at their front of the mouth, where they form a bristly looking grasping apparatus when the jaw closes. Although the teeth are generally rather large, *Dorygnathus* retains a number of smaller teeth in the posterior region of its lower jaw. This and other features link *Dorygnathus* to other early pterosaurs and is one reason why it is thought to represent an early grade of rhamphorhynchine evolution.

The neck vertebrae of rhamphorhynchines have unusually high, pointed neural spines and, in some taxa, the anterior cervicals are somewhat enlarged (fig. 4.5; Wellnhofer 1975). These vertebrae, along with the anterior dorsals, shoulder girdle, and sternum, are extensively pneumatized (Bonde and Christiansen 2003). Their torsos are slight and compact, and retain the primitive pterosaur condition of being longer than their skulls. Their tails are long and stiffened (except for *Bellubrunnus*, which does not bear stiffened, elongated zygapophyses), and contain unusually high numbers of vertebrae: *Bellubrunnus*, *Rhamphorhynchus,* and *Nesodactylus* each have around 40 (Colbert 1969; Wellnhofer 1991a; Hone, Tischlinger, et al. 2012). Their pectoral girdles are relatively slender, and their glenoids possess an abbreviated articulatory surface that seems to have restricted ventral motion of the forelimb. Their sterna are generally large with prominent cristospines, though the sternum of *Dorygnathus* seems to be particularly small. It is likely that this, as with other pterosaurs with small sterna, was extended by cartilage in life (Padian 2008a).

Rhamphorhynchine forelimbs are among the longest of all early pterosaurs, their lengths only

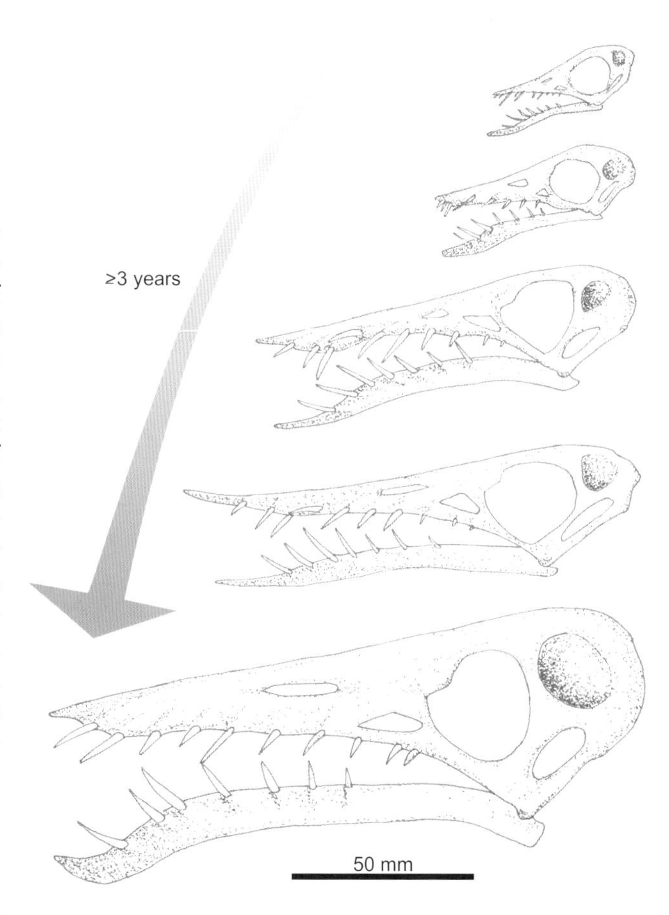

FIG. 13.7. The cranial growth series of *Rhamphorhynchus*. Modified from Bennett (1995).

challenged by the "campylognathoidids." Their humeri are easily recognized by their large, hatchet-shaped deltopectoral crests. The first three digits of their hands are delicate, though their claws are large and very recurved. *Dorygnathus* again proves itself as distinctive by possessing antungual sesamoids on all three of its manual digits. Rhamphorhynchine flight fingers are elongated to as much as 67 percent of their wing length (Unwin 2003; Hone, Tischlinger, et al. 2012), and *Nesodactylus* and *Rhamphorhynchus* bear wing phalanges with a shallow groove along their posterior margin. What function these grooves had is not clear. In contrast with their forelimbs, rhamphorhynchine hindquarters are quite under-developed, comprising small pelves, delicate prepubes, and short, slender legs. Reflecting this, their foot bones rather are rather fragile-looking with only weakly developed claws.

SCAPHOGNATHINAE

With their chunky skulls and stout neck vertebrae, scaphognathines are not as elegant in appearance as the rhamphorhynchines (fig. 13.8; see Sharov 1971 and Wellnhofer 1975 for their comprehensive descriptions). They range considerably in size, with *Fenghuangopterus* and *Sordes* being the smallest with wingspans of 0.7 m and *Pterorhynchus* a little larger at 0.85 m, but *Cacibupteryx* and *Harpactognathus*—fragmentarily known as they are—clearly represent larger animals with wingspans of 2.5 m or more (Carpenter et al. 2003). These wingspans, coupled with the bulky scaphognathine skeleton, make *Cacibupteryx* and *Harpactognathus* among the largest of all non-pterodactyloid pterosaurs. Scaphognathine skulls are characteristically robust, with blunt jaw tips and relatively deep, angular mandibles. Their jaws are rather broad for much of their length and do not taper to a sharp point. Their teeth are long and slender and, unlike rhamphorhynchines, are orientated vertically along the jaw.

Scaphognathine bodies are of fairly typical proportions for early pterosaurs, but their neck vertebrae are rather stocky, being wider than long and anchoring elongate cervical ribs. Only *Sordes* and *Pterorhynchus* preserve complete tails, showing strongly contrasting proportions. *Sordes* is equipped with a tail of around twice its trunk length, but the tail of *Pterorhynchus* measures a whopping 3.5 times the same dimensions. This tremendously long tail *is* comprised of 45–50 caudals (Czerkas and Ji 2002), giving it the highest caudal count, and perhaps the longest tail with respect to body size, of any pterosaur.

Scaphognathine shoulder girdles feature very small sterna, but as with *Dorygnathus* it seems likely that these structures are missing large cartilaginous regions. The rest of their pectoral girdle is well developed, and their forelimb skeleton is notable for possessing an especially elongate radius and ulna. Unlike rhamphorhynchines, scaphognathine wing fingers are not especially long. Neither, for that matter, are their hindlimbs, though they are not as diminutive as those of rhamphorhynchines. Special mention should be given to their manual claws however, which are not as recurved as those of their sister group, but are considerably more robust.

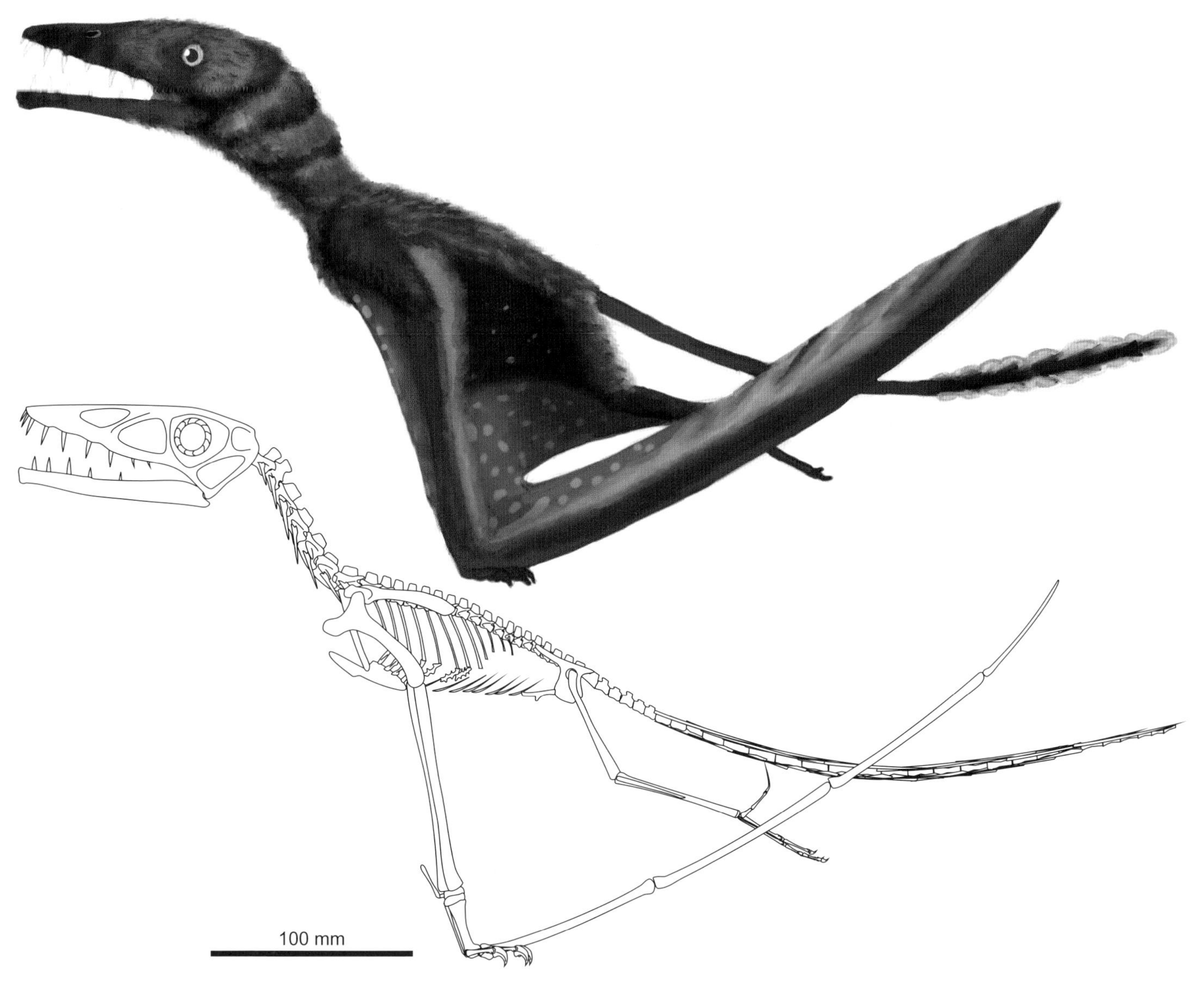

FIG. 13.8. Skeletal reconstruction and life restoration of a launching *Scaphognathus crassirostris.*

SOFT TISSUES

We probably have more data on the soft tissues of rhamphorhynchids than for any other pterosaur group, and particularly so for *Rhamphorhynchus* (see chapter 5). Their cranial soft tissues are very well known, with beaks, crests, and brains represented across different species and specimens. The beaks of *Rhamphorhynchus* extended the bony prows of their jaws and sometimes adopted shapes that would not be predicted from their jaw skeletons (fig. 5.11). The heads of some scaphognathines were adorned with large, rounded soft-tissue crests (Czerkas and Ji 2002), which are marked by a small bony rise on the rostrum. From this, we can assume that *Harpactognathus* possessed such a structure, though the shape and size of its crest remains mysterious. Detailed endocasts are known for *Rhamphorhynchus* and partially for *Dorygnathus* (Newton 1888; Wellnhofer 1975; Witmer et al. 2003), revealing superficially birdlike brains. Although later pterosaurs would assume greater neurological similarity to birds, the brains of rhamphorhynchids were already close

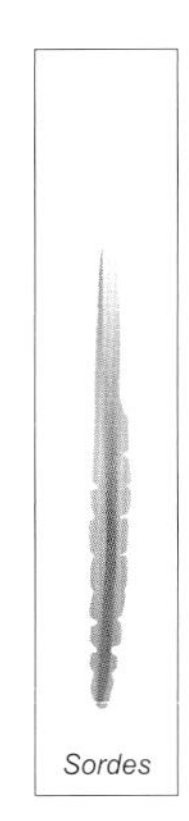

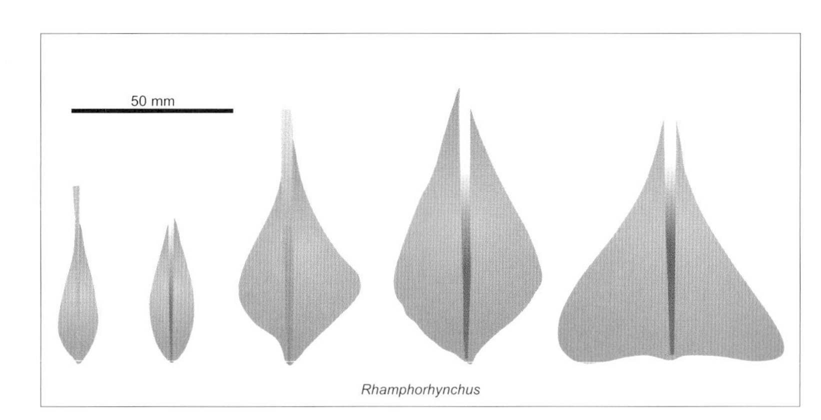

FIG. 13.9. Rhamphorhynchid tail vanes. Note the changes in fin profile with size and age in *Rhamphorhynchus*. *Rhamphorhynchus* tail vanes after Bennett (1995).

enough to assume that they had sufficient computing power for dynamic flight behaviors.

Pycnofibers have been reported from specimens of *Rhamphorhynchus* and *Dorygnathus* (Broili 1927; Wellnhofer 1975), but these interpretations have proved somewhat controversial among pterosaur workers. By contrast, the preserved pelts of *Sordes* and *Pterorhynchus* are inarguable examples of pterosaur fuzz. In *Sordes* they densely cover the face, neck, and torso (Sharov 1971; Bakhurina and Unwin 1995a; Czerkas and Ji 2002). Each fiber is about 6 mm long and often curved, suggesting they would be flexible and soft in life. Some have identified fur-like fibers in *Sordes* wings, but these were more likely to be short, flexible, structural fibers within the wing tissue itself (Unwin and Bakhurina 1994). Flexible wing fibers are certainly known from specimens of *Rhamphorhynchus*, along with stiffened structural fibers in the distal wings, soft-tissue fairing of the wing bones, blood vessels, trailing edge structures and wing attachment sites (Padian and Rayner 1993; Unwin and Bakhurina 1994; Frey, Tischlinger, et al. 2003; Tischlinger and Frey 2010; also chapter 5).

Tail vanes are also known from a number of rhamphorhynchid species. These features are assumed to be present in all long-tailed pterosaurs but are poorly represented in the pterosaur fossil record (fig. 13.9). Two types are known: long, lobed varieties that occupy much of the tail length, and relatively small structures situated at the very tip of the tail. The latter type is found in *Rhamphorhynchus*, and it seems that their shape alters with age from a diamond shape in young animals to a more triangular profile in adults (Bennett 1995). Because these vanes are not symmetrical, it is assumed that they were orientated vertically on the tail to maintain efficient flight dynamics. *Sordes* and *Pterorhynchus* suggest that scaphognathines sported the lobed-fin variant (Czerkas and Ji 2002), but the tail lobes of *Pterorhynchus* extend much further along the tail than those of *Sordes*. Combined with the diversity in tail vane shapes seen throughout the growth of *Rhamphorhynchus*, it seems possible that these organs had a greater function as display devices than as flight rudders (as suggested, for example, by Wellnhofer 1991*a*; and Frey, Tischlinger, et al. 2003).

Locomotion

FLIGHT

Many of the proportional contrasts between the two rhamphorhynchid bauplans can probably be attributed to different adaptations to flight. The long, narrow wings of rhamphorhynchines are widely considered suitable for gull-like, soaring flight (Wellnhofer 1975; Hazlehurst and Rayner 1992; Chatterjee and Templin 2004; Witton 2008a), and the abundance of rhamphorhynchines in marine or marginal marine sediments corroborates this flight style. Interestingly, their sterna are weakly developed compared to those of "campylognathoidids," suggesting their downstroke musculature was comparatively small. Flap gliding and soaring flight may therefore have been employed in these pterosaurs more than continuous flapping. Some modern soaring birds (e.g., albatross, frigate birds) also possess reduced sterna because of their infrequent flapping, which lends some credence to this idea.

Opinions on the flight of scaphognathines are not so clear. Some (Hazlehurst and Rayner 1992; Chatterjee and Templin 2004) consider them to have been adept soarers like rhamphorhynchines, but other analyses have argued for a more generalized flight style (Witton 2008a). The latter would certainly suit the apparent scaphognathine preference for richly vegetated terrestrial environments (Bakhurina and Unwin 1995). Soaring and gliding are excellent methods of flying in very open settings (such as the coastlines that rhamphorhynchines seem to have haunted), but can be difficult to sustain in cluttered, obstacle rich terrestrial environments. In these circumstances, high-angled launching, sharp maneuverability, and short wings that can be neatly tucked away when on the ground are preferable (Rayner 1988). The wing proportions of scaphognathines meet these criteria, with the large proximal wing region maximizing lift and the comparatively short distal wings increasing aerial agility, not to mention tidy wing stowing. Though scaphognathines would have sacrificed their gliding ability to achieve this flight style, they may not have missed it if they frequented the woodlands and forests indicated by the depositional settings of their fossils.

ON THE GROUND

Rhamphorhynchids have been used as principal evidence for non-pterodactyloid pterosaurs being rather shoddy terrestrial animals, because their fossils clearly show that the long fifth toes of early pterosaurs were employed in supporting an extensive uropatagium (e.g., Unwin 2005). Conversely, Wellnhofer (1975) and Padian (2008a) argued for efficient bipedality in rhamphorhynchids, at very least during takeoff. As discussed in chapters 5 and 6, there's some reason to think otherwise on both these fronts, with effective quadrupedal launching and walking apparently quite feasible for all pterosaurs. Indeed, the presence of antungual sesamoids on the hands of *Dorygnathus* indicates that it had a grappling manus adapted for climbing, which supports the idea of the forelimbs being used in rhamphorhynchine terrestrial locomotion. Rhamphorhynchine glenoids do seem to have restricted their ability to bring their forelimbs directly beneath the body however, so their forelimbs may have sprawled somewhat when moving around on the ground. This, combined with their rather short hindlimbs, suggests that rhamphorhynchines may not have been the most proficeint of terrestrial locomotors. The slightly longer limbs of scaphognathines may have fared a little better when walking and running, and their heavier claws probably gave more purchase when scuttling about. This ties in nicely, of course, with the bias of scaphognathine fossils toward terrestrial settings.

Paleoecology

The vast majority of *Rhamphorhynchus* specimens represent immature individuals, with only a handful of adult individuals known. As such, attempts to find sexual dimorphism in this species have been rather inconclusive (Bennett 1995), although some gender differences may be reflected in skull size (Wellnhofer 1975). The growth patterns of rhamphorhynchines have been studied in detail, revealing that their anatomy changed dramatically as they grew (figs. 8.7, 13.3, and 13.7; Bennett 1995). With age, they transformed from snub-nosed, small-toothed animals to slender-skulled forms with narrow teeth and long jaw prows, and finally to robustly skulled forms with tusk-like teeth and a hooked mandibular tip. This has been taken as evidence that pterosaurs occupied different ecological niches as they grew (Unwin 2005), their changes in skull and tooth shape allowing differently aged individuals to feed on prey of varying sizes and types. This would have enabled many different individuals to coexist without competing for food resources (so called niche partitioning; Bennett 1995).

It is well established that *Rhamphorhynchus* were piscivores thanks to the remains of fish in the guts and throat regions of two medium-sized specimens (fig. 8.10; also Wellnhofer 1975; Tischlinger 2010; Frey and Tischlinger 2012), and this diet seems likely for most large rhamphorhynchines. Opinions on the foraging strategy of *Rhamphorhynchus* differ however. Some imagine it as a skim feeder, plowing the tip of its mandible through water like the modern skimming bird, *Rynchops* (Wellnhofer 1991a; Hazlehurst and Rayner 1992), but others argue the opposite, noting the lack of necessary skim-feeding adaptations across the skull and neck (Chatterjee and Templin 2004; Humphries et al. 2007; Witton 2008b: also see chapter 24). A more plausible idea may be that *Rhamphorhynchus* and its kind dip fed, swiping fish from the water surface while on the wing. This notion is supported by the large size and articulations of the anterior cervicals of *Rhamphorhynchus*, indicating the anterior neck region was well muscled and capable of considerable movement. Coupled with the extended reach of the jaws and teeth, slender jaw tips, and apparent ability for stable flight, *Rhamphorhynchus* and its kin may have been well suited to snatching food on the wing, albeit without skim feeding. If so, this would vindicate the seabird-like lifestyle advocated for *Rhamphorhynchus* since its discovery.

The relationship between *Rhamphorhynchus* and its fishy prey was often reversed. Several specimens of the large, spear-nosed Solnhofen fish *Aspidorhynchus* are known to have become entangled with this pterosaur in death, a relationship thought to have developed through numerous failed predatory attempts by this species of fish (Tischlinger 2010). Pterosaurs are otherwise never associated with potential predators in this way, which may indicate that botched attacks on pterosaurs by *Aspidorhynchus* carried relatively high risks of fatality. A more successful predation effort has been demonstrated by a specimen of fossil gut regurgitate, bearing hallmarks of a fishy owner, that is riddled with *Rhamphorhynchus* remains (Schweigert et al. 1998). How these *Rhamphorhynchus* were apprehended, of course, remains unknown, but it is perhaps most conceivable that they were grabbed by fish while alighted on the water surface.

Unfortunately, the foraging habits of scaphognathines are not as clearly dictated by their fossils as those of rhamphorhynchines. Some researchers have proposed a piscivorous or insectivorous diet for these animals (Wellnhofer 1991a; Ősi 2010), while Bakker (1986) proposed that they would hawk other pterosaurs out of the sky in a raptor-like fashion. Aerial foraging seems unlikely because they lack suitable wings and shoulder anatomy for steady flight over water or chasing aerial prey (see chapter 11). I wonder if scaphognathines were specialized for any foraging methods more than others, and were perhaps instead generalized, omnivorous foragers akin to medium- and large-sized corvids. Their robust claws and snouts may have been good for rummaging through leaf litter or shallow digging, and their widely spaced, piercing teeth may have let them even subdue relatively large prey from time to time. Interestingly, the skulls of juvenile scaphognathines (demonstrated by *Sordes* and *Scaphognathus*) are not as different from their older brethren as those of *Rhamphorhynchus*, suggesting these pterosaurs did not hop through as many distinct foraging phases in their life histories as their slender-snouted colleagues.

14
Wukongopteridae

PTEROSAURIA > MONOFENESTRATA > WUKONGOPTERIDAE

The twenty-first century has already furnished pterosaurologists with two of the most exciting pterosaur discoveries in living memory. The first was the recovery of pterosaur eggs (chapter 8), and the second was the discovery of *Darwinopterus*, a small pterosaur from Liaoning deposits of China (fig. 14.1). This humble looking pterosaur was unveiled in 2009, but was not formally published until 2010 (Lü, Unwin, et al.) and has caused a real stir among pterosaurologists. *Darwinopterus* incontrovertibly fills a long-standing gap in pterosaur evolution, bridging the morphological distance between early pterosaurs and Pterodactyloidea. As we will see in subsequent chapters, pterodactyloids are anatomically very different from their ancestors, and until *Darwinopterus* was found, the evolutionary steps between these lineages were not clear. The manner in which *Darwinopterus* links these anatomies is remarkable. Rather than demonstrating a bauplan with a smattering of pterodactyloid and non-pterodactyloid features across the entire skeleton, it possesses the characteristic skull and neck of a pterodactyloid while retaining a body very similar to those of rhamphorhynchid pterosaurs (fig. 14.2). The distinction between these configurations is so marked that fakery was a real concern for the first-found specimens (Lü, Unwin, et al. 2010), but detailed examination and subsequent discoveries have confirmed their aberrant anatomy. This has seen *Darwinopterus* hailed as an example of "modular" evolution, where large, genetically linked blocks of anatomical components change while others remain unaltered. This is reflected in its full name, *Darwinopterus modularis*, which honors both the father of modern evolutionary theory, Charles Darwin, and the mechanism of evolution it may demonstrate.

There May Be Others

Darwinopterus comes from the Tiaojishan Formation of northeast China (fig. 14.3), and was kept company by the rhamphorhynchids *Pterorhynchus*, *Qinglongopterus,* and *Fenghuangopterus* (chapter 13). The age of this unit is somewhat controversial, but it seems to sits around the boundary of the Middle and Upper Jurassic (Callovian-Oxfordian; 165–156 Ma). This suggests *Darwinopterus* existed just before the first pterodactyloids, which can be reliably dated to the Oxfordian (161–156 Ma) (Buffetaut and Guibert 2001; note that some controverisally identified pterodactyloid specimens may be older). *Darwinopterus* has proved to be a relatively abundant fossil with over 10 good specimens recovered since 2009, and it is likely to become one of the best understood pterosaurs (fig. 14.4; Lü, Unwin, et al. 2011). One particularly amazing specimen was found in association with an egg and, as discussed below and in chapter 8, this has significant implications for our understanding of pterosaur lifestyles, taxonomy, and reproduction.

Weeks after *Darwinopterus* was announced, another closely related pterosaur was described from the Tiaojishan Formation: *Wukongopterus lii* (Wang et al. 2009). This specimen is missing most of its skull, and the authors interpreted it as being several phylogenetic miles away from Pterodactyloidea. It was not until 2010 that this animal was appreciated as being a close relative of *Darwinopterus* (Wang et al. 2010). Then, simultaneously, two new *Darwinopterus*-like taxa were also announced: a second *Darwinopterus* species (*D. linglongtaensis*) and *Kungpengopterus sinensis*. These animals, also from the Tiaojishan Formation, comprise part of Wukongopteridae, the recently named group that houses *Darwinopterus* and its kin (Wang et al. 2010).

This was not the end of the stream of *Darwinopterus*-like forms from the Tiaojishan Formation, however. Reappraisal of the small Tiaojishan pterosaur *Changchengopterus pani* suggests it is rather *Darwinopterus*-like. The only known specimen of this pterosaur is missing its skull and most of its neck (Lü 2009b), but it seems to have wukongopterid-like cervicals and may well represent another species of this new group (Wang et al. 2010). Yet another *Darwinopterus* species (*D. robustodens*, named after its relatively large teeth) and the

FIG. 14.1. The early evolution of nagging. *Darwinopterus,* shown here in an Upper Jurassic woodland, is likely to have been sexually dimorphic with large-crested males and entirely crestless females.

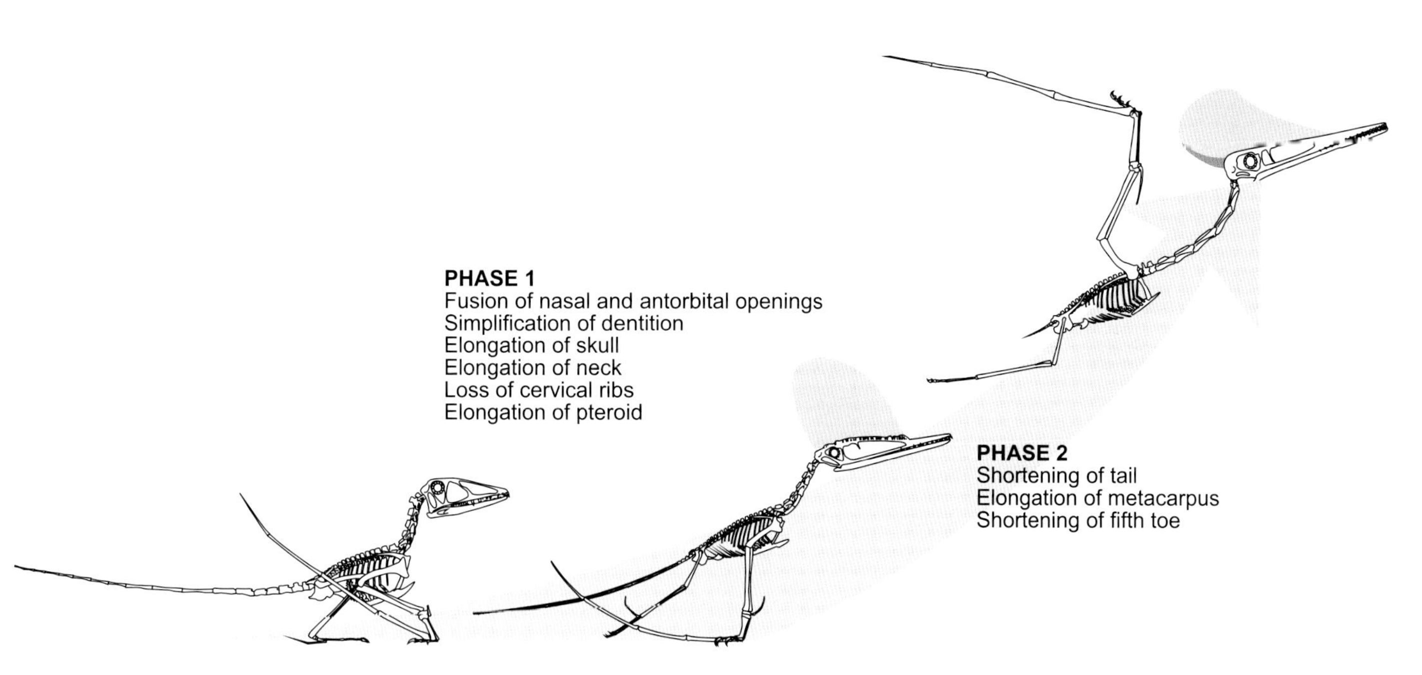

FIG. 14.2. How to make a pterodactyloid. Take one early pterosaur (*Preondactylus*, left), add some monofenestratan characteristics (*Darwinopterus*, middle), and simmer for a few million years with an emphasis on postcranial selection pressures. A pterodactyloid (*Pterodactylus*, right) is the result.

FIG. 14.3. Distribution of wukongopterid taxa.

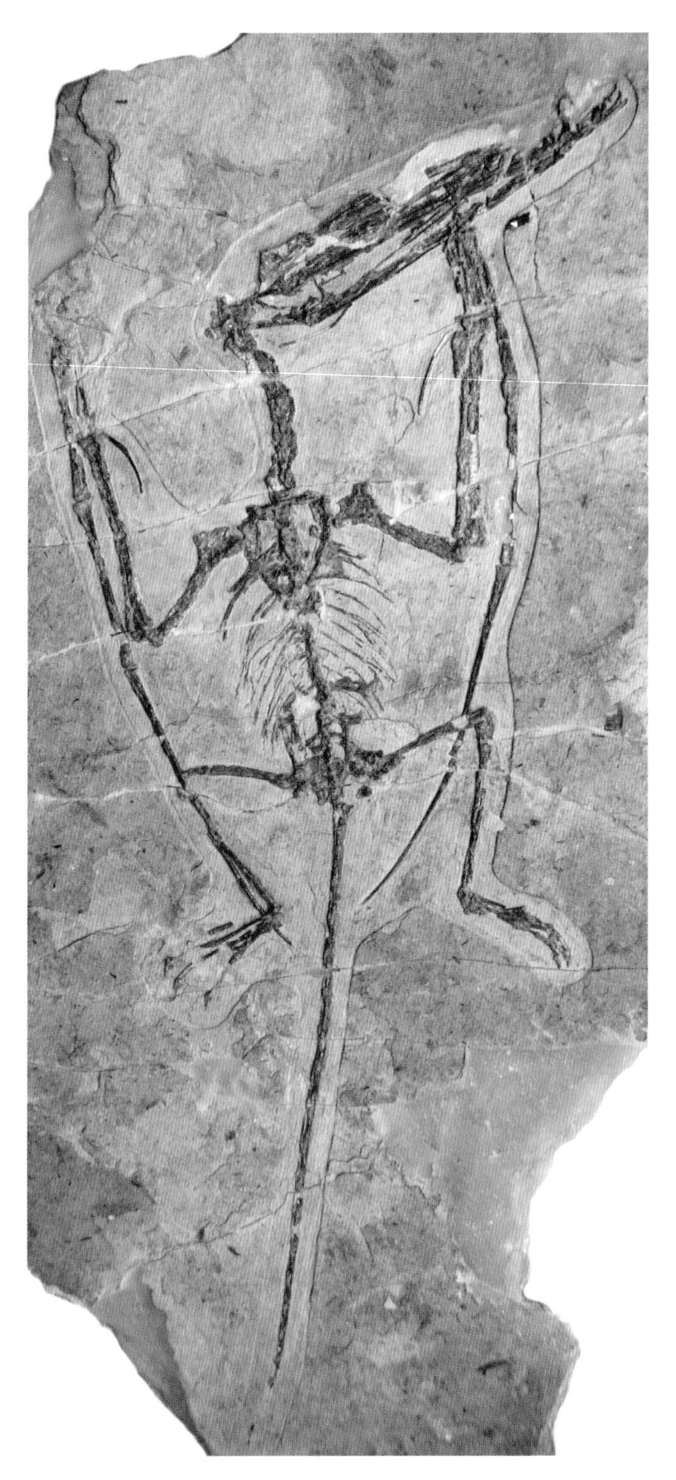

FIG. 14.4. An almost complete skeleton of *Darwinopterus modularis*, of the Callovian-Oxfordian Tiaojishan Formation, China, missing only details of its skull. Photography courtesy of David Unwin.

unfortunately named *Archaeoistiodactylus linglongtaensis* (which almost certainly has nothing to do with the ancestry of istiodactylids [chapter 15], despite its name) represent further wukongopterid species, bringing the total of Tiaojishan wukongopterids to eight (Lü and Fucha 2010; Lü, Xu, et al. 2011). As noted in chapter 13, the alleged Tiaojishan scaphognathine *Jianchangopterus zhaoianus* may also represent a young wukongopterid, which would raise this number to nine. This has been noted as an inordinate number of species considering the very minor morphological differences between them (Lü, Unwin, et al. 2012); quite unbelievably, we're already in need of a detailed taxonomic review of this group only two years after its discovery!

Away from the Tiaojishan *Darwinopterus* factory, wukongopterid remains have been identified in Britain, in the Jurassic (Bathonian; 168–165 Ma) Stonesfield Slate (Steel 2010; Andres et al. 2011) and possibly in the slightly younger (Kimmeridgian; 156–151 Ma) Kimmeridge Clay (Martill and Etches, in press). The latter, named *Cuspicephalus scarfi,* is represented by an incomplete skull and its affinities to wukongopterids are only tentative, but it may well represent an unusually large member of this group. The ages of these English wukongopterids are noteworthy; if both reports are valid, they tell us that wukongopterids existed as a lineage for at least 10 million years and were spread across what is now Asia and Europe. This offers a new perception of wukongopterids as a successful lineage in their own right, and not simply a step into evolution's changing room to put on some pterodactyloid garb.

Anatomy

Proving that good things really do come in small packages, wukongopterids are not big animals. The largest known Tiaojishan individuals are 0.65–0.8 m across the wings (Wang et al. 2009; Lü, Unwin, et al. 2010), and *Changchengopterus,* represented by the remains of a juvenile, is particularly tiny with a wingspan of only 45 cm. *Cuspicephalus* dwarfs its potential cousins by some margin, however, with a predicted wingspan of over 2 m.

Wukongopterid skulls are their most striking feature, demonstrating the first occurrences of anatomies that would become "standard" for pterosaurs throughout the next 100 million years. A particularly noticeable development is their enlarged cranial proportions. The skulls of wukongopterids are over 1.5 times longer than their torsos (fig. 14.5), a proportion unseen in non-pterodactyloids but very typical of

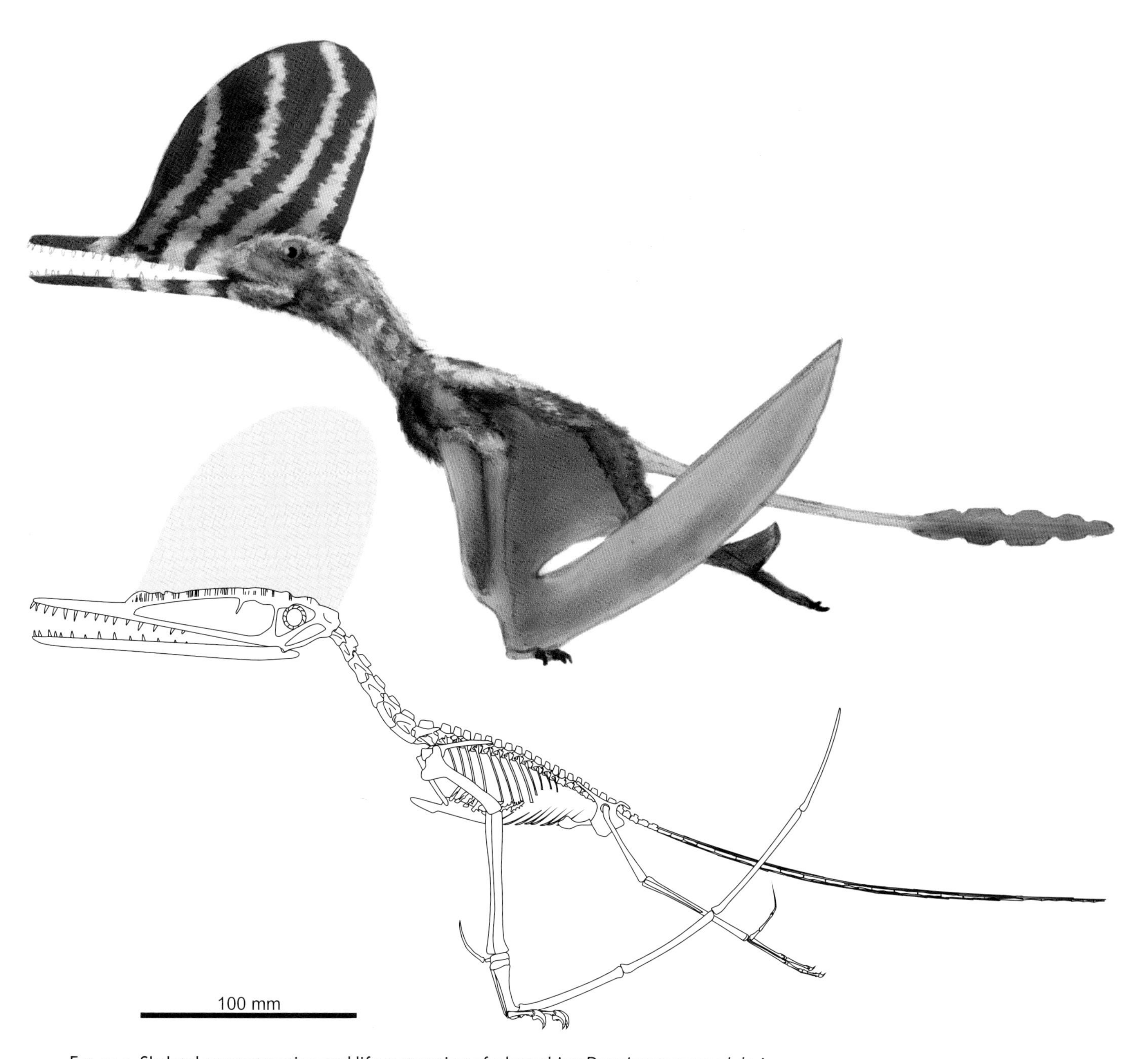

FIG. 14.5. Skeletal reconstruction and life restoration of a launching *Darwinopterus modularis*.

pterodactyloids. Equally obvious is the large opening occupying much of the skull length, the result of the nasal and antorbital openings fusing into the confluent nasoantorbital fenestra (fig. 14.6). This structure, previously only known in pterodactyloids, unites wukongopterids and pterodactyloids in their own group, the Monofenestrata (literally meaning "one opening").

Many other features of the wukongopterid skull are also rather pterodactyloid-like. Much of its dorsal margin is occupied by a low, fibrous bony crest that almost certainly supported a large soft-tissue extension in life. Cranial crests may be a dimorphic feature of wukongopterids as some osteologically and sexually mature wukongopterids seem to have lacked them (Lü, Unwin, et al. 2011; also see below). Their occipital faces are strongly inclined and their jaws are long and slender, with the mandible particularly so. This fuses along the front 20 percent of its length to

FIG. 14.6. Skulls of *Darwinopterus*. A, *D. modularis*; B, *D. robustodens*. Photographs courtesy of David Unwin.

give wukongopterids a short mandibular symphysis, another primarily pterodactyloid feature. Their dental array is comprised of relatively few, simple teeth which are limited to the front half of the jaws. Some dental variation seems to exist among species, with *Wukongopterus* and *Darwinopterus robustodens* having more robust, triangular teeth than other wukongopterids.

As with pterodactyloids, the cervical vertebrae of wukongopterids are twice as long as wide, producing a much longer neck than those of earlier pterosaurs. Their vertebrae also lack cervical ribs and have low neural spines, which are additional pterodactyloid characteristics (although the bizarre anurognathids also lack cervical ribs, so this is not strictly a monofenestratan characteristic). An enlarged pteroid bone is their final pterodactyloid-like feature, and is far longer than anything you'll find in other non-pterodactyloids. It is also strangely bowed and may indicate that these animals had relatively broad propatagia.

The rest of the wukongopterid body is classically non-pterodactyloid, sharing a number of attributes with the bodies of rhamphorhynchids. Their tails are long, consisting of at least 20 elongate caudals wrapped up in stiffening rods. Their sterna are relatively well developed and their humeri are distinguished by small, subtriangular deltopectoral crests. The proportions of their radius, ulna, and metacarpals are comparable to those of rhamphorhynchine pterosaurs, but the wing finger is not particularly elongate. Their manual digits I–III and hindlimbs are also proportionally similar to those of rhamphorhynchids, and their feet are also gracile and slender. Like rhamphorhynchids, wukongopterid fifth toes possess a strongly bent second phalanx. The degree of curvature in this phalanx is quite variable, being relatively gentle in most taxa but sharply angled in *Wukongopterus*.

Locomotion

FLIGHT

Because wukongopterids are a relatively new discovery and have primarily generated attention for their evolutionary and taxonomic significance, virtually nothing has yet been deduced about their locomotory abilities. It has been suggested that *Darwinopterus* was an aerial predator, requiring it to be an agile, skilled flier to pursue its prey (Lü, Unwin, et al. 2010), but this idea does not stand up to scrutiny in my opinion (see below). A provisional assessment offered here notes that wukongopterid wings are relatively short and broad, a morphology well suited to strong, flapping flight and tight turning circles. Their long, robust, and curving pteroids suggest that the propatagium was an important component of wukongopterid flight. Controlling the shape of the forewing during flight can be an excellent aid to maneuverability as ventral deflection of the propatagia allows for use of the wings at higher angles of attack and generates more lift. Such traits would also enable high-angle launching. Taken together, this may mean that wukongopterids were well suited for flapping flight in densely vegetated settings, an idea corroborated somewhat by the diverse and abundant paleoflora in the Tiaojishan Formation and Stonesfield deposits. Of course, these ideas are very provisional, and we can say very little about their flight ability with certainty until detailed investigations into wukongopterid mechanics are performed.

ON THE GROUND

Wukongopterid forelimb:hindlimb ratios are not as skewed as those of some pterosaurs, and their limb bones are not especially slender or overly developed. As such, we can imagine them as relatively competent, if unspecialized terrestrial locomotors. If, as suggested above, their large pteroids developed through strong and frequent launching, they may not have moved very far on the ground anyway. Launches and short flights were probably not too energetically demanding for small animals like wukongopterids, and like many small song birds, they may have preferred to move from place to place using bursts of flight rather than walking, hopping, or running.

Paleoecology

REPRODUCTIVE BIOLOGY AND DIMORPHISM

Wukongopterids are the only pterosaur group in which we have identified a definitively female and mother individual: a complete skeleton of *Darwinopterus* associated with her egg preserved alongside her pelvis (fig. 2.8; Lü, Unwin, et al. 2011). This association reveals that *Darwinopterus* eggs had relatively low masses relative to their parents, a ratio most comparable to lizards when contrasted with modern egg layers. Along with their thin shells, this supports the supposition that pterosaur eggs were buried in moist environments and absorbed large quantities of water during their long incubation time (Lü, Unwin, et al. 2011, also see chapter 8). The skeleton of this female *Darwinopterus* is indistinguishable from those of other *Darwinopterus*, except for her total lack of a headcrest and particularly wide pelvic canal. Other *Darwinopterus* show the opposite configuration (narrow pelves and headcrests), but no specimens yet found show a hypothetical third combination of these characters (broad pelves with headcrests). Thus, it is assumed that these are sexually dimorphic characters and that, based on pelvis width, only males developed the cranial ornament (Lü, Unwin, et al. 2011). Thus, *Darwinopterus* joins *Pteranodon* (chapter 18) in strongly supporting the notion of pterosaur headcrests primarily being sexual display devices.

FORAGING STRATEGIES

Given that wukongopterids have provided an insight into macroevolutionary processes, filled a gap in pterosaur phylogeny, and present a very unique pterosaur bauplan, expectations may be high that their proposed foraging strategies will also be rather amazing. Fittingly, some have proposed that wukongopterids were pterosaur top guns, their newly evolved long necks and oversize heads being used to prey upon dinosaurs, other pterosaurs, and even gliding mammals in midair (Lü, Unwin, et al. 2010). Such acts would be rather remarkable because, with even a generous mass estimate, the biggest Tiaojishan wukongopterids would not weigh much over 300 g (extrapolating data from Witton 2008a), which is about the same as a modern feral pigeon.

At that size, tackling squirrel-sized mammals or crow-sized dinosaurs on the wing would be a feat earning praise from even the hardiest modern raptors, and wukongopterid skeletons would have to be brimming with offensive weaponry for this purpose. Vertebrate-hawking birds are renowned for their talons, incredibly strong feet, robust skulls, and powerful beaks (e.g., Hertel 1995; Fowler et al. 2009), while bats that subdue large vertebrates in flight are also armed with formidable teeth and powerful jaws (Ibáñez et al. 2001). These adaptations provide the means to immobilize their prey quickly and efficiently, and are obvious advantages for animals grappling with large prey while in flight. Vertebrate hawkers are also powerful fliers that can chase down their quarry and, once immobilized, carry the prey to a safe spot to eat. Pterosaurian equivalents would therefore require equally powerful flight musculature to permit the same tasks.

Unfortunately for the *Darwinopterus* raptor hypothesis, wukongopterids do not possess any of these requirements. None of their appendages bear the chunky digits and talons ideal for subduing large aerial prey items, and their long, comparatively delicate skulls and unimpressive teeth are ill suited to this task. Nor, for that matter, do they have the expanded shoulder regions indicative of the powerful flight muscles needed to chase and eventually carry their prey. With this in mind, raptorial pursuits look doubtful for wukongopterids. They appear far better adapted to the considerably more mundane practice of eating small invertebrates, with their long jaws enabling them to investigate and probe cavities for foodstuffs (Lü, Xu, et al. 2011). Their dentition, too, is ideally shaped for gripping small prey items. Noting the relatively robust teeth of *Darwinopterus robustodens*, Lü, Xu, et al. (2011) suggested that larger-toothed wukongopterids may favor harder-shelled beetles over softer prey, while the latter was preferred by more slender-toothed species. This would make them, with their apparent adaptations to flitting around Jurassic woodlands, somewhat comparable in their habits to modern songbirds, such as thrushes and bulbuls. While perhaps not as exciting as terrorizing airborne dinosaurs, mammals, and their fellow pterosaurs, this idea is probably more in keeping with their anatomy.

15
Istiodactylidae

PTEROSAURIA > MONOFENESTRATA > PTERODACTYLOIDEA > ORNITHOCHEIROIDEA > ISTIODACTYLIDAE

At some point in the Upper Jurassic, a lineage of wukongopterid-like pterosaurs ditched their long tails and fifth toes, thinned their bone walls, elongated their metacarpals, and became the first members of Pterodactyloidea. This major pterosaur lineage dominated pterosaur evolutionary history from the end of the Jurassic onward, replacing the preceding pterosaur lineages with new forms that existed until the close of the Cretaceous. It seems that pterodactyloids were generally larger animals and more taxomically and ecologically diverse than their predecessors, which may explain their superior fossil record. In the scheme being followed here, pterodactyloids are divided into 11 groups within two broad categories: Ornithocheiroidea[7] and Lophocratia. The two groups are separated by a suite of anatomical features pertaining to specializations for different lifestyles. Ornithocheiroids were generally highly adapted for flight, often to the detriment of their terrestrial capabilities, whereas lophocratians were adapted for diverse lifestyles in terrestrial settings. We will meet the lophocratians from chapter 19 onward but, for now, will focus on their aerially adept sister branch.

Ornithocheiroidea is a large group that includes the razor-muzzled istiodactylids, the spear-toothed ornithocheirids, bizarre-looking boreopterids and the famous, toothless pteranodontians (Unwin 2003; Lü, Unwin, et al. 2010). As we'll discover over the next few chapters, these groups are characterized by their extremely long, highly modified wings, oversize heads, and relatively tiny bodies and legs (Unwin 2003). Istiodactylids (fig. 15.1), the first representatives of this group we will meet, stand out from other ornithocheiroids by features of their skulls, including tightly packed, razor-edged teeth set into broad snouts. At times, they have been called "duck-billed pterosaurs" (Wellnhofer 1991a; Unwin 2005), a peculiar label considering there is really nothing "duck-like" about the jaws or likely habits of these animals at all. Istiodactylid remains are presently limited to Early Cretaceous deposits of the Northern Hemisphere (fig. 15.2), though a purported Late Cretaceous istiodactylid from Canada suggests that they may have existed a tremendously long time, from at least the Barremian to the Campanian (60 million years, from 130–70 Ma; Arbour and Currie 2010). However, there is a good chance that this find does not represent an istiodactylid (see below), reducing their known evolutionary range to 10–20 million years.

[7] This group is sometimes called "Pteranodontoidea" (Kellner 2003; Andres and Ji 2008; Wang et al. 2009), but it seems that "Ornithocheiroidea" was coined first and under rules of zoological nomenclature takes priority (Unwin 2003).

The Slow-Burning Research History of Istiodactylids

The first istiodactylid fossils were found in the lower Aptian (ca. 120 Ma) Wealden deposits on the Isle of Wight, a small island off the coast of southern England that is famous for its fossils of Early Cretaceous dinosaurs. They were first mentioned, albeit briefly, by Richard Lydekker (1888) under the now defunct name of "*Ornithochirus* [sic] *nobilis*" and were later discussed in more detail by the outspoken pterosaurologist Harry Seeley (1887, 1901). Seeley separated these remains into two species: *Ornithodesmus cluniculus* (named in 1887) and *Ornithodesmus latidens* (1901). The latter indicates that Seeley was somehow aware of the unusual dentition that characterizes istiodactylids (the specific name *latidens* means "broad tooth"), but how he was aware them remains unclear. It is assumed that he knew of skull material that is now lost, as all other istiodactylid skulls from these deposits were recovered after this time (C. Davies, pers. comm. 2012). One of these skulls, along with much of a skeleton, was found after

Fig. 15.1. Recycling, Mesozoic style. When Early Cretaceous dinosaurs died, istiodactylids may have been first on the scene to clean up the remains. Here, three *Istiodactylus latidens* from the British Wealden Group have wasted no time in devouring a recently deceased stegosaur.

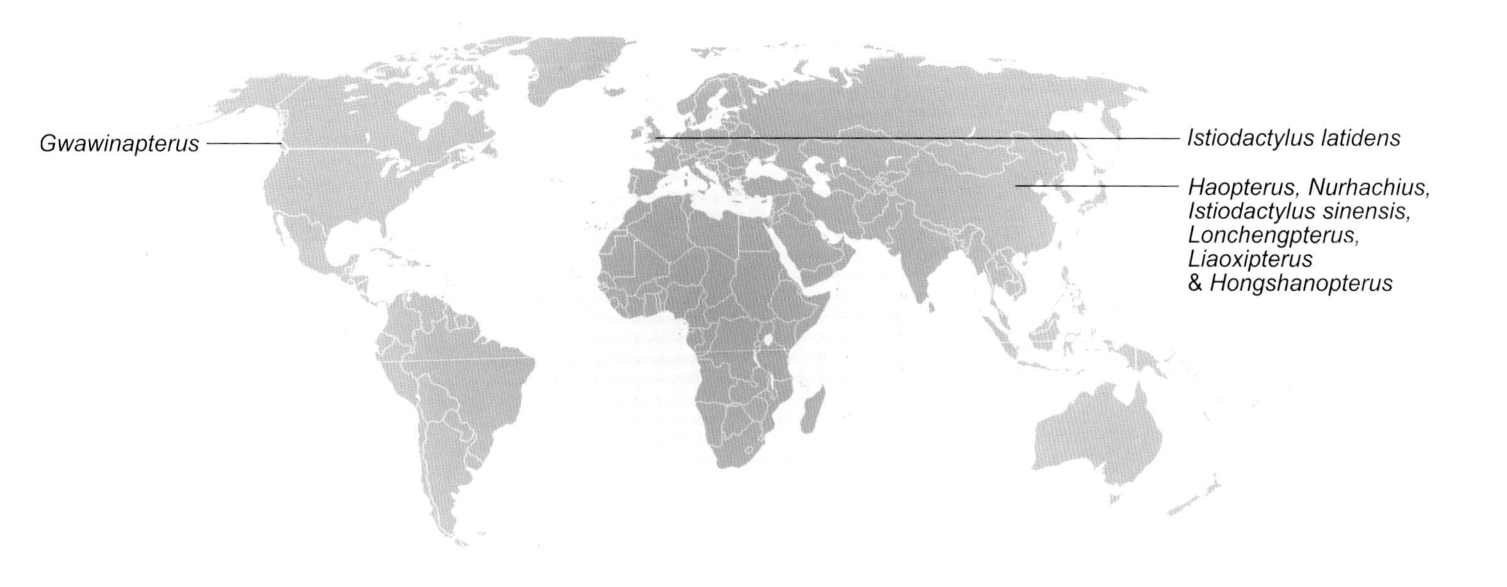

Fig. 15.2. Distribution of istiodactylid taxa.

a Wealden rock fall in 1904. This specimen formed the subject of an extensive monograph by the eminent amateur paleontologist Reginald Walter Hooley (1913), and over one hundred years later, it remains the only substantial, three-dimensional remains of an istiodactylid from anywhere in the world (fig. 15.3A).

Not much was said of *Ornithodesmus* after Hooley's monograph, and it remained the only pterosaur of its kind known for many decades. Indeed, during the rest of the twentieth century only one other pterosaur fossil was favorably compared to it—a very small mandibular fragment from the Upper Jurassic Morrison Formation of Wyoming (Kimmeridgian-Tithonian; 156–145 Ma; Bakker 1998). However, this specimen is unlike other istiodactylids in bearing strongly recurved teeth, and probably has affinities to another pterosaur group.

It wasn't until the twenty-first century that developments in istiodactylid research occurred, and the first was a strictly nomenclatural one. Howse and Milner (1993) realized that Seeley's *Ornithodesmus cluniculus*, only known from a fragmentary sacrum, actually represented a predatory dinosaur. This obviously required separation of the dinosaur and pterosaur material held under the *Ornithodesmus* name. Since *O. cluniculus* had been named first, the pterosaurian *latidens* material was referred to a new genus, *Istiodactylus* (Howse et al. 2001). *Istiodactylus latidens* remained the only istiodactylid known for a few more years until, suddenly, new istiodactylid species began pouring out of Cretaceous deposits of China at a high rate. With retrospect, we may consider the first of these to be *Haopterus gracilis* from the Yixian Formation (likely Barremian; 130–125 Ma) (Wang and Lü 2001), a form represented by a partial skeleton initially thought to represent an animal like *Pterodactylus* (chapter 19). Subsequent analyses have lumped *Haopterus* closer to or within Istiodactylidae (e.g. Lü, Ji, Yuan, and Ji 2006; Lü, Ju, and Ji 2008; Andres and Ji 2008; Lü, Unwin, et al. 2010), suggesting it may represent a relatively early grade of istiodactylid evolution (but see Witton 2012 for a contrary opinion).

Shortly after, the Jiufotang Formation (probably Aptian; 125–112 Ma) began to reveal istiodactylids as fast as they could be described (Andres and Ji 2003). The first Jiufotang form to be named was *Liaoxipterus brachyognathus*, an animal only known from a lower jaw and initially identified as a ctenochasmatoid (Dong and Lü 2005; see chapter 19 for more on ctenochasmatoid pterosaurs). It was quickly reassigned to Istiodactylidae by a host of authors (e.g., Lü, Ji, Yuan, and Ji 2006; Lü, Ju, and Ji 2008; Wang et al. 2007). Only months later, a far more substantial istiodactylid specimen from Jiufotang sediments was named *Nurhachius ignaciobritoi*, a species that remains one of the

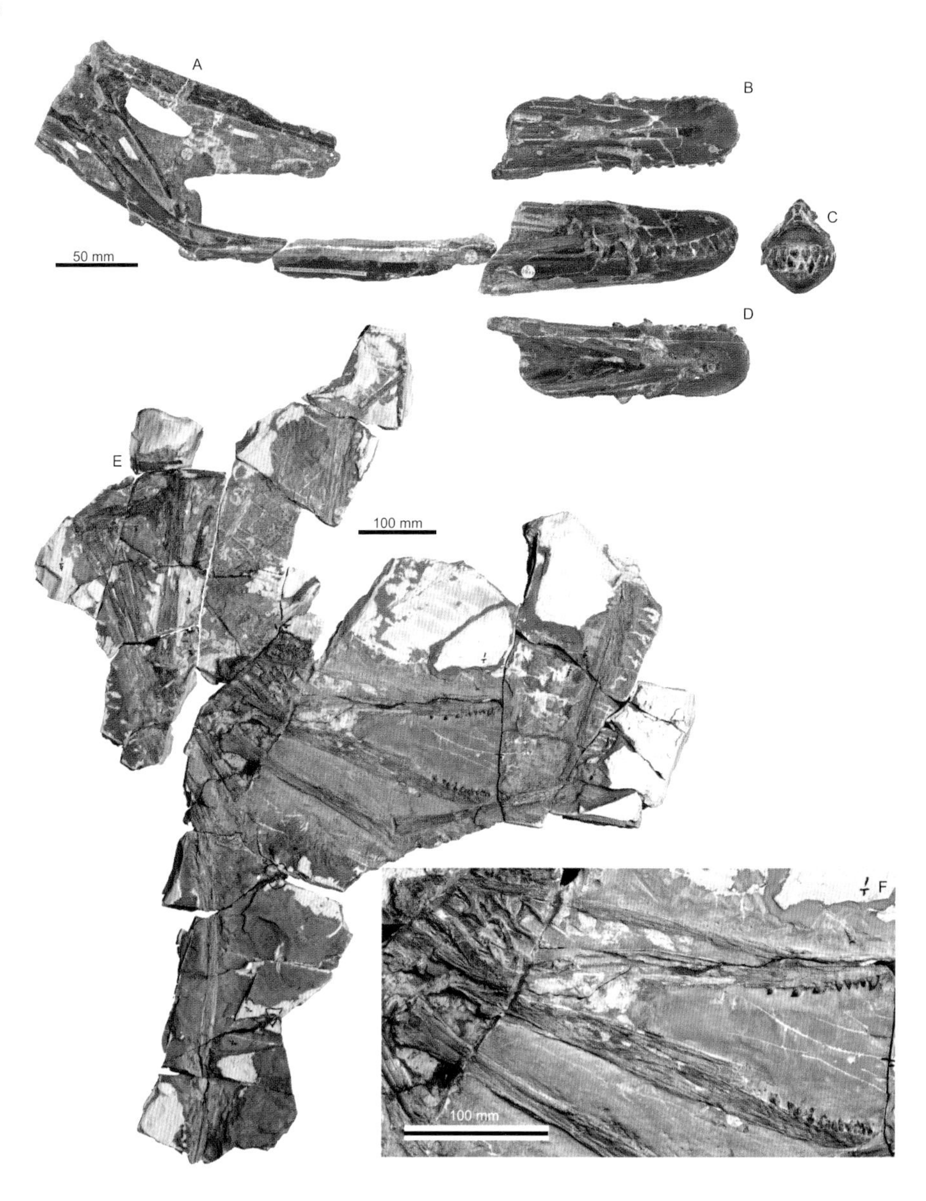

FIG. 15.3. *Istiodactylus* fossils. A–D, cranial remains of *Istiodactylus latidens* from the Barremian Wessex Formation of England. A, right lateral view; B–D depict the rostrum and mandibular symphysis in dorsal (B), anterior (C), and ventral (D) views; E, incomplete skeleton of *Istiodactylus sinensis* from the Aptian Jiufotang Formation, China; F, detail of the *I. sinensis* skull. Photographs A–D, copyright Natural History Museum, London; E–F courtesy of Brian Andres.

most completely known Chinese istiodactylids (Wang et al. 2005). The third Jiufotang istiodactylid species was described soon after this, and was so similar to the Wealden *Istiodactylus* that its describers classed it as a new species of the genus: *Istiodactylus sinensis* (fig. 15.3E; Andres and Ji 2006). One month later, a fourth taxon from these deposits, *Longchengpterus zhoai*, was described (fig. 15.4; Wang et al. 2006). Shortly after, a fifth was announced: *Hongshanopterus lacustris* (Wang, Campos, et al. 2008). The last is perhaps the most distinctive Jiufotang istiodactylid yet described, bearing an unusual dental configuration that may represent a relatively early grade of istiodactylid evolution (Wang, Campos, et al. 2008). Witton (2012), by contrast, suggests that *Hongshanopterus* lacks the features necessary for inclusion into Istiodactylidae, instead considering it an indeterminate ornithocheiroid.

The distinctive nature of istiodactylid teeth has enabled their identification in other parts of the world in recent years. This includes occurrences in Lower Cretaceous deposits of Galve, Spain (Sánchez-Hernández et al. 2007) and in other parts of the British Wealden (Sweetman and Martill 2010), indicating the presence of European istiodactylids as early as the Barremian (130–125 Ma). Even more excitingly, a mysterious jaw fragment from late Campanian (ca. 75 Ma) deposits of British Columbia has been suggested to represent the first occurrence of istiodactylids in the New World (Ar-

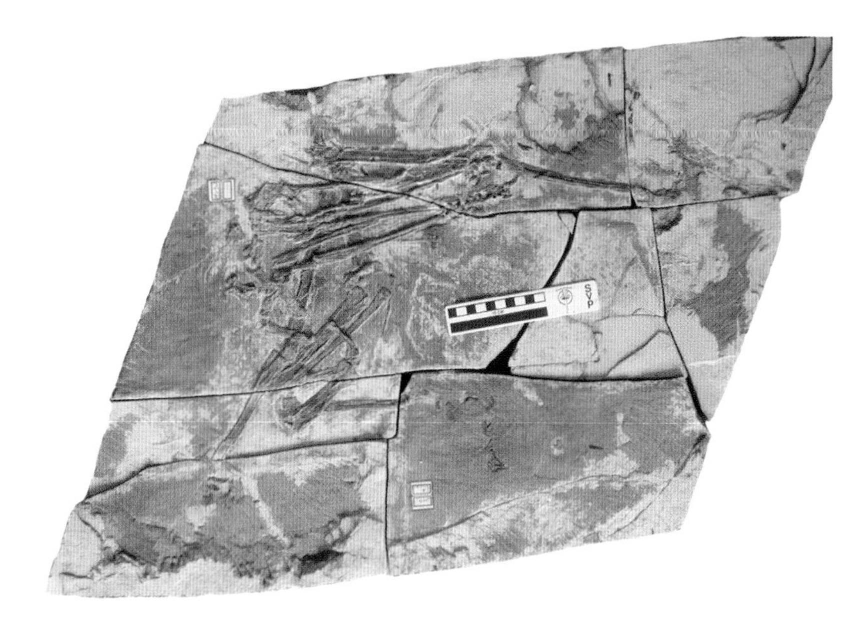

Fig. 15.4. The incomplete skeleton of *Longchengpterus zhoai*, from the Aptian Jiufotang Formation, China. Image courtesy of Lü Junchang.

bour and Currie 2010). This specimen, named *Gwawinapterus beardi,* is not a pretty fossil and comprises a small fragment of rostrum with well-preserved teeth. These are proportionally smaller than those of istiodactylids and bear longer roots, but are otherwise of similar tooth crown morphology. *Gwawinapterus* also bears a very high tooth count for an istiodactylid, packing over 25 tooth sockets into the preserved region of its rostrum compared to less than 15 in most istiodactylid species. If its identity as an istiodactylid is correct, *Gwawinapterus* is pretty durned exciting; not only is it an unusual istiodactylid, but it represents the only toothed pterosaur known from the uppermost Cretaceous and extends the known range of istiodactylids by 40 million years. Unfortunately, neither istiodactylid nor pterosaurian affinities for this specimen are incontrovertible. Its replacement teeth grow directly beneath its existing ones, whereas the replacement teeth of pterosaurs always sprout *behind* their older ones (see chapter 4). The tooth roots are also far longer than those of other pterosaurs, and the tooth crowns lack the "spalling" wear (that is, where the tooth apex is worn rounded) seen on other istiodactylid teeth. Thus, *Gwawinapterus* does not make a particularly strong case for the existence of istiodactylids or even toothed pterosaurs in the Upper Cretaceous (Witton 2012). Support for this idea was published recently, with strong similarity noted between the remains of *Gwawinapterus* and the jaws and teeth of a spear-jawed saurodontid fish (Vullo, Buffetaut, et al. 2012).

It seems quite likely that stormy clouds also lie ahead for other aspects of istiodactylid taxonomy. The sudden discovery of so many new species in China, most of which are extremely similar morphologically, has raised the eyebrows of a number of researchers. Lü, Ju, and Ji (2008) addressed this head on, suggesting that *Longchengpterus* is actually a second specimen of *Nurhachius*. Witton (2012) noted that this claim was strongly contradicted by Lü, Ju, and Ji's own data, but also suggested that the Jiufotang forms *Liaoxipterus* and *Istiodactylus sinensis* may be the same species. Additionally, the "rediscovery" of an overlooked skull piece of *Istiodactylus latidens* (which had apparently been neglected and pushed to the back of a specimen drawer for one hundred years) indicates that it and *I. sinensis* differ in a number of ways, suggesting they are unlikely to represent the same genus. Clearly, something of a reappraisal of istiodactylid taxonomy is needed.

Anatomy

At first glance, istiodactylid anatomy (best documented by Hooley 1913; Howse et al. 2001; Andres and Ji 2006) is a little drab compared to many ornithocheiroids (fig. 15.5). They do not have enormous headcrests, nor do their jaws brim with menacing, tusk-like teeth. Their necks and limbs are of fairly typical proportions for ornithocheiroids and in terms of overall wingspan, they are fairly middling for pterodactyloids. *Istiodactylus latidens* was the largest with a wingspan of 4.3 m, but *Nurhachius* and *I. sinensis* only managed spans of 2.4 m and 2.7 m, respectively (Wang, Campos, et al. 2008; Andres and Ji 2008).

This mediocre appearance is only superficial, however. Closer inspection reveals a unique skull construction with very unusual dentition. Derived istiodactylids have short, laterally compressed and triangular teeth restricted to the anterior third of their jaws. Each side of the jaws only possesses 12–15 teeth, limiting overall tooth counts to only 48–60 teeth in total. Uniquely for pterodactyloids, their teeth have virtually no space between them and neatly slot between each other to form a continuous cutting edge when the jaws were closed. This dental rosette is completed by a triangular process of bone rising up from the tip of the lower jaw to nestle between the two anteriormost upper teeth (fig. 15.3C). Derived istiodactylids enhanced the shearing ability of their dental array by developing carinae, or blade-like cutting edges, along the sides of each tooth (Hooley 1913; Andres and Ji 2006). The possible basal istiodactylids *Haopterus* and *Hongshanopterus* show only partial development of this unusual dentition, however, bearing larger numbers of relatively long, recurved, and widely spaced teeth typical of other early monofenestratans (Wang and Lü 2001; Wang, Campos, et al. 2008).

Istiodactylids have rather blunt jaw tips, and in dorsal or ventral view they assume a rounded, muzzle-like appearance (Hooley 1913; Howse et al. 2001). Their rostra, when closed, have a circular cross section that contrasts strongly with the more angular jaw profiles of most pterosaurs. Both their rostra and mandibular symphyses are rather short at only 30 percent of their jaw lengths but, by contrast, the retroarticular processes are unusually long. Their nasoantorbital fenestrae are enormous—some of the largest of any pterosaurs—and in *Istiodactylus latidens*, they are bordered ventrally by an extremely slender bar of bone (the maxilla), which is only 6 mm tall in a skull approximately 450 mm long (Witton 2012). The posterior skull regions of derived istiodactylids are reclined and lengthened, with that of *I. latidens* also being rather tall. Istiodactylid orbits are also stretched and reclined, and are almost split into two distinct regions by bars of bone situated toward the top of the eye socket in some species (presumably, these occur immediately beneath where the eyeball was situated). The skull of *I. latidens* is wider than most other istiodactylids, spanning 30 percent of the jaw length across the jaw joints. This ratio is only surpassed by the short-faced tapejarids (chapter 22) and anurognathids (chapter 11) (Witton 2012).

Beyond the skull, the istiodactylid skeleton is rather similar to those of other ornithocheiroids, although its "ornithocheiroid" features are somewhat less developed by comparison. Istiodactylids possess a fully pneumatized vertebral column, forelimb, and trunk skeleton (Claessens et al. 2009), and their cervical vertebrae bear tall neural spines and broad articulator faces. They also possess a stout notarium with a broad supraneural plate comprised of the anteriormost six dorsal vertebrae (Hooley 1913; Wang et al. 2005; Andres and Ji 2006). Their supraneural plates are tallest at the midlength where the scapulae articulate with oval depressions in each side. The rest of the istiodactylid vertebral column is poorly known, but *I. latidens* preserves at least another six dorsals and four sacrals. Their tails remain entirely unknown.

A complete set of limbs remains elusive in all known istiodactylid fossils, but many specimens preserve enough of their limb skeletons to provide a basic outline of their proportions and form. Their scapulae are proportionally short and robust, and would have articulated with the notarium at a perpendicular angle. This shoulder configuration seems limited to ornithocheiroids, because other pterosaur scapulae appear to lie diagonally across their backs. In contrast to their short scapulae, their coracoids are relatively long, which would position the shoulder high in the body (Frey, Buchy, and Martill 2003). As with all pterodactyloids, the glenoid is spread across the fused ends of the scapula and coracoid. Istiodactylid sterna are not well known, but seem relatively robust in *Haopterus* and *Nurhachius* (Wang and Lü 2001; Wang et al. 2005) and deep in *I. latidens* (Hooley 1913). Their forelimbs are substantially sized and, including the length of the wing finger, are predicted to stretch 4.5 times further than their legs (Witton 2008a). Their humeri are stout and blocky with a "warped" deltopectoral crest, an ornithocheiroid characteristic that sees the distal end of this process wrapping toward the humeral shaft. Their forearms are relatively elongate and, like virtually all ornithocheiroids, bear proportionally slender radii and widened carpals. As with all pterodactyloids, their metacarpals are proportionally elongate, but two of the smaller metacarpals (the identity of which is uncertain) do not reach the carpal block (Wang et al. 2005; Andres and Ji 2006). The wing fingers of istiodactylids, as typical for ornithocheiroids, are elongate, and from what we can ascertain of their incomplete remains probably occupied over 50 percent of the wing skeleton. Their other fingers, by contrast, were proportionally small.

Istiodactylid hindlimbs are short and slender compared to their oversize forelimbs. This trait is exaggerated even further in other ornithocheiroid species. Istiodactylid pelves are not well known but the relatively large size of the proximal muscle attachment sites on the femur of *Istiodactylus latidens* suggest they could be more substantially developed than those of other ornithocheiroids (Hooley 1913). Their femur and tibiotarsus are subequal in length (Wang et al. 2005), but the proportions of their fibulae aren't known. Another or-

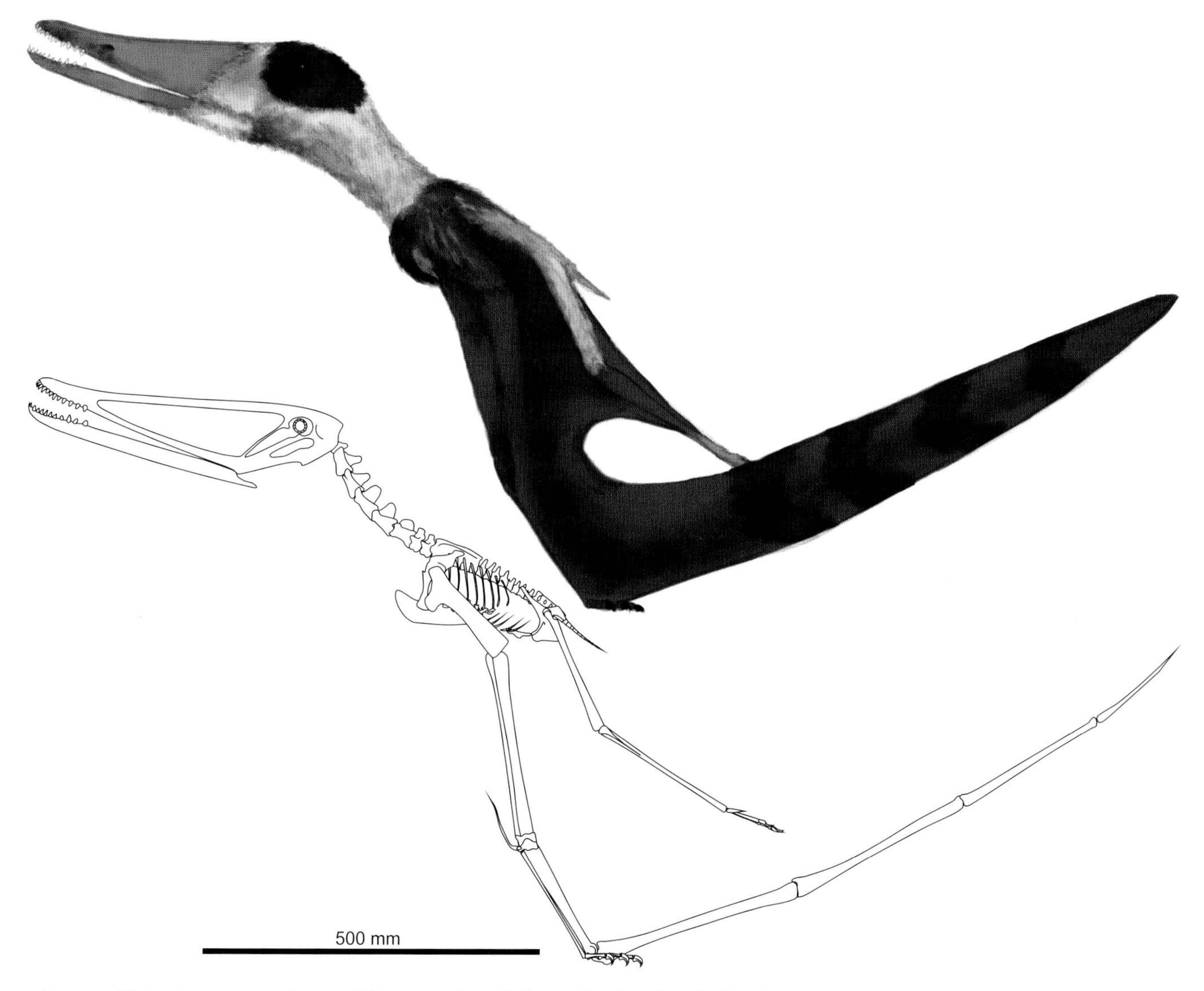

Fig. 15.5. Skeletal reconstruction and life restoration of a launching *Istiodactylus latidens.*

nithocheiroid characteristic is seen in the istiodactylid femoral head, which extends almost straight from the femoral shaft. Both *Haopterus* and *Nurhachius* demonstrate that istiodactylid feet were extremely small, their entire length being comparable to the length of their diminutive third finger.

Locomotion

FLIGHT

While the flight kinematics of some ornithocheiroid groups have been analyzed on multiple occasions (see chapters 16 and 18), istiodactylid flight remains largely uninvestigated. Forelimb elements of *I. latidens* were once used to model pterosaur wing joint mechanics (Bramwell and Whitfield 1974), and more recently, the mass, wing area, and basic flight style of *Nurhachius* were modeled (Witton 2008a), but little else has been specifically said of their flight. Like most ornithocheiroids, the membranes of istiodactylids were coupled to shortened bodies and legs and very long forelimbs, presumably creating relatively large, high-aspect wings with low wing loadings (Witton 2008a). This condition is comparable to the wings of modern shorebirds and is well suited to soaring flight. Further evidence of strong flight capabilities is seen in their well-developed shoulder, chest, and upper arm skeleton (Hooley 1913). With such developed flight anatomy, it seems likely that istiodactylids spent much of their time in the air (fig. 9.1), though it is of in-

FIG. 15.6. The evolution of wing shapes, perhaps toward increasing flight efficiency, within Ornithocheiroidea.

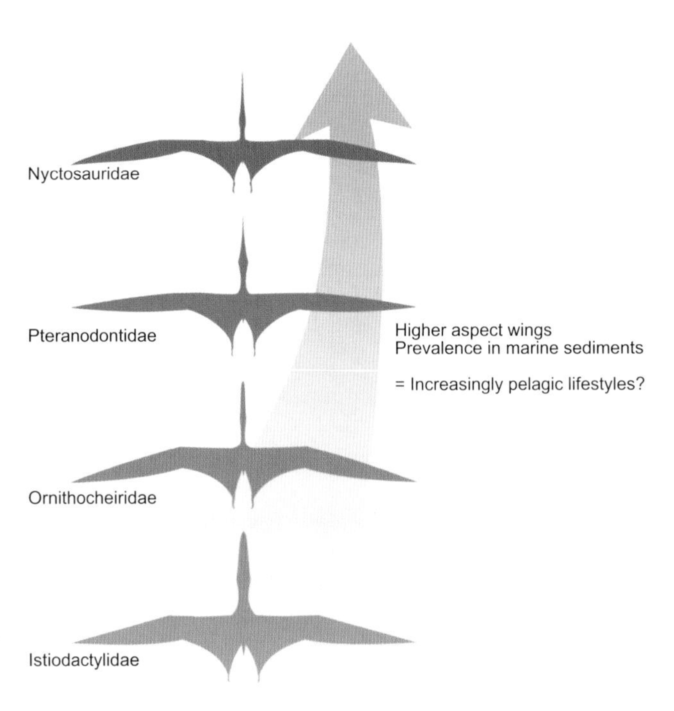

terest that their wings seem somewhat shorter than those of most other ornithocheiroids (fig. 15.6). The latter tend to have slightly higher aspect wings and may represent forms supremely adapted for oceanic soaring (see chapters 16 and 18), while the somewhat greater depth and brevity of istiodactylid wings would permit more efficient takeoff and landing to the detriment of their soaring ability. A similar relationship is seen in the wing shapes of modern soaring birds: those that soar inland have shorter and deeper wings than those that soar oceanically (Pennycuick 1971). This may reflect a preference for terrestrial habitats in istiodactylids, which may explain why their fossil remains have consistently been found in freshwater and lagoonal settings. This is consistent with istiodactylids being unusual among ornithocheiroids in lacking some adaptations for launching from deep water (Witton 2012; also see Habib and Cunningham 2010; and chapter 16).

ON THE GROUND

It seems unlikely that any ornithocheiroids, including istiodactylids, were particularly proficient terrestrial animals. Their disproportionate limbs and tiny appendages look ill suited for walking or running, and even though istiodactylid hindlimbs are more proportionate to their forelimbs than those of other ornithocheiroids, they remain offset enough to have probably hindered terrestrial locomotion. The femur of *I. latidens* has a more prominently developed greater trochanter than those of ornithocheirids or pteranodontians (compare those figured in Hooley 1913 with those in Williston 1903; Kellner and Tomida 2000; Bennett 2001; Veldmeijer 2003). This may indicate that relatively large muscles attached to the femur in istiodactylids compared to other ornithocheiroids. Nevertheless, they were still probably proportionally smaller than those of most pterodactyloids.

Some have suggested that istiodactylids could hang from their hindlimbs in a bat-like fashion (Hooley 1913; Wang and Lü 2001). For reasons discussed in chapter 7, this was almost certainly not the case for any pterosaur, but it would be particularly problematic for istiodactylids with their miniscule feet. Indeed, their appendages were probably too small relative to their body size for use in any sort of climbing or suspensory activities.

Paleoecology

A handful of authors (Hooley 1913; Wellnhofer 1991a; Wang and Lü 2001; Wang and Zhou 2006a) have suggested that fish formed the primary diet of all istiodactylids. This is perhaps likely as part of a generalized diet for purported basal forms with their pointy, well-spaced dentitions, but there seems little reason to link the wide, shearing muzzles of derived istiodactylids to the role of grabbing fish; the spear-like teeth of rhamphorhynchines (chapter 14) or ornithocheirids (chapter 16) are far more likely fish grabbers. Likewise, the duck-like filter-feeding lifestyle for istiodactylids proposed by Fastnacht (2005a) seems unlikely: Why would an animal that dabbles food from water have interlocking, sharpened teeth (Witton 2012)?

Howse et al. (2001) proposed the more plausible notion that the razor muzzles of derived istiodactylids were the foraging tools of vulture-like scavengers. A shearing, interlocking dental rosette is of obvious use to anything trying to remove chunks of meat from a carcass, and it may be no coincidence that the teeth of istiodactylids closely resemble those of cookiecutter sharks, specialized pseudoparasitic fish that remove chunks of flesh from the flanks of much larger marine animals. Witton (2012) built on the Howse et al. (2001) hypothesis and noted that, like modern avian scavengers, istiodactylid skulls are a mosaic of

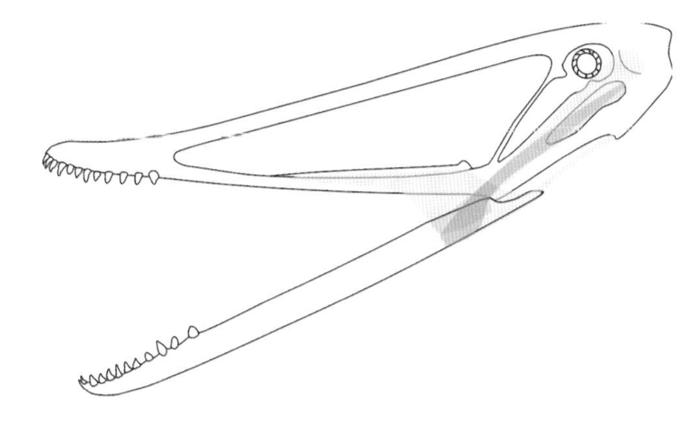

FIG. 15.7. Tentative jaw muscle reconstruction of *Istiodactylus latidens*. Note the relatively great posterior extension of *m. pterygoideus posterior* relative to that of *Quetzalcoatlus*, with its more typical jaw myology depicted in figure 5.6.

mechanically strong and weak features. Scavengers need strong enough skulls to remove flesh from cadavers, but can economize on other aspects of their skull construction because their preferred foodstuffs are immobile and unlikely to generate unpredictable stresses during feeding. Thus, they do not need to reinforce their skull as completely as predatory animals, which have to ensure their skulls and jaws are strong enough to withstand fighting and struggling prey items (Hertel 1995).

Some aspects of cranial strengthening in istiodactylid skulls are reflected in their expanded jaw muscles (fig. 15.7). The long istiodactylid retroarticular processes may have anchored a relatively large *m. pterygoideus posterior*, the same muscle that provides crocodiles their extremely strong bites and characteristically bulging posterior jaws (chapter 5). Strong bites would enable those shearing teeth to slice through viscera and muscle tissue, though it seems unlikely that istiodactylid jaws and teeth were strong enough to bite through bone. The tall and wide posterior skull, circular rostral cross sections, and the narrowing and partial closure of their orbits are other indications of cranial reinforcement and resistance to feeding forces (some aspects of istiodactylid orbit reinforcement mirror those of strong-jawed predatory dinosaur orbits; see Henderson 2002). Their expanded posterior skull is likely to have anchored large neck musculature, which would lend itself to twisting and pulling the head during feeding.

Many of these attributes are consistent with the skull morphology of modern avian scavengers, which also have large attachment sites for neck muscles and reinforcement against feeding-induced stresses (e.g., Spoor and Badoux 1986; Hertel 1995), as is another, perhaps unexpected trait: rather weak cheek bones. *Istiodactylus latidens* and vultures share proportionally slender maxillary bones, with additional thin bones comprising their posterior skull regions. Other derived istiodactylids share similar lightly built posterior skull constructions. These slender bones would be very prone to injury if handled roughly, suggesting that some delicacy in foraging was required. This contrasts, of course, with the aspects of istiodactylid skulls that are more strongly built and seem indicative of powerful biting. Put together, these traits perhaps suggest that istiodactylids fed on large prey that required powerful jaws to process, but was inert enough to allow complete control of the strain on their jaws and skull during feeding. Animal carcasses are probable sources of large but inert prey items, indicating that derived istiodactylids may have habitually scavenged for food.

Several aspects of istiodactylid anatomy are consistent with a scavenging hypothesis. Their lack of killing adaptations—large teeth or talons— implies that they did not have any effective means of immobilizing large prey, despite their apparent abilities to eat large animals. Their ability to soar efficiently is of obvious use to a scavenger, as it would enable them to effectively find and travel to carcasses much more rapidly than ground-based animals. Istiodactylid orbits are also a lot smaller than those of likely predatory pterosaurs (such as those of ornithocheirids; see chapter 16), implying a lessened requirement for the high visual acuity necessary for efficient predation, just as in modern scavenging birds (Hertel 1995). As with modern scavengers, it seems likely that istiodactylids would have to step back from a carcass should a larger, more powerful carnivore also take an interest in their quarry, but once those animals had their fill, istiodactylids would almost certainly be there to finish the remains.

16
Ornithocheiridae

PTEROSAURIA > MONOFENESTRATA > PTERODACTYLOIDEA > ORNITHOCHEIROIDEA > ORNITHOCHEIRIDAE

It seems that more pages of pterosaur literature are devoted to ornithocheirids (fig. 16.1) than any other group, a reflection of the abundance and quality of their fossil remains. The fossil record of these primarily marine, snaggle-toothed animals is expansive with 55 million years of evolution through the Cretaceous (specifically, Valanginian to Cenomanian; 140–93 Ma) and occurrences on every continent except Antarctica (fig. 16.2; Barrett et al. 2008). Much of the ornithocheirid record consists of isolated teeth and fragmentary bones, and in some localities, such remains number in the hundreds or thousands. Other ornithocheirid fossils are among the highest quality pterosaur fossils ever found. They are, without question, one of the best known pterosaur groups, and their anatomy has been documented in great detail in the last few decades (e.g., Campos and Kellner 1985; Wellnhofer 1985, 1987a, 1991b; Kellner and Tomida 2000; Fastnacht 2001; Veldmeijer 2003; and others), as has their functional morphology (e.g., Wellnhofer 1988, 1991b; Witmer et al. 2003; Frey, Buchy, and Martill 2003; Bennett 2003a; Veldmeijer et al. 2006; Wilkinson 2008; Habib and Cunningham 2010).

Among pterosaur workers, ornithocheirids are infamous for their enormously controversial and confused taxonomy. Although there is general agreement that these animals are related to the likes of istiodactylids and pteranodontians (e.g., Unwin 2003; Kellner 2003; Andres and Ji 2008), there is virtually no consensus over the exact content and interrelationships of the group. Literally dozens of ornithocheirid species—most of them suspect—have been named since their remains were first found in the 1840s. Thanks to convoluted Victorian nomenclatural schemes and differing attitudes on the taxonomic utility of some historically important but scrappy ornithocheirid fossils, it seems that anyone who has ever studied this group has a different opinion on its taxonomic composition (compare Hooley 1914; Kuhn 1967; Wellnhofer 1978, 1991a; Kellner and Tomida 2000; Fastnacht 2001; Unwin 2001, 2003; Kellner 2003; Unwin and Martill 2007; Rodrigues and Kellner 2008; and Andres and Ji 2008 for contrasting ornithocheirid taxonomies). Even the name of the group is debated. Ornithocheiridae is sometimes referred to as "Anhangueridae," a term coined by Campos and Kellner in 1985, but Unwin (2001, 2003) has argued that Seeley's (1870) term "Ornithocheiridae" refers to an identical group and should have nomenclatural priority. At a conservative estimate, there may be 12–16 valid ornithocheirid genera and between 20–30 species, but as we'll see in the following glance at their research history, plenty would disagree with these figures. In the interests of brevity and readability, we will not discuss the naming of each ornithocheirid species in the following review, and will deal with genera only. Interested readers should seek the publications listed above for a more detailed overview. And probably a headache.

Small Fossils and Big Problems

The root cause of the confusion in ornithocheirid taxonomy can be traced to the earliest discovery of their fossils. Fragmentary pterosaur material recovered in the 1840s from the upper Cretaceous Chalk of southern England (Cenomanian to Turonian; 100–90 Ma) may be the first records of ornithocheirids, but definitively identified remains of this group were not reported until the 1850s (Bowerbank 1851). This material, also from the British Chalk, comprised a fairly large partial rostrum named *Pterodactylus cuvieri*. Shortly after, another British deposit, the upper Albian (ca. 100 Ma) Cambridge Greensand, also began yielding ornithocheirids, but in much greater quantities. Over 2000 pterosaur bones are known from this deposit, 90 percent of which represent ornithocheirids. Alas, they are of notoriously poor quality (Unwin 2001). Though three dimensional, Cambridge Greensand pterosaur fossils are fragmentary, worn, disasso-

Fig. 16.1. *Anhanguera*, a pterosaur highly adapted for flight in marine settings, faces a turbulent Cretaceous sea over what will become southern Britain.

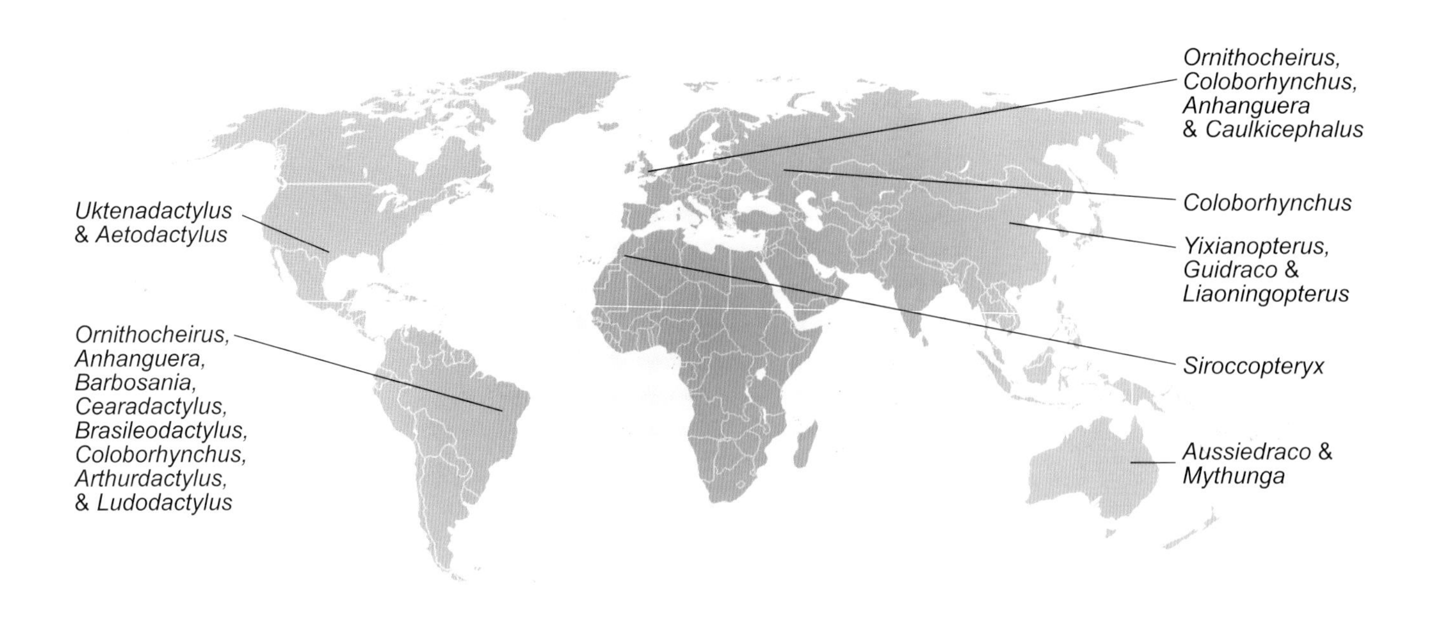

FIG. 16.2. Distribution of ornithocheirid taxa.

ciated, and devoid of any delicate features (fig. 16.3C–D, also fig. 2.3). These fossils represent the remains of pterosaurs that had died or drifted far out to sea, were bored into and encrusted by invertebrates, buried, and phosphatized before being exhumed later, as fossils, in ancient storms and then buried again. It was not until the early Cenomanian (ca. 98 Ma) that these fossils were finally laid to rest and, as may be expected, this turbulent history took its toll on their already fragile skeletons (for an excellent overview of this pterosaur assemblage and its depositional context, see Unwin [2001]).

Nowadays, it is likely that little taxonomic importance would be placed on Cambridge Greensand pterosaurs because of their rather awful condition, but Richard Owen, Harry Seeley, and Reginald Walter Hooley—aforementioned giants of early pterosaurology—strove to understand these fragments throughout the late nineteenth and early twentieth centuries. In the process, they laid foundations for our modern understanding of Ornithocheiridae but also created a taxonomic nightmare, naming dozens of species based on woefully scrappy bits of pterosaur bone. Modern pterosaur workers have tried to integrate this taxonomy into our new understanding of ornithocheirid anatomy and diversity, but opinions differ as to how much utility and significance the Cambridge Greensand material has in our modern concept of the group. This has become such a problem that some modern workers more or less ignore the Cambridge Greensand material altogether, and have established their own schemes based on more substantial finds. Others—particularly pterosaur guru David Unwin—have argued that the Cambridge Greensand material is still taxonomically relevant and useful to modern pterosaur workers. Unwin, Lü, and Bakhurina (2000) and Unwin (2001) argued that, of the dozens of species named from the Cambridge Greensand, only three genera and six species could be recognized, including *Ornithocheirus* (fig. 16.3C–D; two species), *Anhanguera* (fig. 16.3E; including Bowerbank's "*P.*" *cuvieri* and one other species), and *Coloborhynchus* (two species). Each of these genera can be recognized from their jaw tips alone. Another genus of Cambridge Greensand ornithocheirid, "*Criorhynchus*," was resurrected from the taxonomic bin by Kuhn (1967) and Wellnhofer (1978, 1991a) and continues to crop up to this day (Fastnacht 2001; Veldmeijer 2003), despite the eight "*Criorhynchus*" species being referred to other ornithocheirid genera in other schemes (Unwin 2001).

With so little to go on, it is unsurprising that some early restorations of ornithocheirids were a bit zany. Those of G. von Arthaber (1922) are especially

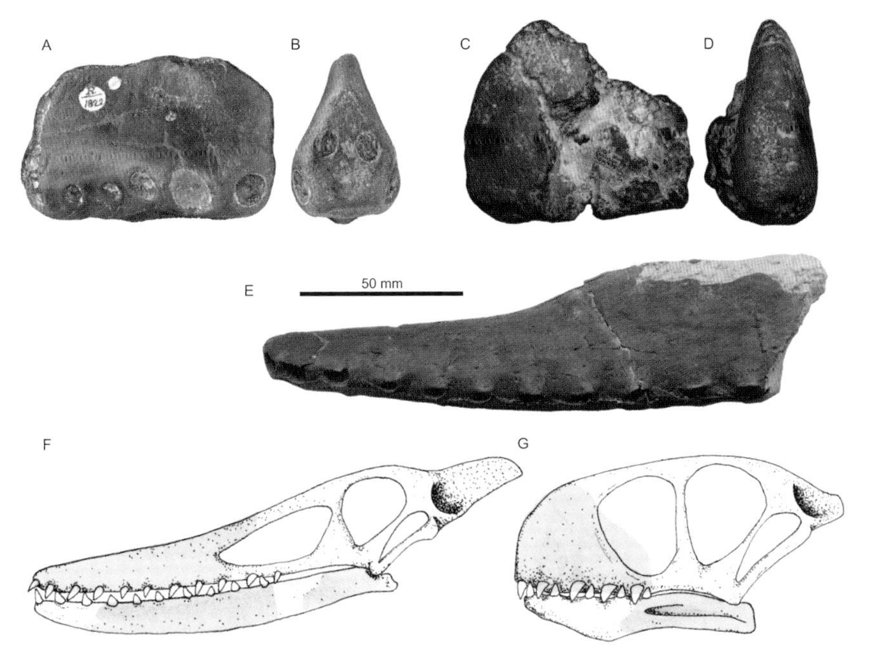

Fig. 16.3. Ornithocheirid jaw tips from Britain, the only skull remains of this group known until the latter decades of the twentieth century. A–B, *Colobo-rhynchus clavirostris* of the Berriasian-Valanginian Hastings Group, in right lateral and anterior view, respectively; C–D, the Albian Cambridge Greensand form *Ornithocheirus simus* in left lateral and anterior view; E, the Turonian Chalk Group form *Anhanguera cuvieri*, left lateral view; F Arthaber's 1922 reconstruction of "*Ornithocheirus*" (= *Anhanguera*) *cuvieri*; G, Von Arthaber's 1922 reconstruction of *Ornithocheirus simus*. Grey shading in F and G indicates remains known to Arthaber when reconstructing the skulls of these pterosaurs. Photographs A–D, courtesy of Andre Veldmeijer and Erno Endenburg; E, courtesy of Bob Loveridge; A–B and E, copyright Natural History Museum, London; F, redrawn from Arthaber (1922).

fun as they portray various ornithocheirids with tub mouths and snub noses (fig. 16.3F–G), conditions we now know as very different to their actual appearance. Their true skull shape was mysterious until the 1980s, when several complete skulls were unearthed in what were then newly opened pterosaur-bearing rocks of Brazil (fig. 16.4; Leonardi and Borgomanero 1985; Campos and Kellner 1985; Wellnhofer 1987a). These showed ornithocheirids in their true guise, as long-snouted animals with robust, fang-like teeth and large, rounded crests on both jaw tips. These skulls were recovered from the Lower Cretaceous Santana Formation of Brazil (probably Aptian-Albian; 125–100 Ma; see Martill 2007), an important pterosaur fossil Lagerstätte that continues to provide abundant and exceptionally preserved ornithocheirid remains. Detailed study of these skulls, and the extensive skeletons often found with them, has given a greater understanding of ornithocheirid anatomy than achieved for the majority of other pterodactyloid groups (e.g., Wellnhofer 1991b; Kellner and Tomida 2000; Veldmeijer 2003; Veldmeijer et al. 2009).

Unfortunately, nomenclatural issues also dog the Santana Formation ornithocheirids. The first Santana ornithocheirid genus, "*Araripesaurus*," was based on limb bones that do not bear sufficient diagnostic features to distinguish it from other ornithocheirids (Price 1971), and is now considered an "invalid" name (Unwin 2003). Similarly, Wellnhofer's (1985) "*Santanadactylus*" is now of doubtful validity according to most authors (but see Kellner and Tomida [2000] for a contrary view). *Brasileodactylus*, another of the earliest named genera (based on a fragmentary mandible; Kellner 1984), is thought to be so closely related to *Anhanguera* (fig. 16.4B; Campos and Kellner 1985) that Unwin (2000) raised concerns that they may represent the same animal. *Anhanguera* is otherwise generally accepted as valid, though the number of species it contains is highly contentious. The Cambridge Greensand taxon *Coloborhynchus* has also been reported from the Santana Formation (Fastnacht 2001) and many species have been plundered from *Anhanguera* to sit in this genus. "*Tropeognathus*," a genus based on a large, complete skull (fig. 16.4A; Wellnhofer 1987a), has been referred to another taxon from the Greensand, *Ornithocheirus* (Unwin 2003), but this has been questioned (Kellner and Tomida 2000; Fastnacht 2001; Veldmeijer 2003; Andres and Ji 2008). The most recently named Santana genus, *Barbosania* (Elgin and Frey 2011a), has escaped taxonomic revision so far, but it is probably a matter of time before the nomenclatural parade of this animal is also rained upon.

Cearadactylus, another Santana ornithocheirid, has also provided confusion to pterosaurologists but for different reasons than most other members of its group. First known from a skull and named as an indeterminate pterodactyloid by Leonardi and Borgomanero (1983), it was considered an ornithocheirid by Dalla Vecchia (1993) and Kellner and Tomida (2000), but Unwin (2002) argued that it was a bizarre cteno-

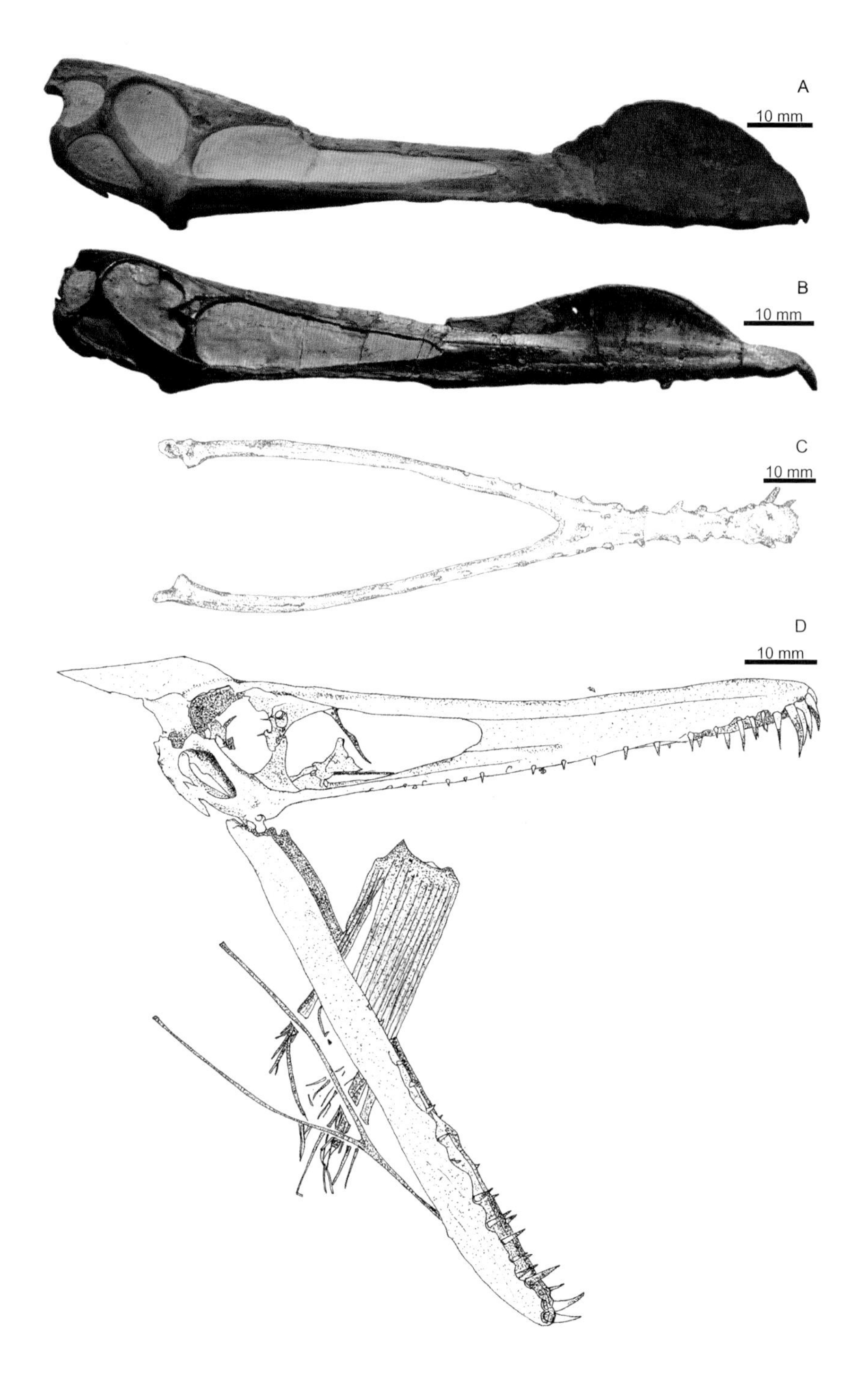

Fig. 16.4. Complete ornithocheirid skull material, stemming from the probable Albian Santana Formation (A–C) or Aptian Crato Formation (D). A, skull of *Ornithocheirus mesembrinus* (housed in the Bavarian State Collection of Palaeontology and Geology, Munich; used with permission), right lateral view; B, *Anhanguera blittersdorffi*, right lateral view; C, mandible of *Coloborhynchus spielbergi*, dorsal view; D, the complete skull and mandible of *Ludodactylus sibbicki*, complete with a possibly deadly plant frond (see chapter 8), right lateral view. Photographs A–B courtesy of Andre Veldmeijer and Erno Endenburg; C, redrawn from Veldmeijer (2003); D, redrawn from Frey, Martill, and Buchy (2003b).

chasmatoid (see chapter 19). *Cearadactylus* stood out among other pterosaurs because it seemed to possess enormous anterior teeth that dwarfed even those of ornithocheirids, as well as a crocodile-like smile to its shallow jaws. The rear of the skull otherwise seemed quite ornithocheirid-like, despite the jaw tips and teeth being very atypical of this group. Recent preparatory work has revealed why this was the case. The abnormal front end was a fabrication by fossil dealers attempting to make the specimen more complete and marketable (Vila Nova et al. 2010), and it is now clear that *Cearadactylus* had a more "standard" ornithocheirid jaw morphology. Unfortunately, such fakery is not uncommon with pterosaur fossils, but in most cases it's detected early on and does not reach the scientific literature. *Cearadactylus* is thankfully a rare example of one that slipped through the net.

And the Rest

Ornithocheirids from other localities have been less affected by the taxonomic confusion rampant in the Cambridge Greensand and Santana Formations and, generally speaking, are more recent discoveries. The Lower Cretaceous Crato Formation (probably Aptian; 125–112 Ma), a pterosaur Lagerstätte that neighbors the Santana Formation, has yielded further remains of *Brasileodactylus* (Sayão and Kellner 2000), as well as a virtually complete ornithocheirid skeleton christened *Arthurdactylus conandoylei* (Frey and Martill 1994). The latter was named after Arthur Conan Doyle, the *Sherlock Holmes* scribe who also popularized pterosaurs in his 1912 novel *The Lost World*. A complete ornithocheirid skull and mandible, named *Ludodactylus*, represents a third Crato genus (fig. 16.4D; Frey, Martill, and Buchy 2003b). In a recent review of the Crato ornithocheirids, Unwin and Martill (2007) suggested that *Ludodactylus* may represent further material of *Brasileodactylus*, but more fossils of both pterosaurs are needed before this can be ascertained.

Somewhat surprisingly for a nation known for its vast fossil deposits and abundant pterosaurs, China has yielded relatively little ornithocheirid material. That said, one of the four ornithocheirids known from China is an embryo within an egg (Wang and Zhou 2004), which makes China an important ornithocheirid-bearing locality despite its low yield. *Yixianopterus*, *Liaoningopterus*, and *Guidraco* are the only named Chinese ornithocheirids (Wang and Zhou 2003a; Lü, Ji, Yuan, Gao 2006; Wang et al. 2012), and they all stem from the famous Liaoning region. The first is known from a partial skeleton and, as the name suggests, comes from the Early Cretaceous Yixian Formation, as does the aforementioned egg and embryo (likely Barremian; 130–125 Ma). *Liaoningopterus* and *Guidraco* were uncovered in the slightly younger (probably Aptian) Jiufotang Formation and are represented by a set of jaw tips and a complete skull with several associated vertebrae, respectively. These Jiufotang forms are generally very similar to each other (they both, for example, are notable for their incredibly large anterior teeth), and there is a strong probability that they represent the same species.

Elsewhere in the world, most ornithocheirids are only known from jaw tips. This includes *Uktenadactylus*, an ornithocheirid from Albian (112–100 Ma) deposits of Texas (Lee 1994; Rodrigues and Kellner 2008), which was once thought to represent a North American species of *Coloborhynchus*. Actual *Coloborhynchus* material, from the Valanginian (140–136 Ma) Weald of Britain (fig. 16.3A–B; Owen 1874) and Cenomanian (100–94 Ma) deposits of the Saratov region of Russia (Bakhurina and Unwin 1995), also comprise jaw tips. The latter is one of the youngest ornithocheirids known and was roughly contemporary with the Texan ornithocheirid *Aetodactylus* (known from another fragmentary rostrum; Meyers 2010). The English Weald also holds an endemic ornithocheirid *Caulkicephalus*, represented by a fragmentary brain case and—you guessed it—a jaw tip (Steel et al. 2005). Ornithocheirids have also recently turned up in Australia, represented by the obligatory jaw tips of the Albian forms *Mythunga* (Molnar and Thulborn 2007) and *Aussiedraco* (Molnar and Thulborn 1980; Kellner et al. 2010, 2011). Morocco also has its own ornithocheirid genus represented by a partial rostrum, *Siroccopteryx*, from the Cenomanian Kem Kem Beds (Mader and Kellner 1999). Some suggest that this animal is not a distinct genus, but is another species of *Coloborhynchus*, however (Unwin 2001).

Anatomy

Most named ornithocheirids are poorly known, with only the Santana and Crato Formation forms represented by substantial skeletons. They appear to have been fairly large pterosaurs with wingspans regularly achieving 4 or 5 m (e.g., Wellnhofer 1985, 1991b; Frey and Martill 1994; Kellner and Tomida 2000), and some species of *Ornithocheirus* and *Coloborhynchus* may achieved proportions of 6 m or more (fig. 16.5; Dalla Vecchia and Ligabue 1993; Martill and Unwin 2012). There are claims of Santana ornithocheirids with wingspans up to 9 m (Dalla Vecchia and Ligabue 1993), but they are based on such fragmentary remains that their ornithocheirid identity is not certain. Accordingly, wingspans of 6 to 7 m seem to be the most reliable size estimates for ornithocheirids at present (Martill and Unwin 2012).

Ornithocheirid skulls are proportionally large with long jaws and toothrows (fig. 16.6; see Campos and Kellner 1985; Wellnhofer 1987a; Kellner and Tomida 2000: Veldmeijer 2003; and Wang et al. 2012 for some of the best examples). Around half, if not more, of the skull length is positioned anterior to the nasoantorbital fenestra to form an elongate rostrum. Their skulls often possess large, rounded crests at the front of both

FIG. 16.5. No one messes with the 6 m wingspan *Ornithocheirus*, not even two 4 m wingspan *Anhanguera* shown on the left on this image.

jaws, the exact position of which is often used to diagnose different genera. Some ornithocheirids lack these crests on one or both of their jaws (e.g., *Brasileodactylus*, *Ludodactylus*), and others may sport supraoccipital crests of variable size instead of (*Ludodactylus*), or along with (*Ornithocheirus* and *Caulkicephalus*), their rostral ornament. *Guidraco* sports a type of crest unlike that of any pterosaur, with a tall, rounded blade projecting anterodorsally from the region above its eyes. The crest margins of ornithocheirids lack the fibrous bone textures we see supporting soft-tissue crests in other pterosaurs, indicating that their crests may have been entirely comprised of bone. Their skull widths are quite variable with most species possessing fairly narrow skulls (e.g., *Anhanguera*, most *Coloborhynchus*, *Cearadactylus*), but others demonstrating considerable broadening at the jaw joints (e.g., some *Coloborhynchus, Ornithocheirus*). Most taxa have long, narrow mandibular symphyses that extend at least 30 percent of the jaw length, and perhaps considerably more. Their eye sockets are large and somewhat forward-facing, while their temporal openings are relatively broad. The excellent preservation of some ornithocheirid skulls has permitted detailed reconstruction of their brain cavities with CT scanning, so although no fossil endocasts are known, ornithocheirid neuroanatomy is well understood (Witmer 2003; also figs. 2.10 and 5.5).

The rather savage-looking teeth of ornithocheirids are one of their defining characteristics (Kellner 2003; Unwin 2003). The anterior teeth of both jaws form a distinctive "fish grab," a structure seen at its most fearsome in *Guidraco* and *Liaoningopterus*, where the teeth are so long that that they extend well beyond the upper and lower snout margins when the jaws are closed. Fish grabs are comprised of a similar dental arrangement in both jaws. Moving from the tips backward, we see three increasingly large, recurved tusk-like teeth, two or three much smaller teeth, and then three progressively larger teeth. Beyond this are smaller, well-spaced, and slightly recurved teeth that extend to the back of the jaws. The jaw tips are often slightly expanded to accommodate the larger anterior teeth. The orientation at which the anterior teeth project from the jaw margin is variable, and seems to characterize different ornithocheirid species.

As with the istiodactylids, the entire vertebral column, forelimb bones, and trunk skeleton of ornithocheirids are pneumatized so extensively that only pteranodontians (chapter 18) and azhdarchids (chapter 25) can claim greater degrees of skeletal pneumatization (Claessens et al. 2009). Ornithocheirid necks are long in comparison to their trunks and are comprised of large vertebrae with tall, complex neural spines (Wellnhofer 1991b). Their torsos, by contrast, are relatively tiny and shorter than both the neck and skull. Mature individuals possess a notarium comprised of six dorsals and, beyond this, no more than seven "free" dorsals (fig. 16.7A; Veldmeijer 2003). This figure may change depending on how many dorsals are incorporated into the sacrum, which may comprise four to seven vertebrae (fig. 16.7B; Wellnhofer 1991b;

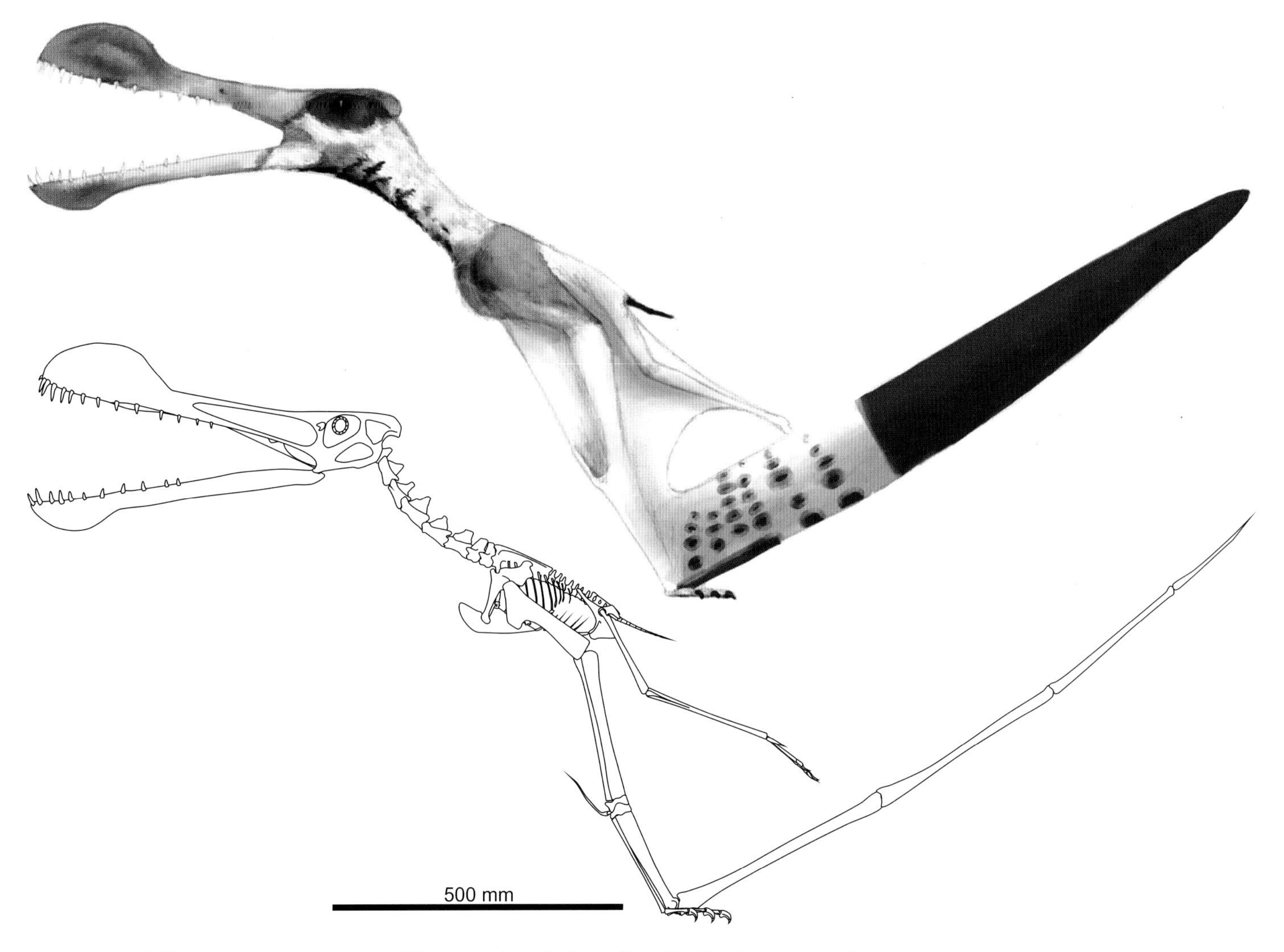

Fig. 16.6. Skeletal reconstruction and life restoration of a launching *Ornithocheirus mesembrinus*.

Kellner and Tomida 2000; Veldmeijer 2003). In adults, the sacrum also develops a supraneural plate atop its neural spines. Ornithocheirid tails are poorly known but appear to be comprised of at least 11 short vertebrae, which become relatively circular in cross section toward the end of the series.

Ornithocheirid forelimbs are proportionally enormous. They are around five times longer than their legs and much stronger under biomechanical testing (Habib 2008). Such mighty arms require substantial anchorage on the body, and ornithocheirids accordingly have robust scapulocoracoids and stout, deeply keeled sterna to house their substantial forelimb muscles (Bennett 2003a). Their pectoral girdle is of typical construction for ornithocheiroids, and is set at a perpendicular angle to the spine with the coracoids much longer than their scapulae (Bennett 2003a; Frey, Buchy, and Martill 2003). Their humeri possess the characteristically "warped" deltopectoral crest of most other ornithocheiroids, though these features are much larger than those we saw in istiodactylids (fig. 16.7C). Their carpals are also enlarged in the manner typical of all ornithocheiroids. Ornithocheirid wing fingers occupy over 60 percent of the wing length, making them among the longest possessed by any pterodactyloids. Their first three fingers are rather less impressive however, as can be said of their rather small pelves and hindlimbs. Their pelves appear to take longer to form a solid ischiopubic plate during growth than those of other pterosaurs, and are notable for the large angle between the preacetabular process and the posteriorly deflected ventral pelvic bones (Hyder et al., in press).

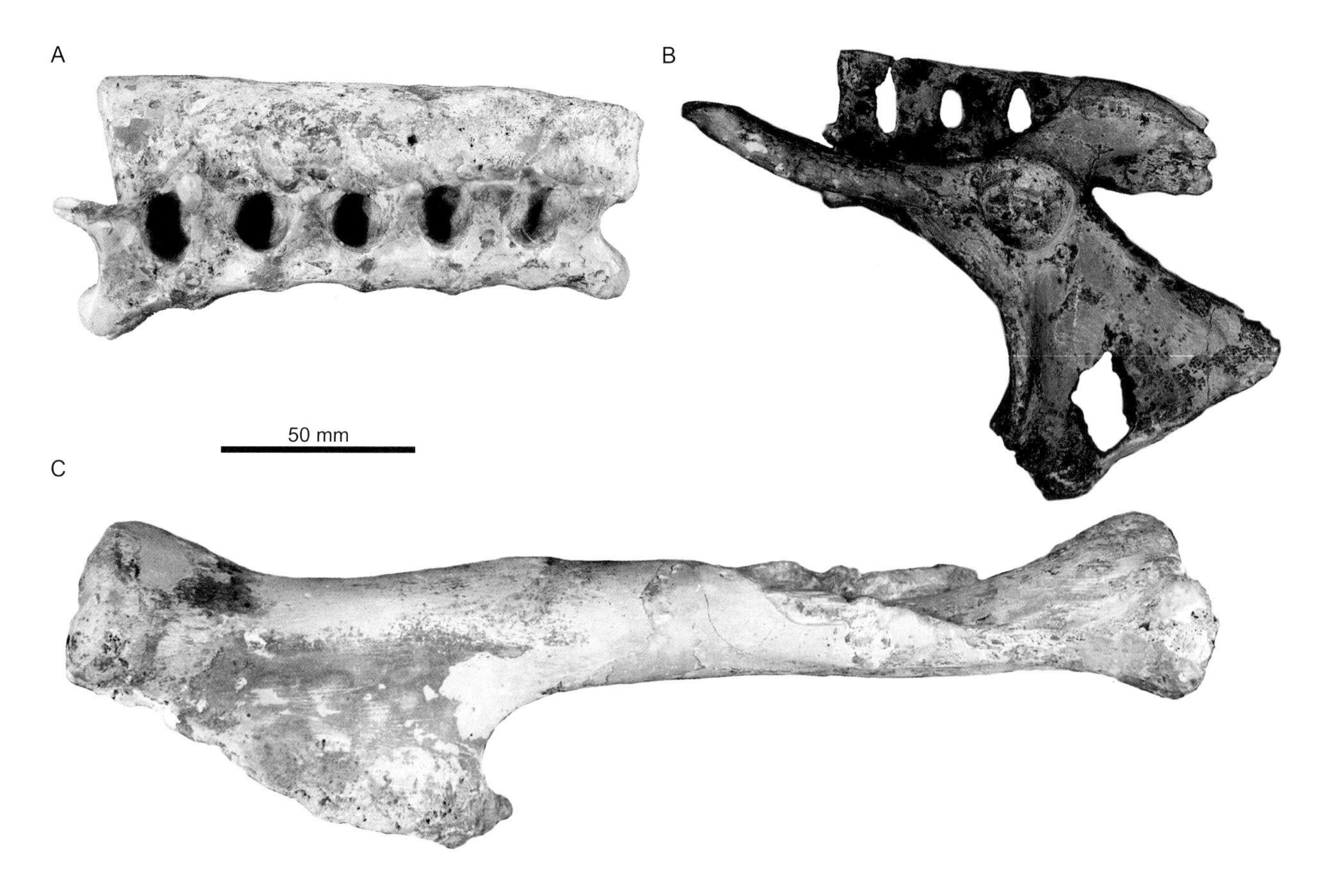

FIG. 16.7. Skeletal elements of *Coloborhynchus spielbergi*. A, notarium in left lateral view; B, pelvic girdle in left lateral view; C, left humerus in dorsal view. Images courtesy of Andre Veldmeijer and Erno Endenburg.

Their slender femora have, like istiodactylids, femoral heads that project almost in line with the femoral shaft, but they seem to lack prominent processes anchoring their hindlimb muscles. Their shin bones are similarly developed and of equal length to the femur. Their feet are poorly known, but seem relatively small and gracile with undeveloped claws and a hooklike fifth metatarsal (Kellner and Tomida 2000).

Locomotion

FLIGHT

The diminutive legs but enormous wings of ornithocheirids betray their preference for life in the air. Indeed, ornithocheirid anatomy is best summarized as a pair of wings with a head and neck bolted on, a configuration that produces relatively low body masses for their wingspans, high-aspect wings, and reduced wing loading. These attributes strongly indicate that long-distance soaring was their preferred mode of flight (Chatterjee and Templin 2004; Frey, Buchy, and Martill 2003; Witton 2008a). Their wing shapes are quite like those of oceanic seabirds (Witton 2008a) and suggest they were better suited to dynamic soaring over oceans and seas than to flight in terrestrial settings. The development of their shoulder girdles and sterna indicate a strong flapping ability with an emphasis on high downstroke power output, which was presumably used to power their own flight while looking for external sources of lift.

Some confirmation that ornithocheirids were marine fliers stems from their fossil record, which is largely biased toward marine settings. Osteological correlates of their skeleton for water launching also suggest they spent a lot of time around water, and frequently—deliberately or not—landed on it (fig. 16.8; Habib and Cunningham 2010). Quite how pterosaurs took off from floating starts has long puzzled pterosaurologists, but recent work from the same scientists who generated the quadrupedal launching hypothesis has found that the same technique, somewhat modified, also works

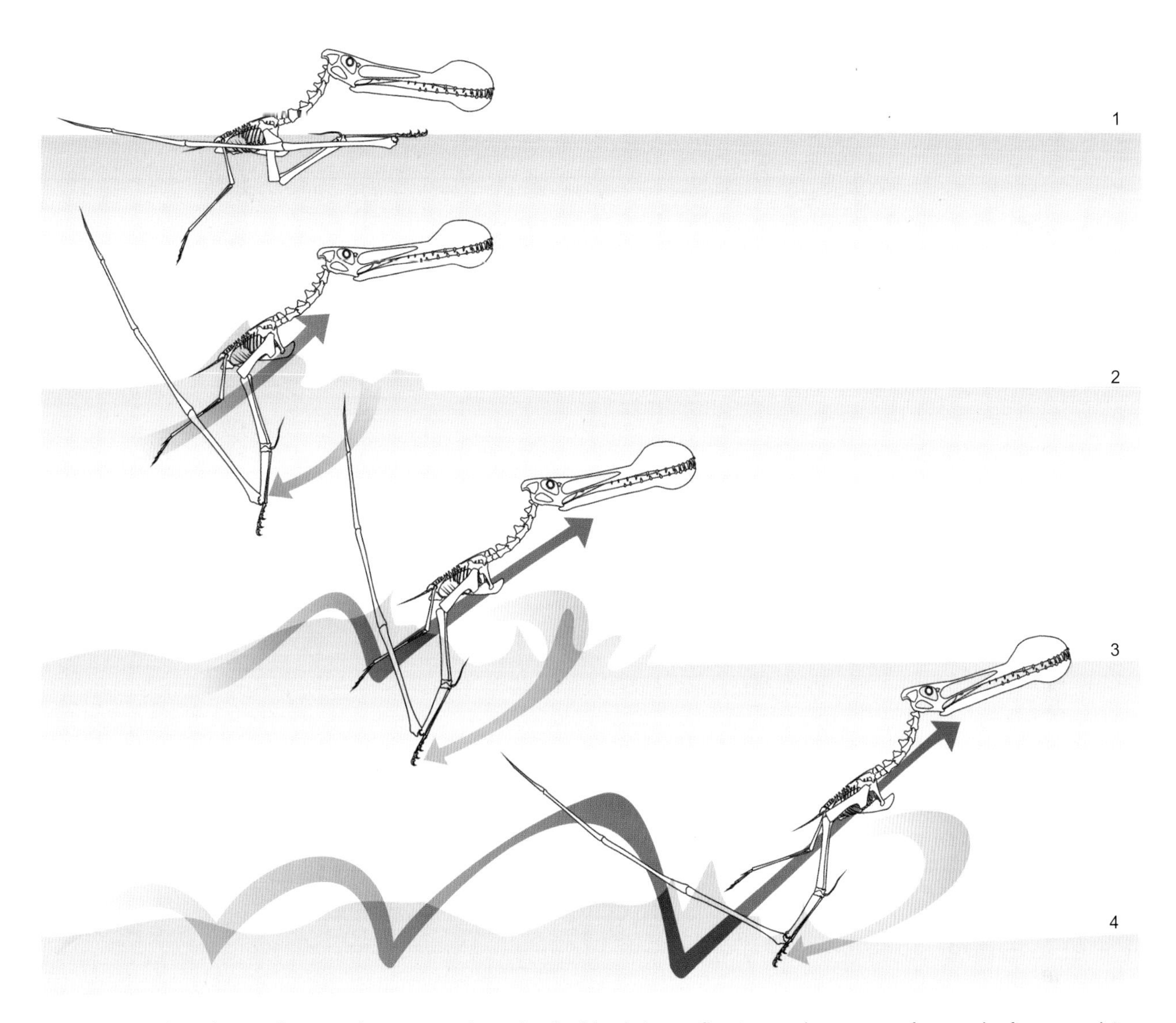

Fig. 16.8. Water launching in four simple steps, as shown by *Ornithocheirus*. 1, floating on the water surface; 2, the first propulsive phase, where the limbs are thrust into and posteroventrally through the water; 3, the hopping phase, in which the body is propelled from the water surface with increasing height in every bound; and 4, the escape phase, in which enough velocity and clearance from the water has been achieved so that full launch is possible. See Habib and Cunningham (2010) for more details.

in water. Escaping water requires overcoming both its surface tension and suction, but several aspects of ornithocheirid anatomy suggest they were better equipped than most pterosaurs for these challenges (Habib and Cunningham 2010). Their expanded forelimb muscles, reinforcement of their shoulder girdles, and reconfiguration of certain muscle orientations around the warped deltopectoral crest permitted ornithocheirids to deliver large power outputs in the required vectors for water escape, though rising from the surface still required several quadrupedally powered "hops." These bounding motions would successively gain more power and momentum and eventually lift the bulk of the ornithocheirid out of the water. Once sufficiently clear, their unusually broad wing finger joints would provide enough contact area with the water surface to generate propulsive thrust for a final, fully powered shove that would launch them skyward (Habib and Cunningham 2010). The reduced torso and hindlimb proportions of ornithocheirids were probably helpful in such launches, as they would create less drag than those of more typically proportioned pterosaurs. Note

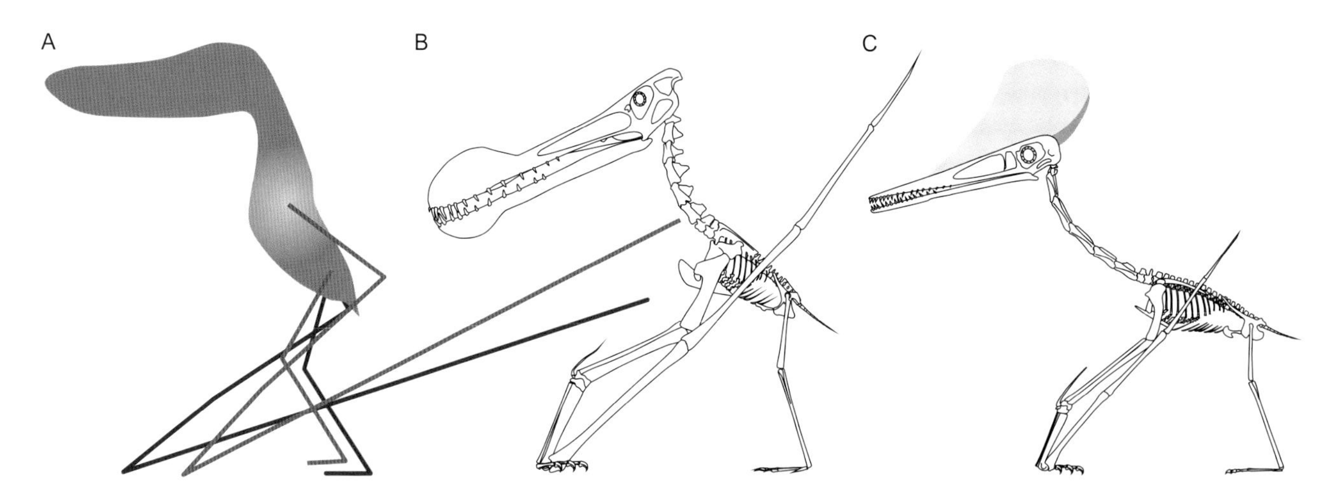

FIG. 16.9. Possible terrestrial stances of ornithocheirids. A, walking posture for *Anhanguera* as suggested by Unwin and Henderson (1999) and Unwin (2005). Note the lack of flexion of the carpal joints; B, standing *Ornithocheirus* with a slightly flexed wrist and relatively erect torso, the ornithocheirid stance preferred here; C, standing *Pterodactylus*, a nonornithocheiroid pterodactyloid. Note how this more typically proportioned pterodactyloid has a relatively low-angle torso when standing, due to its more equal limb proportions.

that other pterosaurs were also capable of launching from water using similar techniques (see chapter 25 for an example), but most would not have the same launch efficiency as ornithocheirids (Habib and Cunningham 2010).

ON THE GROUND

The excellent condition of some ornithocheirid fossils has inspired several dedicated studies into their terrestrial locomotion (e.g., Wellnhofer 1988; Unwin and Henderson 1999; Unwin 2005; Chatterjee and Templin 2004; Wilkinson 2008) with the results often implicated to other pterodactyloids. The extent to which such comparisons are fair is debatable, given how different the ornithocheirid bauplan is to virtually all other pterodactyloids (fig. 16.9). Ornithocheirids have generally been considered quadrupedal animals, but some controversy exists over their likely stance. Wellnhofer (1988) and Wilkinson (2008) assumed that they held their limbs in a sprawling, somewhat crocodile-like fashion (fig. 7.2D), but others (Unwin and Henderson 1999; Chatterjee and Templin 2004; Unwin 2005) concluded that their limbs more or less extended vertically in an avian- or mammal-like configuration (fig. 16.9A–B). Only Chatterjee and Templin (2004) supposed that any flexion was possible in their wrists; Unwin (2005) and Unwin and Henderson (1999) maintained that the wrist was virtually immobile. In the latter model, the absence of wrist flexion renders the arms largely useless for propulsion, and would mean they did little more than stop the animal from toppling forward while the hindlimbs provided forward momentum.

To my mind, the model proposed by Chatterjee and Templin (2004) seems most likely. The ornithocheirid glenoid, though different from those of other pterosaurs, still permits the humerus to swing vertically beneath the shoulder when walking, and the acetabulum also allows the hindlimb to move vertically. As discussed in chapter 4, there is now good evidence that pterosaur carpal joints were fairly mobile (e.g., Bennett 2001; Wilkinson 2008). Such a configuration allows ornithocheirids to walk in a broadly similar way to other pterosaurs, but with a more steeply elevated spine to accommodate their hypertrophied forelimbs (Chatterjee and Templin 2004). The flexible wrist in this model enables the forelimb to fold up somewhat more when walking however, thus lessening the elevation of the spine and taking some of the weight off the hindlimbs. A slightly elevated vertebral column seems unavoidable in animals with such disparate limb proportions, and the reconfiguration of ornithocheirid pubes and ischia appears to reflect this. The configuration of more typical pterosaur pelves provides inefficient anchorage for hindlimb muscles when walking with elevated torso angles (Fastnacht 2005b), but posterior rotation of the ischiopubic plate may relocate the pelvic hindlimb muscles into a more optimal position for walking (Hyder et al., in press).

If the above description is accurate, ornithocheirids may not have been especially unusual in their approach to terrestrial locomotion, but their actual terrestrial capabilities were probably still limited. Their pace was dictated by their short legs, and their forelimbs could presumably only shuffle along so as not to outpace their hindquarters. This would not permit particularly fast terrestrial locomotion, but there does not seem any reason to assume that ornithocheirids could not also bound from each limb set to cover ground quickly (Witton and Habib 2010). When imagining this, bear in mind that pterosaurs probably incorporated elements of this motion into quadrupedal launching; bounding simply removes the flighted end of this action, and uses the powerful forelimb muscles to propel the animal primarily forward, rather than skyward. Ornithocheirid limb dichotomy seems to restrict the use of intermediate speeds between explosive bounding and a shuffling walk, however, suggesting they only had two "gears" for terrestrial locomotion. This, in conjunction with their strong flight adaptations, may indicate they spent a fairly limited amount of time on the ground. We can assume that they were capable of swimming relatively well, however. Ornithocheirids must have been at least competent and stable in water (fig. 7.10) to ensure that they could take off from water surfaces in the manner discussed above.

Paleoecology

Many aspects of ornithocheirid paleoecology—their growth regimes, sexual dimorphism, and the like—have not been studied despite the wealth of fossils referred to this group. To my mind, this is something pterosaurologists could, and perhaps should, address soon. Studies into possible dimorphism and growth within ornithocheirid collections may help resolve some of the taxonomic problems dogging this clade. It seems possible, and perhaps likely, that many alleged ornithocheirid "species" are actually differently aged, or gender variant individuals, but only detailed studies of their remains will shed light on this.

Given that ornithocheirids seem suited to flight over marine settings, it's unsurprising that they are also considered to have dined at sea. The manner in which they gathered their food has not been researched in detail, but it is generally thought that they either skim fed, pushing their lower jaw tip through the water to snap up food upon impact (Nessov 1984), or gleaned food from the water surface like some terns and frigatebirds (e.g., Wellnhofer 1991a; Unwin 1988a, 2005; Veldmeijer et al. 2006). The ornithocheirid skim-feeding hypothesis has been discounted in recent appraisals of pterosaur skim-feeding (Humphries et al. 2007; Witton 2008b, also see chapter 24), but dip-feeding is supported by a number of anatomical features. Their elongate rostra are ideal for reaching into water to grab swimming animals, as are the "fish grabs" lining their jaw tips. Wellnhofer (1991a) and Veldmeijer et al. (2006) suggested that the rostral crests of ornithocheirids would work well as stabilizers for jaw tips being plunged into water in search of food. This may be so, but we should then consider that a number of ornithocheirids lack rostral crests altogether, and that modern dip feeders can perform this action without analogous structures. A neck of goodly length, strength, and flexibility does, however, seem critical for this feeding method, and these characters all fit the ornithocheirid cervical series (Wellnhofer 1991b). Indeed, the anterior neck vertebrae of *Anhanguera* are more robust than those behind them, suggesting the muscles anchoring the head (and therefore those battling water drag during dip feeding) were proportionally large and suited to this role (Witton 2008b). Large, forward-facing eyes and developed flocculi (see chapter 5) are ideal for dip feeding too, permitting effective spotting of prey and judgment of distances when striking at said foodstuffs (Witmer et al. 2003). As such, it seems likely that at least some ornithocheirids were efficient dip feeders, though we cannot rule out the possibility that more sedate foraging methods—reaching for food while alighted on the water surface, or shallow surface dives—were also used. Indeed, the diversity in ornithocheirid tooth morphology is perhaps a good sign that some variety in foraging methods may have existed.

17
Boreopteridae

PTEROSAURIA > MONOFENESTRATA > PTERODACTYLOIDEA > ORNITHOCHEIROIDEA > BOREOPTERIDAE

By now, I'm sure we're all becoming accustomed to how downright strange pterosaurs look. Even species of mundane appearance have natty looking teeth and preposterous body proportions, while more extreme variants would not look out of place in a Tim Burton movie. The bizarre appearances of even these have been superseded by a newly discovered ornithocheiroid group, however, one characterized by chunky-looking headcrests, enormously stretched jaws, and incredible numbers of teeth that are so long they project well beyond the bounds of the upper and lower jaws. These animals are the boreopterids (fig. 17.1), a Cretaceous group that become known to paleontologists as recently as 2005 (Lü and Ji 2005a). To date, only two definitively identified boreopterid specimens have been found, both from Lower Cretaceous deposits of northern China. Fortunately, these specimens are complete enough that much of their anatomy is already clear.

Boreopterids have generally been placed within Ornithocheiroidea (Lü and Ji 2005a; Unwin 2005; Lü, Ji, Yuan, and Ji 2006; Lü, Unwin, et al. 2008; Lü, Unwin, et al. 2010), although affinities to ctenochasmatoids (chapter 19) have also been suggested (Andres and Ji 2008). The former is almost certainly more likely. The skulls of boreopterids may be some somewhat ctenochasmatoid-like, but their postcranial skeletons are almost identical to those of ornithocheiroids, with particular similarities to members of Ornithocheiridae (chapter 16). Indeed, some components of the boreopterid skeleton are so ornithocheirid-like (such as their humeri) that we may want to reappraise the identity of some isolated "ornithocheirid" limb bones: they may well belong to boreopterids. In addition, boreopterids also have the tiny feet seen in some istiodactylid ornithocheiroids (chapter 15), which are a stark contrast to the massive, wading feet of most ctenochasmatoid species.

Presently, boreopterids have only been identified from one stratigraphic unit: China's Lower Cretaceous Yixian Formation (likely Barremian; 130–125 Ma) (fig. 17.2). The first of their fossils was unveiled in 2005 when the skeleton of a virtually complete, if somewhat imperfect and rather immature individual, was described (Lü and Ji 2005a) and named *Boreopterus cuiae*. Lü and Ji thought that the tooth morphology of *Boreopterus* suggested affinities to Ornithocheiridae, and other authors generally agreed (Unwin 2005; Lü, Ji, Yuan, and Ji 2006; Lü, Unwin, et al. 2008; Lü, Unwin, et al. 2010; but also see Andres and Ji 2008). *Boreopterus* became the namesake of Boreopteridae, a distinct group of ornithocheiroids, when Lü, Ji, Yuan, and Ji (2006) considered it to have a close evolutionary relationship with another Chinese pterodactyloid *Feilongus* (see below, and chapter 19).

The second boreopterid specimen, named *Zhenyuanopterus longirostris*, presented a substantially better preserved skeleton that dismisses any ideas of housing boreopterids outside of Ornithocheiroidea (Lü 2010b). The preservation of the only known specimen of *Zhenyuanopterus* is exceptional (fig. 17.3), especially for such a large pterosaur (3.5 m wingspan). Barely a bone is out of place and, although somewhat squashed, almost every component of the skeleton can be observed in detail. The sole-known *Zhenyuanopterus* is also more osteologically mature than *Boreopterus*, giving us a good idea of what adult boreopterids looked like. Indeed, there is a very good chance that both *Zhenyuanopterus* and *Boreopterus* represent the same species, with the latter merely being a subadult version of the former. Most of the differences between them (e.g., size, presence of a headcrest, and number of teeth) can be explained as effects of age (see chapter 8), and their occurrence in the same depositional unit heightens this possibility further.

One animal that is definitely distinct from *Boreopterus*, however, is a third alleged boreopterid: *Feilongus youngi*. Roped into Boreopteridae by Lü, Ji, Yuan,

Fig 17.1. Two *Zhenyuanopterus* float on a lake in Lower Cretaceous China.

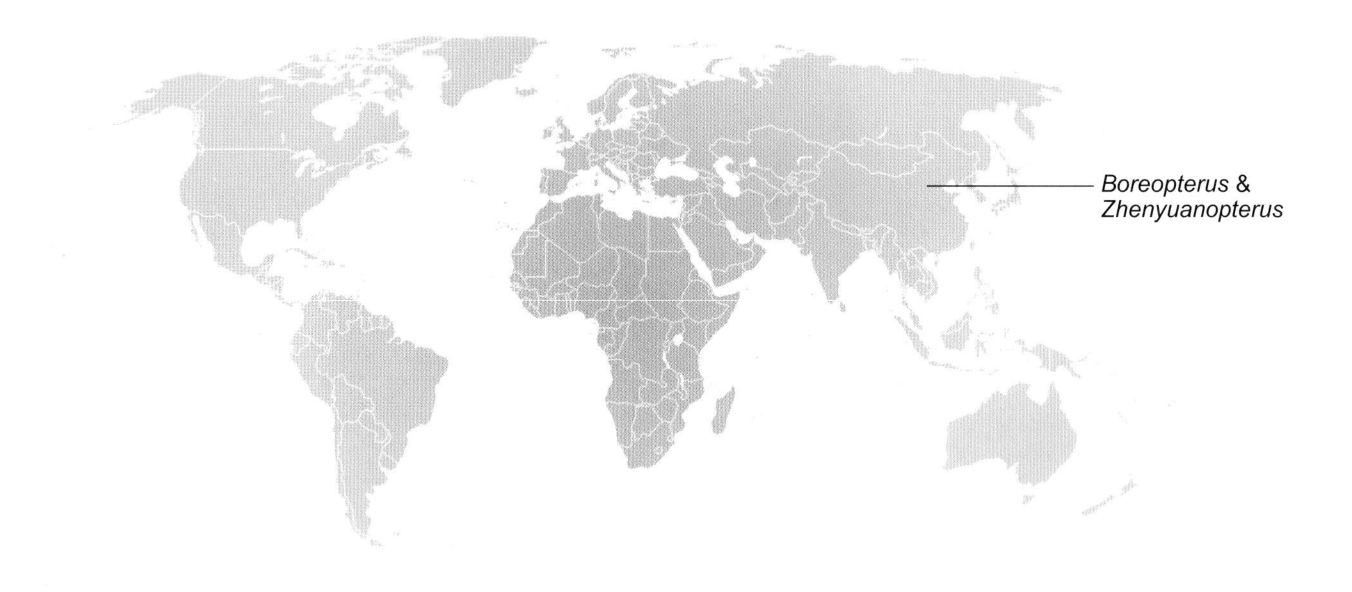

FIG 17.2. Distribution of boreopterid taxa.

and Ji (2006) and Lü (2010b), this animal is known only from a skull that, it must be said, looks out of place in a boreopterid lineup. *Feilongus* does possess the long rostrum and needlelike teeth of boreopterids, but their number and position, nature of its headcrest, inclination of its posterior skull bones, and elongate cervicals are more like those of ctenochasmatoids than of ornithocheiroids (Wang et al. 2005; Wang et al. 2009; Andres and Ji 2008). A rather similar, recently uncovered Yixian species *Morganopterus zhuiana* was also proposed to be a boreopterid on the same grounds as *Feilongus* (Lü, Pu, et al. 2012), but its placement among boreopterids is equally problematic. Because these species appear to show greater affinities to ctenochasmatoids than boreopterids, we will discuss them in more detail with the other ctenochasmatoids in chapter 19.

Anatomy

Boreopterid anatomy (fig. 17.4) is unusually well known for such a new pterosaur group, with descriptions of most bones from the two recognized boreopterid taxa given by Lü and Ji (2005a), Lü, Ji, Yuan, and Ji (2006), and Lü (2010b). They are much smaller than other ornithocheiroids: *Boreopterus* has a wingspan of only 1.9 m, while that of *Zhenyuanopterus* attains 3.5 m. This was likely not the full size of *Boreopterus*, seeing as it is only known from a young individual, but the sole known *Zhenyuanopterus* specimen possesses a notarium and fused skull bones, suggesting it had probably finished growing.

Boreopterid skulls are their most striking and defining feature (fig. 17.3B). Like those of most other ornithocheiroids, they are proportionally large (twice the length of the torso) with particularly long rostra that occupy over two-thirds of the skull length. *Zhenyuanopterus* possess a large crest that extends from the skull midlength until level with the posterior region of the nasoantorbital fenestra. Another small crest is found at the back of the skull. *Boreopterus* lacks both of these features, but the osteological immaturity of the specimen suggests it may have been too young to begin crest development before it died. The temporal openings and orbits are small compared to those of other ornithocheirids, suggesting that neither their eyes nor jaw muscles were large. Their mandibles are long, crestless, and fused along the anterior two-thirds of their length. Both the upper and lower jaws are brimming with equal numbers of teeth that occupy most of the jaw length. In total, *Boreopterus* packs 116 teeth into its mouth, while *Zhenyuanopterus* possesses 184. The teeth are particularly long and slender, being 10 times taller their wide, and have a slight posterior curve. Unlike virtually all other pterosaurs, the teeth

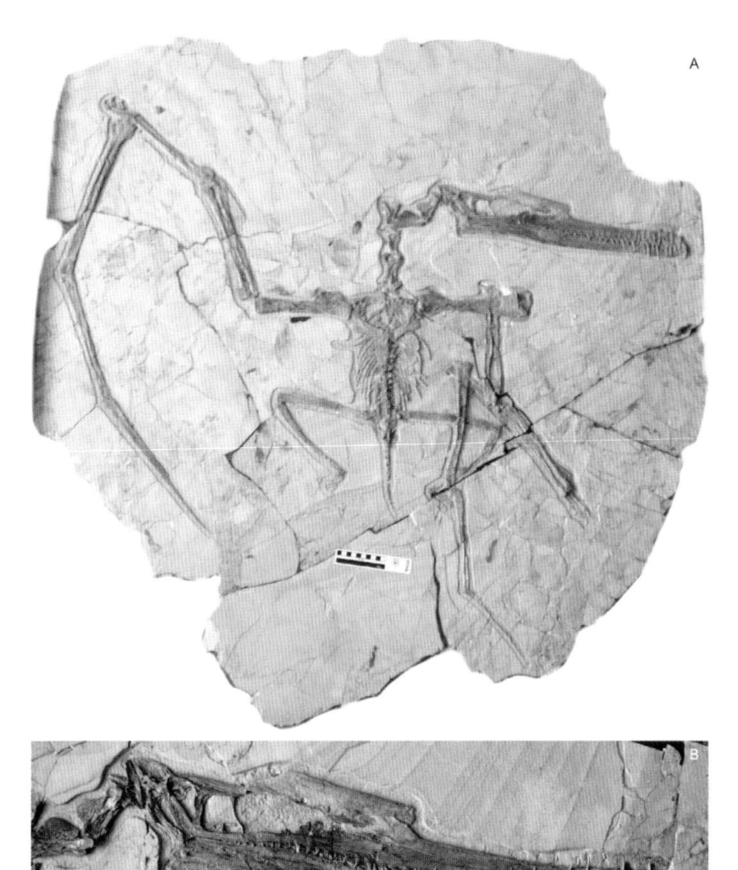

FIG 17.3. An exquisitely preserved complete skeleton of *Zhenyuanopterus longirostris,* from the Barremian Yixian Formation of China. A, the complete and fully articulated skeleton; B, detail of the skull and mandible. Photographs courtesy of Lü Junchang.

are long enough that they extend well above and below the opposing jaw bones. The teeth are longest at the jaw tips and steadily reduce in size posteriorly until, at the back of the mouth, they're a mere tenth of the length of those at the front.

Much of the postcranial skeleton of boreopterids is like that of other ornithocheirids. The cervical series is long compared to the torso, with particularly large, rounded neural spines. Twelve dorsals are preserved in *Zhenyuanopterus*, three of which have fused into the beginnings of a notarium. The sacrum contains four vertebrae and the tail is comprised of at least 13 elongate caudals. Boreopterid tails are among the longest of all pterodactyloids, a feature that they share with another ornithocheiroid group, the pteranodontians (chapter 18).

Boreopterid forelimbs are overdeveloped in the manner typical of ornithocheiroids and, with the exception of the wing finger, are proportionally typical for this group. The scapulocoracoids are large and reinforced, anchoring on to broad sterna with pronounced cristospines. Their humeri possess elongate, warped deltopectoral crests and stout shafts. The radius and ulna are similarly both robust and, unusually for ornithocheiroids, of similar diameters. The pteroid is relatively slender and around 40 percent of the forearm length. Some details of the wrist and metacarpals are not well understood, as neither boreopterid specimen has well preserved or fully fused carpal blocks, and it cannot be seen if some metacarpals have lost contact with the carpals. What is clear, however, is that boreopterid wing fingers are proportionally more like those of istiodactylids than ornithocheirids, occupying half of the wing length compared to the 60 percent or more of ornithocheirids. Their first three fingers are small but capped with relatively robust claws. Boreopterid pelvic girdles are poorly known, but their hindlimbs are very diminutive. They are generally similar in length and proportions to those of ornithocheirids, but their feet are tremendously small, with their maximum length a mere 10 percent of the total leg length.

Locomotion

With perhaps one exception, diddly-squat has been said in print about the biomechanics and functional biology of boreopterids, including their locomotory abilities. Their proportional similarity to ornithocheirids suggests that some generalities about ornithocheirid locomotion will also apply to boreopterids, however. For example, the elongate, warped deltopectoral crest and reinforced scapulae that we see in boreopterids probably signifies frequent launching from deep water, just as it does in ornithocheirids (Habib and Cunningham 2010). Similarly, the tiny hindlimbs of boreopterids dictate that their wings bore a correspondingly shallow wing, although their abbreviated wing fingers would make the wing somewhat shorter and lower aspect than those of other ornithocheiroids. Perhaps, as we discussed for istiodactylids in chapter 15, this reflects a preference for inland habitats where short wings would be benefi-

Fig. 17.4. Skeletal reconstruction and life restoration of a launching *Zhenyuanopterus longirostris*.

cial for takeoff (Rayner 1988). The discovery of boreopterid specimens in a freshwater deposit fits with this idea but, of course, a much larger sample size is needed to ascertain a depositional bias in their record.

The miniscule feet of boreopterids suggest that they did not worry too much about moving around on the ground, and their heavily skewed limb proportions would have probably limited them to shuffling or bounding around in the fashion we described for ornithocheirids in the previous chapter. We may assume, given their apparent adaptations for taking off from water, that they were at least stable when swimming or floating, and their possible adaptations for aquatic feeding suggest they would be at home in watery settings.

Paleoecology

To date, perhaps the only comments about boreopterid dietary preferences stem from Wang and Zhou (2006a), who suggested that they habitually dined on fish. This agrees with the elongate jaws and necks of boreopterids, which are well suited to being plunged into deep water to gather fish or other swimming prey, but exactly how they apprehended their food has not been established. Boreopterid teeth look so slender and delicate that it is easy to imagine large, feisty prey items damaging them. Their eyes are also small, which does not suit a predator that is out to spot prey from a distance. Perhaps, instead of feeding on large prey, boreopterids used their jaws as a cage to trap numerous, small swimming critters in one go. The close spacing and extremely long nature of their teeth is certainly consistent with this action, and we can imagine boreopterids combing water for food before lifting their heads to strain consumables from the water through their tightly packed teeth, or perhaps transporting food items to their throats with their tongues. Their neck morphology seems consistent with this idea, with the large and complex nature of their cervical vertebrae indicating a flexible and well-muscled neck.

Straining water for food would probably require feeding boreopterids to either be on the water surface or standing in shallow water, but this brings their stunted feet and hindlimbs into question. Perhaps they preferred rather still waters where they didn't need large, flipper-like feet to punt themselves around, or perhaps they were able to use their forelimbs to shunt themselves about while their bodies were buoyed up by water. Of course, all of these ideas are provisional, and there is clearly a lot left to know about this new group of pterosaurs. We can only hope that further discoveries revealing more details of their paleobiology, evolutionary history, and global distribution are not too far away.

18 Pteranodontia

PTEROSAURIA > MONOFENESTRATA > PTERODACTYLOIDEA >

ORNITHOCHEIROIDEA > PTERANODONTIA

Pteranodontians are probably the most famous of all pterosaurs. Their toothless, crested forms grace films, television shows, paintings, museum halls, books, toys, and anything else with which you would want to associate a pterosaur. Fame probably found these pterosaurs because they were the first truly gigantic and spectacular flying reptiles to be discovered. The very first discoveries of pteranodontian fossils left no question that they were seriously big animals—much larger than virtually any other extinct flying creature known at that time—and they were represented by far more complete and impressive specimens than the shrapnel-like remains of large pterosaurs then known from Europe. Their elaborate headcrests were also a novel feature for nineteenth-century paleontologists, and make them easily identifiable even for people with no special interest in paleontology. Their size and elaborateness has been superseded by a good number of other pterosaurs in the last few decades, but pteranodontians remain, in the eyes of many, the quintessential pterosaurs.

In the phylogeny followed here, Pteranodontia is considered a particularly derived lineage of Late Cretaceous ornithocheiroids and comprised two relatively small groups: Pteranodontidae and Nyctosauridae (Lü, Unwin, et al. 2010; Lü, Unwin, et al. 2011). They are united by their complete lack of teeth; the position of their headcrests; the particularly deep profiles of their mandibular rami; and characters of their wing skeletons, most obviously the elongation of their wing metacarpals (Unwin 2003). Each group contains one famous genus. Pteranodontidae houses *Pteranodon*, the first genuine pterosaur known from the United States and also the first giant pterosaur known from substantial remains(fig. 18.1). Nyctosauridae holds *Nyctosaurus*, an animal well known for its truly outrageous antler-like headcrest. As we'll see later, pteranodontians and nyctosaurids differ in some aspects of their shoulder and forelimb anatomy, but they otherwise seem so similarly constructed that shared ancestry does not seem unreasonable. This is only one of several ideas about the placement of these animals among pterodactyloids, however. Some consider nyctosaurids to be an independent offshoot of pterodactyloids, an individual branch of the tree that split from other pterosaur stock before a broad radiation of the other ornithocheiroids, dsungaripteroids, and azhdarchoids (Bennett 1989, 1994; Kellner 2003; Wang et al. 2009; Wang et al. 2012). They are retained within an "ornithocheiroid" group by others, but are considered the most primitive members of the lineage (Andres and Ji 2008). Lü, Ji, Yuan, and Ji (2006) place both *Pteranodon* and *Nyctosaurus* in another position, at the foot of Azhdarchoidea.

Pteranodontia: All-American Leathery-Winged Heroes?

The discovery of pteranodontians is intimately associated with the recovery of the first pterosaur fossils from the United States, which were also the first records of pterosaurs found outside of European sediments (see Bennett 1994; Everhart 2005; and Witton 2010 for overviews). The tale of their discovery has all the right ingredients for a great paleontological campfire story—a famous paleontologist, uncharted fossil deposits, spectacular fossil animals—and goes thus. American pterosaurs were found by the famous paleontologist Othniel Charles Marsh, the man who named and described the well-known dinosaurs *Apatosaurus*, "*Brontosaurus*," *Allosaurus*, and *Stegosaurus*, along with innumerable other fossil species. Marsh and his team found his pterosaurs in the Santonian/Campanian (85–70 Ma) Smoky Hill Chalk of the Niobrara Formation, Kansas (fig. 18.2). These chalky deposits are the remains of a shallow sea that bisected North America during the Late Cretaceous and are

Fig. 18.1. *Pteranodon* dives for a bait ball of panicked fish in the Late Cretaceous Western Interior Seaway. Following the logic of Bennett (1992), the large *Pteranodon* with the oversized headgear represents an assumed male, while the smaller individuals with more sensible headcrests are presumed females.

thors including Williston (1897), Eaton (1910), and most recently, Bennett (2001). Chris Bennett's monumental work on *Pteranodon* is particularly important to modern researchers because it not only highlights several flaws in the "classic" view of *Pteranodon* anatomy (which was outlined by George Eaton in 1910), but also used the vast sample size of *Pteranodon* fossils to gain insight into pterosaur growth (Bennett 1993), sexual dimorphism (Bennett 1992), ailments (Bennett 2003c), and detailed anatomy (Bennett 2007a). Some recent work has challenged a number of these ideas, suggesting that Bennett's interpretations of *Pteranodon* diversity are overly conservative and that we can split *Pteranodon* into four species: *Pteranodon longiceps*, *Geosternbergia sternbergi*, *Geosternbergia maiseyi*, and *Dawndraco kanzai* (Kellner 2010). I must admit a certain amount of skepticism about this recent revision. These new species seem entirely diagnosed by crest and rostral morphology, both of which have been demonstrated as being subjected to strong sexual and individual variation in *Pteranodon* (Bennett 1992; Tomkins et al. 2010). Ergo, using those features to diagnose new species may be like applying taxonomic significance to subtle differences in deer antlers or elephant tusks. Erring on the side of caution, the following chapter follows Bennett's *Pteranodon* taxonomy.

Various remains referred to *Pteranodon*, but probably only representing indeterminate pteranodontids, demonstrate that they achieved a global distribution in the closing stages of the Cretaceous (Barrett et al. 2008). Nyctosaurids, by contrast, are rarer fossils that only occur in Coniacian-Maastrichtian strata (89–65 Ma) of the Americas. The first nyctosaurid fossils were recovered by Marsh from the Niobrara Formation in 1876, and were initially considered to represent a species of *Pteranodon* (*P. gracilis*), but it was not long before the two were recognized as significantly different and the name *Nyctosaurus* was established. At one point, Marsh incorrectly assumed that *Nyctosaurus* was a preoccupied name and later suggested *Nyctodactylus* as a replacement, but Samuel Williston (1903) identified this as a mistake and restored *Nyctosaurus* as the correct moniker. Williston produced a series of detailed anatomical papers on this pterosaur, which meant that *Nyctosaurus* was almost as well known as *Pteranodon* by the end of the twentieth century (Williston 1902a, 1902b, 1903; and others). Other nyctosaurid material from the Niobrara Formation has been suggested to represent different species of *Nyctosaurus*, including *N. nanus* (Marsh 1881) and *N. bonneri* (Miller 1972), but it seems likely that the latter, or perhaps even both, are synonymous with *N. gracilis* (Bennett 1994; Unwin 2005). Some adult *Nyctosaurus* possess one of the most elaborate cranial crest morphologies known for any pterosaur, with a huge, bifurcating "antler" projecting dorsally and somewhat posteriorly from the back of their skulls (fig. 18.4; Bennett 2003b). Kellner (2010) considered these crested animals to belong to a dis-

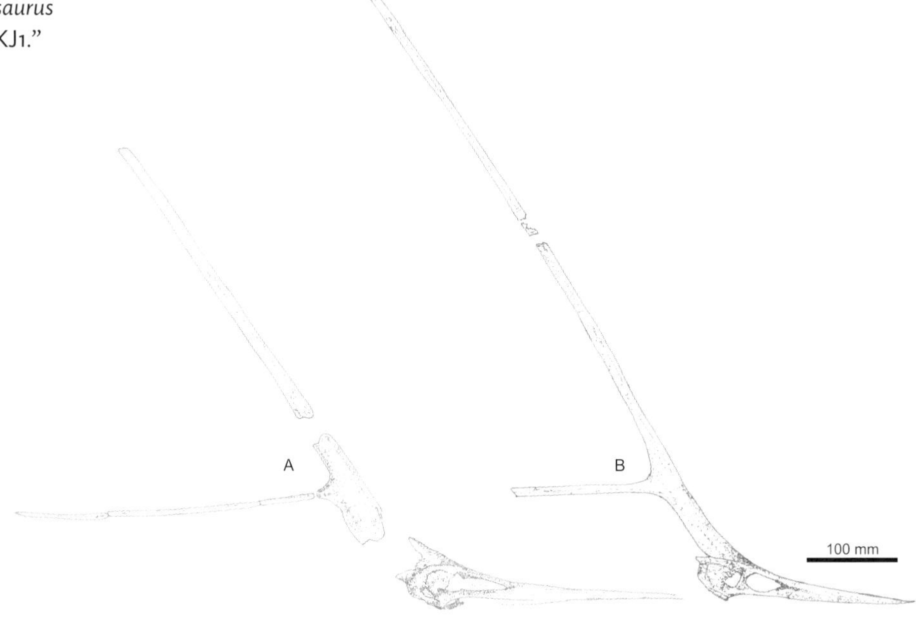

FIG. 18.4. Privately owned, antler-crested *Nyctosaurus* sp. skulls. A, the specimen known as "KJ2"; B, "KJ1." Redrawn from Bennett (2003b).

tinct (unnamed) species, but with good evidence that crestless *Nyctosaurus* were also not fully grown, this may be unlikely given the patterns of cranial crest growth we see in other pterosaurs.

Additional nyctosaurids are known from further south in the Americas. Frey et al. (2006) named a second nyctosaurid from a fairly complete skeleton found in Coniacian (89–85 Ma) deposits of Mexico *Muzquizopteryx coahuilensis*. This specimen occupied an ornate wall stone in a quarry manager's office for some time before it was handed over to Mexican scientists. Heading even further south leads us to a solitary nyctosaurid humerus from Maastrichtian (70–65 Ma) deposits of northeast Brazil, which may be one of the most significant nyctosaurid fossils yet found despite its unassuming nature (Price 1953). This specimen (named "*Nyctosaurus lamegoi*," but probably too fragmentarily represented to warrant a name) is currently the only record of a pteranodontian from the end of the Mesozoic, and thus makes them one of only two pterosaur groups that lived until the end of the Cretaceous. As discussed in chapter 26, this has an important bearing on the mystery surrounding pterosaur extinction.

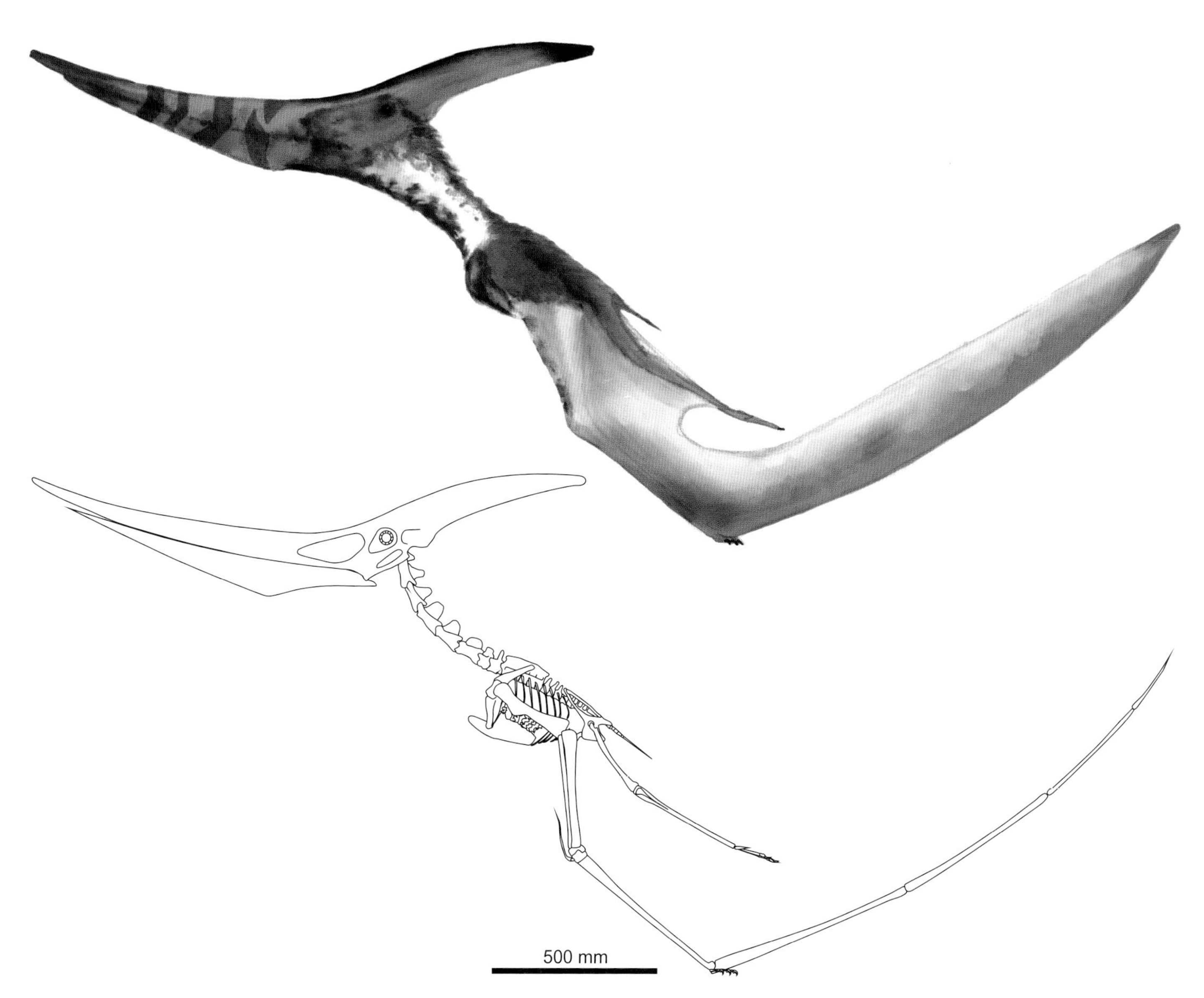

Fig. 18.5. Skeletal reconstruction and life restoration of a launching *Pteranodon longiceps*.

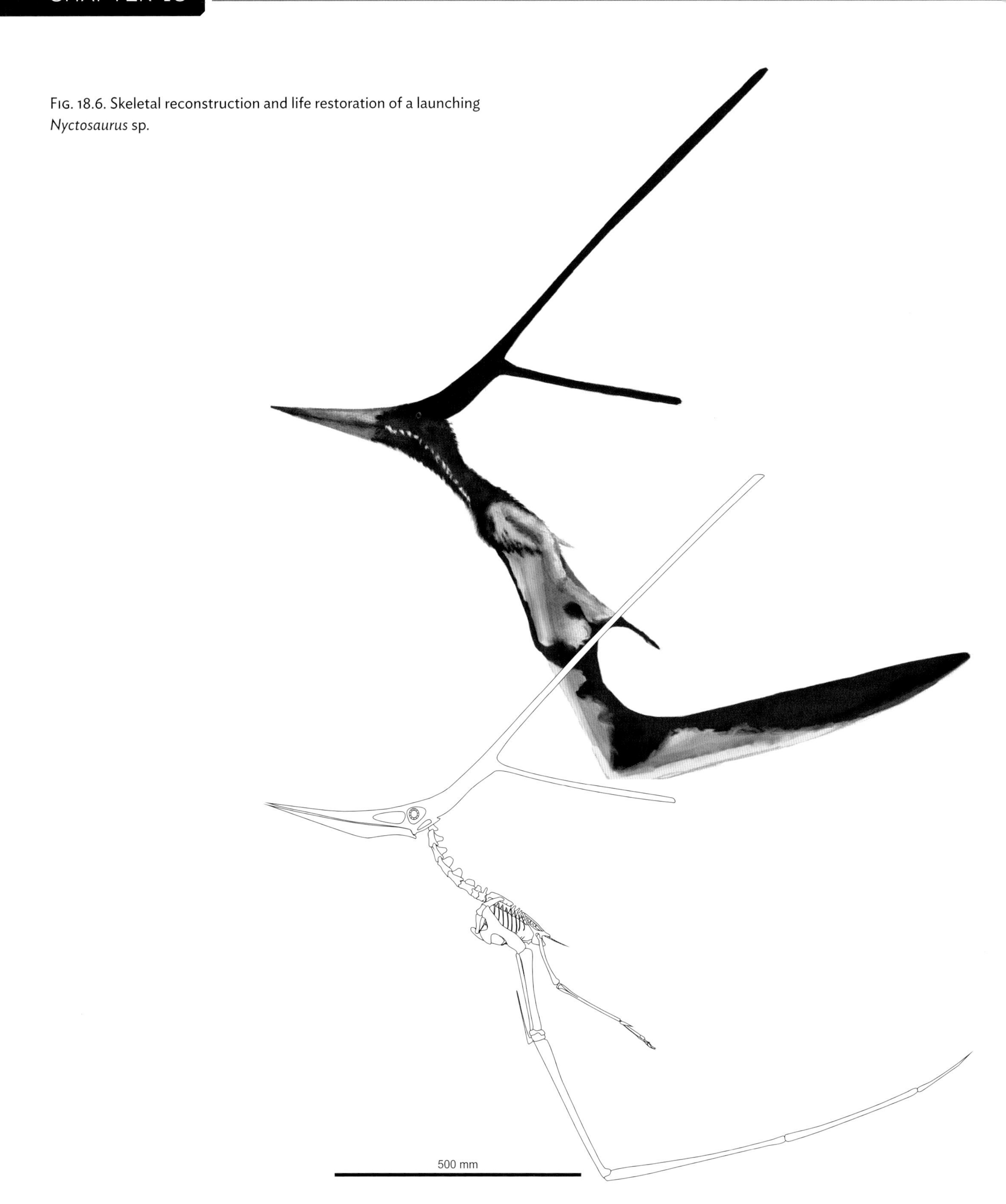

FIG. 18.6. Skeletal reconstruction and life restoration of a launching *Nyctosaurus* sp.

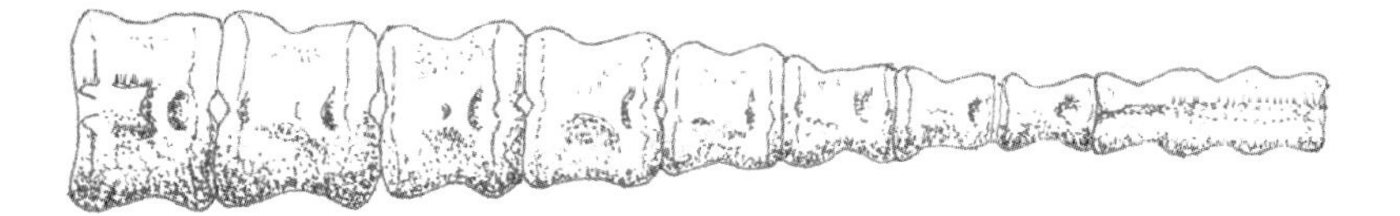

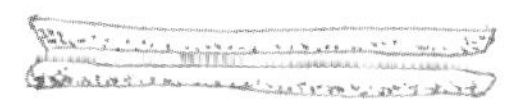

Fig. 18.7. The bizarre caudal series of *Pteranodon* in dorsal view. Note the wide "duplex" centra and long, paired tail rods. Redrawn from Bennett (2001).

Fig. 18.8. *Nyctosaurus* with the wind in its pycnofibers, a sunset, and some dramatic cloudscaping.

Anatomy

Pteranodontians have the same basic long-winged, big-head bauplan as ornithocheirids, but have injected enough of their own character into it that much of their anatomy is recognizable even in isolation (figs. 18.5–18.6). Pteranodontian anatomy is also very well known, and is described in extensive detail in monographs by Williston (1903) and Bennett (2001), with additional details by Bennett (2003b) and Frey et al. (2006).

Pteranodontian wingspans are quite variable. Nyctosaurids were smaller than many ornithocheiroids with wingspans of 2 m in *Muzquizopteryx* (Frey et al. 2006) and up to 2.9 m for *Nyctosaurus* (Wellnhofer 1991a). "*Nyctosaurus lamegoi*" may be the largest nyctosaurid, with a wingspan of up to 4 m (Price 1953). Pteranodontids, by contrast, were giants. They were the largest of all ornithocheiroids and, excepting the mighty azhdarchids, were the largest pterosaurs of all. Wingspan estimates for *Ornithostoma* are not possible because of its fragmentary nature, but Bennett (2001) found that *Pteranodon* regularly attained wingspans over 6 m, and the largest known individuals spanned 7.25 m. It seems, however, that only purported males grew this large; females, in comparison, were more modestly sized with 4 m spans (fig. 8.8; Bennett 1992).

Pteranodontian skulls are characterized by very long, generally tapering jaws devoid of teeth. The snouts of male *Pteranodon* show subparallel dorsal and ventral margins that lead to a relatively blunt, robust upper jaw tip. Their upper jaws were noticeably longer than the lower, a trait that was more exaggerated in males than females. The upper jaw tip of *Nyctosaurus* seems to have been quite delicate, breaking off in many specimens and inspiring many artists to incorrectly restore it with an underbite (Bennett 2003b). The mandible of *Pteranodon* is characteristically deeper at the fusion of the mandibular rami than anywhere else along its jaw length (Unwin 2003), and nyctosaurids show a similar, if less pronounced, version of this mandibular morphology. In all pteranodontids, the mandibular symphysis is long and occupies at least 75 percent of the lower jaw length.

The nasoantorbital fenestrae and temporal openings of pteranodontians are relatively small, but their orbits are large and positioned high in the skull. While *Muzquizopteryx* and some *Nyctosaurus* lack headcrests, all *Pteranodon* and some adult *Nyctosaurus* bear supraoccipital crests, the largest of which have bases extending from the posterodorsal region of the skull to immediately anterior of their orbits. All purported female *Pteranodon* have short, posterodorsally directed crests with rounded margins, but different species of male *Pteranodon* show at least two distinct crest morphs; *P. longiceps* have narrow, posterodorsally projecting crests that may almost double the length of the skull, while *P. sternbergi* possess dorsally directed, anteriorly bulbous crests that are four times taller than the underlying skull (fig. 18.3). Crested *Nyctosaurus*, as mentioned above, possess thin, bifurcating rami that erupt from the posterodorsal region of their skulls to comprise a crest almost three times the longer than the skull itself (fig. 18.4). The resemblance of this to a boat sail mast has not been lost on engineers, and two teams of authors have modeled the aerodynamic properties of the *Nyctosaurus* crest with a sail-like membrane extending between its rami (Cunningham and Gerritsen 2003; Xing et al. 2009). The idea of the *Nyctosaurus* crest supporting a sail—which is very prevalent in many *Nyctosaurus* illustrations—is almost entirely unfounded. As we have seen in other chapters, pterosaurs with extensive soft-tissue components to their crests typically (but not always; see chapter 5) have fibrous margins where the bony crest anchored the soft-tissue component (e.g., Campos and Kellner 1997; Bennett 2002). Like all ornithocheiroids, the crest of *Nyctosaurus* is completely smooth, so there is no reason to think that this pterosaur possessed a soft-tissue "sail" (Bennett 2003b).

The postcranial skeleton of pteranodontians is extensively pneumatized with even components of the hindlimb bearing hallmarks of pneumaticity (Claessens et al. 2009). The anterior seven cervicals of pteranodontians are generally similar to those of other ornithocheiroids, being relatively complex and robust structures with high, triangular neural spines. The nature of the cervical articulations and the preservation of some articulated *Pteranodon* necks suggest that pteranodontians—and probably all ornithocheiroids, based on their cervical similarity—could hold their necks in a relatively tight arc (Williston 1902b; Bennett 2001). The two posteriormost cervicals, as in all pterosaurs, are "dorsalized" and sit in front of their fused notaria, which are in turn comprised of up to six (*Pteranodon*) or seven (*Nyctosaurus*) anterior dorsals. Approximately six unfused dorsal vertebrae exist behind the notarium in most *Pteranodon* individuals, and their sacral vertebral counts are of similar number (Bennett 2001). *Muzquizopteryx* and mature *Nyctosaurus* have sacral regions comprising eight vertebrae, and a particularly well-ossified *Pteranodon* shows a synsacrum comprised of 10 vertebrae, including one "sacralized" caudal. The 11 or more free caudals of *Pteranodon* are very distinctive, thanks to the distal elements forming fused, elongate, "duplex" centra (fig. 18.7). In anterior or posterior view, these two rounded bodies assume a figure of eight profile but become separate distally to form two parallel-sided "caudal rods." The length of this tuning fork-like structure gives *Pteranodon* an unusually long tail for a pterodactyloid, comparable in length with those of boreopterids (chapter 17). The function of these duplex centra remains unclear, however. Bennett (2001) proposed that it supported and controlled a small uropatagium stretching between the thighs, but this is inconsistent with pterosaur fossils suggesting that their tails were not bound into their uropatagia (e.g., Sharov 1971; Wellnhofer 1970). It is entirely possible that *Pteranodon* bucked this trend of course, but other functions (such as a swimming aid, anchoring a soft-tissue display device, or a tail vane) cannot be ruled out. In contrast, the tails of nyctosaurids are either short and comprised of fairly unremarkable caudals (*Muzquizopteryx*), or possess a single, rather than a "tuning fork," caudal rod at their distal end (*Nyctosaurus*).

The main distinctions between pteranodontids and nyctosaurids are seen in their forelimb morphology, largely due to nyctosaurids developing wing anatomy quite unlike that of any other pterosaur. In each case, their scapulocoracoids are broad and robust, but those of nyctosaurids are odd in not articulating with their notaria. Instead, the proximal end of the nyctosaurid scapula is enlarged and rounded. The sterna of both *Pteranodon* and nyctosaurids are similar, however, with both being proportionally large and deeply dished. Pteranodontid humeri are very much like those found boreopterids and ornithocheirids with long, warped deltopectoral crests and a generally chunky profile, but nyctosaurids differ again in bearing unwarped, "hatchet-shaped" deltopectoral crests that project prominently from their humeral shafts. In this respect, they resemble those of derived rhamphorhynchines (chapter 13), but are proportionally larger and bear a unique set of muscle scars, even compared to those of pteranodontids (Bennett 1989). The wing metacarpals of all pteranodontians are relatively long, more than twice (*Pteranodon*) or 2.5 times (nyctosaurids) the length of the humerus (Unwin 2003). Such proportions are only seen in one other pterosaur group: the distantly related azhdarchids (chapter 25). Many of the tendons associated with the upper arm and forearm are mineralized in nyctosaurids (most extensively in *Muzquizopteryx*, but also in *Nyctosaurus*), a trait not seen elsewhere in Pterosauria. Pteranodontian wing fingers are long, occupying over 60 percent of the wing skeleton in *Pteranodon* and 55 percent of the wing in *Nyctosaurus*. The wing fingers of *Nyctosaurus* are further distinguished by only having three phalanges, a configuration that is not unheard of in other pterosaurs (see chapters 11 and 19), but is rare nonetheless. Perhaps one of the most striking differences between pteranodontian groups concerns their smaller metacarpals and fingers. The first, second, and third metacarpals of *Pteranodon* have lost contact with the carpus, but they still support three tiny fingers at their distal ends. Nyctosaurids, by contrast, have lost metacarpals I–III and their corresponding digits. Thus, excepting the flight digit, nyctosaurids had no fingers whatsoever.

In contrast to their unusual forelimbs, pteranodontian hindlimbs are fairly conservative, with rather standard ornithocheiroid hindlimb and pelvic bauplans. Like ornithocheirids, their pelves bear posteriorly deflected ischiopubic plates with dorsally curving preacetabular processes. The postacetabular processes are dorsally prominent in both groups, but they are more complex and proportionally larger in *Pteranodon*. Pteranodontian hindlimbs are among the shortest, relatively speaking, of any pterosaur group, with *Pteranodon* hindlimbs a mere 20 percent of a single wing length and those of *Nyctosaurus* even shorter at 16 percent. This gives *Nyctosaurus* the most piddling pterosaur legs relative to body size yet known for any pterosaur. Nyctosaurid feet are poorly known, but those of *Pteranodon* seem more akin to the larger feet of ornithocheirids than the extra small feet of boreopterids and istiodactylids. Like most pterodactyloids, *Pteranodon* feet have four toes with quite weakly developed claws, and they lack a fifth digit entirely.

Locomotion

FLIGHT

The flight ability of *Pteranodon* has been researched more than extensively than any other pterosaur. Pioneering studies into pterosaur flight were performed on this animal, and subsequent generations of researchers have simulated its flight mechanics with increasing sophistication (Hankin and Watson 1914; Kripp 1943; Bramwell and Whitfield 1974; Stein 1975; Brower 1983; Hazlehurst and Rayner 1992; Chatterjee and Templin 2004; Elgin et al. 2008; Witton 2008a; Sato et al. 2009; Witton and Habib 2010). Nyctosaurid flight has not been as consistently investigated, but a small body of research into its flight does exist (Brower and Veinus 1981; Chatterjee and Templin 2004; Witton 2008a; Xing et al. 2009).

Generally speaking, pteranodontians are considered to have been oceanic soarers *extraordinaire*, with their long, narrow wings ideally suited to effortless gliding over seas and oceans, exploiting air currents for lift, and only rarely flapping. In this respect, pelagic seabirds like albatross and petrels are the best modern analogues to pteranodontians. The only opposition to this hypothesis is that *Pteranodon* was too large and heavy to fly, a proposal made after scaling flapping rates and launch strategies of modern albatross to giant *Pteranodon* (Sato et al. 2009). Pterosaur workers have not been too kind to this idea, with numerous arguments raised against the flightless *Pteranodon* hypothesis (Witton and Habib 2010). Most fundamentally, it is argued that albatross and pterosaur takeoff cannot be treated interchangeably given the likelihood of very different launch strategies in both, and that use of quadrupedal launch in *Pteranodon* would permit far more powerful takeoffs than those of a comparably sized, bipedal bird. Moreover, *Pteranodon* has some of the most developed flight anatomy of any flying reptile, which can only be explained through its retention of flight at wingspans well beyond the size limits of modern birds.

Further evidence that *Pteranodon* was flight worthy stems from the occurrence of most of their fossil remains in deposits hundreds of kilometers from the nearest paleoshoreline, right in the center of the Western Interior Seaway. With little in the way of strong swimming adaptations, the only way thousands of pteranodontians made it so far from shore was through powerful flight. The once popular idea of pteranodontians using their headcrests as rudders or airbrakes during flight (e.g., Bramwell and Whitfield 1974; Stein 1975) has fallen out of favor in modern times after wind tunnel and digital experiments with variably crested model pteranodontians found their crests had negligible aerodynamic effects (Elgin et al. 2008; Xing et al. 2009).

Like most ornithocheiroids, pteranodontids appear well adapted for takeoff from water, possessing the robust shoulders and musculature required for

Fig. 18.9. Suggested postures of *Pteranodon* (A) and *Nyctosaurus* (B). Note that the forelimbs of *Nyctosaurus* are so long that they are of little propulsive use during walking, and were likely used as stability aids alone.

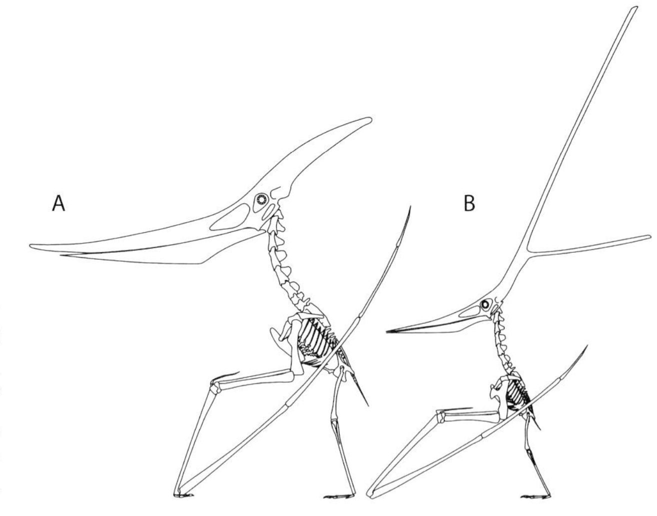

this job (Habib and Cunningham 2010). It's doubtful that nyctosaurids would have been quite as adept at this trick, however. Two key characteristics of habitual water launchers—expansion and reconfiguration of their downstroke muscles and reinforcement of the scapula/notarial brace—are lacking in these forms. By contrast, nyctosaurids seem to have been particularly happy in the air, perhaps more than any other pterosaur group (fig. 18.8). The development of ossified tendons across their proximal wings is a good reason to suspect this, as these structures only develop when constant strain on muscle and tendons encourages their reinforcement to the point where they mineralize. The same effect happens in many bird hindlimbs, and sometimes even in our own legs should we overindulge in high-heeled shoes. Ossified wing tendons in nyctosaurids probably developed through constant contraction of their wing muscles, perhaps by adopting their flight postures for very, very long periods of time. The maladaptation of nyctosaurid anatomy to terrestrial locomotion (see below) adds further credence to this idea, as does the astoundingly high-aspect shape of their lightly loaded wings, a clear hallmark of extremely efficient soaring abilities (see fig. 6.4O). In fact, nyctosaurid wing shape hints that they may have possessed the highest glide ratio of any creature, ever (Michael Habib, pers. comm. 2012); perhaps, like modern frigate birds, they were capable of effectively infinite flight time, only landing occasionally to roost and breed.

ON THE GROUND

A number of different suggestions have been made for the preferred methods of terrestrial locomotion in *Pteranodon*. Hankin and Watson (1914), and Bramwell and Whitfield (1974) considered *Pteranodon* to have hardly any terrestrial competency at all, suggesting they pushed themselves around with their legs while lying on their bellies (fig. 7.2B). Wellnhofer (1991*a*) restored *Pteranodon* as a sprawling quadruped, while Bennett (1990, 2001) considered it a very upright biped, suggesting its arms were too long for effective movement on all fours (fig. 7.2C). To my eyes, there seems no reason to assume that *Pteranodon* couldn't shuffle or hop about in the manner discussed for the long-limbed ornithocheiroids in chapter 16, though it does seem that its forelimbs were approaching the point of being too long for effective use in quadrupedal locomotion (fig. 18.9A).

That threshold has almost certainly been surpassed by *Nyctosaurus*. As noted by Bennett (1996), the forelimbs of this animal are so long that they could not be used for movement over the ground in concert with the hindlimbs. This does not necessarily mean that nyctosaurids were bipedal, but their arms may have been utilized more like walking sticks than propulsive limbs, using their arms to aid balance while their hindlimbs propelled them along. It is quite possible that the loss of their smaller fingers, and the increased traction they would offer, reflects limited use of their forelimbs in terrestrial locomotion. This configuration may bestow upon nyctosaurids the prize for the Worst Pterosaur Terrestrial Locomotors, but they probably would not have minded too much. Their overwhelming adaptations to flight suggest they landed infrequently and could choose their landing spots, carefully ensuring they were not at risk from predators or challenged by the terrain.

Paleoecology

Thanks to the enormous sample size of *Pteranodon*, more details about its life history are known than those any other pterosaur. Indeed, many of the paleobiological insights gleaned from *Pteranodon* have shone light into the lifestyles of other pterosaurs, and particularly with regard to their reproductive mecha-

FIG. 18.10. The most unfriendly of all pterosaurs: a big male *Pteranodon sternbergi* and his harem.

nisms (Bennett 1992) and growth strategies (Bennett 1993). Interestingly, all *Pteranodon* individuals known to date represent osteologically mature, or very nearly mature, individuals (Bennett 1993). With all of these specimens also occurring well out to sea, this perhaps suggests that only mature *Pteranodon* flew over the middle of the Western Interior Seaway. Where the juvenile indiviudals were is not clear, but they may have been living in separate environments and perhaps filling different ecological niches (an idea most notably championed in Unwin 2005). The same does not seem quite so true for *Nyctosaurus*, however, which is known from the same open marine deposits as *Pteranodon*, but from a range of differently aged individuals (Bennett 2003b). We may see this as evidence that nyctosaurids began their pelagic existence earlier in life than *Pteranodon*.

The correlation between pterosaur maturity and headcrest growth was first identified in *Pteranodon*, and the patterns seen in this pterosaur also seem true of other pteranodontians (Bennett 1992). The crest development seen in mature *Nyctosaurus*, when contrasted with their crestless juvenile forms, represents the most extreme cranial modification recorded in any pterosaur species. As discussed in chapter 8 and described above, the crests of *Pteranodon* are known to vary in a manner that correlates with their pelvic morphology and overall body size, all of which are strong indicators of pronounced sexual dimorphism (fig. 18.10). Bennett (1992) interpreted this as evidence of lek mating strategies in *Pteranodon*, in which males would compete for harems of females. This would require the offshore-feeding *Pteranodon* to reproduce during congregations in short-lived rookeries, at which males would compete intensely with each other for reproductive access to females before the rookery dispersed again. Such a situation would strongly favor the biggest, most decorated, and probably worst-tempered males. This level of interpretation may sound farfetched, but the extent of dimorphism seen in *Pteranodon* is mirrored in numerous, unrelated modern birds and mammals that also use lekking social strategies such as seals, deer, grouse, and birds of paradise. Bennett's inferences should not be casually dismissed (Bennett 1992). It remains to be seen whether the headgear of *Nyctosaurus* reflects similar sexual selection pressures and intense male competition, but it is interesting to note that the most pelagic of pterosaurs also has one of the biggest pterosaur headcrests known. Maybe, as proposed for *Pteranodon*, *Nyctosaurus* individuals rarely met to breed, leading to windows of very intense competition to impress prospective mates and strongly sexually selected headgear.

The occurrence of pteranodontians in marine deposits, not to mention the regurgitated gut content of one *Pteranodon* individual (fig. 8.10A; Brown 1943), both suggest that pteranodontians dined on small fish and other nektonic animals. A number of different foraging strategies have been proposed for pteranodontians including skim feeding (e.g., Marsh 1876; Eaton 1910; Bramwell and Whitfield 1974; Wellnhofer 1980; Cunningham and Gerritsen 2003), dip feeding (Wellnhofer 1991a), diving (Brown 1943; Kripp 1943; Bennett 2001), or alighting on the water to grab passing prey (Bramwell and Whitfield 1974). The lack of skim-feeding adaptations in all pteranodontians (Humphries et al. 2007; Witton 2008b) suggests that they could not forage in this manner. The relatively small anterior cervicals and known preference for small prey of *Pteranodon* may indicate that habitual dip feeding was also not in the cards for *Pteranodon* (see chapters 13 and 16 for discussions of this feeding strategy). Instead, these same features may have favored the more sedate foraging strategy of snatching fish from a floating position on the water surface (Bramwell and Whitfield 1974). This would also suit the particularly slender, transversely weak *Pteranodon* jaw tips. There also seems no reason to assume that *Pteranodon* could not perform shallow dives from floating starts like many modern seabirds (fig. 18.1). Whether they dived or not, the adaptations of the *Pteranodon* shoulder girdle for aquatic launching are consistent with frequently landing on the water surface, and feeding is perhaps the most obvious reason for them to regularly get their feet wet.

The apparent exclusivity of nyctosaurid fossils in open marine deposits, and their lack of terrestrial adaptations, suggest they must also have fed in open water. A lack of adaptations for aquatic launching argues for frequent feeding on the wing. Because of their lack of skim-feeding adaptations, it seems likely that they dip fed, swooping across the water surface to snatch fish or squid with their jaw tips. This would complete the frigatebird analogy already made for these animals, as these birds obtain most of their sustenance in this manner (alternatively, they plunder freshly caught food from other birds) and are also reluctant (though capable) swimmers.

19 Ctenochasmatoidea

PTEROSAURIA > MONOFENESTRATA > PTERODACTYLOIDEA > LOPHOCRATIA > CTENOCHASMATOIDEA

The ctenochasmatoids are the first of the lophocratian pterosaurs that we will meet. Comprising seven major pterodactyloid groups, Lophocratia is the more terrestrially adapted side of the pterodactyloid coin, contrasting with the strongly flight adapted ornithocheiroids we discussed in previous chapters.[8] Two of the characteristic features of Lophocratia, their long hindlimbs and robust extremities, are obvious adaptations for a more terrestrially based existence, while the "simple" shape of their deltopectoral crests adds a further distinguishing feature (Unwin 2003). Their name—translating to "crested heads"—reflects the propensity for lophocratian pterosaurs to sport cranial crests occupying most of their skull lengths. These typically erupt in the rostral region and extend to their posterior skulls, their extent marked by low bony ridges that were enlarged in life with enormous soft-tissue extensions (fig. 19.1). This was once thought to be a unique feature of this group, but the discovery of similar crests in wukongopterids and perhaps some "campylognathoidids" suggests that this crest construction is not exclusive to lophocratians. The full extent, and sometimes even the very existence, of these crests can often only be determined under ultraviolet light, and as this technique has only become established recently, the almost universal distribution of crests among lophocratians is a relatively modern concept (Frey and Tischlinger 2000; Frey, Tischlinger, et al. 2003; Tischlinger 2010).

The earliest and most "primitive" of the lophocratians are the aforementioned ctenochasmatoids, a large, long-lived, and morphologically diverse group of pterodactyloids with a prominent place in pterosaur research history. A well-known member of this clade, *Pterodactylus*, was the first pterosaur known to science (fig. 2.2) and, rather obviously, is the namesake of the major pterosaur group Pterodactyloidea. The name "*Pterodactylus*" dominated pterosaur nomenclature for much of the nineteenth century (see preceding chapters for examples), and because the anatomy of this animal has been well documented for over 200 years it maintains a prominent place in phylogenetic and functional research. Ctenochasmatoids also played an important role in breathing life back into pterosaur research in the latter part of the twentieth century, being the principle focus of Peter Wellnhofer's extensive 1970 monograph on pterodactyloids from the Solnhofen Limestone, and thus they were thus a major part of a significant publication in the modern reboot of pterosaur studies.

Most ctenochasmatoids seem to have been pterosaur equivalents of wading shorebirds, using extremely specialized jaws and dentitions to feed on small prey items suspended in water. Others seem more terrestrially adapted, but have truly bizarre jaws and teeth that defy easy explanation. Ctenochasmatoids are well represented from Upper Jurassic to Lower Cretaceous (150–105 Ma) deposits from Asia, Europe, and South America, with some fragmentary remains also known from Upper Jurassic deposits of Africa and North America (fig. 19.2; Barrett et al. 2008). They are particularly well represented in the Kimmeridgian (156–151 Ma) Nusplingen Limestone and Tithonian (151–145 Ma) Solnhofen Limestone of southern Germany, and much of the taxonomic and descriptive work on these ctenochasmatoids underpins our modern understanding of this group (e.g., Wellnhofer 1970, 1978; Bennett 1996b).

The content of Ctenochasmatoidea is fairly uniform across different models of pterosaur evolution.

[8] Not all pterosaur workers agree on the existence of Lophocratia (e.g., Kellner 2003; Andres and Ji 2008; Wang et al. 2012), instead considering Ctenochasmatoidea to represent the first offshoot of Pterodactyloidea. In these alternative schemes, ornithocheiroids group closer to dsungaripterids and azhdarchoids. Happily, these interpretations do not affect our consideration of pterosaur diversity too much, as the contents of the broader groups remain fairly similar, and only their relative positions on the pterosaur tree are different.

FIG. 19.1. A portrait of the Late Jurassic pterosaur *Cycnorhamphus suevicus*, complete with goofy teeth, a divergent anterior jaw, and a headcrest that looks like a silly hat. This is not a joke; a real animal once looked something like this. See figure 19.10 for the proof.

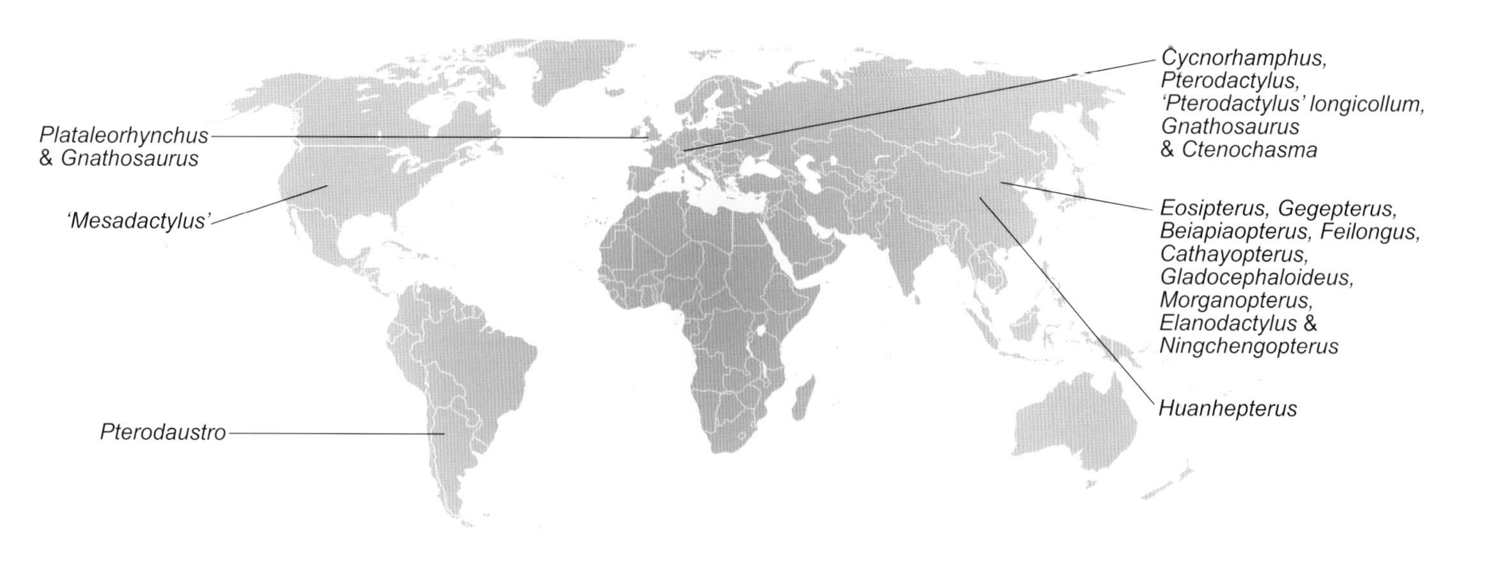

Fig. 19.2. Distribution of ctenochasmatoid taxa.

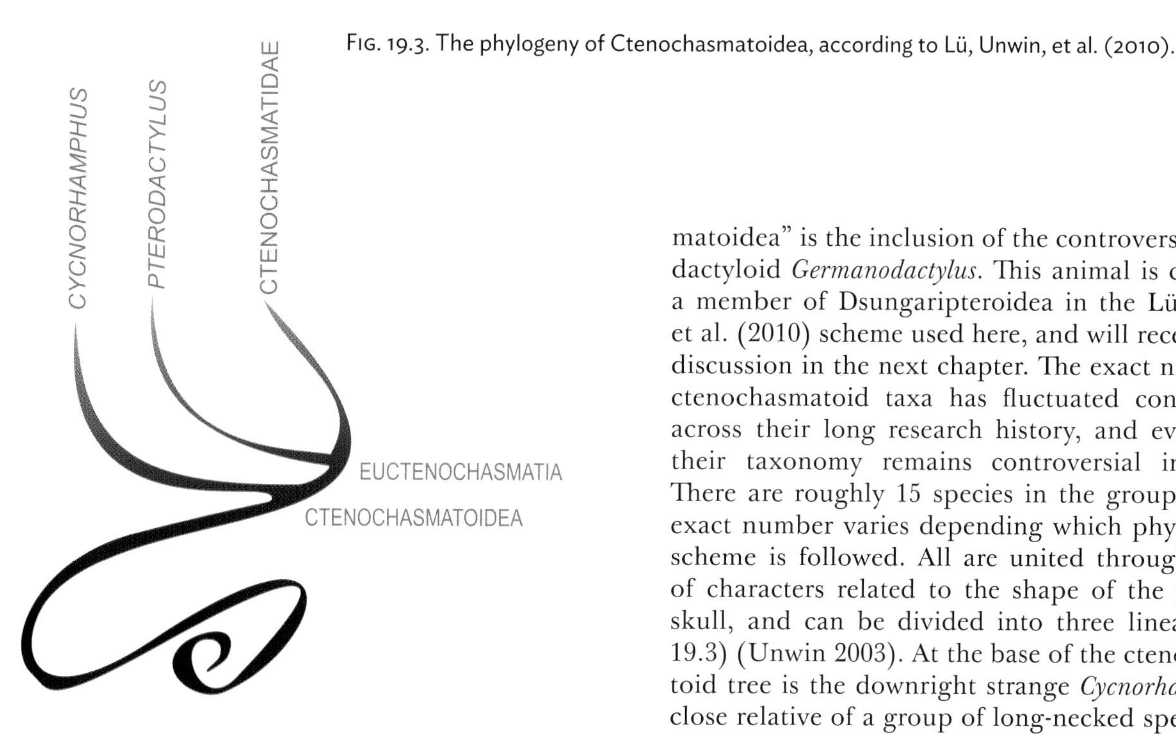

Fig. 19.3. The phylogeny of Ctenochasmatoidea, according to Lü, Unwin, et al. (2010).

Some schemes have labeled a practically identical group "Archaeopterodactyloidea" (Kellner 2003; Lü, Ji, Yuan, and Ji 2006; Andres and Ji 2008; Wang et al. 2009), but its only difference from "Ctenochasmatoidea" is the inclusion of the controversial pterodactyloid *Germanodactylus*. This animal is classed as a member of Dsungaripteroidea in the Lü, Unwin, et al. (2010) scheme used here, and will receive more discussion in the next chapter. The exact number of ctenochasmatoid taxa has fluctuated considerably across their long research history, and even today their taxonomy remains controversial in places. There are roughly 15 species in the group, but the exact number varies depending which phylogenetic scheme is followed. All are united through a suite of characters related to the shape of the posterior skull, and can be divided into three lineages (fig. 19.3) (Unwin 2003). At the base of the ctenochasmatoid tree is the downright strange *Cycnorhamphus*, a close relative of a group of long-necked species, Euctenochasmatia. This group includes the well-known *Pterodactylus* and a further set of derived species, the Ctenochasmatidae, which are characterized by an unusually high number of slender teeth packed into elongate jaws.

Back to the Beginning

The research history of ctenochasmatoids, as mentioned briefly in chapter 2, dates back to at least 1784 when Cosimo Collini described a perfectly preserved, fully articulated ctenochasmatoid specimen from the Tithonian Solnhofen Formation of Bavaria, Germany. The revered French naturalist Georges Cuvier noted its probable volant nature and in 1809 gave it a now iconic generic name: "Ptero-dactyle." In 1812, a species name was allocated to the specimen by Thomas von Soemmerring, through the erection of "*Ornithocephalus*" *antiquus*. Cuvier's "Ptero-dactyle," formulated correctly into *Pterodactylus*, had priority over Soemmerring's "*Ornithocephalus*" by three years, so the two names were combined in 1819 to form *Pterodactylus antiquus*. In the coming century, innumerable specimens and dozens of species were housed within *Pterodactylus* (see below, and chapters 10, 13, 16, 17, 20, and 21 for examples). These have since been recognized as unique genera belonging to disparate parts of the pterosaur tree, so *Pterodactylus* is no longer the taxonomic waste bin it once was (Wellnhofer 1970; Bennett 1996b).

Some question still exists over how many *Pterodactylus* species we can recognize, however. Only one *Pterodactylus* species, *P. antiquus*, is discerned by some (Bennett 1996b; Atanassov 2000; Jouve 2004), but a second species, *P. kochi*, is maintained by many (e.g., Unwin 2003, 2005; Andres and Ji 2008; Wang et al. 2009). There is some agreement regarding a third purported *Pterodactylus* species, however: the Nusplingen form "*Pterodactylus*" *longicollum*. This animal is widely considered to be in need of its own genus and reallocation to another position within Ctenochasmatoidea (e.g., Unwin 2003; Andres and Ji 2008). One of the most recently named Solnhofen pterosaur species, the alleged azhdarchoid "*Auroazhdarcho primordius*" (Frey et al. 2011; see chapters 22–25 for more on azhdarchoid pterosaurs), is almost certainly synonymous with "*P.*" *longicollum*. Not only is "*A. primordius*" riddled with ctenochasmatoid traits (and chronically lacking in any unambiguous azhdarchoid characteristics), but it matches the distinctive proportions and anatomy of "*P.*" *longicollum* perfectly (D. M. Unwin, pers. comm. 2011).

Pterodactylus was only the first of many ctenochasmatoids that would be found in European sediments. The next was an isolated lower jaw, also from the Solnhofen Limestone, that was documented in 1832 and named *Gnathosaurus subulatus* two years later (Meyer 1834). *Gnathosaurus* was considered to be some sort of crocodile until somewhat recently, and it is not hard to see why. Its elongate jaws bristle with long, curved teeth that approximate the condition of a slender-toothed gharial-like crocodile. It was only when a complete skull of *Gnathosaurus* was found in 1951 that its pterosaurian nature became apparent (Mayr 1964). Bennett (1996b) suggested that a tiny, alleged *Pterodactylus* species from the Solnhofen Limestone, "*P.*" *micronyx*, may represent a juvenile stage of this animal, but this synonymy has not been accepted by all (e.g., Unwin 2005). *Gnathosaurus* may also be known from the Berriasian (145–140 Ma) Purbeck Limestone Formation of southern Britain, albeit as a second species, *G. macrurus* (Howse and Milner 1995). Another British fossil, a fragmentary jaw from the Middle Jurassic Stonesfield Slate (Bathonian; 168–165 Ma), may also represent a *Gnathosaurus*-like animal (Buffetaut and Jeffrey 2012). If correctly identified, the Stonesfield specimen is one of the oldest pterodactyloid fossils in the world. However, in a reversed situation to the first discovery of *Gnathosaurus*, a number of pterosaur workers suspect these remains actually represent a type of marine crocodile.

The next ctenochasmatoid taxon to be discovered and named was a close relative of *Gnathosaurus*, the comb-toothed *Ctenochasma* (fig. 19.4). Initially found in "Purbeck" Tithonian deposits of Saxony, Germany (Meyer 1851), it has since emerged in the Solnhofen Limestone and contemporaneous deposits in eastern France (Taquet 1972; Jouve 2004). Following a recent review, three species of *Ctenochasma* are currently recognized: the Saxony form *C. roemeri*, the Solnhofen *C. elegans* (which includes the recently synonymized *C. porocristata*; see Bennett 1996b), and the French species *C. taqueti* (Bennett 2007c). This long snouted pterosaur genus is characterized by extremely fine, intermeshing teeth set into a long jaw, a morphology that strongly contrasted with that of the next ctenochasmatoid to be uncovered, *Cycnorhamphus*. A recently discovered skull and mandible has revealed how astonishingly weird *Cycnorhamphus* is (fig. 19.1)—sporting strangely curving jaws with bizarre oral soft tissues and a very unusual dentition. Many of these features were unknown to nineteenth and twentieth century paleontologists however, as the two *Cycnorhamphus* specimens known at this time were either a juvenile individual lacking the bizarre cranial morphology of adults, or had an incomplete skull. *Cycnorhamphus* was originally reported from the Nusplingen Limestone of Germany and christened *Pterodactylus suevicus* (Quenstedt 1855), but was then transferred to its own genus *Cycnorhamphus* by Harry Seeley in 1870. *Cycnorhamphus*-like fossils

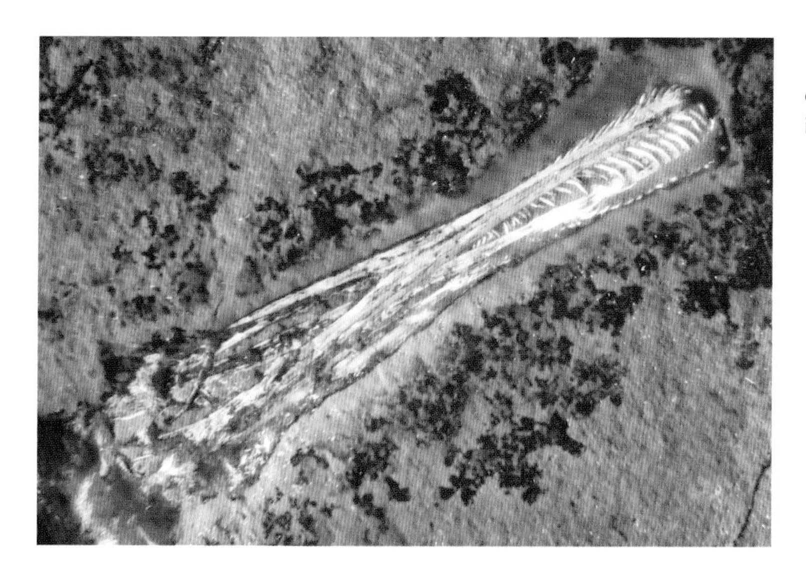

Fig. 19.4. The skull and mandible of a juvenile *Ctenochasma elegans* from the Tithonian Solnhofen Limestone under UV light in ventral view. Photograph courtesy of Helmut Tischlinger.

were found later in Tithonian deposits of France and one, a fairly complete skeleton, was given its own name: *Gallodactylus canjuerensis* (Fabre 1976). Bennett (1996c, 2010) recently swept all material referred to these pterosaurs into a single species *C. suevicus*, assuming that their subtly different anatomy reflects distinct growth phases.

After the plethora of ctenochasmatoid discoveries in Europe throughout the 1800s, the twentieth century proved that they were equally abundant and diverse around the rest of the world. The Argentine *Pterodaustro guinazui* was unveiled in the 1970s and remains one of the most distinctive ctenochasmatoids (and pterosaurs) yet known, with its long, curving jaws packed with hundreds of needlelike teeth that superficially resemble whale baleen (fig. 19.5; Bonaparte 1970). This animal is now known from hundreds of specimens from the Albian (112–100 Ma) Lagarcito Formation of Argentina, including complete skulls (Chiappe et al. 2000), juveniles (Codorniú and Chiappe 2004), and an egg (fig. 8.3; Chiappe et al. 2004). Indeed, *Pterodaustro* remains are so abundant that the site yielding their fossils has been named in their honor: *Loma del Pterodaustro*. *Pterodaustro* also probably occurs in slightly older, Aptian rocks (125–112 Ma), thanks to its likely synonymy with the Argentine ctenochasmatid *Puntanipterus globosus* (Bonaparte and Sánchez 1975; Chiappe et al. 1998).

The year 1989 saw "Dinosaur Jim" Jensen and Kevin Padian name "*Mesadactylus ornithosphyos*," the probably chimeric North American pterosaur from the Upper Jurassic Morrison Formation (Kimmeridgian-Tithonian; 156 –145 Ma) mentioned in chapter 11. Although only some of the "*Mesadactylus*" material is likely to represent a ctenochasmatoid and its taxonomic validity is highly suspect (Bennett 2007b; Smith et al. 2004), it remains significant for being the only substantial ctenochasmatoid fossil yet found in North America. By contrast, the next ctenochasmatoid to be named, *Plataleorhynchus streptorophodon* from the Early Cretaceous (Berriasian; 145–140 Ma) Purbeck Beds of southern England, is more easily diagnosed thanks to its unusually spoonbill-like jaw (Howse and Milner 1995).

The latest chapter in ctenochasmatoid research, as with many pterosaur groups, has been written in Chinese. Zhiming Dong (1982) reported China's first ctenochasmatoid, *Huanhepterus quingyangensis*, from a partially complete skeleton found in the Upper Jurassic Huachihuanhe Formation. Following a short discovery hiatus, Chinese paleontologists then began to discover and name Cretaceous ctenochasmatoids species at an unprecedented rate. Since 1997, *eight* different ctenochasmatoid species have been described from the Yixian Formation (likely of Barremian age; 130–125 Ma). These include *Eosipterus yangi* (Ji and Ji 1997), *Beipiaopterus chenianus* (Lü 2003), *Feilongus youngi* (Wang et al. 2005), *Cathayopterus grabaui* (Wang and Zhou 2006b), *Gegepterus changi* (Wang et al. 2007), *Elanodactylus prolatus* (Andres and Ji 2008), *Ningchengopterus liuae* (Lü 2009a), and *Gladocephaloideus jingangshanensis* (Lü, Ji, et al. 2012). (This species count will increase to nine if *Morganopterus zhuiana* is also a ctenochasmatoid—see below.) Each species is represented by partial skeletons or cranial material, but rarely both, which hinders comparisons among them. Interpretations of specimens of *Eosipterus* and *Beipiaopterus* are also hampered by "improvements"

FIG. 19.5. The skull and mandible of *Pterodaustro guinazui*, of the Albian Lagarcito Formation, Argentina. Note that the majority of the mandibular teeth are missing, but the anterior dentition remains to demonstrate its clear filter-feeding function. Photograph courtesy of Laura Codorniú.

made by fossil dealers (Wang et al. 2007). Where comparable, the anatomy of these animals is really quite similar and there is a very real likelihood that some, and perhaps most, of these species are synonymous with one another.

Special mention should go be given to *Feilongus* and *Morganopterus*, the distinctive pterosaurs we first mentioned in chapter 17. *Feilongus* is considered to be a ctenochasmatoid by some (Wang et al. 2005; Andres and Ji 2008; Wang et al. 2009), but others argue that it, along with *Morganopterus*, should be housed in Boreopteridae (Lü, Ji, Yuan, and Ji 2006; Lü, Pu, et al. 2012; Lü 2010b). There are good reasons to doubt the latter, however. Both *Feilongus* and *Morganopterus* show anatomies that are very distinct from those of boreopterids, but fairly typical for ctenochasmatoids. Their teeth are much shorter than those of boreopterids, and restricted to the anterior region of the jaw rather than occurring almost beneath the orbit. Their headcrests extend along most of the skull and, unlike all ornithocheiroids, possess a fibrous margin that likely anchored a soft-tissue crest in life. Their orbits are rounded, their posterior skull regions are very strongly reclined, and their cervical vertebrae are long and low (Lü 2010c; Lü, Pu, et al. 2012). Such features suggest *Feilongus* and *Morganopterus* are more at home among ctenochasmatoids than with any ornithocheiroids and, in particular, they seem to bear similar jaw tips to those of the Jurassic Chinese ctenochasmatoid *Huanhepterus*.

Anatomy

From species to species, there is probably more variation in the body proportions of ctenochasmatoids than in any other pterodactyloid group (figs. 19.6–19.9). Happily, their anatomy is documented fairly well, and

Fig. 19.6. Skeletal reconstruction and life restoration of a launching *Cycnorhamphus suevicus*.

readers are referred to the likes of Wellnhofer (1970), Fabre (1976), Bennett (1996b, 2007c), and Chiappe et al. (2000) for some of the best detailed descriptions. Despite their diversity, all ctenochasmatoids share an expanded neurocranial region (the portion of skull containing the brain), which results in the posterior skull reclining posteriorly and, in most species, a ventrally projecting occipital face. A typical lophocratian cranial crest is found in most species, and recent discoveries have shown that even skulls without these bony supports can possess substantial soft-tissue crests (fig. 19.7; Frey, Tischlinger, et al. 2003; Tischlinger 2010). Ctenochasmatoids are relatively unusual among pterodactyloids for their pneumatic features

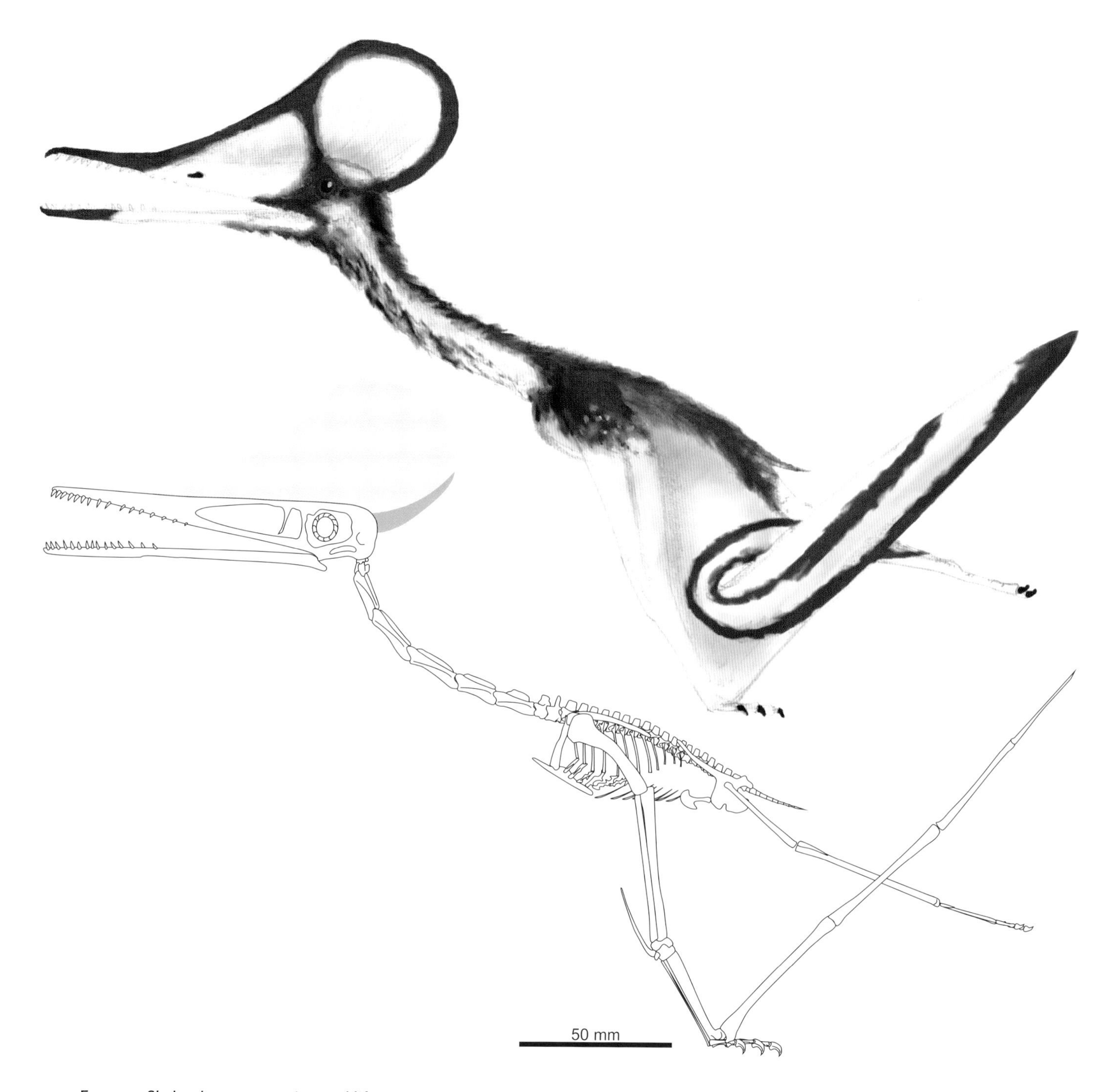

Fig. 19.7. Skeletal reconstruction and life restoration of a launching *Pterodactylus antiquus*.

being restricted to their skulls and axial skeletons, with the exception of *Pterodaustro*, which shows some evidence of pneumatic structures in its wings (Claessens et al. 2009).

Ctenochasmatoids generally failed to attain the giant proportions that many other pterodactyloid clades are famous for, although it is important to stress that many ctenochasmatoid species are not yet known from adult individuals. An exception to this is *Morganopterus*. With a skull 750 mm long (over 950 mm including its supraoccipital crest), it has one of the largest toothed pterosaur skulls known. This easily renders it the largest ctenochasmatoid known, with a wingspan of 4.2 m. (This estimate is consider-

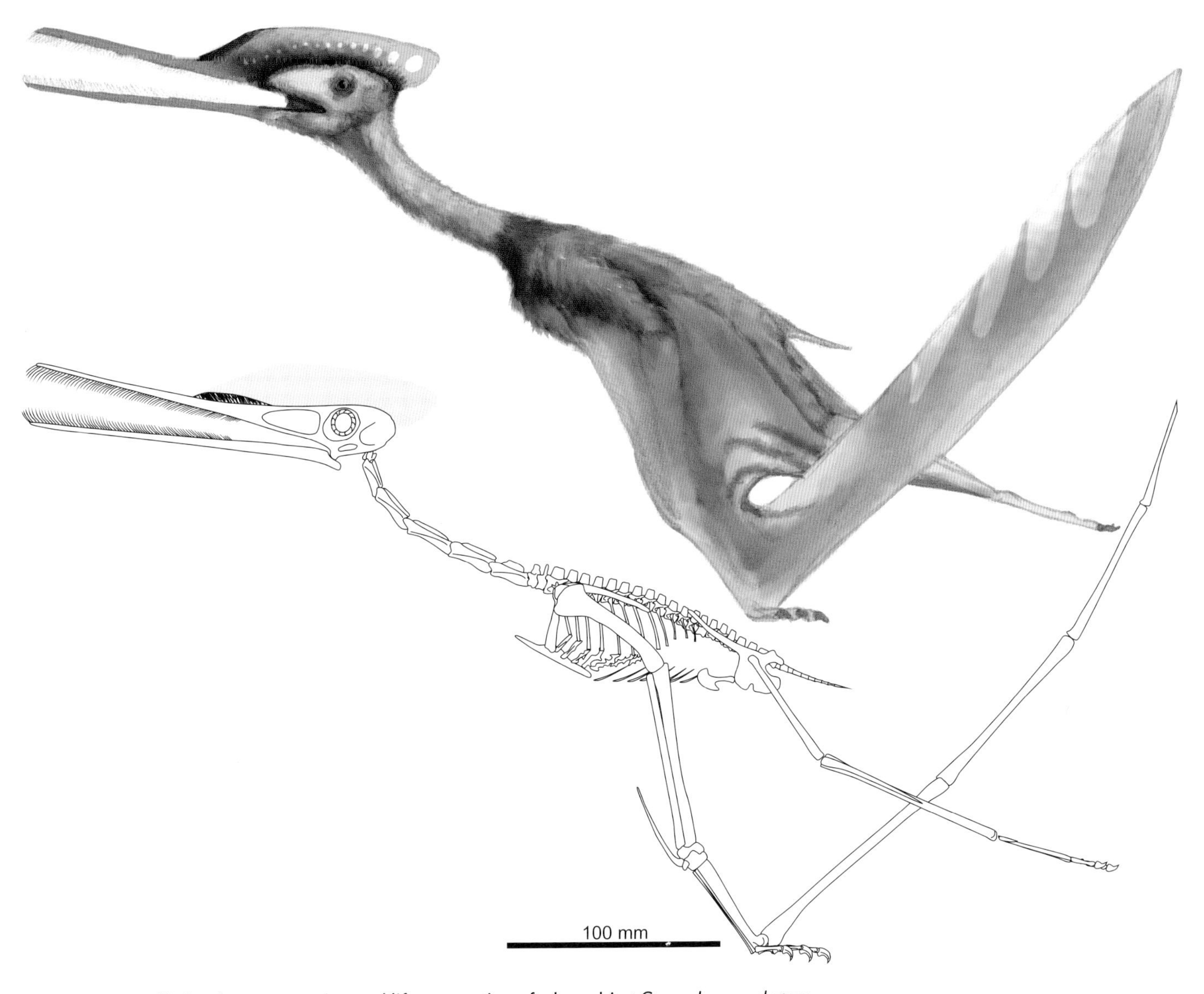

Fig. 19.8. Skeletal reconstruction and life restoration of a launching *Ctenochasma elegans*.

ably less than that of Lü, Pu, et al. [2012], who suggest a span of 7 m for *Morganopterus*. Seven meters is a considerably overinflated wingspan estimate, even if their boreopterid identity of this animal is correct, because a boreopterid with *Morganopterus*-like skull proportions would have a wingspan of around 4.7 m.) The next largest animals were *Pterodaustro*, *Feilongus*, *Cycnorhamphus*, and *Elanodactylus*, all of which had wingspans of around 2.5 m. *Huanhepterus* achieved a slightly lesser wingspan of 2 m. Most other ctenochasmatoids, by contrast, were no more than 1.5 m across the wings. *Pterodactylus* probably represents one of the smallest species of the group, with a wingspan less than 1 m likely.

OSTEOLOGY

Ctenochasmatoid skulls and teeth are their most variable features. Some are not too morphologically dissimilar from those of wukongopterids and dsungaripteroids (chapters 14 and 20), but others possess some of the most "extreme" cranial anatomy of any pterosaurs. The most generalized ctenochasmatoid skull is that of *Pterodactylus*, with a long, gently tapering snout, a bulbous neurocranial region, and large, circular orbits (fig. 19.10A). Its upper and lower jaws each contain around 40 laterally compressed, conical teeth, a rather low number compared to the more derived ctenochasmatoids. Its dentition is more or less

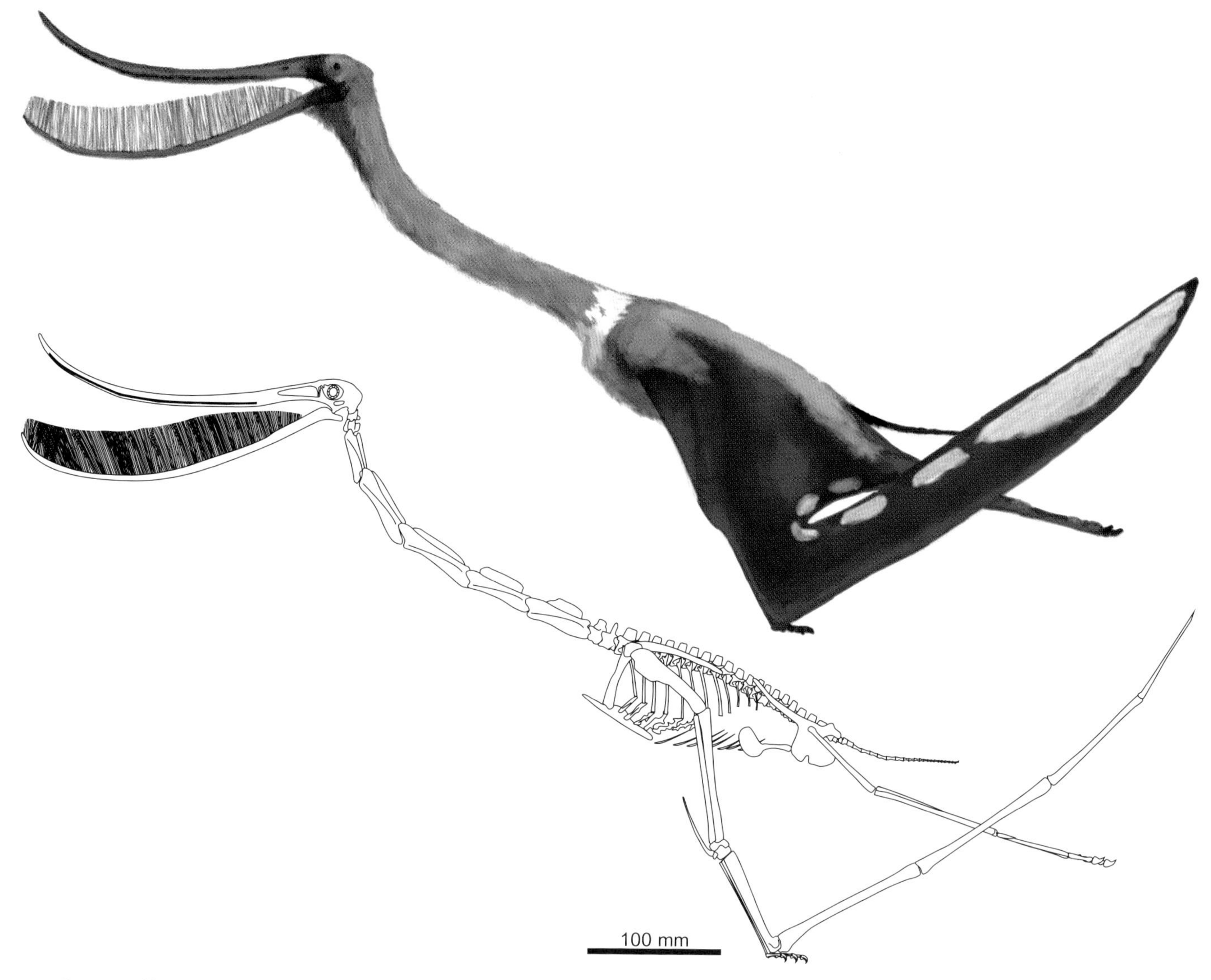

FIG. 19.9. Skeletal reconstruction and life restoration of a launching *Pterodaustro guinazui*.

confined to the front half of the jaws. The skulls of *Feilongus* and *Huanhepterus* (for what little is known of it) are not dissimilar to those of *Pterodactylus*, with comparable tooth counts and teeth confined to the front of their jaws. *Morganopterus* also shows some cranial similarity to *Pterodactylus*, although its skull is considerably more elongate. *Feilongus* and *Morganopterus* both possess particularly long bony crest supports that extend from very near the jaw tips to their supraoccipital crests. The anterior end of a similar crest is seen in *Huanhepterus*, but its posterior extent is unknown, along with the rest of the posterior skull region in this animal. The denition of *Huanhepterus*, *Feilongus*, and *Morganopterus* are more slender and densely packed than those of *Pterodactylus* (Dong 1982; Wang et al. 2005; Lü, Pu, et al. 2012), and *Feilongus* is further characterized by the apparent presence of an overbite. This may well reflect imperfect preservation of its skull and disarticulation of the skull and mandible, however (D. M. Unwin, pers. comm. 2012).

The concept of slender, tightly packed teeth is taken to an extreme by the ctenochasmatids. These animals have skulls of generally similar morphology to those of *Pterodactylus*, but have greatly elongated jaws that can contain hundreds of teeth. Their teeth also splay laterally, creating distinctive dental profiles for each species in dorsal or ventral view (fig. 19.11). Dental counts for the various *Gnathosaurus* species number

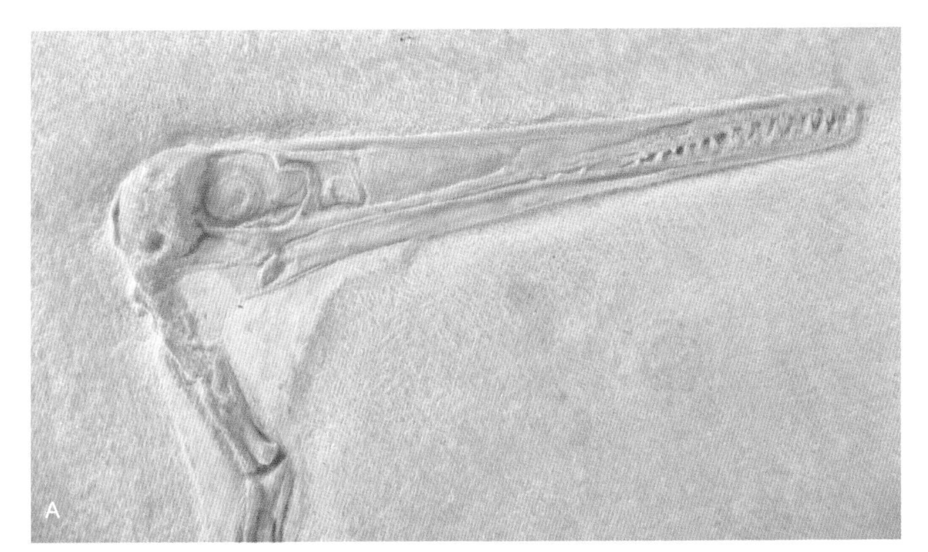

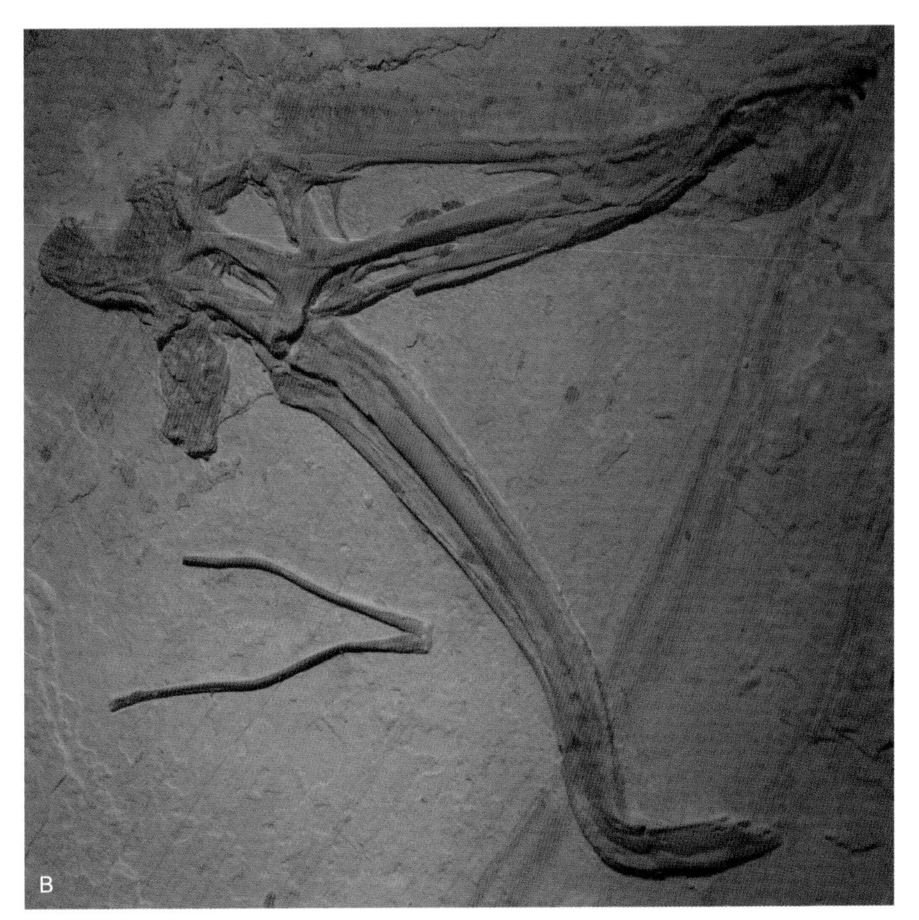

Fig. 19.10. Cranial remains of Solnhofen ctenochasmatoids. A, *Pterodactylus kochi* (the "Vienna specimen"; B, *Cycnorhamphus suevicus*. Note the presence of soft-tissue preservation around the throat and neck of *P. kochi*, and the large crest base and rostral soft tissues of *C. suevicus*.

120–130 (Wellnhofer 1970; Howse and Milner 1995) and a similar figure (150) is known for *Gegepterus*. In dorsal profile, the teeth of *Gnathosaurus* are arranged so that they form a spatulate rosette at the jaw tip, a condition that was probably also present in the spoon-billed *Plataleorhynchus* (which bore at least 62 teeth in its upper jaw alone). The 260 teeth of *Ctenochasma* are arranged in a parallel-sided "comb" when viewed dorsally and are much more delicate than those of most ctenochasmatids. The internal jaw surfaces of ctenochasmatoids can be extremely complex with ridges, abundant foramina, and rugosities (Howse and Milner

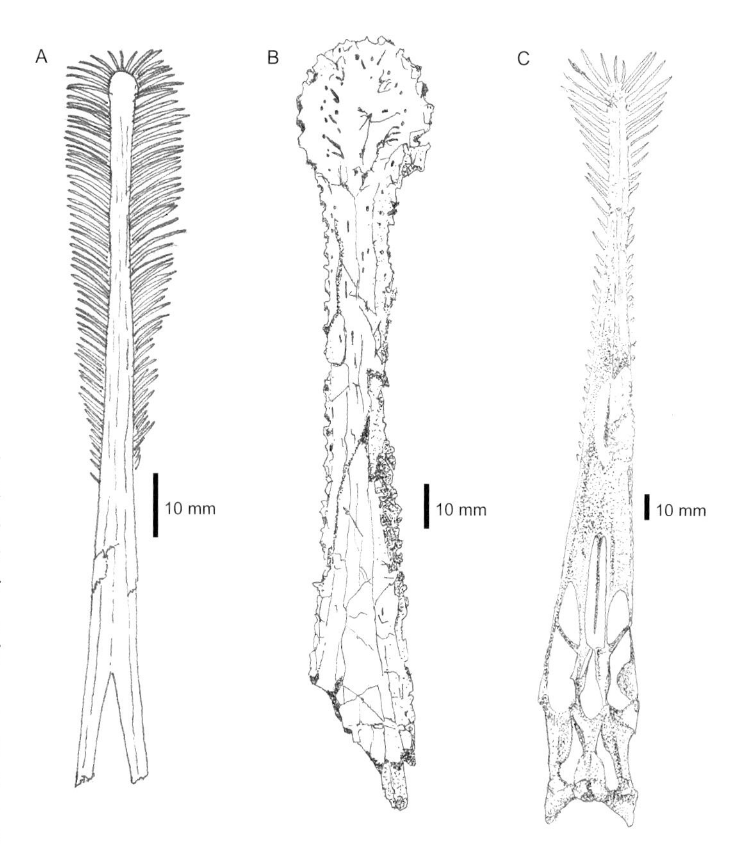

FIG. 19.11. Jaw variation in ctenochasmatids. A, mandible of the Tithonian Solnhofen Limestone form *Ctenochasma elegans*, ventral view; B, upper jaw of the Berriasian Purbeck Beds species *Plataleorhynchus streptorophodon*, ventral view; C, skull of the Solnhofen form *Gnathosaurus subulatus*, ventral view. A and C, redrawn from Wellnhofer (1970); B, redrawn from Howse et al. (1995).

1995) hinting at complex soft tissues lining their palates. Both *Gnathosaurus* and *Ctenochasma* are known to have low bony crest supports atop their upper jaws (Frey and Tischlinger 2010), which implicates the presence of soft-tissue crests in these taxa. *Gegepterus* is also known to have possessed a soft-tissue crest but, like *Pterodactylus* (see below), it lacks a corresponding bony component (Wang et al. 2007).

Perhaps the most derived jaws of all ctenochasmatids are those of *Pterodaustro* (fig. 19.5; see Chiappe et al. 2000 for a thorough description). The rostrum of this pterosaur is the longest known of any pterosaur, being 12 times longer than tall (Martill and Naish 2006). It also shows a distinctive upward curve along its entire length. These jaws are riddled with two types of teeth estimated to number 1000 in total, which collectively occupy 90 percent of the jaw length (Chiappe and Chinsamy 1996). The most obvious teeth are those of the lower jaw, which are long and extremely narrow at 0.3 mm across. Hundreds of these teeth line each side of the mandible so tightly that, for most of the jaw length, they sit in a groove rather than individual sockets. At 20 times longer than wide, these teeth have the greatest width/height ratio of any animal tooth and form a dental battery resembling the baleen of filter-feeding whales. Indeed, some authors (Benton 1990; Wellnhofer 1991a) have suggested that the long teeth of *Pterodaustro* were not teeth at all, but derivatives of keratin or some other protein and thus compositionally convergent with cetacean baleen. However, detailed examination of the *Pterodaustro* mandibular teeth has confirmed that they bear microscopic hallmarks of genuine teeth (pulp cavities, enamel, and dentine; see Chiappe and Chinsamy 1996), making them some of the most remarkable teeth ever to evolve.

In contrast, the teeth of the *Pterodaustro* upper jaw are mere nubbins, teardrop-shaped structures only a millimeter in length that sit within recessed margins. It is thought that they were anchored into these recesses by ligaments or other soft-tissue structures. Like the mandibular teeth, the dental count for the upper jaw stretches into the hundreds, and each tooth is set below a series of ossicles (up to four) arranged in anterodorsally deflected lines. As with the teeth themselves, these seem to be attached to the jaw via soft tissues. The unusual configuration of the upper dentition in *Pterodaustro* has prompted some suggestions that a muscular "lip" extended along the upper jaw (Chiappe and Chinsamy 1996; Chiappe et al. 1998), though an absence of nutritive foramina in the adjacent bone surface, typical hallmarks of fleshy lips and cheeks, may indicate otherwise. What is perhaps more certain is that the jaw muscles powering this unusual set of jaws were fairly large, assuming that the large retroarticular process of the *Pterodaustro* mandible anchored atypically big jaw-closing muscles (see chapter 5). Away from the jaws, the *Pterodaustro* skull is less remarkable, being fairly similar to those of other ctenochasmatids, and so far at least, it does not seem to have a cranial crest.

The skull of *Pterodaustro* has some competition in the oddball stakes from another ctenochasmatoid: *Cycnorhamphus* (fig. 19.10B). The skull of this animal is relatively short and it only possesses teeth at the very tips of its jaws, features that immediately make it an unusual ctenochasmatoid. Its teeth are low in number and rather peg-like, projecting procumbently from very slightly spatulate jaw tips. Some change in tooth morphology seems to occur with growth, with juveniles possessing relatively pointed, conical teeth, and older individuals displaying somewhat blunter, stouter teeth. The strangest part of the skull is found immediately behind the toothrow, where the upper and lower jaws arch away from the biting surface to create a prominent, rounded opening between the closed jaws. Soft tissues preserved in an adult *Cycnorhamphus* skull suggest that this opening was occupied by a mineralized structure projecting from the upper jaw, but the nature of this structure is poorly understood (fig. 19.1, Frey and Tischlinger 2010). A well-developed, fibrous bony crest extends along much of the skull roof and another, rather rounded, crest projects posteriorly from the skull. It is probable that these structures were linked in life by a soft-tissue crest, but no direct evidence of this has yet been reported from any fossils.

The neck of *Cycnorhamphus* is also unusual in ctenochasmatoid circles, being of a "typical" length for a pterodactyloid and comprised of stout, complex vertebrae. All other ctenochasmatoids, by contrast, bear elongate cervical vertebrae that create some of the longest necks seen in Pterosauria. The cervicals of these ctenochasmatoids are similar to those of lonchodectids and azhdarchids (chapters 21 and 25, respectively), but their specific anatomies are distinct enough to suggest that their elongated necks developed independently (Andres and Ji 2008). For example, compared to other long-necked pterodactyloids, the cervicals of ctenochasmatoids remain relatively complex with high centra and obvious, elongate neural spines. These become larger and more blade-like toward the base of the neck and indicate the presence of large neck muscles in this region. The neck vertebrae of *Elanodactylus* also possess cervical ribs—an unusual feature for any pterodactyloid.

Ctenochasmatoid trunk skeletons are unusual among pterodactyloids for two reasons. At over twice the length of the humerus, they are proportionally long, and unlike virtually all other pterodactyloids, lack fusion of their dorsal vertebrae in osteologically mature animals. This lack of fusion may explain the relatively small body size of ctenochasmatoid species, with their unfused torso skeletons not addressing flight stresses as well as those of their relatives. Their particularly elongate, strap-like scapulae (which extend along the ribs to the eighth dorsal in some species [Bennett 2003a]) may have compensated for this somewhat, permitting a larger attachment site for axial and upstroke muscles and helping to bind their trunks with soft tissue. In contrast with their atypically long scapulae however, ctenochasmatoid coracoids and sterna do not seem particularly large.

Where known, most ctenochasmatoid tails are fairly short, unremarkable, and comprised of around 14 simple caudals. That of *Pterodaustro* possesses a hefty 22 caudals however, making it unusually long compared to those of other pterodactyloids (Codorniú 2005). Only the ornithocheiroids *Pteranodon* and *Zhenyuanopterus* seem to have tails of similar length, and the structure of each of these "long tails" is distinctive enough to suggest unique functions.

Compared to most other pterodactyloid groups, the limbs of ctenochasmatoids are only modestly developed. Indeed, some species look rather inelegant compared to most other pterodactyloids thanks to their long trunk skeletons and short limb lengths (fig. 19.9). Their humeri do not bear particularly large or modified deltopectoral crests and, while longer than those of the non-pterodactyloids, their metacarpals are relatively short compared to other pterodactyloid clans. The pteroids of most ctenochasmatoids are fairly stout, but those of *Cycnorhamphus* is remarkably long and slender at almost 70 percent of its ulna length (Bennett 2007a). The metacarpals of "*P.*" *longicollum* and *Cycnorhamphus* are also proportionally longer than those of other ctenochasmatioids. The first three fingers of ctenochasmatoids are fairly well developed and bear stout claws. Their wing fingers are approximately 60 percent of the wing length, but this can only be ascertained for a few species known from complete wing skeletons (*Pterodactylus*, *Huanhepterus*, *Pterodaustro*, and *Ctenochasma*).

The hindlimb lengths of ctenochasmatoids are, for pterosaurs, fairly proportionate to their forelimb length, though *Cycnorhamphus* has disproportionately long forelimbs compared to other members of the group. Euctenochasmatian feet are elongated and robust, with some species taking this to a great extreme. *Pterodactylus* possesses feet measuring 69 percent of its tibial length, while those of *Pterodaustro* are huge, at 84 percent of the same metric (Witton and Naish 2008). The latter also possess especially wide, robust pedal bones. Interestingly, the proportions of the foot bones in these forms are not dramatically different from those of pterosaurs with smaller feet, suggesting that ctenochasmatoids feet enlarged as one unit instead of specific pedal elements lengthening to gain greater foot size.

SOFT TISSUES

Several types of soft tissue are recorded for ctenochasmatoids with the majority, by far, known from *Pterodactylus*. This soft-tissue data gives a more com-

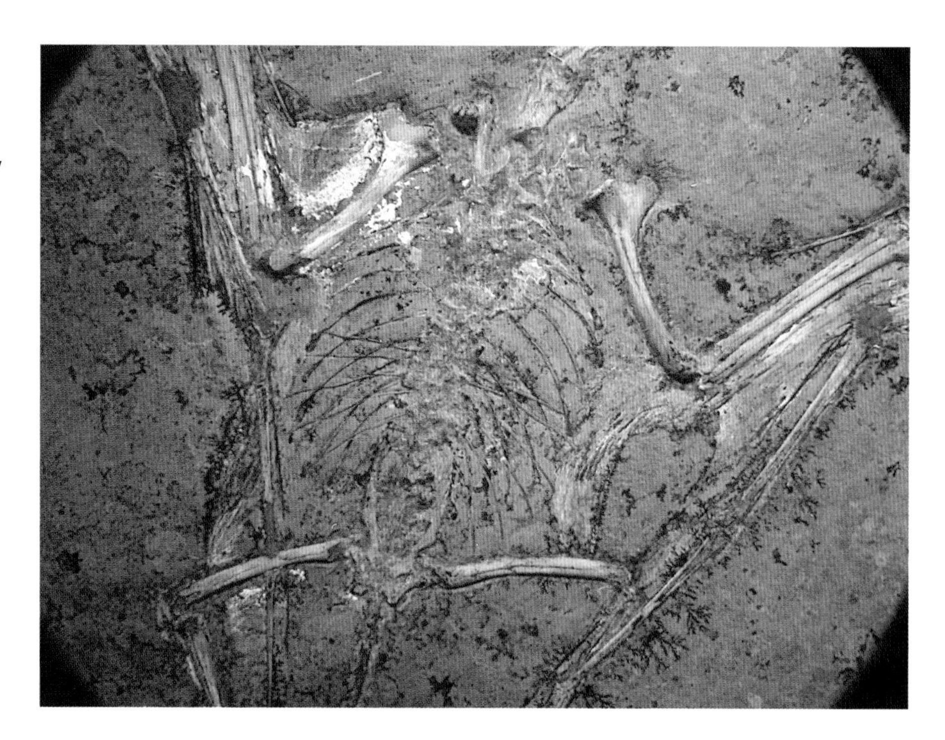

FIG. 19.12. UV photograph of the trunk and propodia and the "Vienna specimen" of *Pterodactylus kochi* in dorsal view. Note the sparse distribution of the brachiopatagia, which is now thought to represent a somewhat desiccated, shriveled condition rather than its full, *in vivo* extent.

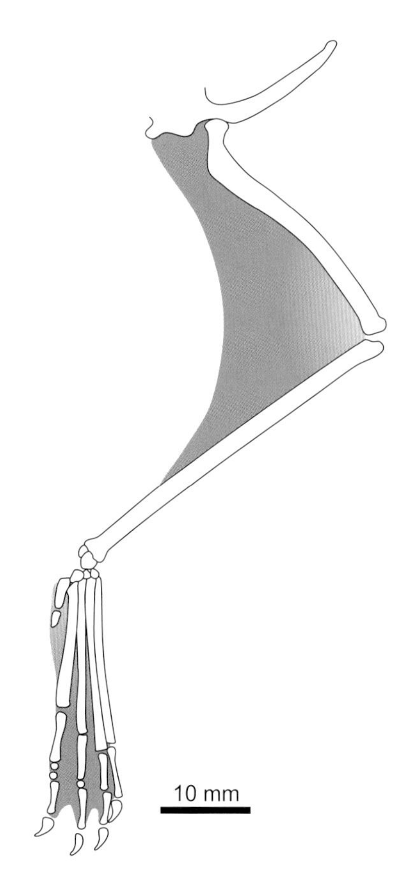

FIG. 19.13. Soft tissues of the right hindlimb of *Pterodactylus kochi* (the "Vienna specimen"). Note the small uropatagium, heel pad, and webbed toes. Redrawn from Wellnhofer (1970).

prehensive picture of the life appearance of *Pterodactylus* than for any other pterodactyloid (fig. 2.1; Wellnhofer 1970, 1987b; Frey and Martill 1998; Frey and Tischlinger 2000; Frey, Tischlinger, et al. 2003; Elgin et al. 2011).

The jaws of *Pterodactylus* bear a single, keratinous "pseudotooth" on each tip and, combined with their interlocking teeth, this makes their jaws resemble a pair of coarsely serrated tweezers (fig. 5.11D). A posterodorsally directed occipital cone projects from the posterior face of the *Pterodactylus* skull and supported an extensive soft-tissue crest. The crest itself extended along the skull to the posterior region of the nasoantorbital fenestra (Frey, Tischlinger, et al. 2003), if not slightly further along the snout (Tischlinger 2010). Unusually, there is no bony support for the crest on the skull, and its presence is only recorded from specimens with exceptional soft-tissue preservation. The pelage of *Pterodactylus* is known in some detail, with the pycnofibers on the back of the neck longer than those on the rest of the body (Frey and Martill 1998). Another ctenochasmatoid, *Gegepterus*, demonstrates rare evidence that pterodactyloids had fuzzy tails (Jiang and Wang 2011). Throat pouches have also been recorded in *Pterodactylus* (Wellnhofer 1987b; Frey and Martill 1998), as have intricately preserved foot pads with tiny (0.2 mm diameter) scales and toe webs (Frey, Tischlinger, et al. 2003).

The preserved brachiopatagium of one *Pterodactylus* specimen (the so-called Vienna specimen) served as the only known example of a preserved pterodactyloid wing membrane for many years (fig. 19.12; Wellnhofer 1987b; also see chapter 5). Although employed as evidence that the brachiopatagium of pterodactyloids was attached at the knee or thigh for many years, reinterpretation of the specimen suggests that the wing tissues of this animal had shriveled before preservation (Elgin et al. 2011). New *Pterodactylus* specimens, along with the Chinese ctenochasmatoid *Beipiaopterus*, suggest that these pterosaurs probably anchored their main flight membranes at their lower legs (Frey and Tischlinger 2000; Lü 2002). The Vienna specimen remains important however, because it preserves some of the only evidence for the shape and extent of the pterodactyloid uropatagium (fig. 19.13; Wellnhofer 1987b). Well-preserved wing membranes, showing features rather typical of pterosaur wing membranes (chapter 5), are also recorded in *Ningchengopterus* (Lü 2009a).

Locomotion

FLIGHT

Despite the long research history on ctenochasmatoids, little work has been performed on their flight mechanics. Frey, Tischlinger, et al. (2003) used the wealth of soft-tissue data known for *Pterodactylus* as a model for a generic flying pterodactyloid, suggesting that their various webs, crests, and membranes could be employed as flight control agents. Witton (2008a) modeled basic wing attributes of *Pterodactylus*, *Pterodaustro*, and *Ctenochasma* and found the flight of the first two animals comparable to wading birds, such as snipes and dunlins, and the latter to powerful fliers like skuas. All of these birds are capable of long-distance flights, at least during periods of migration. Ctenochasmatoids lack the specialized flight anatomy of some pterosaurs, but there seems no reason to exclude such behavior from ctenochasmatoids. Indeed, the elongate scapulae of many ctenochasmatoids may represent particularly large upstroke musculature suited to powerful and sustained flights.

Launch efficiency may have been strikingly different among ctenochasmatoid species. The long limbs and relative short trunk of *Cycnorhamphus* suggests it may have found little difficulty with becoming airborne, but the stretched bodies and shorter limbs of some derived ctenochasmatids may have lowered their launch efficiency somewhat. It seems that the long necks and bodies of ctenochasmatoids were proportionally heavy (Henderson 2010), and because fairly squat limbs would generate relatively lessened leverage during takeoff, launch may have been harder work for these animals than other pterosaurs. *Pterodaustro* probably suffered unusually in this respect due to its particularly stumpy limbs and elongate neck, head, and torso. These combined to create an unusually front-heavy bauplan with a low center of gravity (fig. 19.9). Perhaps a little like swans and geese, *Pterodaustro* was limited to fairly low-angle, rather frantic, and high-energy launch mechanics.

ON THE GROUND

The enormous, robust feet of most ctenochasmatoids are testament to their ability to wander around on foot. Relatively uniform limb proportions, well-developed pelvic regions, and large foot pads also indicate that ctenochasmatoids spent a fair amount of their time on the ground. Their large feet are clearly specialized for a purpose other than just walking, however. Long, plantigrade feet are hard work to bring into the "push off" phase of the walk cycle and are relatively heavy appendages to lift with each step. Thus, they are a burden to animals that spend much of their time walking on firm ground, as they decrease the efficiency of the hindlimb mechanics. Oversized feet are ideally suited, however, to spreading body weight over a broad area, and enable their owners to walk on soft substrates more effectively than short, compact feet. Given that ctenochasmatoids appear to be a clan of pterosaurs suited to finding food in shallow water (see below), such adaptations are of obvious utility. However, it should be noted that their hands do not seem to be similarly expanded, which is all the more confusing since pterosaurs are front-heavy animals (e.g., Henderson 2010). Perhaps the apneumatic hindquarters of ctenochasmatoids needed a little more support during wading than their comparatively buoyant, anterior regions which were filled with air sacs. Another, and perhaps more intriguing hypothesis, is that ctenochasmatoid feet were enlarged to enhance propulsion when swimming. Indeed, the long torsos of these animals may have enhanced their stability when floating on the water compared to other, shorter-bodied pterosaur species.

Paleoecology

Many words have been written on the paleoecology of ctenochasmatoids. The growth series of several species are fairly complete (for *Pterodaustro*, this includes an egg and embryo), which has permitted documentation of their growth rates and skeletal allometry in some detail (e.g., Wellnhofer 1970; Bennett 1996b, 2007c; Chiappe et al. 2004; Codorniú and Chiappe 2004; Chinsamy

Fɪɢ. 19.14. Juvenile ctenochasmatoids from the Tithonian Solnhofen Limestone. A, *Ctenochasma elegans*; B, *Pterodactylus* sp. The latter is especially tiny, with a skull length of only 30 mm. Photograph B copyright Natural History Museum, London.

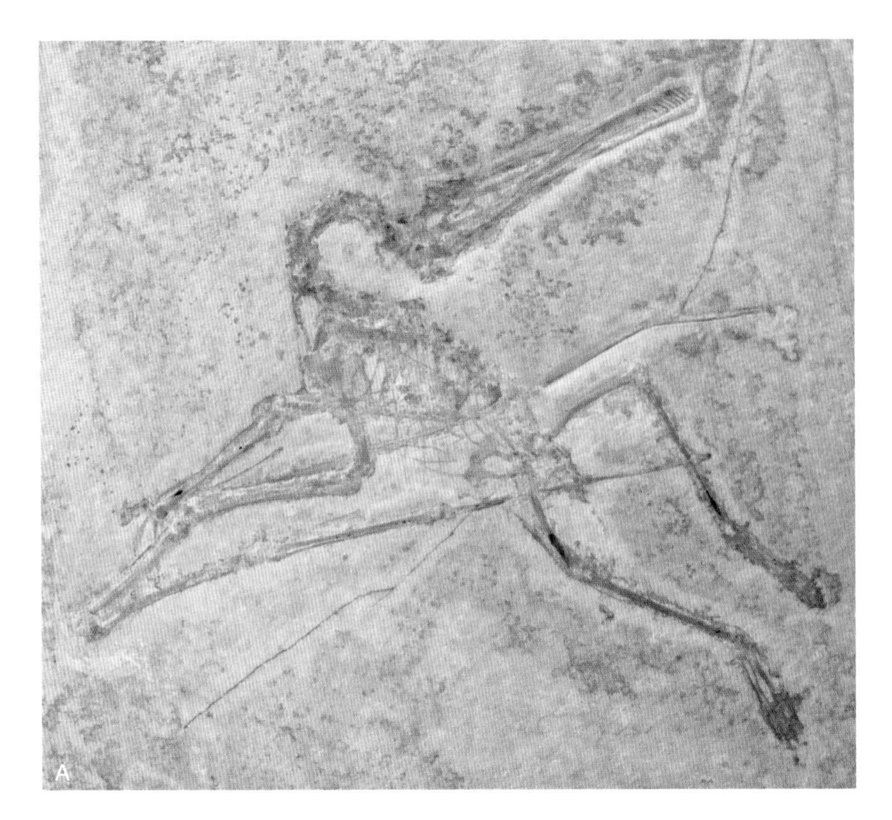

et al. 2009). It seems that their most marked changes in morphology throughout growth relate to the elongation of their jaws and dramatic increases in tooth counts (fig. 19.14). Bennett (1996b) estimated that *Ctenochasma* increased its tooth count and skull length by factors of 6, from hatching to adulthood. Thanks to *Pterodaustro*, we know that this growth may have taken several years, at variable speeds (Chinsamy et al. 2009; also see chapter 8 for more details). Attempts have also been made to uncover sexual dimorphism in *Pterodactylus* (e.g., Wellnhofer 1970), but to date no clear sexually related differences between specimens have been identified (Bennett 1992).

Perhaps the most intriguing part of ctenochasmatoid paleoecology is the function of their unusual teeth and probable wading habits. Differences in their cranial morphology likely reflect varied dietary preferences and foraging strategies, suggesting niche partitioning akin to that of modern wading birds. It seems probable that preferred morsel size decreased with tooth size and spacing, which may have allowed many different ctenochasmatoids to coexist in the same environments without competing for food. The long necks of ctenochasmatoids, as with other long-necked pterosaurs, may reflect adaptations to searching for small prey, enabling them to sweep their jaws around in search of food without expending energy to move their bodies. *Pterodactylus* was probably capable of taking relatively large food items with its comparatively robust, well-spaced teeth and may have been something of a generalist feeder. The same may also be true of *Huanhepterus, Feilongus*, *Morganopterus*, and *Gegepterus*, although their comparably narrow, tightly-packed teeth indicate a preference for somewhat smaller prey items. Bakker (1986) proposed that *Pterodactylus* was a prober of invertebrate burrows on shorelines and tidal flats, but this idea is contradicted by a lack of tactile sensory equipment or a mechanism to extract prey from burrows. Modern probers show a suite of adaptations pertaining to these tasks, but none are found in *Pterodactylus*. Instead, perhaps *Pterodactylus* simply gleaned prey, which may have included invertebrates and small fish, from the water column or the substrate surface.

Derived ctenochasmatids are generally considered to be filter feeders (e.g., Wellnhofer 1991a), but this term is probably applied too often to these forms. True filter feeding can be achieved in many ways but always requires anatomy that can control water flow and strain food. This is frequently associated with complex actions of the jaws to generate flowing water and an ability to transport collected food to the throat, and is therefore a very sophisticated and specialized foraging mechanism (Zweers et al. 1995). Among pterosaurs, only the anatomy of *Pterodaustro* seems to measure up to this feat (fig. 19.15). The specifics of the *Pterodaustro* feeding apparatus have not been researched in detail, but several features of its jaws indicate that it possessed a highly sophisticated filtering mechanism. This includes the upward curvature of its jaws, which decreases the distance between its open jaw surfaces to maximize generation of water currents and filtering efficiency when the jaws were pumped open and closed. Its long retroarticular processes may have anchored lengthened, fatigue-resistant jaw muscles capable of continuously working the mandible for this purpose. The remarkably long and densely packed mandibular teeth are clearly filtering devices, trapping food on the medial surface of the jaw as water was flushed out between them. Finally, the peculiar teeth and rows of ossicles on the upper jaw may have been employed in retaining filtered food in the mouth and moving it toward the throat. These structures may have allowed filtered food items to move back along the mouth when water flowed in, but trapped them when water was pumped out. Alas, we lack knowledge of the oral soft tissues in *Pterodaustro,* which may prohibit us from ever developing a complete understanding of its filtering mechanics. Such soft tissues make crucial contributions to the filtering strategies of modern filter feeders (Zweers et al. 1995), and there seems no reason to assume that *Pterodaustro* did not also employ similarly complex, intricate structures in its filtering anatomy. However it obtained its food, the recent discovery of gizzard stones in the gut cavity of a *Pterodaustro* specimen indicate that powerful grinding actions were required to unlock nutrients from it (Codorniú et al. 2009). Thus, it is thought that *Pterodaustro* primarily sustained themselves with tough food items, such as seeds, planktonic arthropods, or perhaps tiny shelled mollusks (e.g., Chiappe et al. 2000).

Other fine-toothed ctenochasmatids do not seem suited for filtering food like *Pterodaustro* because their jaws lack the specializations required for straining food from the water column. Rather, their elongate, broad jaw tips and meshing teeth seem much better suited for tactile-feeding strategies, where small prey items are snatched from the water column as they contact the sensitive mouth parts of the foraging animal. The similarity between the spatulate jaw and dental profiles of the *Gnathosaurus* and *Plataleorhynchus* to the bills of modern spoonbills is striking; the richly vascularized, complex morphology of the *Plataleorhynchus* palate suggests that it may have been lined with touch-sensitive tissues in a similar manner to these birds (see Swennen and Yu 2004). These pterosaurs

FIG. 19.15. The *Loma del Pterodaustro* at night, complete with a flock of wading, filter-feeding *Pterodaustro guinazui*.

probably swept their slightly open jaws through the water, catching small fish and other swimming animals as they brushed the sensitive innards of their mouths. The same also may have been true of *Ctenochasma*, though its relatively delicate teeth and broader catchment area suggest it was after finer food items.

An elephant in the room of any discussion of ctenochasmatoid paleoecology is the freakishly jawed *Cycnorhamphus*, an animal of mysterious foraging strategy and dietary preferences. Perhaps its dished anterior mandible and rostral "plug" reflect a diet that required very particular jaw morphologies to access, just as open-billed storks and crossbills demonstrate with snails and pinecones today. Just what this food may have been is not clear, but its rounded teeth may betray a fairly coarse diet. Perhaps *Cycnorhamphus* foraged on large insects, shellfish, or other abrasive organisms that would wear their teeth down. The cradling lower jaw may have served as a device to hold this prey in a particular way, allowing the mineralized crest tissues to open, bisect, or crush them. Clearly, there is a lot of work to be performed here, but we may need to be wary of overinterpreting this animal's strangeness, lest we forget how much effort many pterosaurs put into visual display devices. It is just possible that the unusual jaws of *Cycnorhamphus* were nothing to do with feeding and were simply another mechanism for showing off their prowess or status to other pterosaurs.

20
Dsungaripteroidea

PTEROSAURIA > MONOFENESTRATA > PTERODACTYLOIDEA > LOPHOCRATIA > DSUNGARIPTEROIDEA

Morphologically speaking, dsungaripteroids are a country mile away from the slender, gracile pterosaurs we have met in previous chapters. With skulls and teeth that are characteristically chunky, compact necks and trunks, and reinforced limb walls, dsungaripteroids appear to represent animals adapted for a tough diet and a rough lifestyle (fig. 20.1; Fastnacht 2005b). Their remains occur in Jurassic and Cretaceous rocks from Europe, Asia, South America, and Africa (fig. 20.2), making dsungaripteroids a widespread lineage with around 40 million years of evolutionary history spanning the Kimmeridgian (155–150 Ma) to probably the Albian (112–100 Ma). There is some ambiguity over the age of the Chinese beds containing the youngest dsungaripteroid material however, so the upper extent of their temporal range is not certain. Happily, the anatomies of some German and Chinese dsungaripteroid species are well known along with the outlines of a growth series for two species (Unwin 2005; Bennett 2006), but sadly, other dsungaripteroids are known from far less substantial and abundant remains.

Before progressing further, it is important to note that interpretations of the name "Dsungaripteroidea" vary considerably between pterosaur workers. Some authors use it to refer to a relatively exclusive clade of perhaps only eight genera (e.g., Unwin 2003; Lü, Unwin, et al. 2010), but others use it as a label for a major branch of Pterodactyloidea, essentially accommodating all pterodactyloids with a notarium (e.g., Kellner 2003, 2004; Wang et al. 2009). Proponents for the more restricted use of this name argue that pterosaur groups have distinct notarial constructions (Bennett 2001), suggesting that some lineages evolved notaria independently of each other (Unwin 2003). Moreover, it seems that only pterodactyloids achieving large adult size possess notaria, with no known adult pterosaur specimen with a wingspan under 2 m possessing one (Unwin 2003). Size may therefore be the primary driving force behind notarial development, and not taxonomy. We're following the more selective use of "Dsungaripteroidea" here, considering it to include Dsungaripteridae (a group of derived, robustly built pterodactyloids, such as *Dsungaripterus* and *Noripterus*, which is supported by all modern phylogenetic studies) and a few early dsungaripteroids, including *Germanodactylus*, the systematically problematic taxon we met briefly in chapter 19.

The History of Hardy Pterosaurs

The discovery of *Germanodactylus* represented the start of dsungaripteroid research. The first specimen of this pterosaur, a disarticulated skeleton from the Late Jurassic Solnhofen Limestone (Tithonian; 151–145 Ma) (fig. 20.3), was initially described and referred to *Pterodactylus kochi* by Felix Plieninger (1901). Soon after, Carl Wiman (1925) decided that this specimen warranted its own species, *Pterodactylus cristatus*, because of clear differences between its skull and those of *P. kochi*. The eminent Chinese paleontologist C. C. Young thought that specific distinction from *P. kochi* was not enough however, and created a whole new genus for the specimen, *Germanodactylus* (Young 1964; note that Young did not seem aware of Wiman's erection of "*P. cristatus*," and it was not until Wellnhofer's 1970 monograph that the name "*Germanodactylus cristatus*" was fully formulated). It did not take too long before *Germanodactylus* became home to a second species, with Peter Wellnhofer suggesting that "*Pterodactylus*" *rhamphastinus*, a relatively large pterodactyloid from the Mörnsheim limestones (Solnhofen-like deposits slightly younger than the Solnhofen Formation) was congeneric with *G. cristatus* (Wellnhofer 1970). Suggestions that *G. rhamphastinus* should be given its own genus have challenged this opinion recently, however (Maisch et al. 2004). It also seems that *Germanodacty-*

Fig. 20.1. Gnarly skulls, chunky teeth, and tough skeletons suggest that dsungaripteroids, like the Lower Cretaceous *Dsungaripterus weii* shown here, were hardier beasts than other pterosaurs, and correspondingly better candidates for appearing on the front cover of rock music albums.

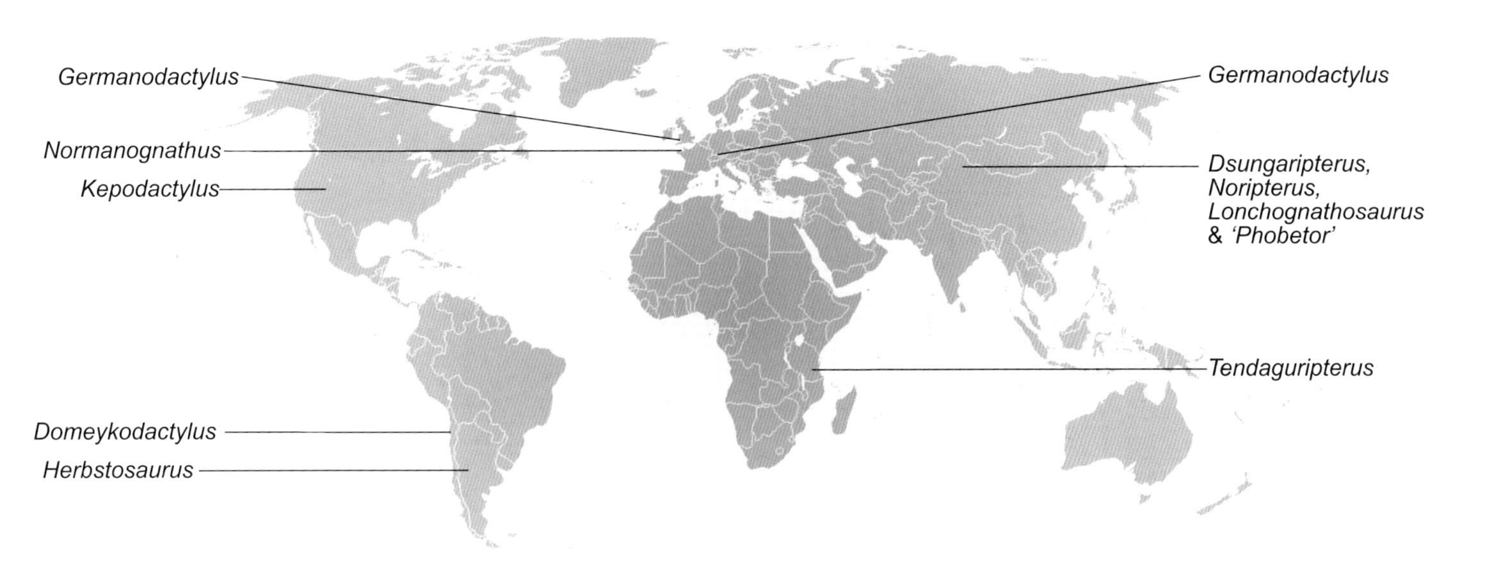

FIG. 20.2. Distribution of dsungaripteroid taxa.

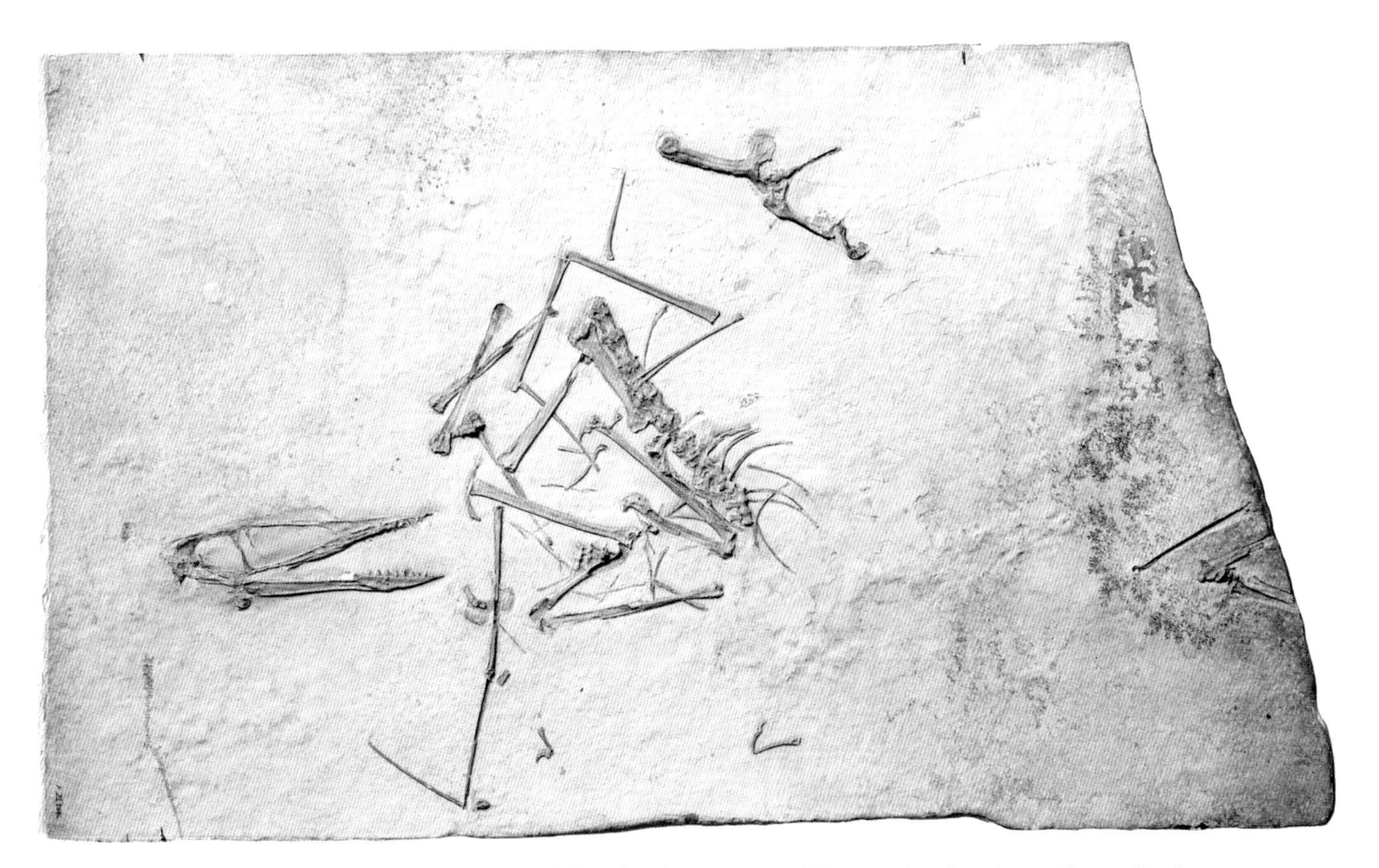

FIG. 20.3. An incomplete skeleton of the Tithonian Solnhofen dsungaripteroid *Germanodactylus cristatus* (housed in the Bavarian State Collection of Palaeontology and Geology, Munich; used with permission).

FIG. 20.4. A recently described incomplete skeleton of *Noripterus complicidens,* from the Lower Cretaceous Tsagan-Tsab Formation, Mongolia. Photograph courtesy of Lü Junchang.

lus is present in Kimmeridgian deposits (156–151 Ma) of Dorset, southern England (Unwin 1988c).

The debate over the placement of *Germanodactylus* in Dsungaripteroidea is tied to the naming of another dsungaripteroid by C. C. Young in 1964. This second animal, the large, Lower Cretaceous dsungaripterid *Dsungaripterus weii* from the Upper Tulugu Group of China, is known from numerous fossils including complete, three-dimensionally preserved skulls and incomplete skeletons (Young 1964, 1973). With its robust cranial morphology, toothless jaw tips, and thickened limb bones, *Dsungaripterus* was so different from almost all other pterosaurs known at that time that Young realized he had discovered an entirely new pathway of pterosaur evolution. Young noted that the only pterosaur then known to show any similarity to *Dsungaripterus* was *Germanodactylus cristatus*, which he considered to be a likely *Dsungaripterus* ancestor because of its blunt teeth and edentulous jaw tips. Many pterosaurologists agree with this idea (e.g., Wellnhofer 1978; Unwin 2003; Lu, Unwin, et al. 2010), but not others, who prefer the original interpretation of *Germanodactylus* as a ctenochasmatoid (e.g., Bennett 1994, 1996b, 2006; Kellner 2003; Andres and Ji 2008; Wang et al. 2009). My vote is cast with the Young camp on this issue, and not just because of *Germanodactylus*' dentition. Like all dsungaripteroids, *Germanodactylus* has a skull with enormously expanded opisthotic processes (laterally situated flanges at the back of the skull that anchor neck musculature), thickened bone walls, and a strongly curved femur. No other pterosaur groups show these features and they suggest, as postulated by Young even before the Beatles went psychedelic, that *Germanodactylus* represents a Jurassic dsungaripteroid pterosaur.

Young's legacy with this group continued when he described another large dsungaripterid from the same deposits as *Dsungaripterus*: *Noripterus complicidens* (fig. 20.4; Young 1973). This was joined a few years later by the smaller "*Phobetor*" *parvus* from Mongolia (Bakhurina 1986; Bakhurina and Unwin 1995b; note that the generic name of this form is occupied by another animal, so a new name is needed). Recent work has controversially suggested that *Noripterus* and "*Phobetor*" are one and the same (Lü 2009c), although they remain considered distinct animals by most. Frustratingly, although it seems likely that these pterosaurs represent the last of the dsungaripteroids, the age of the sediments containing their fossils is poorly constrained and their exact age remains unknown.

Since the uncovering of the Chinese dsungaripteroids, the quality of dsungaripteroid fossil specimens being discovered has dropped substantially. The scatty remains of *Herbstosaurus pigmaeus* are rather typical of modern dsungaripteroid discoveries, and are represented by femoral and pelvic remains that were initially thought to belong to a small dinosaur (Casamiquela 1975). Several authors have noted the pterosaurian affinities of these uppermost Jurassic Argentinian remains (late Tithonian; ca. 145 Ma), including Unwin (1996), who considered it a possible dsungaripteroid. *Kepodactylus insperatus*, from the Upper Jurassic Morrison Formation (Kimmeridgian-Tithonian; 155–145 Ma) in Colorado is also known from insubstantial material, with only a cervical vertebra and a few wing bones to its name (Harris and Carpenter 1996). The Kimmeridgian (155–151 Ma) *Normanognathus wellnhoferi* from northern France is known from even less, solely represented by the anterior ends of its jaws (Buffetaut et al. 1998). The same is true for *Tendaguripterus recki* from Tanzania (Kimmeridgian–Tithonian; Unwin and Heinrich 1999), *Domeykodactylus ceciliae* from Chile (of uncertain age, but likely to stem from the Late Jurassic or Early Cretaceous; Martill et al. 2000), and *Lonchognathosaurus acutirostris* from China (Aptian-Albian; 125–100 Ma; Maisch et al. 2004; note that this species may merely represent remains of *Dsungaripterus* [Andres et al. 2010]). In addition to these finds, unnamed dsungaripteroid material representing various skeletal elements has been reported from around the world, including deposits in Romania, Japan, and additional Chinese and German localities (Barrett et al. 2008).

Anatomy

Several authors have noted that dsungaripteroid cranial anatomy is rather plastic and prone to evolutionary convergence (Buffetaut et al. 1998; Unwin and Heinrich 1999; Maisch et al. 2004). As such, it can be difficult to readily split dsungaripteroids into "basal" and "derived" forms, and especially so for the less completely known species. Generally speaking, *Germanodactylus*, *Tendaguripterus*, and *Normanognathus* are considered to represent relatively early grades of dsungaripteroid evolution, while *Noripterus*, "*Phobetor*," *Dsungaripterus*, and *Lonchognathosaurus* comprise the relatively more derived Dsungaripteridae (Unwin and Heinrich 1999; Unwin 2003; Maisch et al. 2004). Dsungaripteroid fossils have been well described and most descriptive papers on these animals provide a wealth of information, but those in pursuit of particularly comprehensive overviews should seek Plieninger (1901), Young (1964, 1973), Wellnhofer (1970), and Lü (2009c).

Early dsungaripteroids are generally small, with *Germanodactylus cristatus* just about obtaining a wingspan of 1 m (fig. 20.5), and the wings of *G. rhamphastinus* stretching no further than 1.1 m. "*Phobetor*" (fig. 20.6) and *Noripterus* are somewhat larger with wingspans of 1.5 m, while *Lonchognathosaurus* and *Dsungaripterus* are the largest dsungaripteroids known with wingspans of 3 m. As such, even at their largest dsungaripteroids were only moderately sized pterodactyloids. Dsungaripteroids commonly have thickened bone walls in their vertebrae and limb bones that render their skeletal densities more comparable to those of early pterosaurs, mammals, and some birds than to other pterodactyloids. Because dsungaripteroid anatomy is otherwise very derived compared to other pterosaurs, it is thought that these thickened bone walls represent an evolutionary reversal of the thin-walled condition common to other pterodactyloids. With this reinforcement in mind, it is unsurprising that dsungaripteroids do not seem to have pneumatized skeletons beyond their skulls and vertebral columns (Claessens et al. 2009). Interestingly, their reinforced limb bones are not reduced in overall dimensions compared to other pterosaurs, suggesting dsungaripteroids probably bore relatively heavy, but perhaps unusually strong skeletons. Their thickened bone walls would render them particularly strong against buckling forces, which pterosaurs with thinner bone walls were vulnerable to (Fastnacht 2005b). Additional strength was added to their limbs by unusually expanded bone joints that could dissipate forces incurred on them across a relatively wide area (Fastnacht 2005b). This reinforcement makes dsungaripteroid bones rather characteristic among those of other pterodactyloids, even if the specific forces behind the evolution of this feature are not entirely understood (see below).

Dsungaripteroid skulls are characteristically chunky, with stout, thickened bones around their jaws, orbits, and braincases (fig. 20.7). Like some istiodactylids, derived forms have partially or completely closed ventral orbits, due to either a small strut bridging the midheight of the eye socket (as in "*Phobetor*") or by filling the lower half of the eye socket with bone (*Dsungaripterus*). All dsungaripteroids possess fibrous bony crests along their snouts, although their proximity to the jaw tip is rather variable (e.g., Buffetaut et al. 1998). Dsungaripterids also bear thin supraoccipital crests projecting from the posterodorsal regions of their skulls. Both crests almost certainly supported soft-tissue crest components in life, although they have never been recorded in a fossil. Dsungaripteroids uniquely possess strongly expanded opisthotic processes projecting somewhat laterally from the posterior face of their skulls, and are further characterized by toothless jaw tips (excepting *Germanodactylus rhamphastinus*, which bears fully toothed jaw tips). The extent of this edentulous region seems to have increased over time, allowing their jaw tips to become more laterally compressed and more slender than those of earlier

FIG. 20.5. Skeletal reconstruction and life restoration of a launching *Germanodactylus cristatus*.

forms. Most dsungaripteroid jaws are rather straight, but those of *Dsungaripterus* and *Normanognathus* are slightly upturned at their tips. Like their skulls, dsungaripteroid mandibles are robust with thick, deep mandibular rami and mandibular symphyses that occupy between 30–50 percent of the mandible lengths. In most forms, a rather shallow crest occurs along the underside of the mandibular symphysis, with *Dsungaripterus* demonstrating its most pronounced development. *Dsungaripterus* also has a peculiar, knobbly ridge running along the midlength of its palate, roughly in line with the back half of the toothrow.

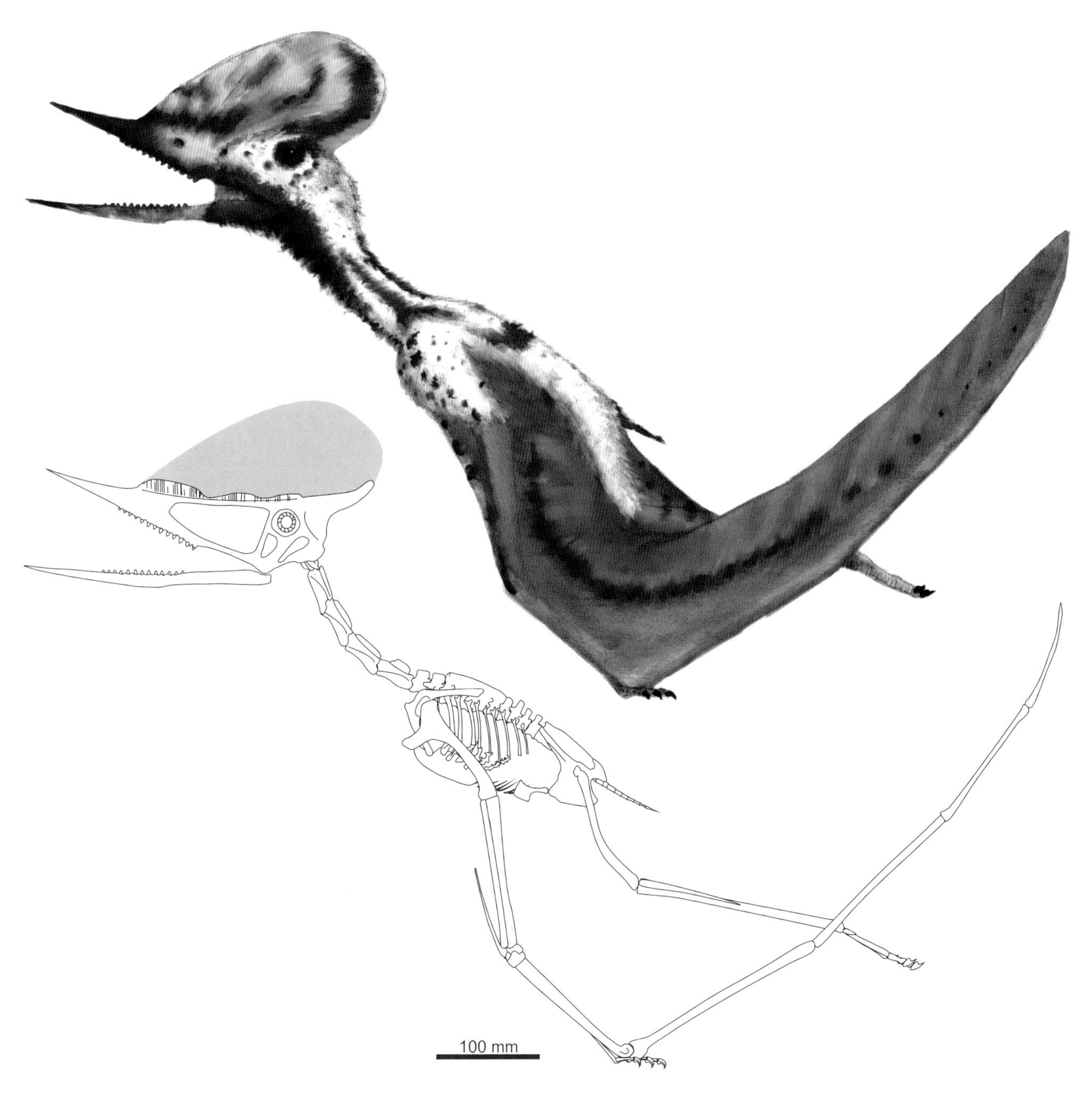

Fig. 20.6. Skeletal reconstruction and life restoration of a launching *"Phobetor" parvus*.

Dsungaripteroid dentition is remarkably odd and, like the rest of their anatomy, seems adapted for great strength. The toothrows of some forms were constricted from the rear as well as the front, forming characteristically short dental regions (Maisch et al. 2004). Their teeth are universally robust, low in number (15 or so along the side of each jaw), well spaced and generally increase in size toward the back of the toothrow. This becomes more marked over the evolution of the group, with only slightly enlarged posterior teeth in *Germanodactylus* but tremendously enlarged posterior teeth in *Dsungaripterus*. *Noripterus* is unusual among dsungaripteroids, however, for possessing the "normal" pterosaur condition where the larger teeth are at the front of its jaws (Lü 2009c). Dsungaripteroid tooth sockets are also remarkable, swelling up around the teeth so that the jaw bone grows around and over the tooth bases. These margins almost overgrow the entire dentition in some species, a condition that raises all sorts of interesting functional questions, not the least

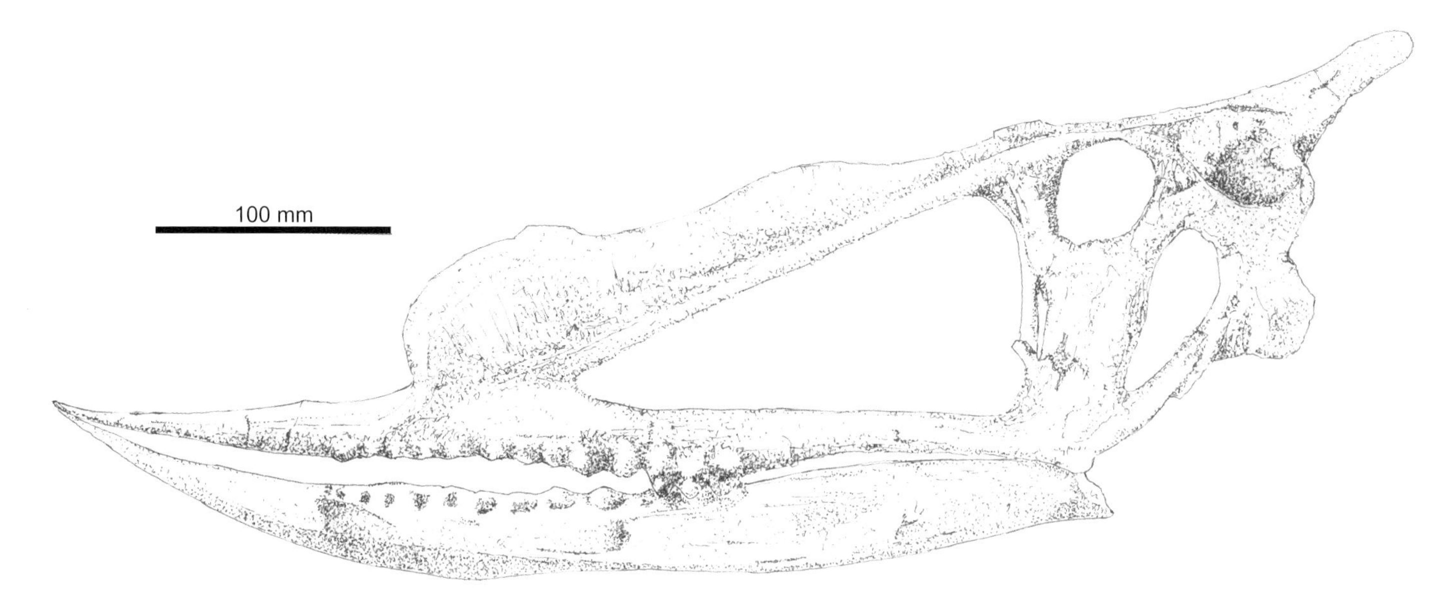

FIG. 20.7. Complete skull and mandible of the large, Lower Cretaceous dsungaripterid *Dsungaripterus weii*. Note the infilled ventral orbit and the general robust construction of the jaws and teeth. Redrawn from Young (1973).

important being how—or even if—dsungaripteroids could shed and replace their teeth throughout their lives like other reptiles.

The axial columns of dsungaripteroids seem to share many common features with those of azhdarchoids (chapters 22–25). Their cervicals are rather square and blocky in dorsal aspect, with prominent zygapophyses that bear large articular surfaces. The neural spines are rather low, and their neural arches are contained within the base of the vertebrae (Buffetaut and Kuang 2010). A complete dsungaripteroid dorsal series remains elusive, but trunk skeletons are partially known from *Dsungaripterus, Noripterus*, and *Germanodactylus*. At least 10 dorsals are present in an excellently preserved, partial dsungaripteroid skeleton from Germany (Fastnacht 2005b), and *Dsungaripterus* and *Noripterus* both show that derived dsungaripteroids fuse their anterior dorsals into notaria. Interestingly, this does not seem to be the case in smaller forms (e.g., *Germanodactylus*), suggesting that dsungaripteroids only fused their dorsals in response to large overall size (Unwin 2003, though see Bennett 2003a; and Kellner 2003 for counterarguments). Dsungaripteroid sacral counts are also poorly known, but seven are found in the fused sacrum of *Dsungaripterus*. The caudals of dsungaripteroids are rarely preserved but seem relatively complex and sculpted in *Dsungaripterus*, as if the tail was unusually strong in this form. The tail of *Germanodactylus*, by contrast, is a fairly straightforward affair comprised of 15 short, simple vertebrae.

Like the axial column, the dsungaripteroid pectoral girdle is not known in entirety. The scapulocoracoid of *Germanodactylus* shows a strap-like scapula that is longer than the coracoid, while the scapula of *Noripterus* seems rather robust compared to those of other pterosaurs. Articulated specimens of this pterosaur show that the scapula lies across the ribcage at an acute angle to the spinal column, suggesting dsungaripteroids retained the "primitive" orientation of the pectoral girdle, rather than rotating it to a perpendicular angle with the spine, as is the case in ornithocheiroids. Dsungaripteroid pelvic girdles are known in a little more detail with several fairly complete girdles preserved (Young 1964; Fastnacht 2005b). These elements show well-developed pelves with elevated postacetabular processes that may, in some forms, have resembled those of the terrestrially adapted azhdarchoids (Hyder et al., in press).

Dsungaripteroid forelimbs show a unique humeral morphology, which lacks openings for the invasion of pneumatic tissues and bears a rather long, medially deflected deltopectoral crest. The forearms are a little shorter than the wing metacarpals in early forms, but this condition is reversed in more derived species. Their pteroids occupy about half of the forearm length, and their first three digits, where known, are comprised of rather robust phalanges. Dsungaripteroid wing finger proportions are only represented completely by *Germanodactylus* and *Noripterus*, both of which suggest that their wing fingers occupy about

60 percent of the wing length. The phalanges of the wing finger generally to decrease by 10–25 percent in each successive element, so the terminal phalange is about 65 percent of the length of the first. Proximal dsungaripteroid wing phalanges seem to be unusually curved (Unwin 2003), a trait that may be diagnostic of this group.

Dsungaripteroid hindlimbs are fairly long and robust, being 75 percent as long as their forelimbs (not including the wing finger). Their femora are often peculiarly bowed, with a marked curvature discernible in lateral view and a slight medial curve in anterior aspect (figs. 2.5 and 20.6). No other pterosaurs seem to have their femora curved in this manner, suggesting a unique biomechanical regime for this bone in dsungaripteroids (Fastnacht 2005b). Their tibiotarsi are around 70 percent longer than the femora. Dsungaripteroid feet are poorly known, but the first metatarsal and pedal digit of *Germanodactylus* suggest they had a fairly compact, robust construction. The foot of *Noripterus* is similarly constructed, being about 30 percent of the tibial length with short toes and slender, relatively straight claws.

Locomotion

FLIGHT

The few studies performed on dsungaripteroid flight mechanics have invariably concluded that they were adapted for flying in terrestrial settings. Computations of the wing attributes of *Dsungaripterus* suggest that this animal was either a rather skilled, agile flier like modern kites (Hazlehurst and Rayner 1992) or, in stark contrast, a somewhat more heavily loaded flier akin to birds that frequent marshlands and rivers such as anhingas and dippers (Witton 2008a). While these interpretations are obviously quite different, they at least agree that dsungaripteroids were not gull- or albatross-like marine soarers and that their relatively short, deep wings seem well suited for flapping around terrestrial environments. Further support for this interpretation comes from the preponderance of dsungaripteroid fossils in continental sediments, as well as a study of their rather unusual femoral morphology. Fastnacht (2005b) demonstrated that the curved femoral shaft, thickened bone walls, and expanded knee joint of the dsungaripteroid femur rendered it an excellent shock absorber, the evolution of which he imagined was caused by repeated and frequent landing on hard ground. This, according to Fastnacht, suggested that dsungaripteroids were primarily adapted for regular, short-lived flights over inland settings.

There is some need to treat all of these conclusions with caution, however. No published studies of dsungaripteroid flight have considered how their reinforced, and probably heavier skeletons may have affected their flight performance. Seeing as skeletal mass seems to be directly proportional to body mass (Prange et al. 1979; also chapter 6), dsungaripteroids may have had to lug greater weights around with them in the air. This may explain why they apparently did not grow to the same sizes as many other pterodactyloid clades, despite possessing similar reinforcement of their trunk skeletons as in all other large pterosaurs. Perhaps their limbs developed shock-absorbing characteristics for the same reason; greater masses will obviously incur greater forces acting on the limbs during landing, not necessarily because they landed more frequently than other pterosaurs. Accordingly, while the basic conclusion that dsungaripteroids were adapted for flight in terrestrial environments may remain sound, there is scope for refining our flight models of these animals with assessments of how their reinforced skeletons affected their overall mass.

ON THE GROUND

The pelves of dsungaripteroids have played some role in untangling just how well pterosaurs could move around terrestrially. The articulated femora of a fantastically preserved *Dsungaripterus* partial skeleton were crossed beneath the pelvis when fossilized, refuting ideas that pterosaur hip joints prohibited movement of their hindlimbs directly beneath their bodies (fig. 2.5, Bennett 1990). Biomechanical studies on another dsungaripteroid pelvis showed that pterosaur pelvic morphology provided little anchorage for hindlimb muscles should pterosaurs stand in an erect, bipedal posture, but worked much more efficiently in the more horizontal configuration of a quadruped (Fastnacht 2005b). These studies aside, little has specifically said of the terrestrial abilities of dsungaripteroids, but we can assume that they were probably fairly adept terrestrial locomotors. With hindlimbs almost as long as their forelimbs, both limb sets could take equally long strides during walking and running. Their short, robust pedal anatomy is also indicative of efficient walking mechanics, as this configuration maximimizes the power output of the foot during the propulsive phase of a stride in plantigrade animals. Their thickened limb bone walls, curved femora, and expanded joints are also of relevance here, making their limbs more resistant to the compressive forces felt when walking or running (Fastnacht 2005b). Such reinforcement may have permitted dsungaripteroids to be rather sprightly on their feet, much as the expanded joints function in rapidly running mammals and birds (see Coombs 1978 for examples). Perhaps it was even energetic terrestrial locomotion that drove the strengthening of the dsungaripteroid skeleton, allowing them to frequently engage in running, bounding, or leaping behaviors without fear of injury.

Another form of nonvolant locomotion may explain the thickened bone walls of these pterosaurs: Were dsungaripteroids aquatic? Some aquatic birds and mammals use thickened bone walls to counteract the buoyant air of their lungs and air sacs, thereby giving them greater control over their position in the water column (Wall 1983). There is no reason to think that aquatic pterosaurs could not do the same, but the lack of specific swimming adaptations in dsungaripteroids suggests otherwise. Even semiaquatic animals develop some features conducive to life in the water (e.g., shortening of limbs to increase their paddling efficiency, development of paddle-like extremities, etc.; Stein 1989), so a lack of such features in dsungaripteroids is probably indicative that they did not regularly swim or even wade into deep or swampy waters. Thus, their greater bone masses are unlikely to reflect adaptations to an aquatic lifestyle.

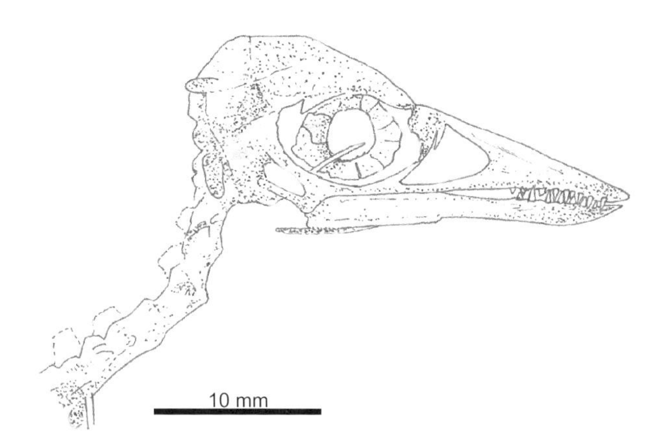

Fig. 20.8. Skull and anterior cervical vertebrae of a juvenile *Germanodactylus cristatus*, from the Tithonian Solnhofen Limestone. Note that the jaws bear the same short toothrow and edentulous jaw tips of older *Germanodactylus*, suggesting they occupied a generally similar lifestyle throughout their lives. Redrawn from Bennett (2006).

Paleoecology

Our discussion of robust bones in the dsungaripteroid skeleton has, thus far, overlooked one particularly reinforced component of their anatomy: their skulls. Even the smallest species have stout-looking crania and teeth, while larger, more derived forms possess the strongest-looking skulls and jaws of any pterosaur. This, in concert with their strange dental configurations, suggests that dsungaripteroids employed powerful bite forces during feeding. The presence of their largest teeth at the back of the toothrows (where the bite forces are maximized) and the closure of the ventral orbits (thereby strengthening the back of the skull) are further obvious adaptations to this habit. The construction of their teeth is particularly suited to crushing and smashing hard, brittle foodstuffs rather than holding soft prey, and we may imagine that the knobbly palatal ridge of *Dsungaripterus* was also involved in a crushing action. Such teeth and jaws are well adapted for breaking open shellfish, such as bivalve mollusks or crustaceans (e.g., Wellnhofer 1991a; Unwin 2005), while smaller dsungaripteroids perhaps also ate hard-shelled insects. This prey may have been acquired with their narrow, edentulous jaw tips, which resemble those of probing birds. Dsungaripteroids probably grabbed, probed, or dislodged shellfish using their pincerlike jaws, before smashing them to pieces once in their mouths. It's quite likely that the variations in jaw tip shapes reflect adaptations to procuring diverse types of food, allowing different species to share foraging grounds without treading on the ecological toes of others (Unwin 2005). The expanded opisthotic processes may present a further indication of shellfish-eating habits, as they anchor strong neck muscles that enabled forceful twists and turns of the head for extracting or dislodging prey (Habib and Godfrey 2010). Because of their small feet and heavy skeletons, we may assume that dsungaripteroids generally found their food on the rocky, or at least firm, shores of lakes or rivers rather than wading out into deep water.

There is some evidence that dsungaripteroids foraged in this manner from the moment they hatched. The fossils of extremely young *Germanodactylus cristatus* (fig. 20.8) show the same characteristic dental patterns, robust jaws, and signs of expanded neck musculature that we see in adults (Bennett 2006). Thus, unlike some other pterosaur species, juvenile dsungaripteroids seem to have been miniature versions of their parents, perhaps their only distinction being an inability to hold and crack the same tough-shelled prey that adults fed on. However, baby *Germanodactylus* do differ from adults by lacking cranial crests, suggesting that dsungaripteroids are yet another pterosaur group where only sexually mature forms developed cranial ornamentation (Bennett 2006).

21
Lonchodectidae

PTEROSAURIA > MONOFENESTRATA > PTERODACTYLOIDEA > LOPHOCRATIA > LONCHODECTIDAE

By and large, pterosaur workers agree that that most of the major groups we are covering in this book exist in some capacity or another. As we've seen, their position in the pterosaur tree, composition, and nomenclatural basis may be disputed, but there is at least a general agreement that they actually *exist*. This is not the case for Lonchodectidae, animals that are championed as mysterious but important components of Cretaceous pterosaur faunas by some (fig. 21.1; Unwin 2005) but as merely a type of ornithocheiroid by others (e.g., Kellner 2003; Andres and Ji 2008; Wang et al. 2009). As such, there is something of a question over whether a unique lonchodectid bauplan actually exists within Pterosauria at all, or if considering them a major group of pterodactyloids is overstating their morphological distinction.

Predictably, at the core of this controversy are some exceptionally scrappy fossils that until recently suggested lonchodectids were exclusively known in Cretaceous (Valanginian-Turonian; 140–90 Ma) rocks of Britain (fig. 21.2). These specimens, principally jaw elements and a few fragmentary limb bones (figs. 21.3–21.4), were distinctive enough to suggest to David Unwin that they represented a distinctive lineage of toothed Cretaceous pterodactyloids. Many of these specimens could be directly tied to an animal named *Lonchodectes* by Reginald Walter Hooley in 1914, a largely overlooked pterosaur from the upper Albian Cambridge Greensand (ca.100 Ma) (Unwin 2001). Hooley (1914) suggested that six species of *Lonchodectes* were present in the Greensand, but Unwin (2001) found only four: *L. compressirostris*, *L. microdon* (which might be synonymous with the former), *L. machaerorhynchus*, and *L. platystomus*. By contrast, some workers (Kellner 2003; Wang et al. 2009) do not even consider *Lonchodectes*, to be a valid, diagnosable genus.

Based on similarities with these scrappy Greensand fossils, additional pterosaur remains have also been referred to Lonchodectidae. Other British deposits, notably the Weald and Chalk groups of southern England (late Berriasian–Valanginian; 142–136 Ma and Cenomanian-Turonian; 100–90 Ma, respectively) have yielded several fragmentary lonchodectid fossils, including the species *L. sagittirostris* (from the Weald) and *L. giganteus* (from the Chalk) (Unwin, Lü, and Bakhurina 2000; Unwin 2001; Witton et al. 2009). The latter remains one of the most significant *Lonchodectes* species known because, unlike virtually all others, it is represented by both cranial and postcranial remains. However, its importance was trumped recently by an incomplete lonchodectid skeleton from the Jiufotang Formation of China (probably Aptian; 125–112 Ma). This animal has yet to be named, but was colloquially christened "Chang-e," or "Moon Goddess," by Unwin et al. (2008). Further potentially important lonchodectid material stems from the lower Aptian Leza Formation of Spain, which has yielded the upper jaw tip, partial mandible, and various postcranial remains of *Prejanopterus curvirostra* (Vidarte and Calvo 2010). The identity of this species was considered mysterious by Vidarte and Calvo (2010), but the long, low nature of its jaw and unusual tooth socket morphology tentatively suggest it might represent a lonchodectid. Likewise, a partial jaw from the Santana Formation of Brazil (likely Aptian-Albian; 125–100 Ma) could record the first occurrence of this group in the Southern Hemisphere. This animal was identified as an aberrant ctenochasmatoid and named *Unwindia trigonus* by David Martill (2011), but again, several features of its jaw and dental morphology are consistent with a lonchodectid identity (D. M. Unwin, pers. comm. 2011). If *Unwindia* is indeed a lonchodectid, it is of further note for representing a form of unprecedented body size within this group.

Fig. 21.1. *Lonchodectes* sp., in full view moments before the picture was drawn, flies in front of the Cretaceous sun to once again obscure its mysterious anatomy.

FIG. 21.2. Distribution of lonchodectid taxa.

FIG. 21.3. Cranial remains of *Lonchodectes* in left lateral view. A, rostral and mandibular tip of the Cenomanian-Turonian Chalk species *L. giganteus*; B, left mandibular ramus of the Berriasian–Valanginian Hastings Beds species *L. sagittirostris*. Photographs courtesy of Bob Loveridge, copyright Natural History Museum, London.

So What, If Anything, Are Lonchodectids?

The specimens and species mentioned above bear several characters that prohibit their identification as members of other pterosaur lineages, suggesting that a lonchodectid bauplan does genuinely exist, even if ascertaining exactly where it fits into pterosaur evolution is not easy. Lonchodectids seem to be best defined by features of their jaws, with low jaw profiles, raised tooth sockets, and uniformly small teeth with constricted bases presenting a unique anatomi-

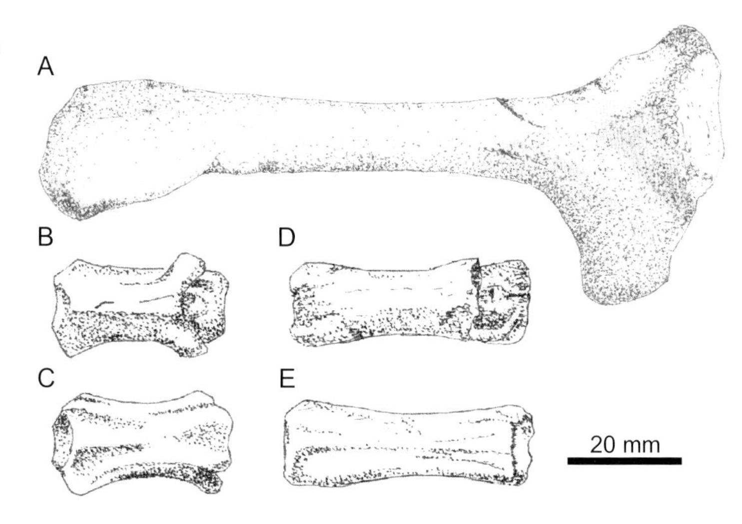

FIG. 21.4. Postcranial remains of lonchodectids. A, right humerus in dorsal view; B–E, associated midseries cervical vertebrae. B–C represent a relatively anterior vertebra to D–E. B and D reflect dorsal views of each element, C and E reflect ventral views. A, after Seeley (1870); B–E after Owen (1851).

cal configuration within Pterosauria. Their humeri, while broadly similar to those of other lophocratians, also bear a characteristic combination of distinguishing features. These attributes are probably enough to distance lonchodectids from ornithocheiroids, as these pterosaurs are known to possess very unusual jaws and humeri (compare figs. 21.3–21.4 with those of chapters 15–18; also see Unwin 2001, 2003) and may suggest a placement somewhere within Lophocratia instead (Unwin 2001). Several different placements have been suggested for lonchodectids within this group (Unwin 2003; Lü, Unwin, et al. 2008; Lü, Unwin, et al. 2010; Lü, Unwin, et al. 2011), but limb bone lengths of the "Moon Goddess," by far the most complete representation of lonchodectid anatomy yet known, hint at a close relationship with azhdarchoid pterosaurs (chapters 22–25; Unwin et al. 2008). A formal description of this specimen, and perhaps the discovery of more lonchodectid material, is needed to verify their position in the pterosaur tree, however.

Anatomy

Lonchodectids are a rather overlooked group of pterosaurs and, to date, no substantial synthesis of their anatomy or taxonomy has been published. At present, the best overview of their anatomy is given by Unwin (2001), but further details can only be found in other, often historic, literature (e.g., Owen 1851; Seeley 1870; Hooley 1914; Witton et al. 2009; Martill 2011). So little is known of their anatomy that even estimating their body size is difficult, but their known humeri do not exceed 100 mm in length and would equate to wingspans well under 2 m if azhdarchoid-like proportions are assumed. *Unwindia* may be an exception, as its preserved jaws indicate a skull length exceeding 300 mm and a wingspan over 3 m.

The jaws of most lonchodectids seem fairly shallow, crestless, and long. *Lonchodectes giganteus* may be an exception to this, bearing what appears to be a particularly short rostrum in front of its nasoantorbital fenestra and shallow crests on both jaws (Martill 2011). *Lonchodectes machaerorhynchus* also bears a mandibular crest, but its shape is not known in detail. Variably formed ridges and grooves adorn the midlines of the biting surfaces of both the upper and lower jaws of all species. Their tooth sockets are regularly spaced, of uniform size, and to greater and lesser extents, elevated from the jaw margins to give the impression that their teeth sit on small pedestals. The toothrows themselves occupy most of the jaw length, extending along much of the mandibular rami in at least *L. sagittirostris*. The teeth are well spaced and characteristically slender but relatively short. They are also rather laterally compressed, often slightly recurved, and may bear slightly "pinched" bases. The third tooth pair in the upper jaw is slightly laterally offset from the rest of the toothrow in some species, and the anterior teeth project somewhat forward.

Until assessment of the "Moon Goddess" is finished, very little can be said of lonchodectid postcranial anatomy. Their neck vertebrae are long with low neural spines, perhaps most resembling those of long-necked ctenochasmatoids in general guise. The glenoid of *L. giganteus* is known, but unfortunately the rest of the shoulder girdle is not. Their humeri are quite typical of most lophocratians, bearing straight, squared-off, and unwarped deltopectoral crests projecting perpendicularly from the humeral shaft. Preliminary studies of the "Moon Goddess" show it has the rather short wing fingers and elongate hindlimbs typical of azhdarchoids (Unwin *et al.* 2008), but the exact proportions of its bones remain to be established.

Locomotion and Paleoecology

With such fragmentary material to work from, it is surprising that anyone has even considered what lonchodectids were like as living animals, let alone published any ideas on it. Despite this, it has not only been proposed that lonchodectids were generalized feeders (Unwin 2005), but that their long legs would grant them terrestrial capabilities comparable to some azhdarchoids (Unwin et al. 2008). Perhaps we could stretch what little we know about these animals even further to suggest that their azhdarchoid-like wing proportions were more suited for flight in terrestrial settings than marine (see chapters 22–25). Of course, all of these ideas are very provisional and cannot be verified until more substantial lonchodectid remains are found and described. After a century of ambiguity over their true form, hopefully these discoveries will not be a long time coming.

22
Tapejaridae

PTEROSAURIA > MONOFENESTRATA > PTERODACTYLOIDEA > LOPHOCRATIA > AZHDARCHOIDEA > TAPEJARIDAE

Tapejarids are relative newcomers to the world of pterosaur research. These Early Cretaceous pterodactyloids have already earned a reputation as some of the most striking pterosaurs known (fig. 22.1), largely thanks to the extreme development of their headcrests and their short, deep skulls with birdlike, downturned jaw tips. In short, they look like the devil himself fashioned them using leftover bits of cassowaries after binging on energy drinks. The first remains of tapejarids were reported by Alexander Kellner in 1989, a time in which the pterosaur community was not particularly aware of the wider group that tapejarids belong to, the Azhdarchoidea. Bits and pieces of azhdarchoids had been discovered throughout the twentieth century but their diversity, abundance, and importance has only become apparent in the last 25 years or so. We now recognize at least four major azhdarchoid bauplans united by a wealth of features, including toothless jaws, very large nasoantorbital fenestrae that extend above their eye sockets, short wing fingers with distinctively truncated distal wing phalanges, and long legs (e.g., Unwin 2003; Kellner 2003; Lü, Unwin, et al. 2008; note that Andres and Ji 2008 do not recognize a group of "azhdarchoid" pterosaurs, however). Azhdarchoids seem to have been a very important pterosaur lineage that appeared early in the Cretaceous and dominated the autumn years of pterosaur evolutionary history. Indeed, they were the only pterosaurs to witness the mass extinction at the end of the Mesozoic in real abundance, with most other lineages dying out well beforehand (see chapter 26).

The relationships of azhdarchoids are the subject of rather hot debate. In the scheme we're following here, tapejarids are the first sprig of the azhdarchoid branch, retaining many "primitive" pterodactyloid characters that prohibit them entry into a group of relatively derived azhdarchoids, the Neoazhdarchia (see chapters 23–25). This finding is supported by a number of studies (Unwin 2003; Lü, Jin, et al. 2006; Lü, Ji, Yuan, and Ji 2006; Lü, Unwin, et al. 2008; Lü, Unwin, et al. 2010; Martill and Naish 2006; Witton 2008b), but in other schemes, Neoazhdarchia does not exist and Tapejaridae is thought to accommodate animals we'll refer to as chaoyangopterids (chapter 23) and thalassodromids (chapter 24) (Kellner 1995, 2003, 2004; Kellner and Campos 2007; Andres and Ji 2008; Wang et al. 2009; Pinheiro et al. 2011). The neoazhdarchian grouping is supported by a number of features including straight jaws, several features of the nasoantorbital region, depression of the orbits entirely into the lower half of the skull, relatively elongate rostra that lack cranial crests, and slender mandibles (Lü, Jin, et al. 2006; Lü, Unwin, et al. 2008); while the idea of an expansive Tapejaridae is only defined by the proportions of the nasoantorbital fenestra, the presence of pear-shaped orbits, and a very thin strut of bone separating the orbit and nasoantorbital openings (Pinheiro et al. 2011). We are following the apparent weight of evidence and using the former scheme here, but this does not mean we have heard the final word on this topic. Debates on this issue continue, and several well-preserved azhdarchoid skeletons await comprehensive description and discussion. Our understanding of azhdarchoid evolution may well change as new data comes to light.

Where, When, and Who

Tapejarid remains are known in some abundance from several locations around the world (fig. 22.2). They are recovered in most abundance from Lower Cretaceous deposits of China and Brazil, but they also occur in sediments of Morocco and Spain (Wellnhofer and Buffetaut 1999; Vullo et al. 2012). Tapejarids were first described from a broken skull found in the Santana Formation of Brazil (likely Aptian-Albian; 125–100 Ma) (Kellner 1989). These remains, though fragmentary, revealed a short-faced, toothless creature sport-

Fig. 22.1. *Tupandactylus imperator* doing his best Clint Eastwood impression on the scrubby hinterland of the Aptian Crato lagoon.

Fig. 22.2. Distribution of tapejarid taxa.

Fig. 22.3. Skulls of the Aptian Crato Formation tapejarids. A, *Tupandactylus imperator*, the pterosaur with the largest cranial crest known; B, *Tupandactylus navigans*, an unusual tapejarid that lacks a supraoccipital crest. A, redrawn from Campos and Kellner (1997).

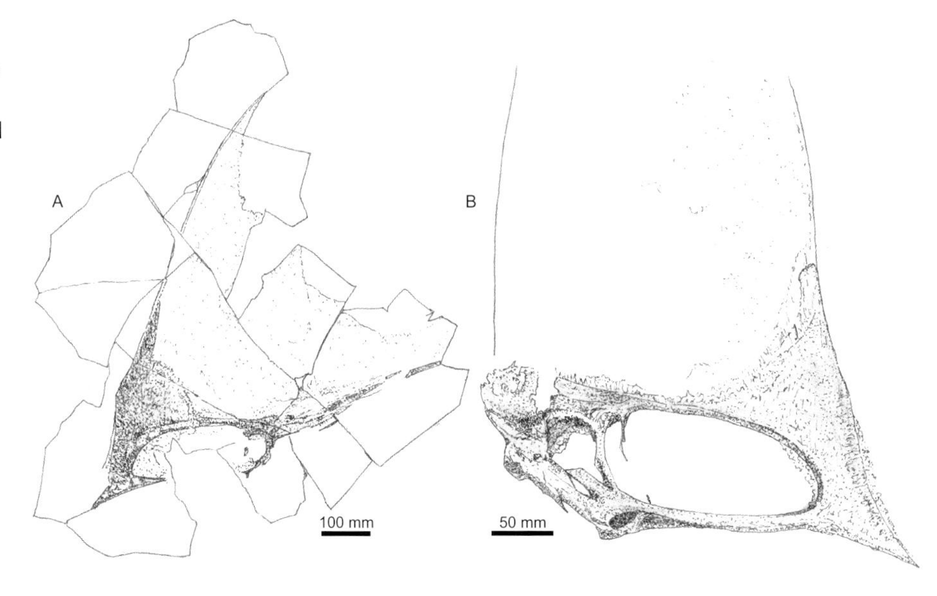

ing a prominent rostral crest. Since then, the discovery of near-complete skeletons of this animal, christened *Tapejara wellnhoferi,* has allowed documentation of its anatomy in detail (Wellnhofer and Kellner 1991; Eck et al. 2011). *Tapejara* is the only tapejarid species represented by three-dimensional remains, with all other tapejarids known from fossils crushed, more or less, to paper-thin proportions. This is certainly true for the most spectacular tapejarid of all, the large, sail-crested *Tupandactylus imperator* (fig. 22.3A; Campos and Kellner 1997; Kellner and Campos 2007). This animal also stems from Brazil, but from deposits that sit beneath the *Tapejara*-bearing Santana beds, the Crato Formation (likely Aptian; 125–112 Ma). Four skulls and one mandible of this species are known (Pinheiro et al. 2011), but none have yet been found associated with any postcranial remains. This quirk of preservation also applies to another Crato tapejarid species, the similarly spectacular *Tupandactylus navigans* (figs. 22.3B, and 5.11–5.12; Frey, Tischlinger, et al. 2003). This pterosaur is mired in a sticky nomenclatural issue thanks to a convoluted series of taxonomic ties with *Tapejara* and *Tupandactylus*. When discovered, *navigans* was considered to be species of *Tapejara*, but recent work suggests it probably belongs to the genus *Tupandactylus* (Unwin and Martill 2007; Kellner and Campos 2007; Pinheiro et al. 2011). While agreeing that *navigans* isn't referable to *Tapejara*, others have found little support for this idea and suggest that *navigans* needs a new generic label (Martill and Naish 2006; Witton 2008b). As we will see below, *navigans* and *imperator* are far more like one another than any other tapejarids, so it does not seem unreasonable to contain them within the same generic label, which we will do here. Additional Crato tapejarid specimens are represented by headless skeletons of varying completeness that cannot be reliably referred to *Tup. imperator* or *Tup. navigans*, but they remain notable for the detailed preservation of soft tissues from their wings and appendages (e.g., fig. 5.10).

The eyes of tapejarid aficionados were turned east in 2003 when *Sinopterus dongi* was described, the first of a collective of tapejarids from China's Jiufotang Formation (likely of Aptian age) (Wang and Zhou 2003b; Lü, Liu, et al. 2006). The Chinese tapejarids are slightly older than their Brazilian relatives and appear to represent an earlier phase of tapejarid evolution. A second *Sinopterus* species, *S. gui*, was named soon after the first (Li et al. 2003) but Kellner and Campos (2007) provided a good case for *S. gui* being a juvenile of *S. dongi*. *Sinopterus* is now represented by numerous specimens that, uniquely among azhdarchoids, provide us with a fairly complete growth series.

Another Jiufotang tapejarid, "*Huaxiapterus*," joins *Tupandactylus navigans* in something of a nomenclatural muddle. The name "*Huaxiapterus*" is based on *Huaxiapterus jii*, a species known from a fairly complete skeleton (Lü and Yuan 2005); and two further species, *H. corollatus* (Lü, Jin, et al. 2006) and the large-crested *H. benxiensis* (Lü et al. 2007). Both Kellner and Campos (2007) and Pinheiro et al. (2011) have suggested that *Huaxiapterus jii* is really not distinct enough from *Sinopterus* to warrant a separate genus, and should instead be labeled *Sinopterus jii*. Echoing the situation with *Tupandactylus navigans*, "*H.*" *corollatus* and "*H.*" *benxiensis* now await a new generic name. However, this issue may bear further thought: "*H.*" *benxiensis* is only really distinguished from *Sinopterus* by the size of its cranial crest, and as we've seen again and again (chapters 8, 14, and 18), pterosaur crest proportions are of questionable use in species-level taxonomy. It is therefore possible, and maybe probable, that the majority of the Jiufotang tapejarid species actually represent one species, *S. dongi* (fig. 22.4).

Further controversy surrounds the identity of another Jiufotang pterosaur linked to Tapejaridae, the tiny *Nemicolopterus crypticus* (fig. 22.5; Wang, Kellner, et al. 2008). This animal is known from a single fossil, a tiny but virtually complete skeleton with a wingspan of approximately 250 mm. This specimen was proposed to represent a unique type of pterodactyloid by Wang, Kellner, et al. (2008), because its anatomy apparently did not match that of any recognized pterosaur group. These authors argued that the ossified phalanges, ribs, and gastralia demonstrated that the animal was not particularly young despite its size, thereby suggesting the differences between it and other pterosaurs were taxonomically significant and not merely growth artifacts. Personally, I'm not so sure about this. Embryonic pterosaurs possess ossified gastralia and phalanges (Chiappe et al. 2004), and the rather simple, unfused bones of *Nemicolopterus* show the same juvenile characteristics and proportions of other baby pterosaurs (e.g., Bennett 1996b, 2006; also see chapter 8). The specimen also demonstrates several tapejarid characteristics (including toothlessness, downturned jaw tips, and some tapejarid features in its humerus), so it seems likely that *Nemicolopterus* is a very young, perhaps even a recently hatched, tapejarid. It even looks a little like the juvenile *Sinopterus* specimen once called "*S. gui*," which may indicate affinities to *S. dongi*. If this assessment is correct, "*Nemicolopterus*" makes the growth series of *Sinopterus* rather complete, comprising near-hatchlings ("*Nemicolopterus*") all the way through to fully grown adults ("*Huaxiapterus*" *benxiensis*).

Tapejarid finds outside of Brazil and China represent the known temporal limits of these pterosaurs. A rather deep, toothless mandibular jaw tip from the Cenomanian (100–94 Ma) Kem Kem beds of Morocco may record the youngest tapejarid occurrence (Wellnhofer and Buffetaut 1999), though its identity as a tapejarid is a little questionable given its very incomplete nature. More faith can be put in a partial tapejarid skull specimen from the Barremian (130–125 Ma) Las Hoyas Lagerstätte of Spain (Vullo et al. 2012). This material, named *Europejara olcadesorum*, is the oldest tapejarid fossil known and demonstrates that the group already possessed their unusual skull morphology very early in the Cretaceous.

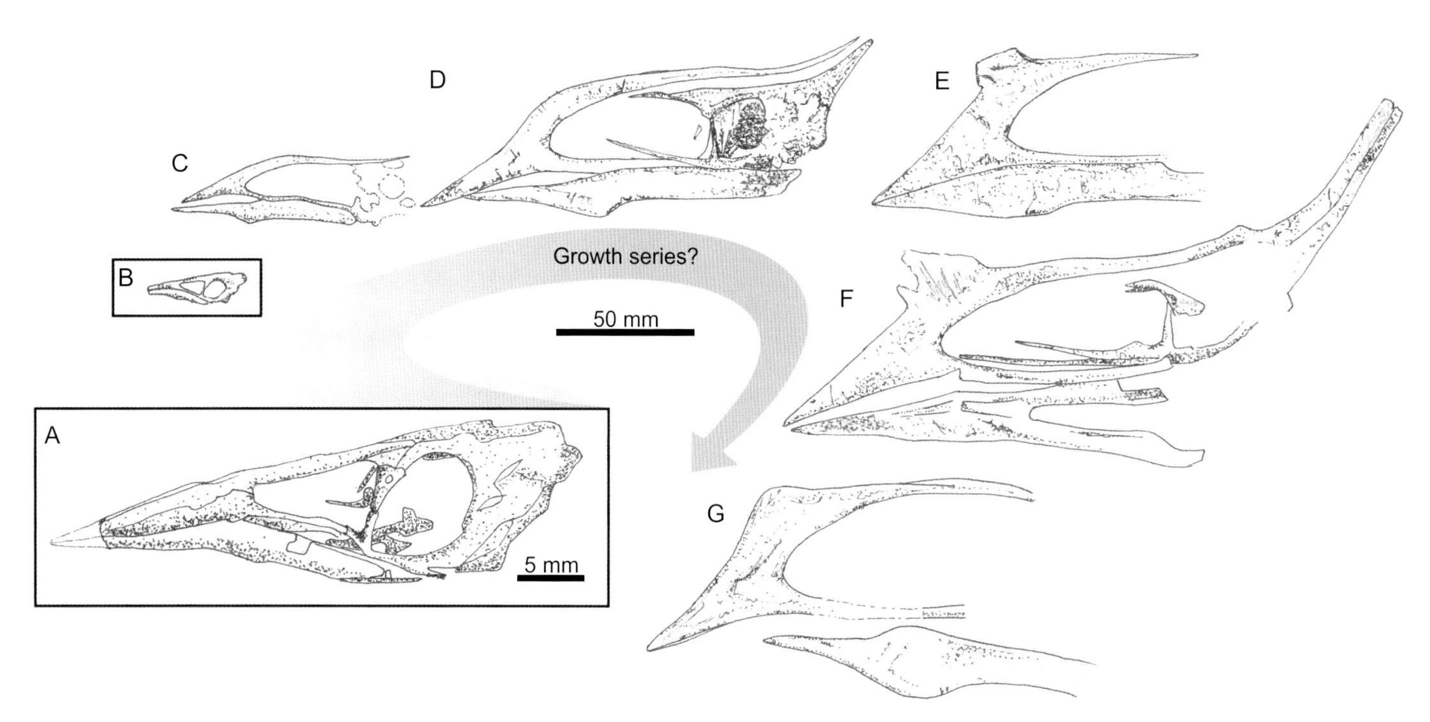

Fig. 22.4. Skulls and mandibles of the Aptian Jiufotang Formation tapejarids, arranged in size order. A, *Nemicolopterus crypticus*, expanded view; B, *N. crypticus* to scale with B–G; C, "*Sinopterus gui*"; D, *Sinopterus dongi*; E, "*Huaxiapterus*" *corrolatus*; F, "*Huaxiapterus*" *benxiensis*; G, *Sinopterus jii*. Note the correlation between size and crest development. A and B, redrawn from Wang, Kellner, et al. 2008; C, redrawn from Li et al. 2003; D, Wang and Zhou 2003a; E, redrawn from Lü, Jin, et al. 2006; F, Lü et al. 2007; G, redrawn from Lü and Yuan 2005.

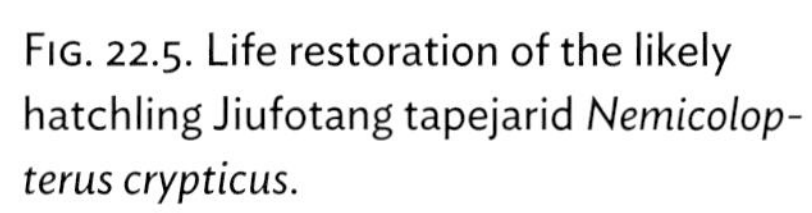
Fig. 22.5. Life restoration of the likely hatchling Jiufotang tapejarid *Nemicolopterus crypticus*.

Anatomy

OSTEOLOGY

Tapejarids have received rather variable treatment in pterosaur literature; some species are very well documented, others less so. The most detailed pictures of tapejarid anatomy are the comprehensive descriptions of *Tapejara wellnhoferi* by Wellnhofer and Kellner (1991) and Eck et al. (2011), along with various descriptions of Crato tapejarids (Frey, Tischlinger, et al.

2003; Unwin and Martill 2007; Pinheiro et al. 2011). Of the Chinese forms, "*H.*" *corollatus* is perhaps the best described (Lü, Jin, et al. 2006). The completeness of the Chinese tapejarids provides us with a handle on how large their poorly known Brazilian relatives were. Despite the proportions of their headcrests, tapejarids seem to have been the smallest of the azhdarchoids. The largest species (*Tupandactylus and* the "Kem Kem tapejarid") probably attained wingspans of no more than 3 m (fig. 22.6). Other species were even smaller. *Sinopterus* grew no more than 1.9 m across its wings, "*H.*" *corollatus* tops 1.5 m, and the largest *Tapejara* do not exceed a 1.5 m wingspan (although no specimens of adult *Tapejara* have been found yet, so it could potentially have grown larger).

Tapejarids are characterized by their short faces, a product of their stunted rostra and downturned jaw tips (figs. 22.3–22.7). The skulls of juveniles are fairly shallow, but their skulls deepened with age and without the significant increases in jaw length we see in other pterosaurs. Tapejarid orbits are rather teardrop shaped and, like all azhdarchoids, sit lower in the skull than the upper margin of the nasoantorbital opening. Their mandibles, like their skulls, are short and deep with downturned jaw tips that mirror those of the upper jaw. Something of a gap exists in the closed jaws of *Tapejara* so that only the posterior half of the jawline and jaw tips touch (Wellnhofer and Kellner 1991). Rounded crests erupt from the mandibular symphyses of mature forms, typically bearing roughly semicircular margins. *Europejara* presents an unusual mandibular crest however, which is backswept (Vullo et al. 2012).

Most mature tapejarid skulls sport two crests: one at the front of the skull, and the other at the back. The soft-tissue components of their crests occupied

Fig. 22.6. Why tapejarids are all show. Despite having the largest cranial crest of any pterosaur, the estimated wingspan of *Tupandactylus imperator* is only 3 m, which equates to an approximate shoulder height of 1.5 m.

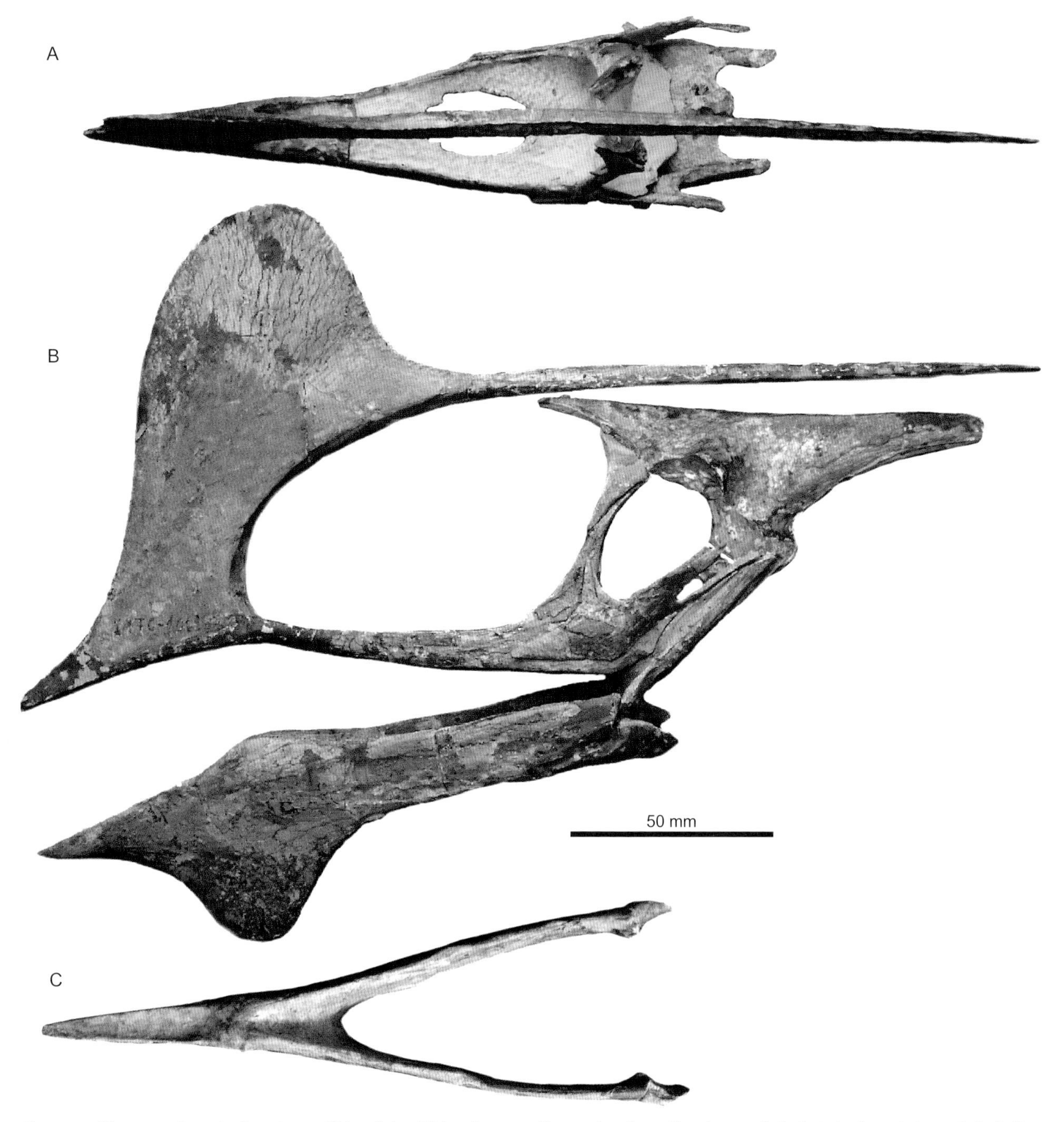

Fig. 22.7. The complete skull and mandible of the Albian Santana Formation form *Tapejara wellnhoferi*. A, dorsal view of skull; B, left lateral view of skull and mandible; C, dorsal view of mandible. Photographs courtesy of Andre Veldmeijer and Erno Endenburg.

the space between these two structures, so it is possible to roughly gauge the approximate size of their headcrests from these elements alone. *Tupandactylus navigans* is an unusual tapejarid, however, for possessing a blunt posterior skull face. The anterior crest projects dorsally or anterodorsally from the rostrum and is of variable shape, ranging from small, angular protuberances in the Chinese species to vastly expanded, triangular-shaped or rounded projections in the Brazilian forms. The anterior margins of the *Tupandactylus* soft-tissue crests are supported by long, bony spines that extend dorsally from the tips

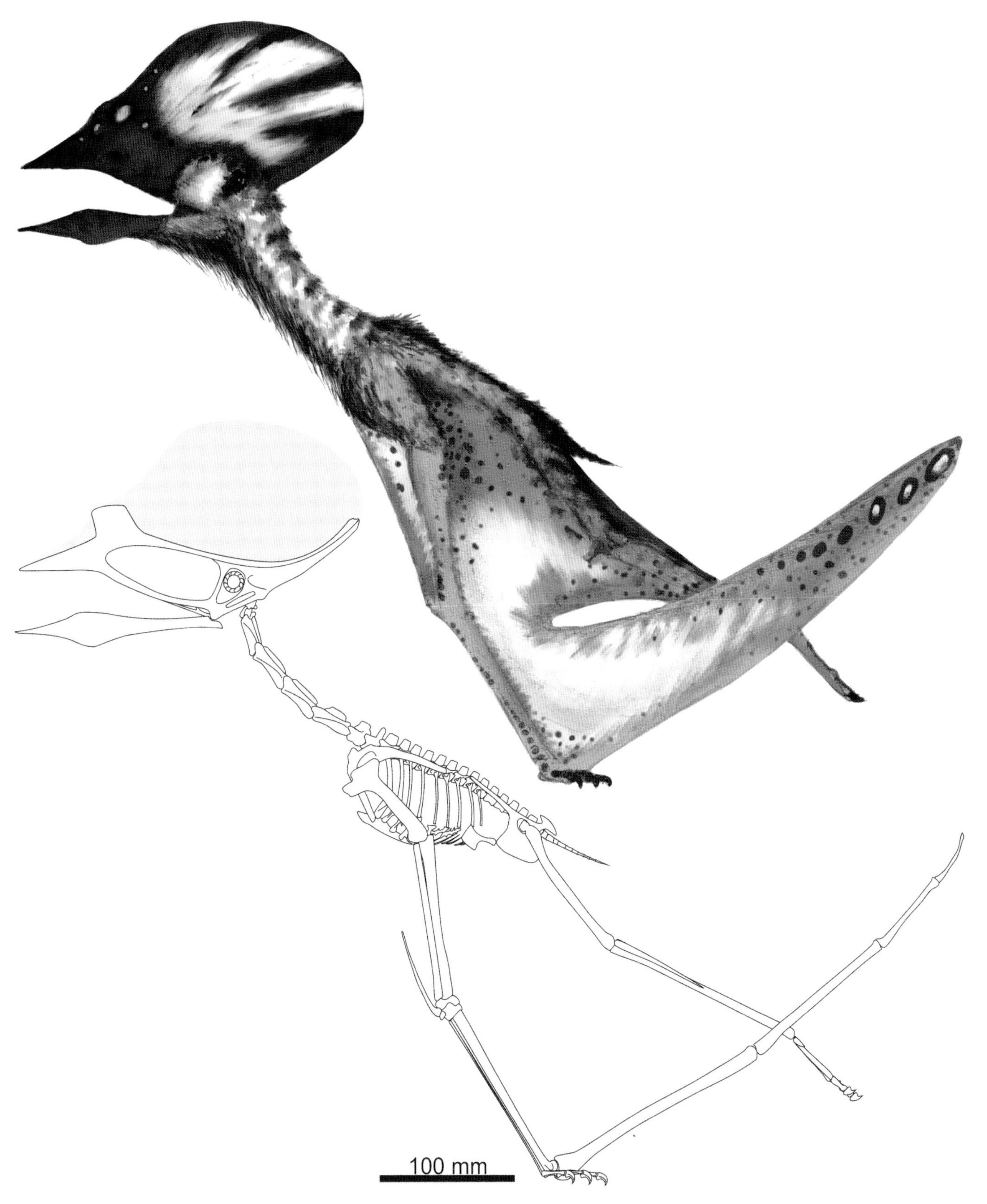

Fig. 22.8. Skeletal reconstruction and life restoration of a launching "*Huaxiapterus*" *benxiensis* (which likely represents a mature *Sinopterus dongi*).

of their rostral crests. The posterior bony crest is partially formed from a rearward projection of the rostral bone—the premaxilla—that overgrows the entire skull to extend beyond the posterior skull margin. This remains separate from the skull roof in juveniles but fuses in adults, combining with a posteriorly projecting supraoccipital crest to form a stout posterior ramus. The full extent of the bony crests in the Crato tapejarids is impressive (figs. 22.3 and 22.6), with their height (and length in *Tupandactylus imperator*) equating to a third of their owner's wingspan. Anchorage of their extensive soft-tissue crests is marked by fibrous bone along the lengths of their skulls that are, texturally speaking, very similar to the fibrous edges of bony crests seen in other lophocratian pterodactyloids.

The tapejarid cervical series is of a fairly typical length for pterodactyloid pterosaurs (fig. 22.8). Like dsungaripteroids, lonchodectids, and other azhdarchoids, their cervicals are low in form with reduced neural spines, and neural arches dropped into their centra. Their dorsal columns are fairly short and comprised of up to 13 vertebrae in adult *Sinopterus* (Lü, Liu, et al. 2006), but in contrast to some other azhdarchoids, their anterior dorsals do not seem to fuse into a notarium. The tapejarid sacral and caudal series are poorly known, but five vertebrae form the sacrum in mature *Sinopterus*. The few caudals recovered of *Tapejara* (Eck et al. 2011), *Sinopterus* (Wang and Zhou 2003), and *Nemicolopterus* (Wang, Kellner, et al. 2008) suggest that, like most pterodactyloids, their tails were little to write home about.

The limb girdles of tapejarids are robust and complex compared to many other pterodactyloids, with unusually long scapulae and coracoids possessing large flanges on their posterior surfaces (Frey, Buchy, and Martill 2003). These connect with rather square-looking sterna that, despite their short cristospines, are rather deep and broad. Their pelves, so far as they are known, suggest that their hindlimb girdles were similarly complex and strongly muscled, with long preacetabular processes and elevated, hatchet-shaped postacetabular processes. Such processes are common only to other azhdarchoids and possibly dsungaripteroids (chapter 20).

The tapejarid scapulocoracoid has been interpreted as rather unique among pterosaurs and possibly indicative of a distinctive flight style (Frey, Buchy, and Martill 2003). Because tapejarid scapulae are a full third longer than their coracoids, articulating their scapulae directly with the anterior dorsal vertebrae in an ornithocheiroid-like fashion positions their glenoids in the ventral trunk region, level with the sternum. Frey, Buchy, and Martill (2003) determined this a "bottom-decker" configuration, while other pterosaurs are either "top-deckers" (ornithocheiroids) or "middle-deckers" (azhdarchids; chapter 25), depending on the height of their glenoids within the body. I have my doubts about the "bottom-decker" interpretation, however, for a number of reasons. Firstly, marks on the ribs, and the nature of the articulation between the scapula and notaria in other azhdarchoids (see chapters 24 and 25), both indicate that their scapulae projected anterolaterally from the vertebral column and lay across the ribs in a fairly standard fashion for pterosaurs, rather than articulating perpendicularly with the spine, as per ornithocheiroids manner. Such marks remain unknown in tapejarids, but their close relatives suggest that their pectoral girdle was not constructed in the unusual ornithocheiroid manner. Secondly, articulated skeletons of pterosaurs with disproportionately long scapulae (campylognathoidids and ctenochasmatoids; chapters 12 and 19) show that analogous scapulae simply lie over the dorsolateral surface of the ribs and extend further down the trunk than those of other pterosaurs, Hence, longer scapulae do not necessarily indicate repositioning of the glenoids. Thirdly, it has also not been demonstrated that a lower decker girdle would fit around the ribs of its owner, a rather important—and often problematic—component of reconstructing extinct animal skeletons (engineers tasked with mounting dinosaur skeletons for museum displays have to frequently contend with this problem). It may, therefore, be safer to assume a "typical" shoulder configuration for tapejarids for the time being, or at least until more substantial evidence for a "bottom-decker" configuration is presented.

The limbs of tapejarids are rather elongate for their body size, a common trait of the azhdarchoid clan. The proportions of both limbs sets are typical of other azhdarchoids, with hindlimb length around 75 percent of the forelimb (excluding the wing finger), elongate wing metacarpals (around 20 percent of the entire wing), but rather short wing fingers (less than 57 percent of the wing length). The first three metacarpals of some tapejarid species (e.g., *Sinopterus*, "*Huaxiapterus*") are particularly short and do not reach the wrist bones (fig. 22.9; as seen in most other azhdarchoids and *Pteranodon* [chapter 18]), but other taxa (*Tapejara*) do not seem to have developed this condition. Tapejarid humeri characteristically bear two pneumatic perforations in their proximal ends (one on the upper surface, another on the lower) and deltopectoral crests that are a little shorter, and possibly more angular, than those of other azhdarchoids. Their femora are generally rather long—around 1.2 times the length of the humerus in adults—but, by

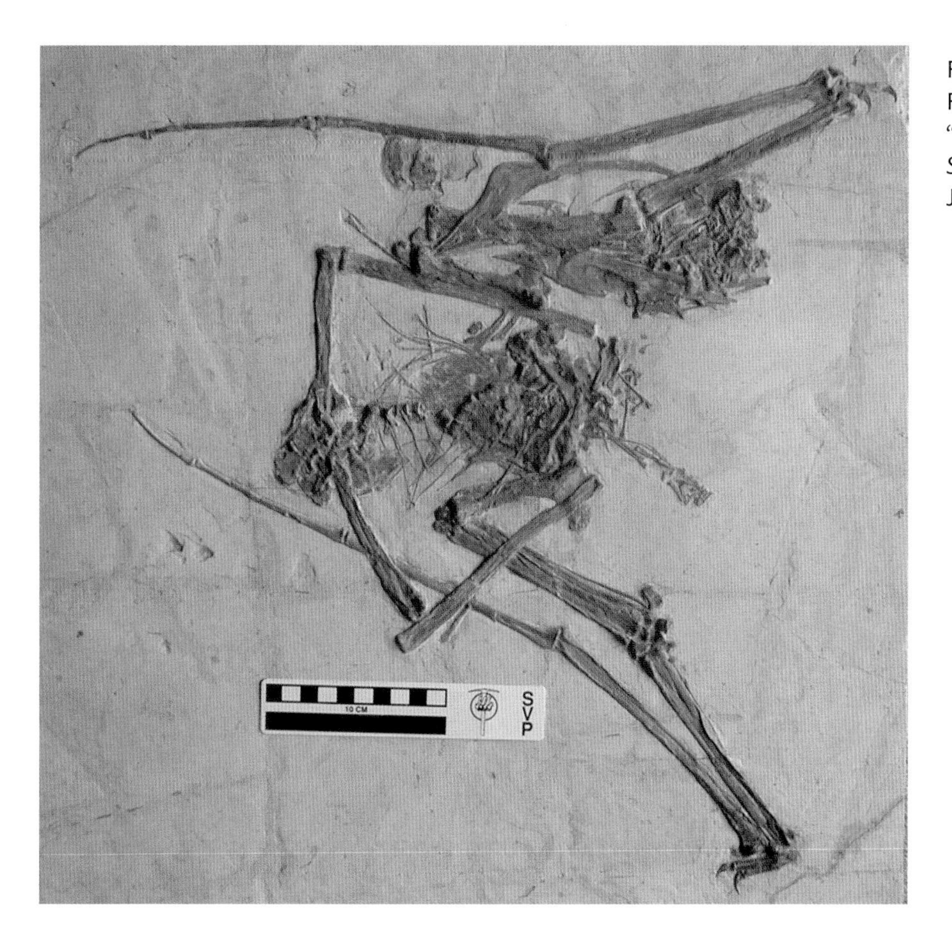

FIG. 22.9. Holotype skeleton of the Jiufotang Formation taxon *Sinopterus jii* (originally called "*Huaxiapterus jii*," but likely representing a large *Sinopterus dongi*). Photograph courtesy of Lü Junchang.

contrast, their feet are short and compact with a tightly bound pedal skeleton and short toes. The pedal digits of *Nemicolopterus* are of note because they possess slightly curved phalanges and, unusually, toes that steadily increase in length from digit I to IV.

SOFT TISSUES

The Crato Formation tapejarids are renowned for the preservation of numerous soft-tissue structures, many of which are not represented by pterosaurs of similar size or derivation (Frey, Tischlinger, et al. 2003; Elgin et al. 2011; Pinheiro et al. 2011). Their crests are undeniably their most arresting soft-tissue features. These seem to have been comprised of fairly inert, keratinous material as their remains show no indication of a blood supply or nerve tissues extending through them. While it is clear that they were huge, their extent is not completely known in either Crato species (fig. 22.3). Those of the four known *Tupandactylus imperator* skulls are ragged and torn (Pinheiro et al. 2011), clearly having been left out in the hot Crato sun a bit too long before they were buried. By contrast, the crest tips of both known *Tupandactylus navigans* skulls are probably missing because of human error, being sawn off by quarrymen unaware of the fossils preserved in the limestone they were cutting into roof tiles (fig. 5.12; Frey, Tischlinger, et al. 2003). Estimating the missing regions of these crests suggests that *Tupandactylus navigans* sported a rather tall number with subparallel anterior and posterior margins, while that of *Tupandactylus imperator* was much broader, being strongly convex along its posterodorsal margin and almost equal in length and height (fig. 22.6; Pinheiro et al. 2011). Each crest is estimated to have been between 3 to 4 times the lateral extent of its corresponding skull, depending on the size of their estimated area. Elements of soft-tissue crests are also known from *Sinopterus* and "*Huaxiapterus*," but their actual extent in life is more mysterious than that of the Crato forms. Additional minor crest tissues are seen along the front of the rostrum in both *Tupandactylus* species, and contribute to the keratinous beak sheath in *Tupandactylus navigans* (see chapter 5; Frey, Tischlinger, et al. 2003). Some soft tissues representing the nasal septum are also known from the front of the na-

FIG. 22.10. Left hindlimb and distal forelimb elements of an Aptian Crato Formation tapejarid, linked by the preserved trailing edge of the wing membrane (the orange minerals between the two limbs). Note the displaced stiffening fibers close to the ankle. Photograph by Bob Loveridge.

soantorbital fenestra of this pterosaur, perhaps indicating that tapejarid nostrils were housed toward the front of their skulls.

The soft tissues of the hands and feet of the Crato tapejarids are also well known (fig. 5.10; Frey, Tischlinger, et al. 2003). Claw sheaths from both extremities occur in several specimens, with those of the pedal claws being particularly long and doubling the length of the bone within. The same specimens show scaly heel and sole pads beneath the foot skeleton, and the heel pads are large enough that they engulf the vestigial fifth digit entirely. Webbing is preserved between tapejarid digits, a feature often suggested by pterosaur trackways but rarely preserved in fossils. Wing membranes occur in several specimens, including one which possesses a particularly good demonstration of an ankle-attached brachiopatagium (fig. 22.10; Frey, Tischlinger, et al. 2003; Unwin and Martill 2007; Elgin et al. 2011).

Locomotion

FLIGHT

As with most azhdarchoids, tapejarids are fairly new discoveries that have yet to receive much dedicated attention from pterosaurologists investigating functional morphology. As such, knowledge of their flight styles is still in its infancy. Witton (2008a) modeled *Sinopterus* and "*Huaxiapterus*" as fairly adaptable, generalist fliers comparable to modern parrots and crows. Their relatively short wings seem consistent with flight around inland settings; with long limbs and large, presumably well-muscled limb girdles, it seems fairly safe to predict that they were adept launchers and flappers.

The effect of tapejarid crests on their flight has yet to be ascertained. Frey, Tischlinger, et al. (2003) speculated that their crests could act as rudders in flight or, in a rather radical idea, as sails when flying close enough to water that their feet could be submerged to aid steering. This notion of pterosaurian catamarans has received little following from other workers, however. There are no modifications seen in tapejarid skeletons for "sailing" habits, which may include larger neck muscles to hold their heads in a desired position relative to the wind, developed adaptations for water launching, and so forth. Moreover, their compact feet with small surface areas, curving claws, and large pads would make lousy rudders. This does not discount the possibility that tapejarid crests had some effect on their flight, however. Xing et al. (2009) found that a hypothetical (and likely nonexistent; see chapter 18) sail-crest of *Nyctosaurus*, perhaps the best experimental analogue for a large-crested tapejarid, would have a substantial impact on the steering and speed of its flight. This suggests that tapejarid cranial crests may have similarly affected the flight patterns of their owners under windy conditions. Given that pterosaur crest function seems primarily related to display and socializing rather than flight (which may also be true for tapejarids; see below), we have to wonder if tapejarids enjoyed the flight advantages that their crests may have incurred or if they found them troublesome and avoided flying in unsettled conditions.

ON THE GROUND

All azhdarchoids seem to have been particularly adept grounded animals, and tapejarids are no exception. Their long, powerfully muscled limbs indicate potentially proficient walking abilities and their compact, padded feet are ideally suited to pacing firm ground. Tapejarids were probably regularly found striding around the woodland floors of Cretaceous China or the open scrubland of contemporaneous Brazil. The curving toes and asymmetrical feet of *Nemicolopterus* have been cited as evidence of a strong climbing ability (Wang, Kellner, et al. 2008), but other, more mature tapejarids do not show the same adaptations. Perhaps hatchling tapejarids spent much of their time hiding in vegetation, only venturing out into open settings once they were big enough that smaller predators would not worry them. Either way, the apparent terrestrial competence of these animals, as with all azhdarchoids, is supported by their biased preservation in continental deposits or, at least, marginal marine sediments that preserve large amounts of fauna and flora from the local hinterland. Recent analysis of tapejarid bones corroborate this habitat preference, as their skeletons are marked with chemicals carrying strong freshwater signals obtained through foraging and drinking in habitats well away from marine influences (Tütken and Hone 2010).

Paleoecology

The short faces and downturned bills of tapejarids are surefire indications that their feeding habits were distinct from other pterosaurs. Wellnhofer and Kellner (1991) proposed that their beak morphology was ideally suited for eating fruits, seeds, and other nutritious vegetative matter, a rather novel niche for pterosaurs. Other authors (e.g., Unwin 2005; Wang and Zhou 2006a; Vullo et al. 2012) have followed this proposal, which links to an idea that pterosaurs may have been important Cretaceous seed dispersers (Fleming and Lips 1991; Vullo et al. 2012). Seed dispersal is most effectively performed by flying animals, setting pterosaurs and early birds as better candidates for this role than herbivorous dinosaurs or other terrestrial Mesozoic animals. Tapejarid crests are thought to be ideally positioned to push and part vegetation should the animals be taking fruiting bodies directly from plants themselves (Wellnhofer and Kellner 1991), and their terrestrial adaptations would have put them in good stead when it came to foraging through leaf litter. I do wonder if seeds and fruits exclusively formed the diet of tapejarids however, as seed-eating birds are often equally prepared to eat small animals. In smaller birds, these may be mere bugs and worms, but larger avians are capable of capturing big invertebrates and small tetrapods. Tapejarids may have been similarly unfussy in their diets.

Tapejarid crest function has been linked to more than simply parting branches. Kellner (1989) proposed that their crests had a thermoregulatory function. While the soft-tissue crest components were probably inert and could do little to change body temperature, the living, bony regions may have been prone to losing or gathering heat as a consequence of their size and shape. This is an unavoidable effect of possessing an appendage with a high surface area to volume ratio, and as we will see in chapter 24, is unlikely to reflect the primary function of azhdarchoid crests. Of relevance here is the crest development of *Sinopterus,* which demonstrates that only older individuals bore large crests (fig. 22.4). This raises the question of how younger, crestless individuals would thermoregulate if tapejarid headcrests were devoted to this function. In contrast, the pattern of crest growth in *Sinopterus* is entirely consistent with the sexual display explanation of pterosaur cranial crests (fig. 22.4) and may indicate that tapejarid crests were not bizarre radiators, sails, or foraging aids, but merely the result of pronounced sexual selection akin to Irish elk antlers and peacock feathers. It is easy to imagine tapejarid crests being more of an encumbrance than a benefit in everyday tasks, including flying, and foraging in confined spaces, avoiding detection by predators, and coping with strong winds in any circumstance. In this respect, their headcrests would be of comparable inconvenience to the extreme sexual ornaments developed by modern animals, which can incur significantly negative impacts on their owners. This explanation may render tapejarids somewhat less mysterious and outrageous, but it is perhaps more in keeping with their anatomy, pterosaur paleobiology, and the evolutionary patterns observed in other animals.

23
Chaoyangopteridae

PTEROSAURIA > MONOFENESTRATA > PTERODACTYLOIDEA > LOPHOCRATIA > AZHDARCHOIDEA > NEOAZHDARCHIA > CHAOYANGOPTERIDAE

It can be hard not to feel a little sorry for chaoyangopterid pterosaurs (fig. 23.1). Not only are they markedly unflashy compared to the other azhdarchoids (there are no spectacular wingspans or oversize headcrests here), they are often rather overlooked by pterosaur workers in discussions of azhdarchoid diversity. However, chaoyangopterids are not only an exciting and novel pterosaur group (Wang and Zhou 2002), but they spread across the globe in their short Lower Cretaceous evolutionary history (Witton 2008c) and are already represented by up to six species. Chaoyangopterids have only recently been recognized as a distinct group of azhdarchoids (Lü, Unwin, et al. 2008; Witton 2008c; Pinheiro et al. 2011) and initially, different chaoyangopterid species were placed in disparate branches of the pterosaur tree. These placements included nyctosaurids (Wang and Zhou 2003a), pteranodontids (Lü and Zhang 2005; Wang and Zhou 2006a; Lü, Gao, et al. 2006), azhdarchids (Lü and Ji 2005b), and tapejarids (Unwin 2005), and some thought chaoyangopterids were so distinct from known pterodactyloids that they could not be placed in any existing group (Dong et al. 2003). This rocky taxonomic history may be more stable from now on however, because their edentulous jaws, large nasoantorbital fenestrae, depressed eye sockets, humeral morphology, long hindlimbs, and short wing fingers have been identified as certain azhdarchoid characteristics (Lü, Unwin, et al. 2008; Witton 2008c; Pinheiro et al. 2011), and links to pteranodontian groups have not been supported by any phylogenetic analyses of their relationships.

Exactly where chaoyangopterids fit into Azhdarchoidea is debated. Most studies find them to be close relatives of azhdarchids, united by features of their cervical vertebrae and possibly their posterior skull (Andres and Ji 2008; Lü, Unwin, et al. 2008; Witton 2008b; Vullo *et* al. 2012). Lü, Unwin et al. (2008) and Witton (2008b) noted that a position alongside azhdarchids may make them members of Neoazhdarchia, the group of straight-jawed azhdarchoids discussed in chapter 22. Pinheiro et al. (2011) offered an alternative interpretation, in which chaoyangopterids are very close relatives of tapejarids, while other analyses have not singled any group to be their closest relatives (Lü, Unwin, et al. 2010). It seems likely that chaoyangopterids have greater affinities to the straight-jawed azhdarchoids for reasons discussed in the previous chapter but, at present, there may not be sufficient data to assess whether they are more closely related to azhdarchids or another neoazhdarchian group, the thalassodromids (chapter 24). Indeed, there are merits to each of the three possible evolutionary configurations within Neoazhdarchia. A close relationship between thalassodromids and azhdarchids is suggested by their mutual possession of unusual wing finger proportions, fusion of their dorsal vertebrae, and several features of their skull construction. On the other hand, azhdarchids and chaoyangopterids have similarly simplified and elongated cervical vertebrae, but thalassodromids and chaoyangopterids share a number of skull and limb bone metrics. Of these three combinations, the features shared by thalassodromids and azhdarchids may just edge close kinship of these lineages into the lead in this dilemma, but this is only a tentative suggestion. As we will see in the next few chapters, there is much to be learned of the anatomy of neoazhdarchian pterosaurs, and new findings will almost certainly alter perception of their evolution.

Definitively identified chaoyangopterid material only stems from Early Cretaceous deposits of China and Brazil (fig. 23.2). They are best represented in the former, with five species described since 2003 from the Yixian (likely Barremian; 130–125 Ma) and Jiufotang (probably Aptian; 125–112 Ma) formations of China's Liaoning region. *Chaoyangopterus zhangi*, the first of those described, is the namesake of the group and is represented by an almost complete skeleton from

Fig. 23.1. They may not have had style, but they had grace. Two *Lacusovagus magnificens* prance around the margins of the Aptian Crato lagoon in a flamboyant courtship dance.

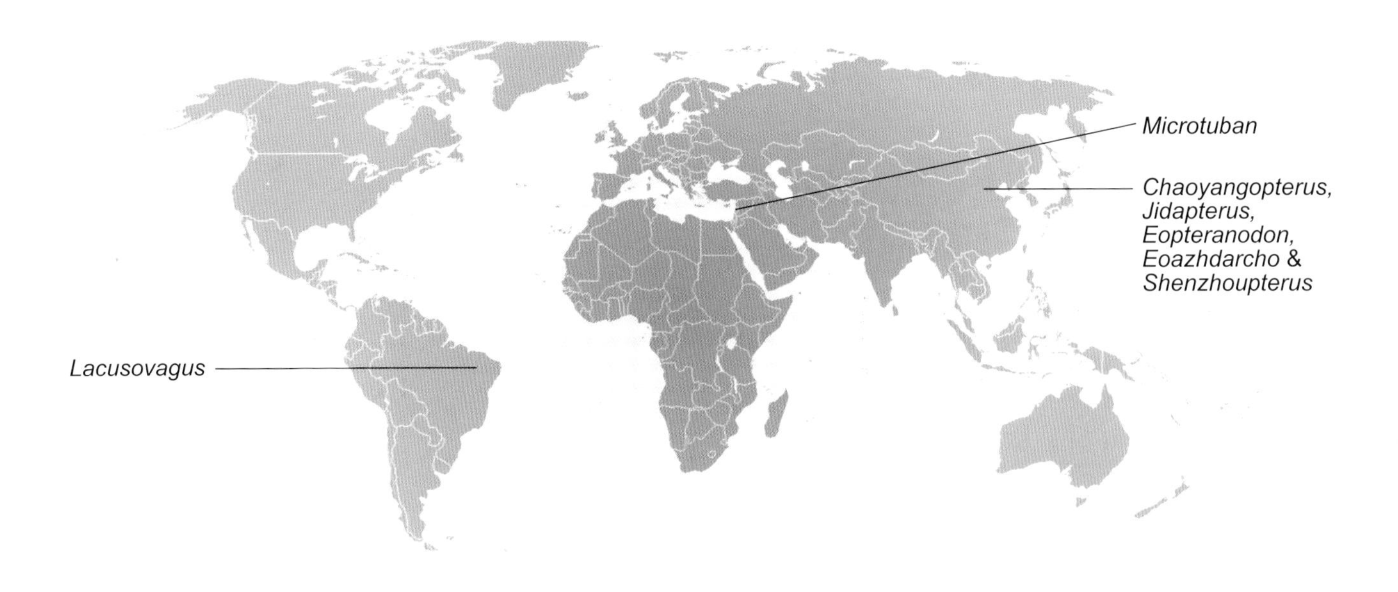

FIG. 23.2. Distribution of chaoyangopterid taxa.

the Jiufotang Formation (Wang and Zhou 2003a). Frustratingly, only the anterior portion of its skull and mandible were recovered, a recurring feature of chaoyangopterid finds. Virtually complete skeletons of the Jiufotang forms *Jidapterus edentus* (Dong et al. 2003) and *Eoazhdarcho liaoxiensis* (Lü and Ji 2005b) are known but, like *Chaoyangopterus and* the Yixian Formation species *Eopteranodon lii*, they lack their posterior skull regions (Lü and Zhang 2005; Lü, Gao, et al. 2006). Thus, only one entire chaoyangopterid skull is known, belonging to a diminutive species from the Jiufotang Formation, *Shenzhoupterus chaoyangensis* (Lü, Unwin, et al. 2008). The solitary specimen representing this animal also bears virtually complete postcranial remains and is, therefore, the most completely known member of its group (fig. 23.3A). Although *S. chaoyangensis* can be easily distinguished from its relatives, it seems likely that the number of other chaoyangopterid species in Chinese deposits is overinflated and in need of detailed taxonomic reassessment.

Outside of China, only one specimen has been referred with confidence to Chaoyangopteridae. Unlike the virtually complete Chinese material, it is only represented by a large but broken snout from the Brazilian Crato Formation (likely Aptian) (fig. 23.3B–C; Witton 2008c). No additional remains of this pterosaur, named *Lacusovagus magnificens,* have come to light since then, though Unwin and Martill (2007) noted the possibility that some headless Crato pterosaur skeletons with neoazhdarchian-like features may belong to this species.

One further pterosaur that may have affinities to chaoyangopterids is the Lebanese form *Microtuban altivolans* (Elgin and Frey 2011b). This species occurs in the Cenomanian (100–94 Ma) Sannine Formation and is only known from postcranial remains that are proportionally similar to those of chaoyangopterids, save for an extremely stunted fourth wing phalanx that constitutes only 1 percent of the wing finger length. Unfortunately, the lack of *Microtuban* cranial material means that its affinities to Chaoyangopteridae are not certain, and it may well represent a thalassodromid (chapter 24) or another, independent azhdarchoid lineage. In any case, its occurrence in Cenomanian deposits is significant because it represents the only upper Cretaceous remains of a nonazhdarchid azhdarchoid pterosaur.

Anatomy

Despite usually being fairly complete, chaoyangopterid fossils are often imperfectly preserved or badly weathered, so many fine details—such as the extent of their

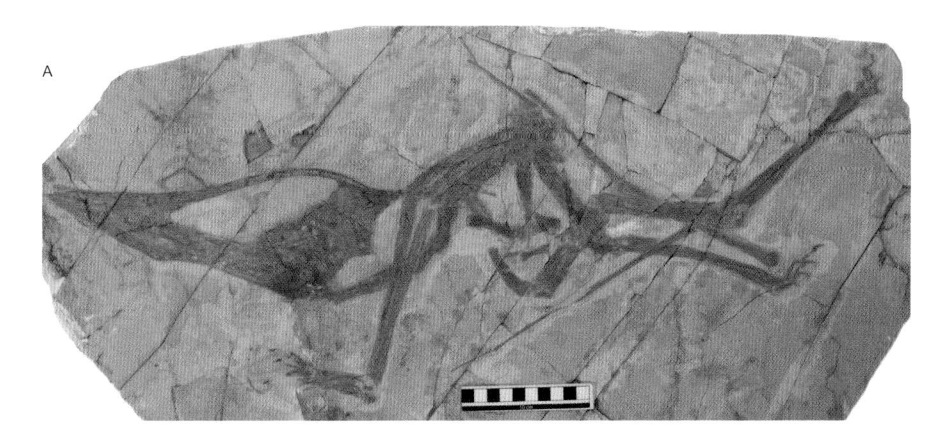

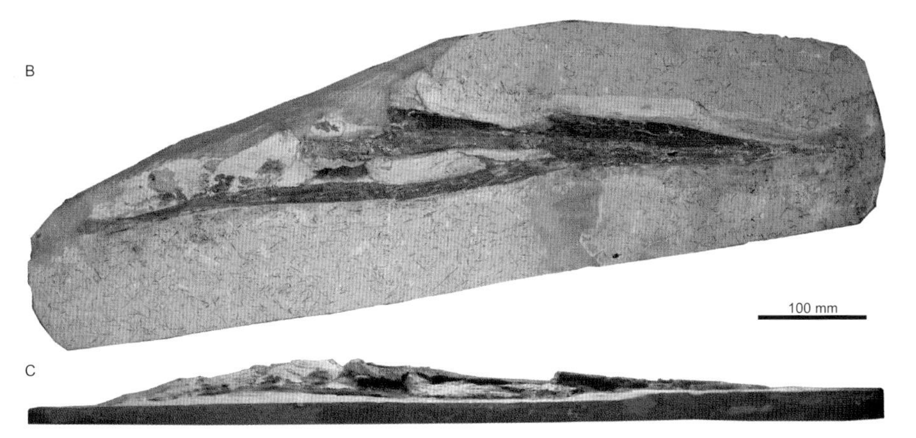

Fig. 23.3. Holotype specimens of two, disparately sized chaoyangopterid taxa. A, the complete skeleton of the Aptian Jiufotang Formation species *Shenzhoupterus chaoyangensis*; B–C, the incomplete skull of the Crato Formation species *Lacusovagus magnificens* in dorsal (B), and left lateral (C) view.

pneumatic system—remain unknown. Most descriptions of their remains are correspondingly rather brief. To date, perhaps the best chaoyangopterid fossil and most extensive description belong to *Shenzhoupterus* (Lü, Unwin, et al. 2008), though even this specimen is rather incomprehensible in places. Most chaoyangopterids were fairly small pterodactyloids, with wingspans ranging from 1.1 m in *Eopteranodon* to 1.9 m in *Chaoyangopterus*. The maximum adult size of *Shenzhoupterus* also seems rather diminutive (1.4 m wingspan). *Lacusovagus* bucks this trend however, spanning over 4 m and representing one of the largest pterosaurs in the Crato pterosaur assemblage (Witton 2008c).

Size aside, the anatomy of chaoyangopterids is fairly uniform (fig. 23.4). Their skulls are proportionally very large (Lü, Unwin, et al. 2008) with long, straight jaws and shallow, crestless rostra. Behind these, their skulls balloon upward with exceptionally large nasoantorbital fenestrae that are characteristically bordered dorsally by very thin, parallel-sided bars of bone. It has been suggested that the slender nature of these bones may explain why the posterior parts of chaoyangopterid skulls are rarely preserved, as the bones holding the skull extremities together seem somewhat fragile (Witton 2008c). *Lacusovagus* possesses a wider jaw and rostrum than the rest of its kin, along with a rather sinuous jaw line (fig. 23.3B). The only posterior skull region known of any chaoyangopterid, that of *Shenzhoupterus*, is rather murkily preserved because the bone minerals appear to have "leached" into the surrounding sediment through weathering. However, careful scrutiny reveals that this region was fairly typical of other neoazhdarchian pterosaurs, with its dorsal half occupied by a broad sheet of bone, and the orbit and both temporal fenestrae situated in the lower half. The posterodorsal region of the *Shenzhoupterus* nasoantorbital fenestra is unusual among azhdarchoids however, as it characteristically stretches beyond the jaw joint and extends into the dorsal cranial region. *Shenzhoupterus* also possess a weakly developed, posterodorsally projecting supraoccipital crest, which (like that of tapejarids and thalassodromids) incorporates elements of a premaxillary bone overgrowing the skull from the rostrum (Lü, Unwin, et al. 2008). More chaoyangopterid skulls are required, of course, to ascertain whether this posterior skull morphology is common to all chaoyangopterids or unique to *Shenzhoupterus*.

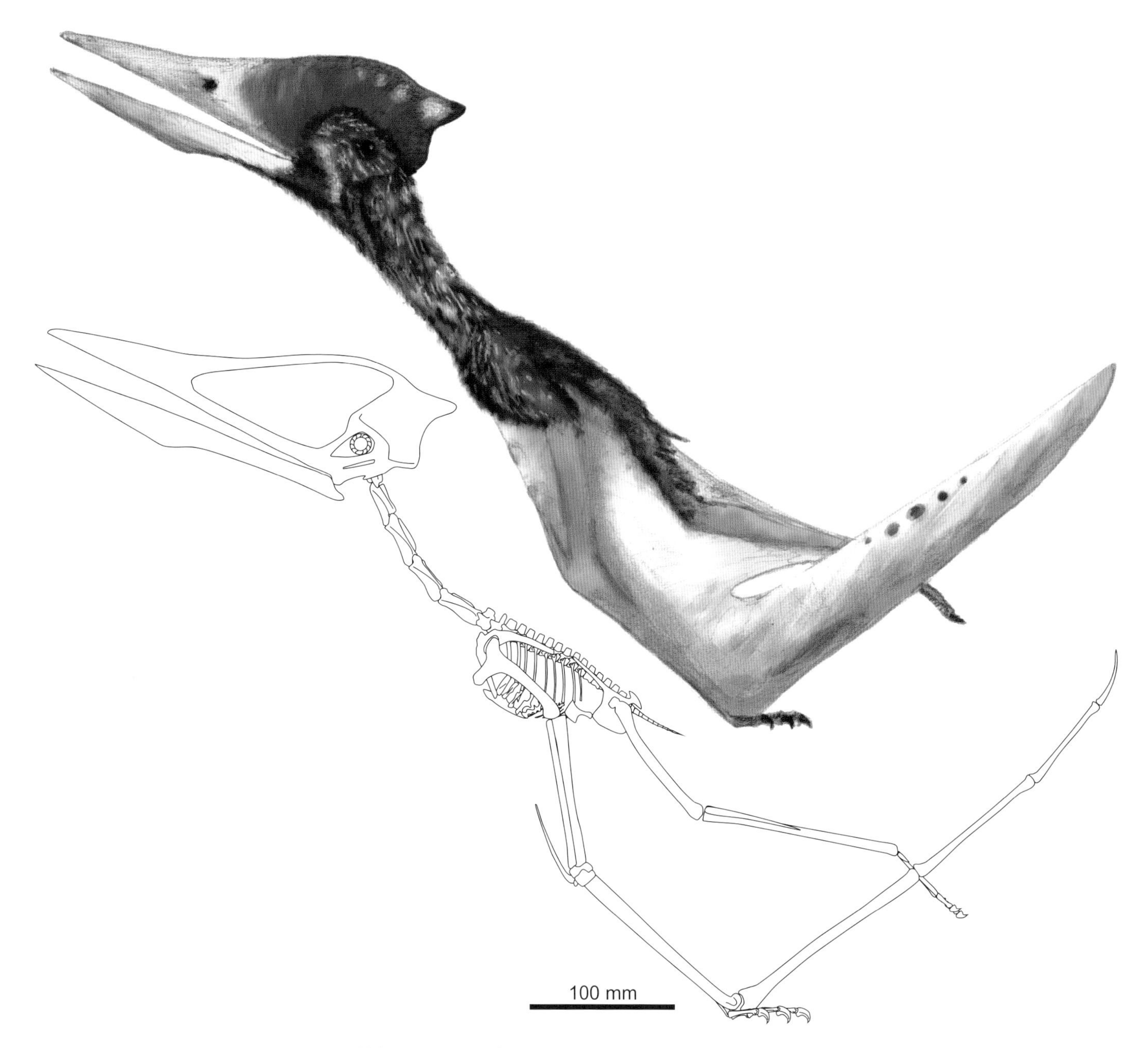

Fig. 23.4. Skeletal reconstruction and life restoration of a launching *Shenzhoupterus chaoyangensis*.

The lower jaws of chaoyangopterids are invariably long and crestless, with elongate mandibular symphyses and shallow mandibular rami. In *Shenzhoupterus* at least, the cervical series articulates with the skull at a rather low angle and may greatly exceed the combined length of the dorsal and sacral series (Lü, Unwin, et al. 2008). The midseries cervical vertebrae are somewhat reminiscent of the distinctive midcervicals of azhdarchids (see chapter 25), with neural arches contained within the laterally compressed centra, well-developed zygapophyses, and low, ridgelike neural spines (Wang and Zhou 2003a; Lü and Ji 2005b, Lü, Unwin, et al. 2008; Andres and Ji 2008). Their dorsal vertebrae do not seem to fuse into a notarium in any species, although most chaoyangopterid individuals we know of are osteologically immature and may not have reached notaria-forming ages at their time of death. Chaoyangopterid caudals are mostly mysterious, but those of

Eoazhdarcho may form a fused tail somewhat like that of pteranodontians (chapter 18; Lü and Ji 2005b).

Chaoyangopterid scapulocoracoids, like those of most azhdarchoids, have scapulae that are longer than their coracoids. The latter have long, prominent flanges on their ventral surfaces, another azhdarchoid characteristic. The chaoyangopterid sternum, currently only represented in *Eopteranodon*, is a rectangular structure with a short cristospine (Lü, Gao, et al. 2006). Frustratingly, chaoyangopterid pelves are, at best, very poorly preserved, so their detailed anatomy remains obscure (e.g., Lü and Ji 2005b; Lü, Gao, et al. 2006; Lü, Unwin, et al. 2008).

Chaoyangopterid humeri demonstrate classic azhdarchoid morphology, with long, parallel-sided deltopectoral crests projecting perpendicularly to their humeral shafts. The detailed structure of their humeri is not clear on any specimens, however, so the taxonomically distinctive configuration of their pneumatic features cannot be determined. The wing metacarpal is the longest bone in the chaoyangopterid forelimb (except for the first wing phalanx), being up to twice the length of the humerus and around 40 percent longer than the ulna. The first three metacarpals are extremely reduced and slender and do not seem to contact the carpus. The first three manual digits are much larger than those of the pes and sport relatively recurved claws. As with all azhdarchoids, the wing finger is proportionally short at only 53 percent of the wing length and is comprised of phalanges that steadily decrease in length distally, successively occupying 40, 30, 20, and 10 percent of the wing finger. Chaoyangopterid hindlimbs are long, between 75–80 percent of the forelimb length, and bear short, compact feet. Their first four metatarsals are pressed tightly together to make a relatively narrow foot terminating in short, slender digits with weakly curved claws. The fifth metatarsal, like most pterodactyloids, is much shorter than the rest and rather spur-like.

Locomotion

Because chaoyangopterids are relatively fresh faces to pterosaurologists, their locomotory biomechanics have not been investigated in any detail. Their limb proportions are very similar to those of tapejarids and we may imagine that their flight and terrestrial capabilities were likewise suited for life in terrestrial settings (see chapter 22). Long hindlimbs and compact feet have been linked to strong terrestrial abilities in other azhdarchoids (Witton and Naish 2008), and the distally shortened wings are well suited to flight in relatively cluttered terrestrial settings (Rayner 1988). We may also interpret their coracoid flanges as anchorage sites for large downstroke musculature, indicating a powerful flapping or launching ability. As with tapejarids, the idea that chaoyangopterids were adapted for life in terrestrial settings is borne out with their occurrences in freshwater deposits (such as the Yixian and Jiufotang Formations) or in brackish environments with high proportions of terrestrially derived fossils (the Crato Formation).

Paleoecology

Little has been said of the likely lifestyles of chaoyangopterids, though it has been suggested that their long jaws and well-developed flight ability may have been suited to fish-eating habits (Wang and Zhou 2006a). Personally, I'm not sure these attributes suggest fish-eating habits more than any other, because chaoyangopterid jaws seem rather unspecialized and would probably suit a variety of foraging strategies. Like other azhdarchoids, the likely terrestrial proficiency of chaoyangopterids suggests they were capable of foraging on land, but as has been noted for azhdarchids (Witton and Naish 2008), their compact feet may not have been particularly useful for wading. We might interpret the long, slender, and probably delicate bones comprising the midregion of their skulls as an indication that they did not generally feed on large, feisty animals, so perhaps smaller prey, nutritious plant matter, or even carrion formed large portions of their diets. If so, this idea may make chaoyangopterid foraging strategies very similar to those of other neoazhdarchians, who also seem to have been omnivores adapted to searching for small game in terrestrial settings.

24
Thalassodromidae

PTEROSAURIA > MONOFENESTRATA > PTERODACTYLOIDEA > LOPHOCRATIA > AZHDARCHOIDEA > NEOAZHDARCHIA > THALASSODROMIDAE

Thalassodromids are among my favorite pterosaur groups, no doubt thanks to several components of my doctoral studies focusing on their taxonomy and paleobiology and, frankly, they look rather cool with their oversize skulls and backswept, bony headcrests (fig. 24.1). This headgear is their most distinctive feature, substantially deepening and expanding the dorsal region of their skulls to create enormous and distinctive sail-like cranial crests. Currently, it seems that thalassodromids were a rather small group comprised of four species from a relatively tiny patch of Brazil. Despite the size of the group, thalassodromids have proved to be a highly contentious research topic and many words have been spilled over their taxonomy and paleobiology since their discovery in the late 1980s.

Perhaps the most contentious issue surrounding thalassodromids concerns their relationships to other pterosaurs. It has been accepted since their discovery that thalassodromids are azhdarchoids (e.g., Unwin 1992), but as we have seen in previous chapters, there is disagreement over whether thalassodromids are more closely related to tapejarids (e.g., Kellner 2003, 2004; Kellner and Campos 2007; Andres and Ji 2008; Wang et al. 2009; Pinheiro et al. 2011) or azhdarchids in the clade Neoazhdarchia (e.g., Unwin 2003; Lü, Jin, et al. 2006; Lü, Ji, Yuan, and Ji 2008; Lü,, Unwin, et al. 2010; Martill and Naish 2006; Witton 2008b). As discussed in chapters 22 and 23, some skull proportions, aspects of axial fusion, and wing bone metrics suggest to me that thalassodromids are probably more closely related to azhdarchids than to any other pterosaur group, but I say this with caution. The anatomy of thalassodromids has been documented in less detail than any other azhdarchoid group (see below), so it is sensible to hedge any large bets on their relationships until we know more of their detailed anatomy.

To date, definitive thalassodromids have only been recovered from one geological unit: the Lower Cretaceous Araripe Group of northeast Brazil (fig. 24.2). This group contains both the frequently mentioned Crato and Santana Formations (of probable Aptian [125–112 Ma] and Aptian-Albian [125–100 Ma] ages, respectively), both of which hold thalassodromid remains. Only fragmentary thalassodromid material has been described from the Crato Formation (Unwin and Martill 2007), but more diagnostic, name-bearing thalassodromid fossils have been recovered from the Santana Formation.

The first thalassodromid to receive a name was *Tupuxuara longicristatus*, a species represented by a fragment of jaw and some associated wing bones (fig. 24.3A; Kellner and Campos 1988). These scraps were enough to reveal aspects of the unusual thalassodromid crest morphology, toothless jaws, and unusual palatal ridges of these animals, features that would be found to characterize the group later on. A second fragmentary jaw specimen, distinguished from the first by details of its palate, was soon used to name a second species *Tupuxuara leonardii* (fig. 24.3B; Kellner and Campos 1994). More complete material of *Tu. leonardii* has since been uncovered, including a virtually complete skeleton of a subadult (Kellner and Hasagawa 1993; Kellner 2004) and a badly smashed skull from a half-grown individual (fig. 24.3C and E; Martill and Witton 2008). A third *Tupuxuara* species, *Tupuxuara deliradamus,* was named from another scrappy specimen for its peculiar eye socket and nasoantorbital fenestra morphology (fig. 24.3D; Witton 2009). More complete remains of *Tu. deliradamus*, representing a fairly large individual, also await further description (fig. 24.3F).

Technically, *Tupuxuara* was not the first thalassodromid to be unveiled to paleontologists however. A large, broken skull fossil found in 1983 was briefly reported in the early 1980s (Bonaparte et al. 1984) before being partially described and illustrated by

FIG. 24.1. The tables are turned: gnarly pterosaur badass *Thalassodromeus sethi* ambushes a baby spinosaur in Lower Cretaceous Brazil (see fig. 8.1 for a contrasting picture).

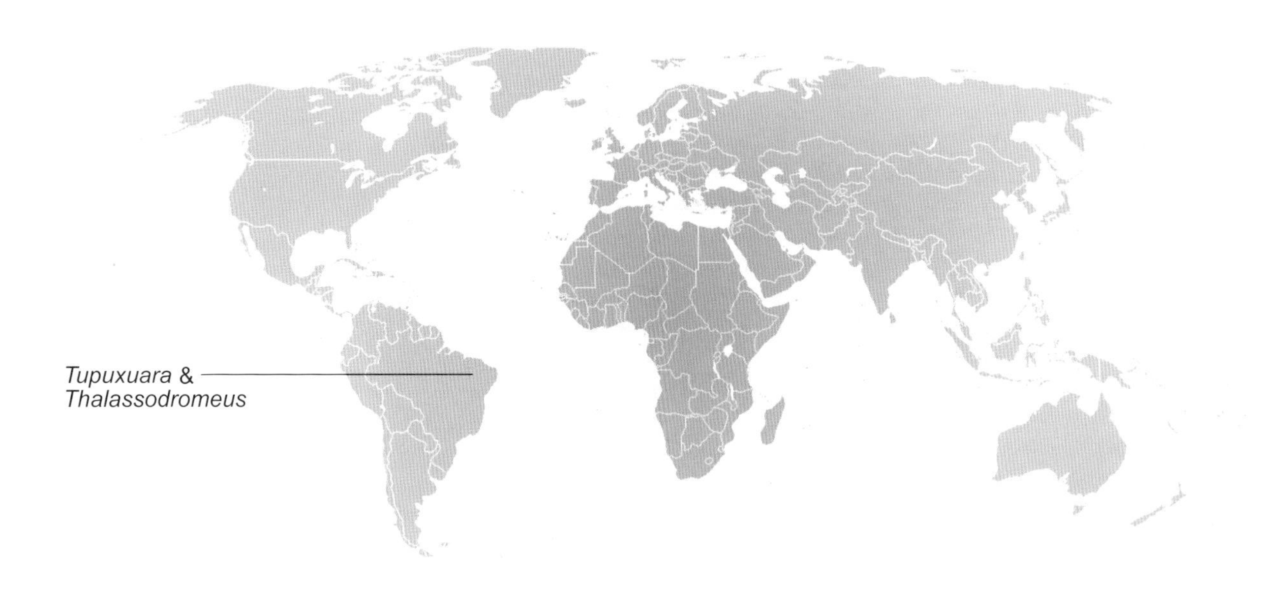

Fig. 24.2. Distribution of thalassodromid taxa.

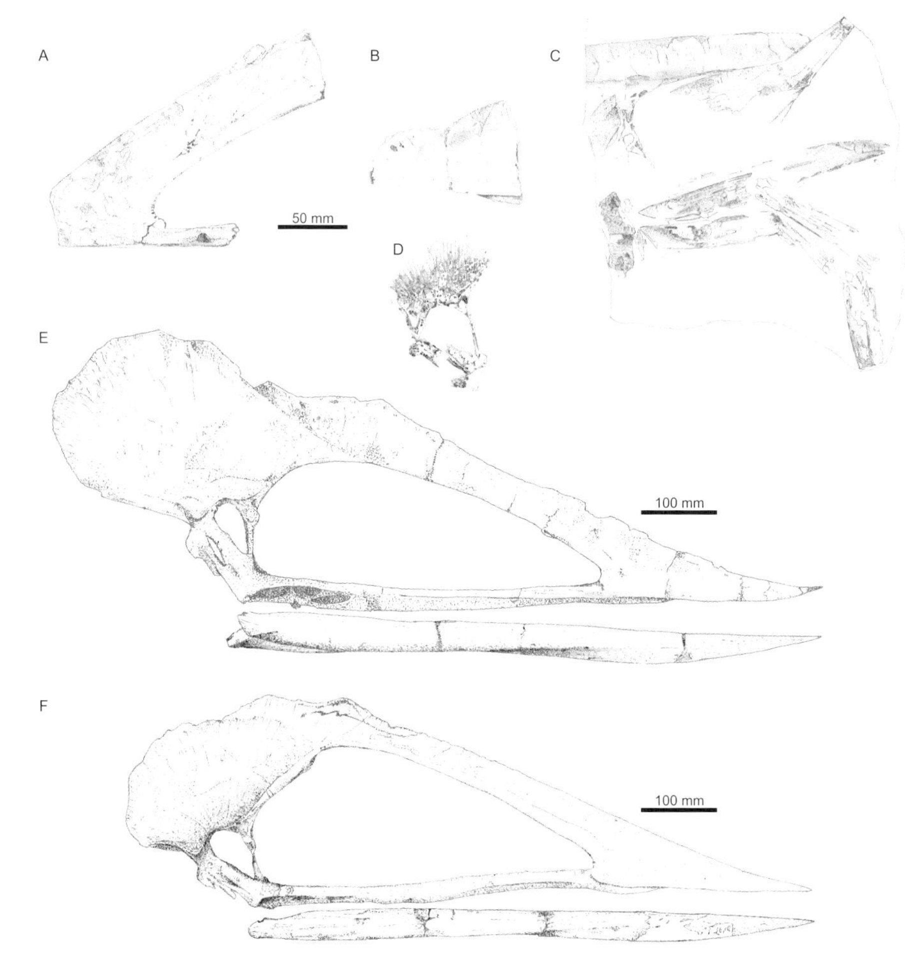

Fig. 24.3. Skulls and mandibles of the three *Tupuxuara* species, all from the Aptian-Albian Santana Formation of Brazil. A, the anterior rostrum of *Tu. longicristatus* (left lateral view); B, holotype rostrum of *Tu. leonardii* (left lateral view); C, broken skull and mandible of a juvenile *Tu. leonardii* (right lateral view); D, posterior skull and mandible of *Tu. deliradamus* (right lateral view); E, complete skull and mandible of *Tu. leonardii* (right lateral view); F, partial skull and complete mandible of *Tu. deliradamus* (the rostrum has been restored; right lateral view). A, B, E, and F redrawn from photographs provided by Andre Veldmeijer and Erno Endenburg.

FIG. 24.4. Skull and mandible of the Aptian-Albian Santana Formation species *Thalassodromeus sethi*. A, referred mandibular symphysis; B, holotype skull and mandible. B was composited from photographs provided by Andre Veldmeijer and Erno Endenburg.

Kellner and Campos in 1990. Somehow, pieces of this skull had been distributed across two museums on different American continents, but were reassembled before 2002 to show the entire skull morphology (fig. 24.4B; Kellner and Campos 2002). Christened *Thalassodromeus sethi*,[9] this animal was clearly *Tupuxuara*-like but differed in numerous details of its jaws and palate, including the possession of a mandibular tip with a compressed, blade-like biting surface. *Thalassodromeus* remains principally known from a single, enormous skull, save for a portion of crest from a half-grown individual and a complete mandibular symphysis (fig. 24.4A; Veldmeijer et al. 2005; Martill and Naish 2006, but see Witton 2008b). For a very brief time, it was argued that all thalassodromids represented a single species (*Tu. longicristatus*) (Martill and Naish 2006), but these ideas have since been firmly rejected (Kellner and Campos 2007).

Other thalassodromid material is thin on the ground. An isolated wrist complex from the Santana Formation ("*Santanadactylus*" *spixi*) may represent a thalassodromid, but its affinities to a named species aren't clear (Kellner 1995; Unwin 2003). Similarly, isolated wings and hindlimbs found in the Crato Formation have the characteristic proportions of thalassodromids, but no features define them as new species or allow their referral to existing taxa. As of yet, no pterosaur remains outside of Brazil can be confidently referred to Thalassodromidae. A large skull

[9] The name *Thalassodromeus sethi*, literally translated, means "Seth's sea-runner," a reference to the alleged skim-feeding habits of the animal (see below) and the purported resemblance of its skull to the crown worn by the Egyptian deity, Set (Kellner and Campos 2002). This has the making of a terrific name but, in actuality, a whole bouquet of nomenclatural oopsie-daisies were picked when this moniker was coined. The idea of this animal skim feeding has been condemned by multiple authors (Chatterjee and Templin 2004; Humphries et al. 2007; Witton 2008b), and the deity Set is not known among Egyptologists for wearing crowns. Veldmeijer et al. (2005) suggested that the plumed crown of another Egyptian god, Amon, was the intended nomenclatural reference.

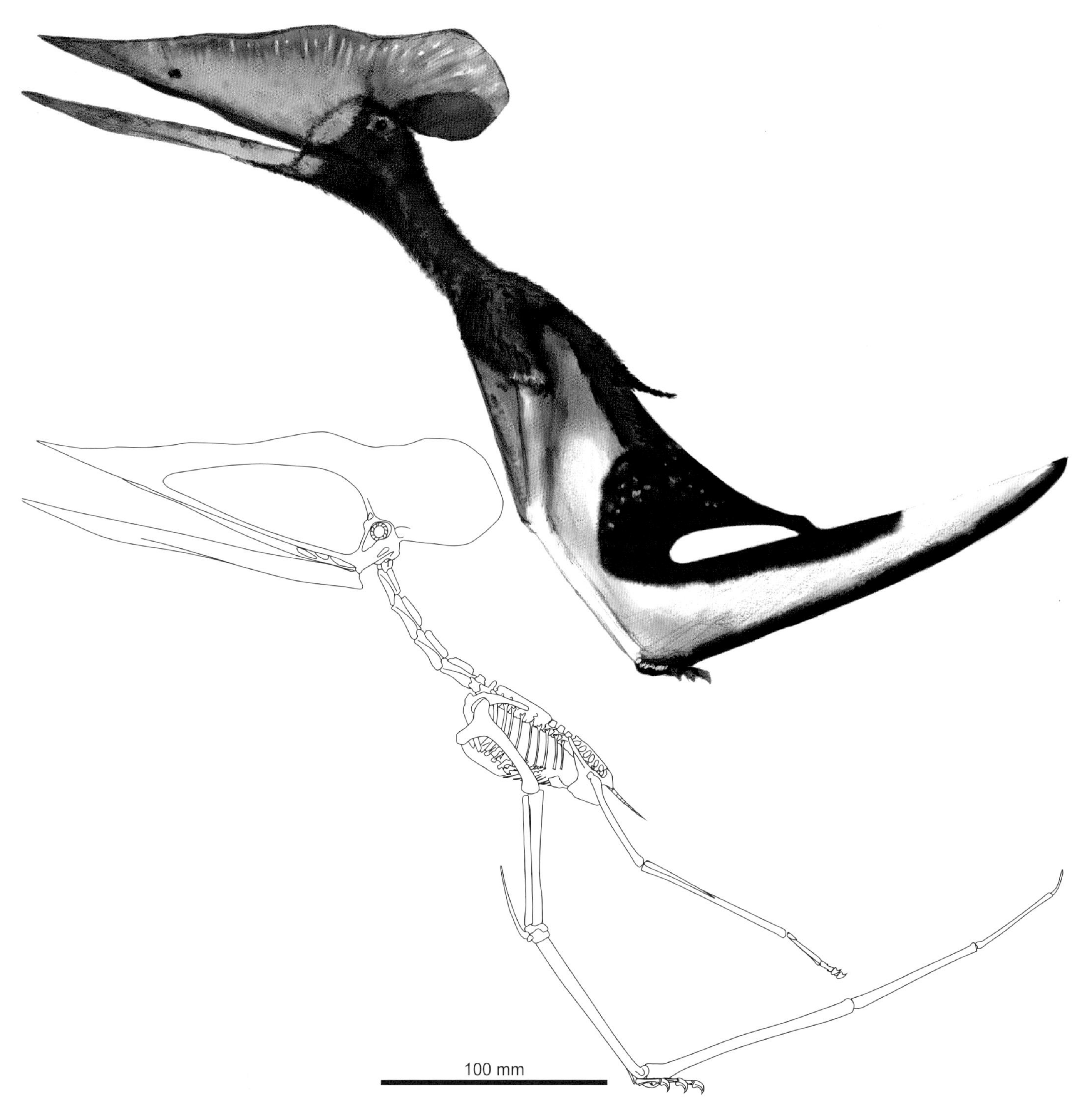

FIG. 24.5. Skeletal reconstruction and life restoration of a launching *Tupuxuara leonardii*.

from Maastrichtian (70–65 Ma) deposits of Texas has been linked to thalassodromids (fig. 25.4H; Kellner 2004; Martill and Naish 2006) but is almost certainly an azhdarchid (see chapter 25; Lü, Jin, et al. 2006; Witton 2008b). Pterosaur humeri recovered from Barremian (130–125 Ma) and Aptian sediments of Britain and Texas also seem to represent nonazhdarchid neoazhdarchians, but cannot be reliably referred to Thalassodromidae or Chaoyangopteridae at this time (Witton et al. 2009). Thus, for the time being, thalassodromids are considered endemic to Aptian-Albian rocks of northeast Brazil.

Anatomy

The reams of literature discussing thalassodromid anatomy have been more concerned with debating their taxonomically important characteristics than describing their overall form, so there are no comprehensive, detailed overviews or illustrations of their anatomy to recommend for pterosaur enthusiasts. The skull anatomy of *Thalassodromeus* is perhaps the best (if briefly) described thalassodromid cranial material (Kellner and Campos 1990, 2002), while their postcrania is briefly outlined in a short abstract on *Tu. leonardii* (Kellner and Hasagawa 1993). Additional published material includes a list of limb proportions (Unwin et al. 2001), and descriptions of the fragmentary Crato forms (Unwin and Martill 2007). Hopefully, some of the more complete thalassodromid material lying on museum shelves will receive more dedicated attention soon.

Thanks to the anatomy of *Tupuxuara* being almost entirely known (fig. 24.5), its wingspan can be accurately measured at 4 m (Unwin et al. 2001), while *Thalassodromeus* is estimated to be a little larger at 4.2–4.5 m (Kellner and Campos 2002). Their Crato cousins were smaller, perhaps only spanning 2–2.6 m (Unwin and Martill 2007). Although sharing similar gross anatomy and proportions, *Thalassodromeus* and *Tupuxuara* are readily distinguished by the heaviness of their builds. The skull of *Thalassodromeus* has large, robust elements with strangely "shielded" orbits, while the skull of *Tupuxuara* is comparatively lightly built. These differences may equate to distinct skull mechanics and foraging strategies in each form, possibly explaining how two otherwise very similar creatures could coexist in the same environment (see below).

The skulls of thalassodromids are proportionally large with jaws 0.7–0.8 m long, which equate to a fifth

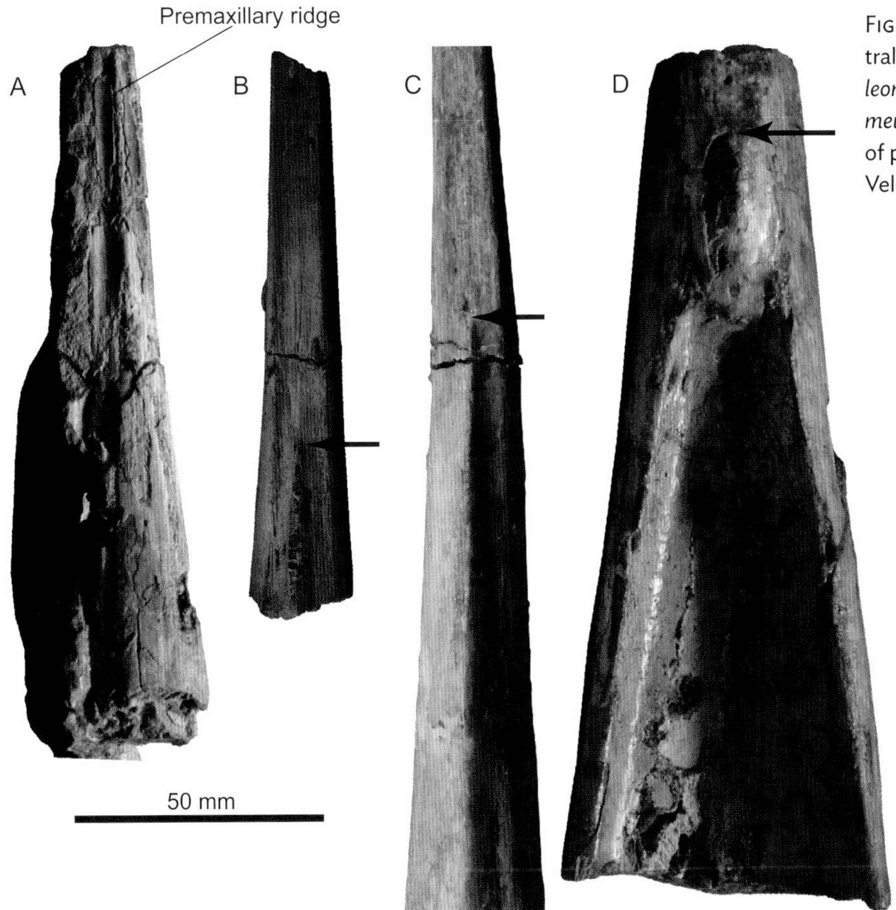

Fig. 24.6. Details of thalassodromid palates in ventral view. A, *Tupuxuara longicristatus*; B, *Tupuxuara leonardii*; C, *Tupuxuara leonardii*; D, *Thalassodromeus sethi*. Arrows mark the anterior termination of palatal ridges. Photographs courtesy of Andre Veldmeijer and Erno Endenburg.

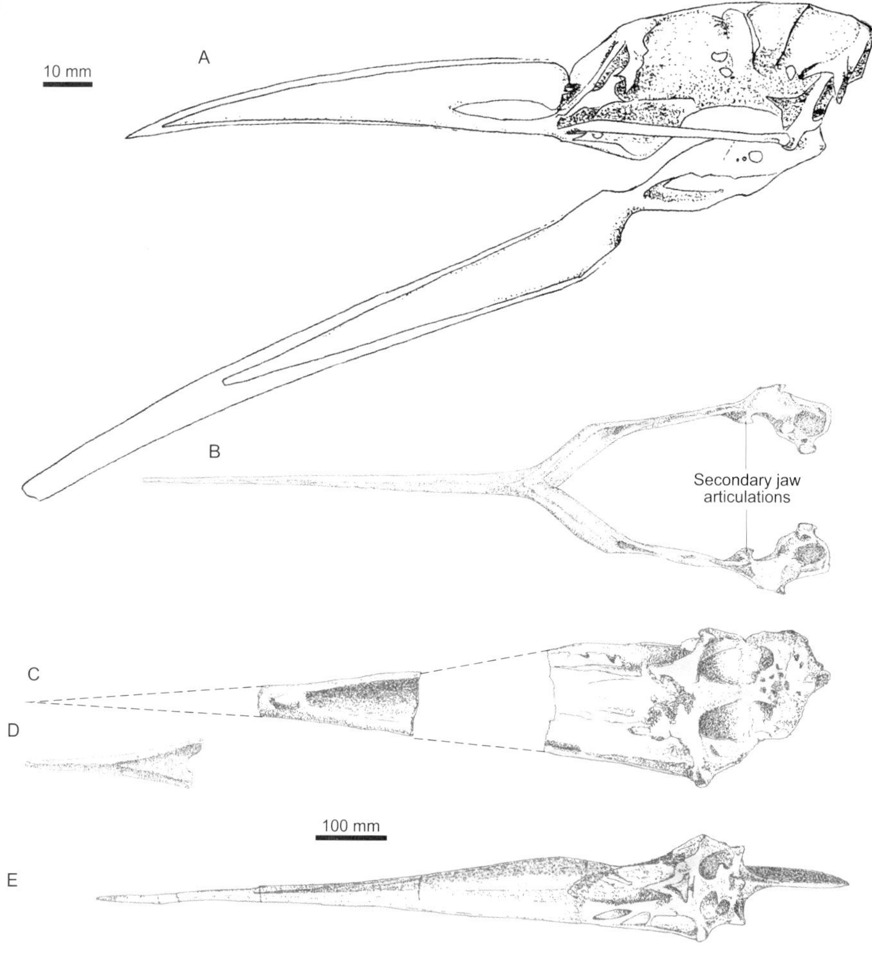

Fig. 24.7. Skull and mandible anatomy of the modern skimmer, *Rynchops niger*, compared with that of thalassodromids. A, skull and mandible of *R. niger* in left lateral view; B, dorsal view of *R. niger* mandible; C, skull of *Thalassodromeus* in ventral view; D, *Thalassodromeus* mandible in dorsal view; E, skull of *Tupuxuara leonardii* in ventral view. The outline surrounding the *R. niger* jaw in A marks the extent of its horny beak, a feature that is lengthened to improve prey-catching potential, but heavily abraded and frequently broken during skim feeding.

of their wingspans. Their nasoantorbital fenestrae are proportionally large compared to other neoazhdarchians but their toothless, straight jaw margins are otherwise in keeping with neoazhdarchian morphology. The orbits are at least 20 percent lower than the dorsal margin of the nasoantorbital fenestrae, positioning their eye sockets very low in the skull. Their crest bases are found almost immediately above the eye sockets, and are formed by broad, laterally compressed sheets of bone that constitute the posteroventral portion of the crest structure.

The bulk of the cranial crest is formed from a single bone—the premaxilla—that overgrows the entire skull from the rostrum (see fig. 8.7). The thalassodromid crest has no obvious anterior termination, instead exaggerating the height of the rostrum so that it is much deeper than those of other pterosaurs. This results in the bones overlying the top of the nasoantorbital region being four times deeper than those lining the jaw, a diagnostic feature of the group (Martill and Naish 2006). The crests of thalassodromids seem to lack the fibrous bone we see in many other pterosaurs with soft-tissue crests, suggesting they were not elaborated with other tissues. Even without these, however, their crests were substantially sized. The headcrest of *Tupuxuara* extends well above and behind the posterior skull face, but even this is dwarfed by the enormity of the *Thalassodromeus* ornament, which almost doubles the length and height of the skull. The apex of the *Thalassodromeus* crest characteristically bears a prominent V-shaped notch between the premaxilla and underlying bone. The lateral surface of this crest recalls the bone textures seen beneath the beaks of large birds in bearing deep, bifurcating channels that probably once held vascular tissues. This indicates that the thalassodromid crest was probably covered with a horny theca continuous with that lining the jaws.

The biting surfaces of thalassodromid jaws are so varied that they can be used to distinguish different species (fig. 24.6). They all possess a ridge along the underside of their rostrum, though it can be very long and shallow (*Tupuxuara longicristatus*), short and deep

(*Tupuxuara leonardii*), or developed into a prominent, rounded protuberance. Behind this, the palatal surface may be markedly swollen and convex all the way to the jaw joint (*Tupuxuara*), or deeply concave (*Thalassodromeus*). At the jaw tips, we find that *Tupuxuara* has flattened biting surfaces, but those of *Thalassodromeus* are compressed into sharpened edges (fig. 24.7D). Their mandibles are equally variable in construction. *Thalassodromeus* bears a slightly upturned mandibular tip and very short mandibular symphysis, but *Tupuxuara* sports a very long, straight symphysis instead.

Tupuxuara possesses neck vertebrae that are unusually short and blocky for pterodactyloids, being about as long as they are wide (Kellner 1995). Broad articular surfaces on each cervical indicates that thalassodromids had rather flexible necks (D. M. Unwin, pers. comm. 2008), and with each cervical sporting fairly tall neural spines along the entire vertebra, there seems plenty of room for cervical muscle attachment. A partially formed notarium comprising four dorsal vertebrae anchors a scapulocoracoid with a scapula that is longer than the coracoid (Kellner and Hasagawa 1993). Articulation surfaces on the notarium and scapula reveal that the thalassodromid shoulder girdle was oriented anterolaterally from the spinal column instead of perpendicularly, as seen in ornithocheiroids (chapters 15–18), and their coracoids possess a prominent process on their ventral surface that appears to have anchored particularly well-developed downstroke muscles. Possible thalassodromid pelves from the Santana Formation suggest that these pterosaurs also had well-developed pelvic elements with hatchet-shaped postacetabular processes, and broad pelvic shields (Bennett 1990; Hyder et al., in press).

Thalassodromid hindlimbs are 80 percent as long as the forelimbs (excluding the wing finger), making their limbs sets among the most equal of all pterodactyloids. Like all azhdarchoids, their feet are rather short and compact. Their humeri are comparatively long and possess a large pneumatic opening on their proximal anterior surface. This represents part of an extensive pneumatic network seen across the thalassodromid skeleton, the extent of which is only rivaled by the highly pneumatized ornithocheiroids and azhdarchids (Claessens et al. 2009). The metacarpals of their smaller manual digits seem to have entirely lost contact with the wrist bones (Kellner and Campos 1988), a feature they share with pteranodontians, other neoazhdarchians, and perhaps the tapejarid *Sinopterus*. Like all azhdarchoids, their wing fingers are rather short (54 percent of the wing length) and feature phalanges that shorten progressively toward their termination. The first wing phalange is proportionally large and occupies virtually half of the entire wing finger; but the fourth, by comparison, is tiny at only four percent of the wing finger length. These wing finger proportions are unheard of elsewhere in Pterosauria, except within Azhdarchidae.

SOFT TISSUE

Thalassodromid soft tissues are virtually unknown, but a controversial set of azhdarchoid body tissues from the Santana Formation of Brazil may tentatively be referable to this group. These are the problematic three-dimensional "wing tissues" briefly discussed in chapter 5, which have been variably interpreted as a piece of proximal wing membrane (Martill and Unwin 1989; Unwin 2005) or a part of the body wall (Kellner 1996b; Hing 2011; this explanation may be most likely). The allocation of these tissues to Thalassodromidae stems from their association with a large Santana azhdarchoid humerus that lacks a pneumatic foramen on its dorsal surface, as well as the presence of rather rotund ribs. These are both features thought to be indicative of a notarium-bearing pterosaur. At present, the only pterosaurs known from the Santana Formation with these features are thalassodromids, suggesting these soft tissues may belong to this group (Hing 2011).

Locomotion

FLIGHT

To date, no thalassodromids have been the subject of locomotory analyses. Hopefully, this situation will change soon given the excellent preservation of some thalassodromid material. Because their limb proportions are similar to the better-studied azhdarchids, their wing shapes and flight styles may be fairly comparable. Like azhdarchids, thalassodromids were probably best suited to flying around inland settings, because their comparatively short, broad wings would have been more maneuverable and less prone to snagging on obstacles than the longer, narrower wings of marine soarers (Rayner 1988). The prominent coracoid flanges of thalassodromids hint at enlarged ventral shoulder muscles that may have aided powerful or frequent downstrokes, along with possibly boosting their takeoff abilities. As with the large-crested tapejarids, it's possible that larger-crested thalassodromids may have had to compensate for their ornament in flight, but the ontogenetic development of this feature (see below) argues against these structures developing purely for aerodynamic use.

ON THE GROUND

As with their flight mechanics, the similarity of limb proportions in thalassodromids and azhdarchids suggests broadly comparable terrestrial abilities, and in the absence of any dedicated study of thalassodromid terrestrial ability, such comparisons are the best assessment currently on hand. If limb proportions are anything to go by, thalassodromids would join azhdarchids in being very proficient terrestrial animals, with elongate limbs capable of long strides and short, compact feet that would provide efficient foot mechanics. The same enlarged ventral shoulder muscles that may have enhanced the thalassodromid launch are also relevant to terrestrial locomotion, perhaps allowing for explosive starts when running or bounding. These attributes suggest that thalassodromids may have been just as competent on the ground as their often giant relatives, although we should remember that these conclusions are based on very limited data, and further anatomical and functional analyses of thalassodromid skeletons may prove otherwise.

Paleoecology

While growth series of thalassodromids can hardly be considered complete, cranial remains of fully grown and half-grown individuals are known for all species except *Tupuxuara longicristatus*, and these provide some insight into the way their skulls altered with age (fig. 8.7B; Martill and Naish 2006; Witton 2009). As with many pterosaurs, thalassodromid jaws increased in length as they grew while their eye sockets, relatively speaking, show little increase in size. The most dramatic difference between young and old thalassodromids, without question, is the extent of their headcrests. The premaxilla overgrew the skull from its starting position on the snout to beyond the orbit when the animal attained half its adult size, and then to the back of the skull at maturity. The premaxilla also deepened so that the rostra and crests of adult thalassodromids are far more substantial than those of juveniles. This is entirely consistent, of course, with the notion that only adult pterosaurs concerned themselves with crest-related activities, suggesting their use was more behaviorally important than physiological.

Despite this, there have been suggestions that thalassodromid headcrests served a more intrinsic function than a display device. The vascular network extending across the posterior region of the *Thalassodromeus* crest has been interpreted as evidence of a thermoregulatory function (Kellner and Campos 2002), a use also postulated for the crest of the tapejarid *Tapejara* (Kellner 1989; chapter 22). The notion of crests as thermoregulatory aids is problematic, however, as pterosaur crest development does not occur in a regular fashion with increasing body size. Instead, they grow in near-adult animals at a pace that far exceeds the requirements predicted for the growth of a thermoregulatory structure (Tomkins et al. 2010), and in any case, pterosaurs probably had no need for such a device. Their enormous, richly vascularized wing membranes would surely provide all the surface area needed for any thermoregulatory requirements. Moreover, the pattern of blood vessels on the *Thalassodromeus* crest does not seem particularly different or denser than those impressed into the bone underlying bird beaks. The vascularization here simply reflects the need to carry nutrients to the beak bones and soft tissues, and not the plumbing for a thermoregulatory device. Bird beaks do loose heat rapidly, however, explaining why long-billed birds often bury their beaks into their feathers when chilled, but this is obviously not their primary function and is only a consequence of their construction. I suspect the same was true of thalassodromid and other pterosaur crests; they may have had a thermoregulatory effect, but there is little reason to think that they were primarily developed for this role.

Although very little has been said of the foraging habits of *Tupuxuara*, the unusual jaws of *Thalassodromeus* have ensured that its diet has been the focus of much discussion. Kellner and Campos (2002) suggested that the relatively wide skull, streamlined mandibular symphysis, and several other features of the *Thalassodromeus* cranium and mandible recall the morphology of the modern skimming bird *Rynchops*, a highly derived species that forages almost entirely by plowing its mandible through water, snagging any prey items it happens to hit (fig. 24.7). The skimming adaptations of *Thalassodromeus* were considered so obvious that the animal was named after its purported habits. We have seen in other chapters that skim feeding has been suggested for plenty of pterosaurs, and favorable comparisons between the jaws of *Rynchops* and these animals date back to at least the 1800s (Marsh 1876). (A nonexhaustive list of purported pterosaur skimmers includes "campylognathoidids," rhamphorhynchines, ornithocheirids, pteranodontians, and azhdarchids; see chapters 12, 13, 16, 18, and 25.)

The idea of skim feeding in *any* pterosaur, even *Thalassodromeus*, has received fair criticism in recent years, however (Chatterjee and Templin 2004; Humphries et al. 2007; Witton and Naish 2008; Witton 2008b). The cause of this backlash is the lack of genuine skim-feeding adaptations in pterosaur skulls. The skull of *Rynchops* is riddled with so many clear-cut skim-feeding adaptations that an entire book has been written on them (Zusi 1962), but pterosaurs lack virtually all of them and do not offer any suitable analogues. For example, the mandibular symphysis of *Thalassodromeus* may have a streamlined cross section compared to those of other pterosaurs, but it is still proportionally much wider than the knifelike mandible of *Rynchops*. This structure is very long and extremely laterally compressed for its entire length in this bird, but that of *Thalassodromeus* is very short and tubby, even for a pterosaur. Many proposed pterosaur skim feeders also have teeth and flattened biting surfaces along their jaw lengths, both of which would cause tremendous drag and turbulence when being pushed through water, thus immediately precluding any skim-feeding behavior. With these features in mind, Humphries et al. (2007) built physical models of several pterosaur jaws (*Tupuxuara*, *Quetzalcoatlus*, *Pteranodon*, and *Thalassodromeus*) to directly measure and compare their streamlining against that of *Rynchops*. None—not even *Thalassodromeus*—came anywhere close to the streamlining of *Rynchops* jaws, and they were found to cut through water so poorly that their owners would struggle to skim feed for long periods. Ironically, the forces exerted on the model *Thalassodromeus* mandible were so high and unstable that they destroyed the aluminum rigging of the modeling apparatus when skimming at high speed. If that doesn't suggest that thalassodromid skim feeding requires a rethink, nothing will.

Comparisons between *Rynchops* and pterosaurs also do not improve elsewhere in their anatomies. *Rynchops* possess enlarged jaw muscles; two reinforced jaw articulations (fig. 24.7B; note that a number of bird groups have two jaw joints, but none as developed as those of skimmers—see Bock 1960); an unusually wide, robust skull; a deepened, stout mandible; and a strong, highly flexible neck. These are unsurprising features for an animal that essentially crashes its jaws into prey to catch it, and again, no pterosaur demonstrates comparable adaptations. *Thalassodromeus* was suggested to fit at least some of these criteria (Kellner and Campos 2002), but its skull is actually no wider than many other long-snouted pterosaurs, the attachment sites for its jaw muscles are not especially big, its jaw articulations are unremarkable, and the skull reinforcement is nowhere near comparable to that of *Rynchops* (Humphries et al. 2007; Witton 2008b). With even the mighty *Thalassodromeus* an unlikely skim feeder, there seems little, if any, reason to think that any known flying reptile was suited to this foraging method.

This opens the question of how else *Thalassodromeus* and its kin avoided going hungry. Skim feeding aside, little has been said about thalassodromid foraging methods. Unwin and Martill (2007) proposed that they may have lived a stork-like lifestyle, an idea that links their anatomical similarities to the potentially stork-like foraging methods of azhdarchids (Witton and Naish 2008; also see next chapter). Like azhdarchids, the length and equal proportions of the fore- and hindlimbs, and rather elongate jaws of thalassodromids seem well suited to roaming over land and opportunistically grabbing small animals and other foodstuffs. It should be noted that their short, relatively flexible necks are very different from the long, stiff necks of their azhdarchid cousins, however. This may infer some differences in foraging strategies between these groups. Perhaps the stouter necks of thalassodromids afforded them more generalized dining habits and sources of food, which are thought to be somewhat restricted for azhdarchids because of their long, stiff necks. The differences in jaw structure and skull construction between *Thalassodromeus* and *Tupuxuara* may also reflect some dietary differentiation, with the rather lightly built skull of *Tupuxuara* perhaps better suited to handling relatively small or weak prey items. By comparison, the more rotund skull of *Thalassodromeus* suggests it was more adept at handling relatively large, feisty animals. Biomechanicist Mike Habib and I once pondered the possibility of *Thalassodromeus* being able to subdue prey rather violently, using its bladed, hooked jaws for delivering strong, crippling bites (fig. 24.1). Perhaps the concave palatal region and short mandibular symphysis provided additional room within its oral cavity for swallowing larger prey items whole. If so, this would make *Thalassodromeus* a real rarity among pterosaurs, a raptor-like predator of small- to medium-sized game. It may be best to not get too carried away with such notions until the anatomy and functional morphology of these forms is studied and documented in more detail, however.

25
Azhdarchidae

PTEROSAURIA > MONOFENESTRATA > PTERODACTYLOIDEA > LOPHOCRATIA > AZHDARCHOIDEA > NEOAZHDARCHIA > AZHDARCHIDAE

It seems fitting that azhdarchids are the last group we will discuss in our tour of pterosaur diversity. Azhdarchids are perhaps the most spectacular of all pterosaurs (fig. 25.1), the sheer size and proportions of many species dwarfing not only all other pterosaurs but every other flying animal known, extinct or not. Recent investigation of their remains, recognized by pterosaurologists since the 1970s (Lawson 1975a), has revealed astounding details of their anatomy, strength, and flight capability. Many azhdarchids were as tall as giraffes and may have had the longest skulls of any nonmarine tetrapod, but they probably weighed no more than a moderately sized domestic pig. They are predicted to have cruised the skies at speeds warranting speeding tickets on most roads, and probably thought little of flying from one side of Earth to the other. There is some evidence that they roamed Cretaceous prairies in large flocks in pursuit of food, devouring fox-sized dinosaurs and other small prey. They were also a genuine evolutionary success, existing across the entire globe (fig. 25.2) and thriving throughout the Upper Cretaceous when other pterosaur groups were dying out. In short, if azhdarchids fail to stir your paleontological loins, you may want to see a doctor.

The taxonomy of Azhdarchidae is relatively uncontroversial. Azhdarchids are a well-defined lineage of azhdarchoid pterosaurs best diagnosed by their hyperelongate, simplified cervical vertebrae (Howse 1986; Unwin 2003; Kellner 2003; Andres and Ji 2008) and several features of their skulls and limbs (Unwin and Lü 1997; Unwin 2003). Exactly where they fit into Azhdarchoidea is, as with other azhdarchoids, a little uncertain. For some, azhdarchids represent all the nontapejarid azhdarchoids (e.g., Kellner 2003; Wang et al. 2009), but others (Unwin 2003; Lü, Unwin, et al. 2010) see them as derived members of Neoazhdarchia, and possibly closely related to Chaoyangopteridae (Andres and Ji 2008; Lü, Unwin, et al. 2008). Over a dozen species have been proposed, although the exact number varies depending on personal consideration of what constitutes a "valid" species. A number of azhdarchids, including some of the most famous forms, are based on remains with questionable diagnostic features (Witton et al. 2010).

The distinctive anatomy that makes azhdarchids easy to recognize for modern eyes was once confusing to pterosaur researchers. The first azhdarchid fossil to receive a name was an incomplete (but still 620 mm long!) cervical vertebra recovered in the 1930s or 1940s (Frey and Martill 1996) from Maastrichtian (70–65 Ma) phosphorous deposits of Jordan (fig. 25.3A). Because this bone was so long and tubular, it was interpreted as a wing metacarpal for many years (Arambourg 1954, 1959) and thought to represent a gigantic animal with a 7 m wingspan. Its describer, C. A. Arambourg, honored this with its name: *Titanopteryx philadelphiae* (Arambourg 1959). It was not until the discovery of complete azhdarchid cervicals in the 1970s that the vertebral identity of the *Titanopteryx* specimen was revealed (Lawson 1975a), with further change occurring when it was realized that its name (meaning "giant wing") was preoccupied by a tiny dipteran fly. The pterosaur "*Titanopteryx*" was subsequently renamed *Arambourgiania* (Nessov et al. 1987). The giant cervical has since had its own adventures through being lost and rediscovered, and has since been redescribed in considerable detail (Frey and Martill 1996; Martill et al. 1998).

The 1970s heralded one of the most important azhdarchid discoveries of all. Between 1972 and 1975 Douglas Lawson and colleagues, prospecting in the Maastrichtian Javelina Formation of Texas' Big Bend National Park, recovered the remains of several incomplete azhdarchid skeletons. Forty kilometers away from these more complete remains, they found an in-

FIG. 25.1. Realizing that the next chapter is about pterosaur extinction, a flock of the Maastrichtian Romanian pterosaur *Hatzegopteryx thambema* tries to fly back to an earlier part of the book to avoid the chop.

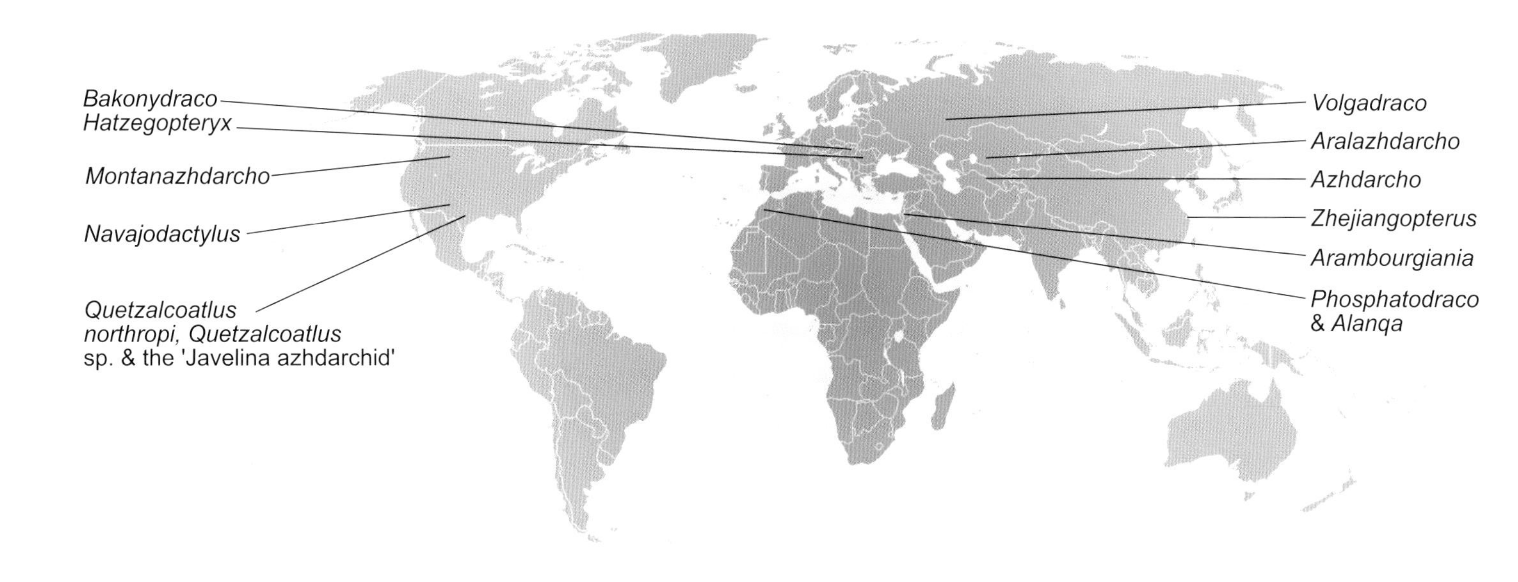

FIG. 25.2. Distribution of azhdarchid taxa.

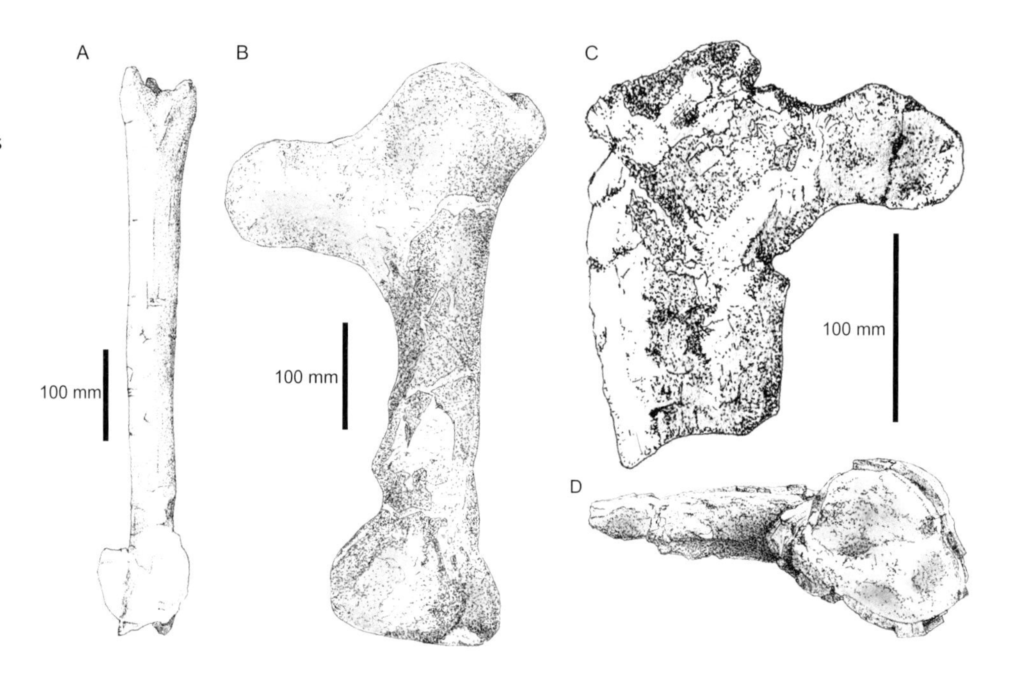

FIG. 25.3. The scant remains of giant azhdarchids. A, cervical vertebra representing *Arambourgiania philadelphiae* from Maastrichtian phosphate sediments of Jordan (dorsal view); B, *Quetzalcoatlus northropi* left humerus from the Maastrichtian Javelina Formation, Texas (dorsal view); C and D, partial humerus of *Hatzegopteryx thambema* from the Maastrichtian Densuş-Ciula Formation, Romania (in C, ventral and D, distal views). A, after Frey and Martill (1996); B, after Wellnhofer (1991a).

complete but gigantic wing that far surpassed the size of any pterosaur wings found previously (fig. 25.4B; Lawson 1975a). These remains constituted *Quetzalcoatlus northropi* (Lawson 1975b), one of the largest and most famous of all pterosaurs. The name *Q. northropi* now exclusively refers to the giant wing remains (Langston 1981) leaving the smaller, more complete material referred to as *Quetzalcoatlus* sp. until it receives a species name (Kellner and Langston 1996). The world is still waiting for a full inventory of the *Quetzalcoatlus* material to be revealed, but what little of this animal has trickled into pterosaur literature

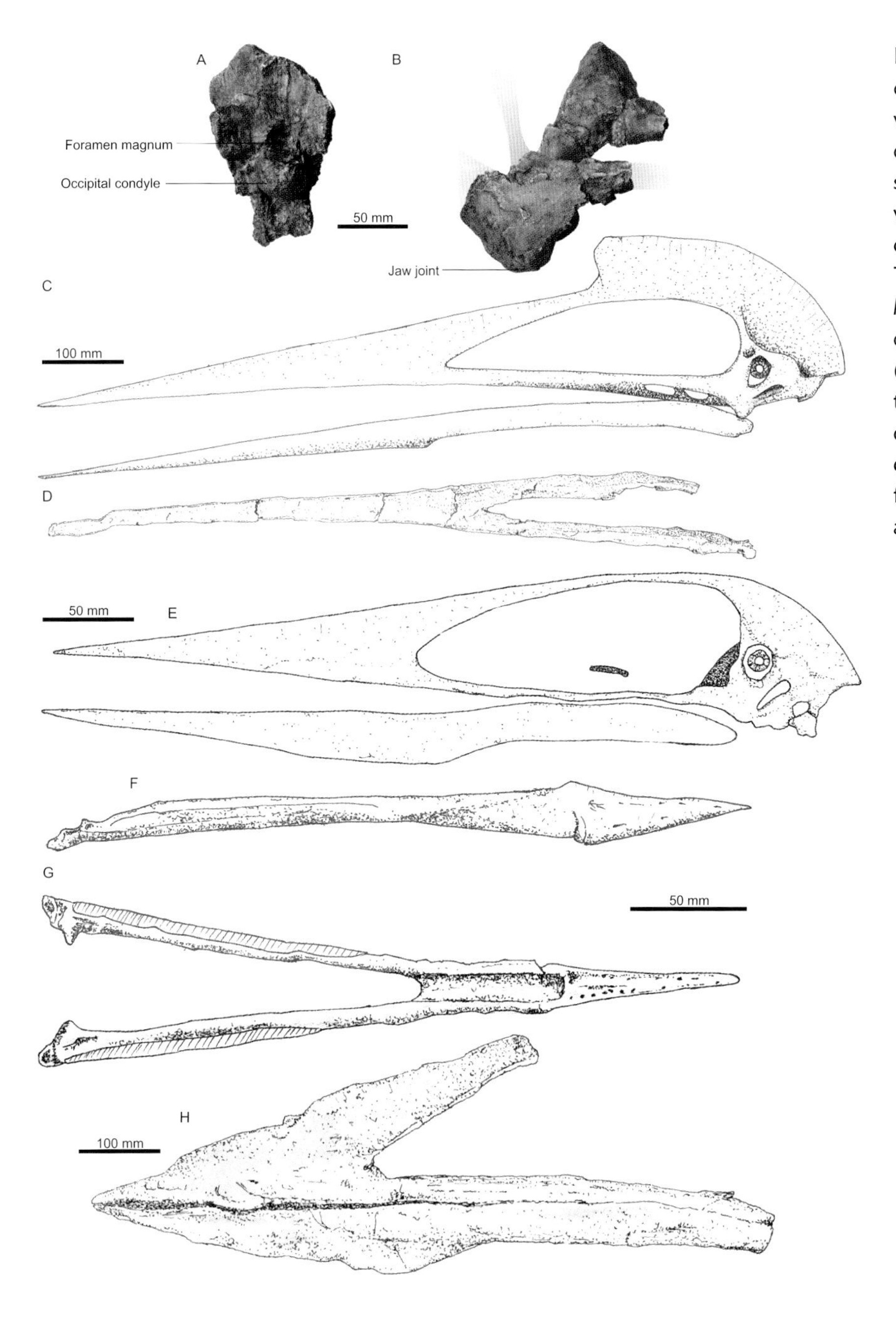

FIG. 25.4. Cranial remains of azhdarchids. A, occipital face of *Hatzegopteryx thambema*, posterior view; B, right upper jaw joint and palatal elements of *H. thambema*, right lateral view; C, reconstructed skull and mandible of *Quetzalcoatlus* sp., left lateral view; D, complete Q. sp. mandible, dorsal view; E, complete skull and mandible of the Campanian Tangshang Formation form *Zhejiangopterus linhaiensis*, left lateral view; F–G, complete mandible of *Bakonydraco galaczi* in right lateral (F) and dorsal (G) view, from the Santonian Csehbánya Formation of Hungary; H, incomplete skull and mandible of the Maastrichtian "Javelina azhdarchid." D, redrawn from Kellner and Langston 1996; E, redrawn from Cai and Wei (1994); F–G, redrawn from Ősi et al. (2005); H, redrawn from Wellnhofer (1991a).

suggests they represent the most complete, highest quality azhdarchid remains known to date (e.g., fig. 25.4C–D). *Quetzalcoatlus*-like fossils have also been reported from as far afield as France (Buffetaut et al. 1997), Montana (Padian and Smith 1992; Henderson and Peterson 2006), and Alberta (Godfrey and Currie 2005), making it the most geographically widespread of the azhdarchids. In recent years, it has become apparent that *Quetzalcoatlus* is not the only azhdarchid occurring in the Javelina Formation. A short-snouted azhdarchid known from an 800 mm set of jaws and some cervical remains represents another, as yet unnamed Javelina species (fig. 25.4H; known hereafter as the "Javelina azhdarchid"). This animal was mistakenly labeled *Quetzalcoatlus* by Wellnhofer (1991a) and considered a relative of *Tupuxuara* by some (Kellner 2004; Martill and Naish 2006), but several features of its anatomy have since been cited as unmistakably azhdarchid and are clearly distinct from those of *Quetzalcoatlus* (Lü, Jin, et al. 2006; Witton 2008b). Thus, at least three azhdarchid species seem to have coexisted in the Javelina paleoenvironment, making it something of a hotspot for azhdarchid diversity.

Azhdarchid remains started to pop out of upper Cretaceous rocks all over the globe in the 1980s, a trick they carry on to this day. Another important species, *Azhdarcho lancicollis*, was unearthed from the upper Turonian (91–89 Ma) Bissekty Formation of Uzbekistan (Nessov 1984). Now represented by over 200 fragmentary bones (Averianov 2010), *Azhdarcho* became the namesake of Azhdarchidae and, in the process, plundered nomenclatural priority for this group from the alternative, tongue-twisting moniker Titanopterygiidae (Padian 1984b) by a matter of months, a blink of the eye in publishing timetables. Other central Asian azhdarchids have also been named from fragmentary fossils, including the Santonian-early Campanian (86–75 Ma) Uzbek species *Aralazhdarcho bostobensis* (Averianov 2007), and *Volgadraco bogolubovi* from Campanian (83–70 Ma) deposits of Russia (Averianov et al. 2008). Another poorly known azhdarchid, "*Bogolubovia orientalis*," comes from the same region (Bogolubov 1914; Nessov 1991) but is of questionable validity. Unfortunately, the uncertainty surrounding this animal is unlikely to ever be resolved, because the sole specimen carrying its name, a portion of posterior cervical vertebrae, is now lost (Averianov 2007).

Yet more azhdarchids were named in the 1990s. Nessov (1991) erected *Bennettazhia oregonensis* for a solitary humerus from the Albian (112–100 Ma) deposits of Oregon (a specimen originally reported by Gilmore 1928), though the lack of pneumatic openings on its proximal palmar surface suggests its affinities lie outside of Azhdarchoidea. Indeed, the diagnosability of this species, which is based on a single limb bone, is highly suspect. Another azhdarchid child of the 1990s is *Zhejiangopterus linhaiensis, a* much better known animal that is represented by several incomplete skeletons from the early Campanian (83–75 Ma) Tangshang Formation of China. Until the *Quetzalcoatlus* material is documented in detail, *Zhejiangopterus* offers the greatest perspective on the general azhdarchid bauplan (figs. 25.4E and 25.5; Cai and Wei 1994). It took some time for the azhdarchid affinities of this species to become apparent, however, because its heavily crushed remains were originally interpreted as belonging to nyctosaurids, and only later were recognized as bearing numerous hallmarks of Azhdarchidae (Unwin and Lü 1997). Campanian deposits in Montana yielded the next named azhdarchid, a diminutive form with a 2.5 m wingspan and a stunted wing metacarpal: *Montanazhdarcho minor* (Padian et al. 1995; McGowen et al. 2002). A key part of identifying *Montanazhdarcho* as a distinct species, and not merely the juvenile of a giant, involved close histological analysis of its bones that suggested it was fully grown, despite its small size.

The turn of the millennium saw the discovery of even more azhdarchids. Another giant, *Hatzegopteryx thambema*, was the first azhdarchid to receive a name in the twenty-first-century (figs. 23.3C–D and 25.4A–B; Buffetaut et al. 2002; Buffetaut et al. 2003). Represented by several very fragmentary skull and limb elements from Romania's Maastrichtian Haţeg basin, *Hatzegopteryx* is among the largest pterosaurs known. It seems to have matched *Quetzalcoatlus northropi* in wingspan (see below) but also bears a very large, robust skull that may infer a heavier overall stature. In fact, the skull remains of *Hatzegopteryx* are so massive that they were initially interpreted as belonging to a large carnivorous dinosaur. The discovery of *Phosphatodraco mauritanicus* followed that of *Hatzegopteryx*, and comprised an incomplete cervical series from phosphatic, Maastrichtian deposits of Morocco (Pereda-Suberbiola et al. 2003). *Bakonydraco galaczi* arrived next and remains one of the more completely known azhdarchids, with a complete mandible (fig. 25.4F–G), numerous jaw tips, and several postcranial remains from the Santonian (85–83 Ma) Csehbánya Formation of Hungary to its name (Ősi et al. 2005; Ősi et al. 2011). Another Moroccan azhdarchid, *Alanqa saharica*, was named by Ibrahim et al. (2010) for a broken mandible and other referred fragments from the Kem Kem Formation (probably Cenomanian; 100–93 Ma). *Navajodactylus boerei* is the latest azhdarchid to join the ranks, and is represented by the proximal end of a first wing phalanx from the Campanian Kirtland Formation of New Mexico (Sullivan and Fowler 2011). I must admit to having considerable doubts as to the validity of *Navajodactylus* and even its classification as an azhdarchid; the proximal ends of pterosaur proximal phalanges are quite variable and have yet to be proven of diagnostic value to broad groups, let alone at generic or species levels.

Along with the named azhdarchid remains, a wealth of indeterminate azhdarchid material is also known from deposits on every continent except Antarctica (a number of overviews of azhdarchid fossil distribution are available: Averianov et al. 2005; Witton and Naish 2008; Barrett et al. 2008 and Ősi et al. 2011). As a whole, azhdarchids appear most abundant in the Late Cretaceous and were still present in great numbers at the very end of the Mesozoic. The frequent occurence of azhdarchid fossils in continental sediments suggests that azhdarchids were abundant, and perhaps even common components of terrestrial eco-

Fig. 25.5. A heavily crushed, headless skeleton of *Zhejiangopterus linhaiensis*, from the Campanian Tangshang Formation of China. Photograph courtesy of Lü Junchang.

systems in the Upper Cretaceous (Witton and Naish 2008). Exactly when the first azhdarchids appeared is poorly constrained, however. Many reports of Early Cretaceous or even Jurassic occurrences have been doubted (see Andres and Ji 2008; Unwin and Martill 2007; and Witton et al. 2009 for details), but convincing, if fragmentary, azhdarchid remains from Berriasian (145–140 Ma) deposits of Romania may indicate an earliest Cretaceous origin for the group (Dyke et al. 2010). This gives azhdarchids an evolutionary history spanning some 80 million years, which is almost half of the Mesozoic. No other pterosaur lineage is known to have similar longevity, with perhaps ornithocheirids being the next longest-lived clan with 50 million years of evolutionary history.

Anatomy

Many azhdarchid fossils are very well described, but their fragmentary nature means they tell us little about the overall anatomy of individual animals. The most complete overviews published to date are the descriptions of *Zhejiangopterus* by Cai and Wei (1994) and Unwin and Lü (1997), and that of the skull of *Quetzalcoatlus* sp. by Kellner and Langston (1996). Witton and Naish (2008) also provide an overview of the generalities of azhdarchid anatomy. Unfortunately, virtually no details have been published of the *Quetzalcoatlus* sp. postcrania despite hints at the preparation of a descriptive manuscript dating back several decades (e.g., Langston 1981; Kellner and Langston 1996). A vague picture of its postcranial anatomy can be gleaned from the *Quetzalcoatlus* sp. limb bone and cervical vertebrae proportions sprinkled through pterosaur literature (Frey and Martill 1996; Unwin, Lü, and Bakhurina 2000; Witton and Naish 2008), but at time of writing, the real details of the best known azhdarchid—and one of the most famous pterosaurs—remain mysterious.

SIZE

Azhdarchids have an extensive size range with the biggest species having wingspans four times those of the smallest. The most diminutive form known is *Montanazhdarcho* with a 2.5 m wingspan, while *Bakonydraco* and *Zhejiangopterus* are only a little larger with spans of 3.5 m. *Azhdarcho* is midsized with a 4–6 m wingspan, a size range that the 5 m wingspan *Quetzalcoatlus* sp., the Javelina azhdarchid, and *Phosphatodarco* also match. Several scrappy remains, including some thought to represent *Alanqa*, suggest that forms spanning between 5 and 9 m were also common (e.g., Padian 1984b; Buffertaut et al. 1997; Company et al. 1999; Ibrahim et al. 2010).

Several azhdarchids were considerably bigger than even these impressively sized species, achieving wingspans and masses so large that they are in their own league of pterosaurian gigantism. These huge animals are represented by so little fossil material that you

Fig. 25.6. The giraffe, the Hatzy, and me. The largest azhdarchids had wingspans between 10–11 m, meaning they were almost as large as the tallest modern giraffes (shown here, a 6 m tall male Masai giraffe, *Giraffa camelopardalis tippelskirchi*).

could quite feasibly fit the entire world stock of giant pterosaur remains onto your kitchen table (fig. 24.3). Accordingly, size estimates of these animals were very poorly constrained when knowledge of azhdarchid anatomy was in its infancy, leading to wingspan estimates of between 11 and 21 m (Lawson 1975a). As understanding of the azhdarchid bauplan has increased, these wingspan estimates were revised to 10.5 m for *Quetzalcoatlus* (Langston 1981); 12 m for *Hatzegopteryx* (Buffetaut et al. 2003); and 11–13 m for *Arambourgi-*

ania (Frey and Martill 1996). Most recently, Witton and Habib (2010) argued that *Hatzegopteryx* had a wingspan matching that of *Quetzalcoatlus* (10.5 m), and that too little is known about the neck growth regimes of azhdarchids to estimate a size for *Arambourgiania*. This suggests that 10.5 m spans are the largest reliable estimates for the giant azhdarchids at present (fig. 25.6), which indicate standing shoulder heights of around 2.5 m, and total standing heights in excess of 5 m.

It is not uncommon for people to question the evolutionary influences behind such large sizes in extinct animals. The simplest and most likely explanation is that giants developed in so many lineages simply because *they could*. There are many advantages to increased body size, including reduced transport costs and predation pressures; the ability to eat larger foods and weather lean times; heightened thermoregulatory efficiency; and increased fecundity in some species. Azhdarchid evolution simply ran with these advantages as far as their biomechanical constraints and ecology would let them, which was clearly much further than other pterosaurs. Interestingly, a wingspan of 10.5 m may not be too far off the assumed size limit for a flight worthy azhdarchid (M. B. Habib and J. R. Cunningham, pers. comm. in Witton 2010). Paleontologists are infamous for having their perceived size limits of extinct animals shattered by fossil evidence of ever more colossal creatures, but our understanding of animal scaling may now be developed enough to predict that the largest azhdarchids could not exceed wingspans of 12–13 m before their skeletal strength and launching capabilities were compromised. This does not mean that flight worthy pterosaurs could not exceed this wingspan limit, but they would have to be very different morphologically from not only azhdarchids, but probably from all other pterosaurs as well.

OSTEOLOGY

Although azhdarchids appear to be fairly conservative in much of their anatomy (fig. 25.7), their cranial morphology is quite varied (fig. 25.4). Azhdarchid skulls and mandibles seem to come in two flavors: extremely long, low skulls that may be 10 times longer than wide, and much shorter skulls with proportions more typical of other pterosaurs. The skulls of *Zhejiangopterus* and *Quetzalcoatlus* sp. reflect the long, low variety, and the long mandible of *Alanqa* suggests that its skull also falls into this group. Their great length is primarily constituted by a long, crestless rostrum that is 4.5 to 7 times longer than tall and occupies over half of the jaw. Behind this sits an expanded nasoantorbital fenestra and a low posterior skull region. Whereas the posterior skull of *Zhejiangopterus* is entirely crestless, *Quetzalcoatlus* sp. sports a short, distinctly humped crest above the posterior half of its nasoantorbital fenestra (Kellner and Langston 1996). This posterior portion of the *Quetzalcoatlus* sp. skull is incompletely known, so its exact shape, and the likelihood of a soft-tissue crest extension, cannot be determined. Indeed, little can be said about the posterior *Quetzalcoatlus* sp. skull region at all, except the fact that its orbit is situated in the ventral portion of the skull and that its jaw joint was unusually robust. The posterior skull of *Zhejiangopterus* is known in entirety, however, and bears a similarly depressed orbit beneath a typically neoazhdarchian, dorsally expanded cranial region. The posterior skull face of *Zhejiangopterus* has rotated to face almost entirely ventrally (Unwin 2003), suggesting that azhdarchid skulls articulated with the vertebral column at a rather sharp angle (echoing the condition of derived ctenochasmatoids; see chapter 19). The mandibles of the long-faced azhdarchids possess very long and shallow mandibular symphyses that may occupy up to 60 percent of the jaw length (Kellner and Langston 1996). The mandible of *Alanqa* is unique for possessing a narrow ridge along the posterior part of its symphysis (Ibrahim et al. 2010), a condition mirrored in other azhdarchids that also show evidence for complex biting surfaces. Witton and Naish (2008) suggested that the posterior palatal region of *Quetzalcoatlus* sp. may have been ventrally bowed, while *Hatzegopteryx* demonstrates a highly vaulted palatal region (fig. 25.4B–C).

The skulls of all "short-snouted" azhdarchids are incompletely preserved, but what little we have of them suggests they were truncated, and perhaps relatively stockier, versions of the long-faced variants. The complete mandible of *Bakonydraco* is five times longer than wide, making it only half as long, for its width, as that of *Quetzalcoatlus* sp. The *Bakonydraco* mandibular symphysis occupies 50 percent of the jaw and bears a distinctly robust termination with a concave and transversely ridged biting surface, a condition that contrasts with the comparatively slender, simple morphology of its recently discovered rostrum (fig. 25.4F–G; Ősi et al. 2011). The overall size of the "Javelina azhdarchid" skull is not known, but it appears to have a comparatively short, deep rostrum compared to other azhdarchids. A similar morphology may also be true of the skull of *Azhdarcho* (Averianov 2010), which possesses a slightly concave dorsal rostral margin that may indicate an unusually short and shallow snout.

The variation of azhdarchid skull morphology raises interesting questions about the proportions of

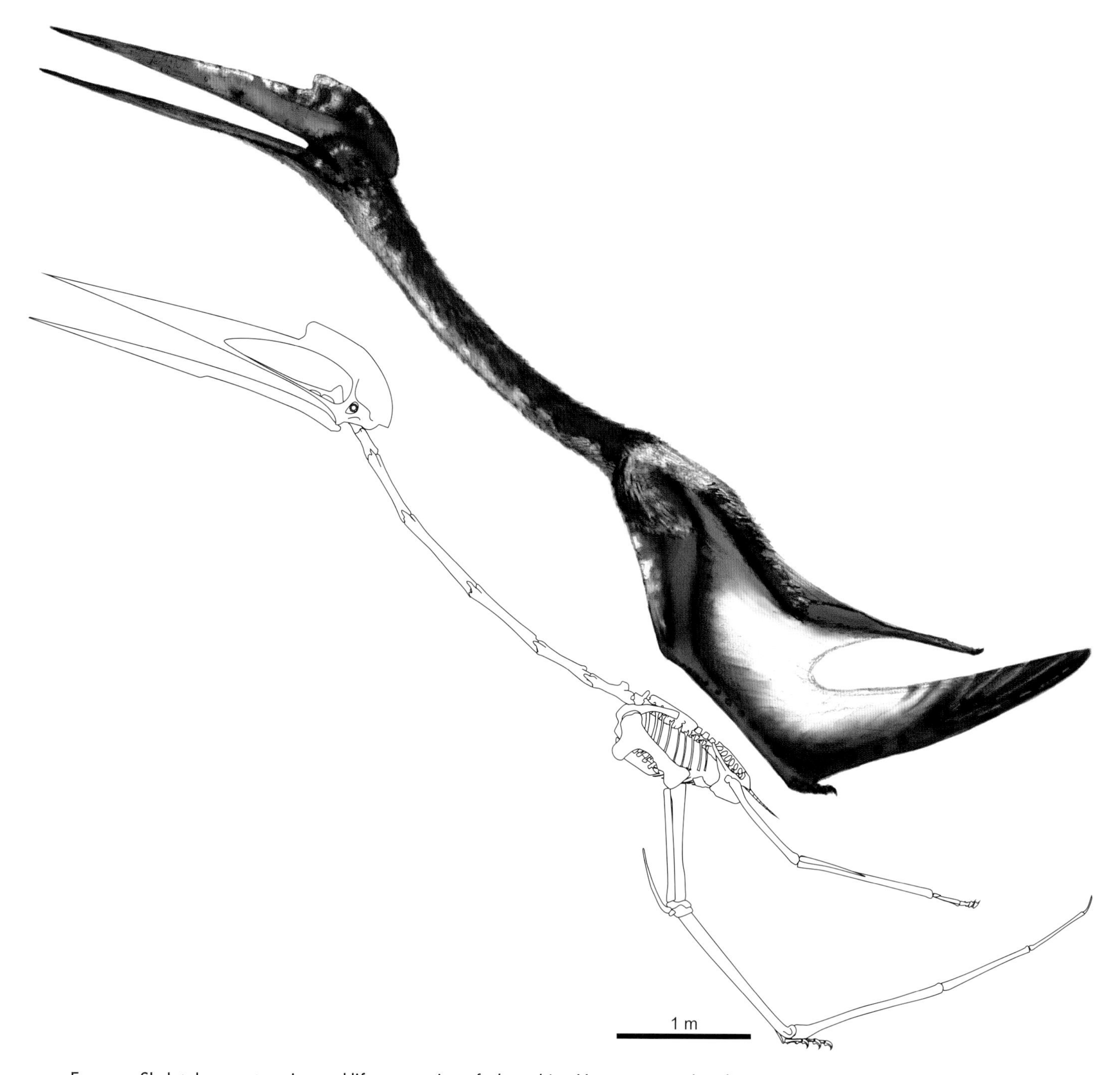

Fig. 25.7. Skeletal reconstruction and life restoration of a launching *Hatzegopteryx thambema*.

the poorly known *Hatzegopteryx* skull. Enough of the posterior skull region of this species is preserved to estimate a jaw width of almost 0.5 m, which could equate to jaws anywhere between 2.5 m (assuming a *Bakonydraco*-like jaw) and 5 m (assuming a *Quetzalcoatlus* sp.-like morphology). The latter has understandably been considered a little outrageous, even for a pterosaur (Buffetaut et al. 2003), but even the more conservative of these estimate remains astonishing. If the jaws are assumed to be 2.5 m long, the overall skull length had to approach 3 m once the posterior skull region is added. If correct, this would give *Hatzegopteryx* one of the longest skulls of any land animal ever, and certainly the longest jaws of any known nonma-

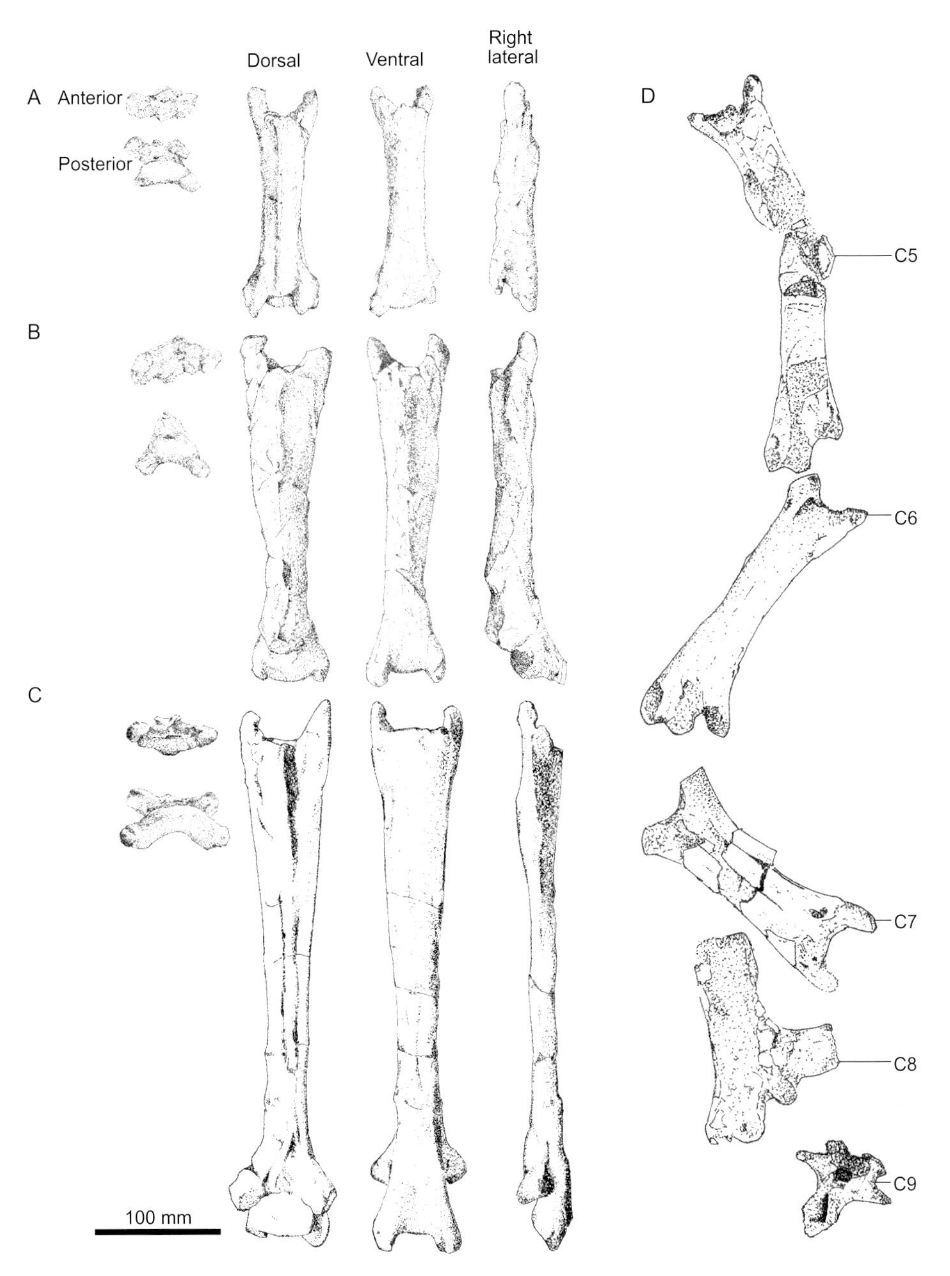

FIG. 25.8. Azhdarchid cervical vertebrae. A–C, cervicals 3, 4, and 5 (respectively) of *Quetzalcoatlus* sp.; D, the cervical series of *Phosphatodraco mauritanicus* from Maastrichtian deposits of Morocco. D, after Pereda Suberbiola et al. (2003).

rine animal. Giant pterosaurs aside, the longest jaws known of any terrestrial vertebrate belong to the giant predatory dinosaur *Spinosaurus aegyptiacus*, but these have a wimpy estimated length of 1.75 m (Dal Sasso et al. 2005). The skull of *Hatzegopteryx* may even give the horned dinosaurs, which hold the record for longest overall skull length for terrestrial animals (thanks to their elaborate frills), a run for their money. *Pentaceratops sternbergi* is the current leader in these stakes, with a skull almost 3 m long (Lehman 1998). If our estimate is correct, *Hatzegopteryx* may well have equaled this even without use of a flashy crest or frill. Of course, the pterosaur skull would be highly pneumatized and far less robust than its dinosaur competitors, but it does not change the sobering fact that one of the largest skulls of any land animal *ever* existed on a *flying* animal.

Azhdarchid necks are also famously large, being proportionally longer than those of any other pterosaurs (fig. 25.7; see Howse 1986; Frey and Martill 1996; and Pereda-Suberbiola et al. 2003 for some of the most detailed studies of azhdarchid cervicals to date). This

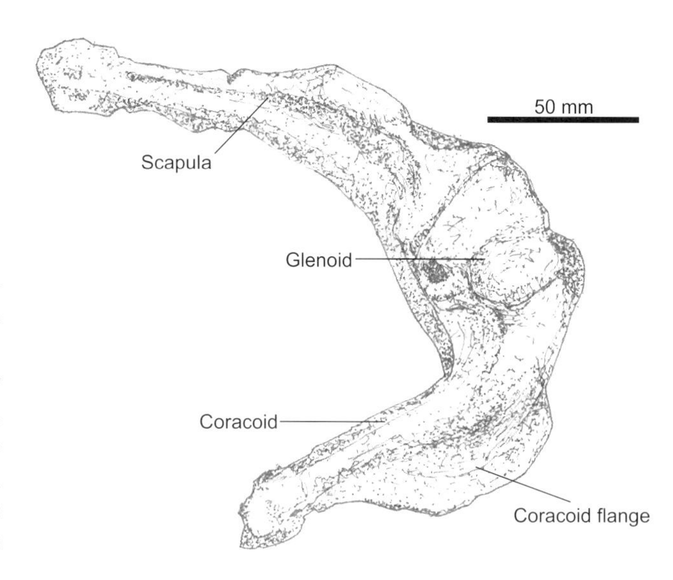

FIG. 25.9. Right *Quetzalcoatlus* sp. shoulder girdle in lateral view. Redrawn from Frey, Buchy, and Martill (2003).

great length is achieved through elongation of cervicals three to eight, with cervical five characteristically stretched to a length eight times its width (fig. 25.8). A complete cervical series is not known for any azhdarchid, but their cervical metrics are documented in enough detail to predict that the giant *Arambourgiania* cervical belonged to a neck approximately 3 m long (Frey and Martill 1996). Like other azhdarchoids and dsungaripteroids, azhdarchid cervical vertebrae are very low with the neural arch contained within the centra and because their midseries neural spines are highly reduced to the point of disappearing, their cervicals have a distinctly "tubular" appearance. Azhdarchid cervical zygapophyses are so large and flared that they have been described as "horn-like by some (e.g., Nessov 1991) and intimated to considerably constrict motion between individual vertebrae (e.g., Lawson 1975a; Wellnhofer 1991a; Martill 1997). This is countered somewhat, however, by the rather bulbous articulatory surfaces of azhdarchid cervical condyles (e.g., Godfrey and Currie 2005), suggesting that more movement was taking place between individual cervicals than previously supposed; they may have been relatively inflexible, but not immobile. As much as we know of the azhdarchid vertebral column is pneumatized, along with a good deal of their skeleton, including their entire forelimb (Claessens et al. 2009).

The azhdarchid trunk skeleton is fairly small, and is around 50 percent longer than their humeral length (Cai and Wei 1994). This means that the giant azhdarchids, despite their long necks, large heads, and ginormous wingspans, probably bore bodies of only around 70 cm in length, which is approximately the trunk length of a typical adult man (Witton and Habib 2010). The robust construction of the giant azhdarchid trunk skeleton compensates for their small size, however. If the large areas for muscle attachment around the shoulders are anything to go by, azhdarchid bodies would be packed with muscle. Paul (2002) estimated that some 50 kg of muscles were anchored around their shoulders, as indicated by their tremendously large coracoid flanges and long, widened deltopectoral crests (figs. 25.3 and 25.9; see Padian 1984b for a discussion of the distinctiveness of the latter feature). Their shoulder bones are correspondingly chunky, with robust scapulae and coracoids of subequal length, large glenoids, and anterior cervicals bound into a notarium (Nessov 1984).

Azhdarchid humeri are well known, and perhaps are their most frequently found remains after cervical vertebrae. They are generally similar to those of thalassodromids with fairly simple, if somewhat distally displaced, deltopectoral crests. As may be expected, the humeri of the giants are tremendously solid-looking, with 80 mm shaft widths (comparable in width to the humeral shafts seen in an adult, 2 tonne hippopotamus!) and massive, chunky articulatory surfaces. Their wing metacarpals are almost 2.5 times the length of these bones, making them the largest wing metacarpals, relatively speaking, of any pterosaur group, and the longest bones in the azhdarchid wing. *Montanazhdarcho* bucks this rule by possessing a distinctly stunted wing metacarpal that is actually shorter than its ulna (McGowen et al. 2002). The first three metacarpals of all azhdarchids are very short and located entirely at the end of the large fourth metacarpal. Their wing fingers characteristically occupy less than 50 percent of the overall wing length and, as with other azhdarchoids, the first wing phalanx is considerably larger than the rest (the wing phalange proportions represent 55, 30, 15, and 5 percent of the wing finger, respectively). The second and third wing phalanges have a curious T-shaped cross section that may have reinforced the torsional motions of the distal wing during flapping (Habib 2010). The first three manual digits of azhdarchids have not been reported

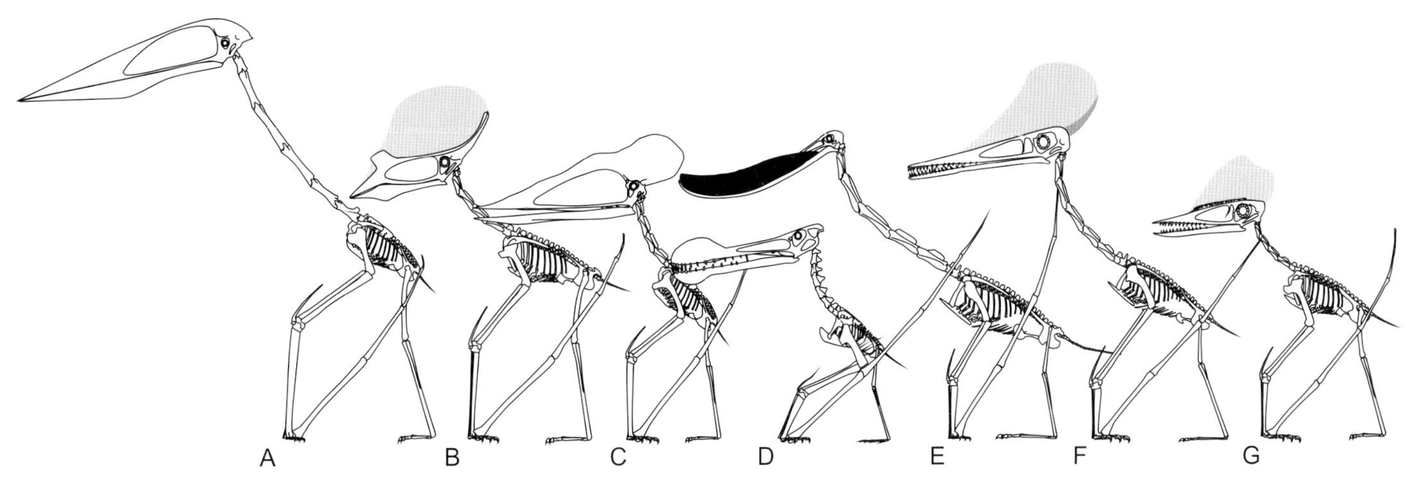

Fig. 25.10. There's nothing quite like an azhdarchid. Pterosaur skeletons in standardized standing postures and scaled to the same humerus length. A, *Zhejiangopterus linhaiensis*; B, *"Huaxiapterus" benxiensis*; C, *Tupuxuara leonardii*; D, *Ornithocheirus mesembrinus*; E, *Pterodaustro guinazui*; F, *Pterodactylus antiquus*; G, *Germanodactylus cristatus*.

in any detail, but those of *Zhejiangopterus* are unusually short in comparison to their limb lengths (Cai and Wei 1994).

Azhdarchid pelves are poorly known, but the crushed pelvic remains of *Zhejiangopterus* suggest they were rather like those of other azhdarchoids (Cai and Wei 1994): relatively robust with large, probably hatchet-shaped postacetabular processes. *Zhejiangopterus* also reveals that the azhdarchid hindlimb skeleton is very long, and comprised of a femur that is 80 percent of the elongate tibiotarsus (Unwin 2003). The combined long legs and elongated wing metacarpal gives azhdarchids longer limbs and taller frames, relatively speaking, than other pterosaurs (fig. 25.10). By contrast, their feet are narrow and very short at no more than 30 percent of their tibial length (Cai and Wei 1994). Despite their size, they may have been quite strong, with Bennett (2001) noting that the feet of *Quetzalcoatlus* were much more robust than those of *Pteranodon*. Azhdarchid footprints (see below) suggest that their feet possessed footpads similar to those seen in tapejarids (chapter 22).

Locomotion

FLIGHT

Azhdarchid flight has been discussed at some length in recent years, with the principal focus placed on whether the largest species could fly at all. It has been thought that giraffe-sized azhdarchids weighing as little as 200–250 kg were simply too heavy to take off (assuming bipedal launch strategies [see chapter 6]; Chatterjee and Templin 2004; Sato et al. 2009) and, in one instance, that giant azhdarchids weighed over 500 kg and were simply too heavy to become airborne (Henderson 2010). These notions have been challenged with arguments that 200–250 kg is a more likely mass estimate for a giraffe-sized pterosaur, largely because of their relatively small torso proportions (Witton 2008; Witton and Habib 2010), and that giant azhdarchids bear the same anatomical hallmarks of flight as their smaller brethren, despite their size (Buffetaut et al. 2002; Witton and Habib 2010). Moreover, newly proposed quadrupedal launch strategies would give giant pterosaurs more than enough power to become sky bound (Habib 2008). Witton and Habib (2010) noted that azhdarchid humeri could launch animals approaching 500 kg before mechanical failure became likely. Accordingly, there is less question of *if* giant azhdarchids could fly, but how well.

It is generally thought that azhdarchids were suited to soaring flight (Lawson 1975a; Nessov 1984; Chatterjee and Templin 2004; Witton 2008a; Witton and Habib 2010), though some have suggested that their large flight musculature would allow them to flap continuously like flying swans and geese (Paul 2002; Frey, Buchy, and Martill 2003). Recently, it has been argued that the short, potentially broad wings of azhdarchids were suited for flight in terrestrial settings, and the finding of comparable wing shapes between azhdarchids and large, terrestrially soaring birds lends credence to this idea (Witton 2008a; Witton and Naish

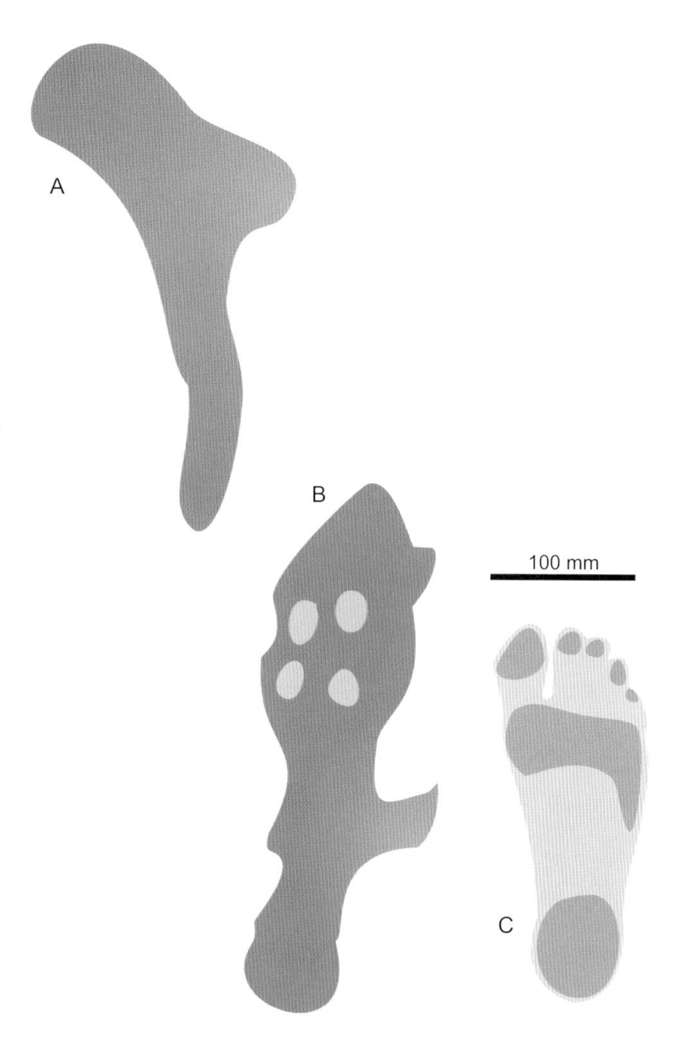

FIG. 25.11. The Late Cretaceous azhdarchid ichnogenus *Haenamichnus uhangrensis,* from the upper Cretaceous Uhangri Formation of South Korea, drawn to scale with a size 10.5 human footprint. Redrawn from Hwang et al. (2002).

2008; Witton and Habib 2010). Albatross-like soaring over marine environments has also been considered as likely (Nessov 1984; Chatterjee and Templin 2004), but the strong terrestrial bias of azhdarchid fossils, not to mention their likely adaptations for ground-based foraging, argue against this.

Detailed flight models of azhdarchids return impressive statistics. A freshly launched, 200 kg azhdarchid with a 10 m wingspan seems capable of sustaining flapping flight under anaerobic power for around two minutes at optimal speeds exceeding 30 m/s (108 k/h, or 67.5 mph) This allows for almost two kilometers in which the pterosaur could seek additional sources of lift—rising thermal updrafts, or winds deflected by elevated ground—to gain additional altitude before soaring effortlessly onward at speeds approaching those achieved in flapping flight (Witton and Habib 2010). Note that this model is based on a conservative estimate using a relatively broad wing shape. Narrowing the wing (which may not be unreasonable given what we know of pterosaur wing membranes) would allow for even faster flight. Traveling at such speeds, and assuming a healthy store of fuel (i.e., fat and muscle tissue to digest along the way), our azhdarchid could potentially travel 16,000 km (10,000 miles) without touching down (Habib 2010). To paraphrase Mike Habib, the chap behind these calculations, geographical barriers would mean *nothing* to these guys.

ON THE GROUND

Azhdarchids are the only pterosaurs to which specific trackways have been reliably referred. The large, compact footprints of *Haenamichnus* (fig. 25.11) are of the right shape, age, and in some instances the right size, for azhdarchids. The last factor is of tremendous help in identifying *Haenamichnus* as azhdarchid tracks. Some *Haenamichnus* hand and foot prints are 35 cm long, and azhdarchids are the only pterosaurs known that could produce tracks of that size. To date, *Haenamichnus* has only been definitively identified from Korea (Hwang et al. 2002), but another series of pterosaur prints from the Campanian of Mexico (Rodriguez-de la Rosa 2003) may also represent *Haenamichnus*. One 7 m long Korean *Haenamichnus* trackway is the longest continuous pterosaur trackway in the world (fig. 7.3) and reveals that azhdarchids walked with their limbs held directly beneath their bodies. This, in association with their long limbs and compact, padded feet, has been taken as evidence for particularly proficient terrestrial locomotion in azhdarchids compared to most other pterosaur groups (Witton and Naish 2008).

By contrast, azhdarchids may not have been so great at swimming (Witton and Naish 2008). Their long extremities, very front-heavy frames, and tiny bodies seem ill suited for creating a stable waterborne posture, and their tiny, compact appendages would probably make lousy flippers. Despite this, it has been predicted that *Quetzalcoatlus* sp. could take off from a floating start (Habib and Cunningham 2010), although it may not have been as graceful at aquatic launching as the ornithocheirids were (chapter 16).

Biomechanical models suggest that *Quetzalcoatlus* sp. could launch from deep water with one enormous push from their forelimbs, but such heft incurred greater energetic costs and biomechanical loading than the water hops possibly practiced by the more aquatically adept ornithocheirids (fig. 16. 8; Habib and Cunningham 2010). This less explosive hopping method would also work for *Quetzalcoatlus* sp. (Habib and Cunningham 2010), but it is testament to the sheer power of azhdarchid limbs that water launching could be performed in one deft move. In any case, it seems unlikely that azhdarchids would have to employ aquatic launching frequently, given their apparent preference for firm ground.

Paleoecology

Just about every conceivable foraging strategy has been suggested for azhdarchids at one time or another. Scavenging (Lawson 1975*a*), sediment probing (e.g., Langston 1981; Wellnhofer 1991a), wading (e.g., Paul 1987; Bennett 2001), skim feeding or to a lesser extent, aerial fishing (e.g., Nessov 1984, 1991; Kellner and Langston 1996; Martill 1997; Martill et al. 1998; Prieto 1998), pursuit swimming (Nessov 1984), aerial predation (Nessov 1984), and terrestrial foraging (Witton and Naish 2008) have all been suggested as possible azhdarchid lifestyles. The disagreements over their habits have been fueled by the strangeness of azhdarchid anatomy, and compounded by conflicting interpretations of their functionality (a full listing of relevant literature on this subject is beyond our means here, but Witton and Naish [2008] provide a comprehensive review of proposed azhdarchid lifestyles).

Some suggested azhdarchid ecologies have received little following. Aerial predation and pursuit swimming, in particular, have not been widely discussed or cited by pterosaur workers. The scavenging hypothesis also never really became widely favored, though the propensity of virtually all modern carnivorous and omnivorous animals to scavenge suggests some opportunistic carrion feeding cannot be ruled out for any azhdarchid. Indeed, their flying ability would be very useful for finding and reaching carcasses before terrestrial carnivores and their frequently large body sizes would ensure first place in most Cretaceous pecking orders. However, there is nothing about azhdarchid anatomy to suggest that they were habitual scavengers in the way that some pterosaurs seem to have been (chapter 15), and in all likelihood they probably scavenged no more or less than most other pterosaurs. Sediment probing also never became a fashionable hypothesis, largely because many aspects of azhdarchid anatomy are quite inconsistent with sticking their beaks into sediment to retrieve burrowing animals (Martill 1997; Witton and Naish 2008).

Azhdarchid skim feeding, by contrast, became the foraging hypothesis of choice in the mid-1990s. Their long necks and jaws were considered to be adaptations for soaring well above the water surface while pushing their mandible tip through the water, allowing them to flap their wings without striking the water surface (Martill 1997). Substantial evidence for this idea was otherwise lacking, leading Humphries et al. (2007) and Witton and Naish (2008) to be rather damning of this concept. Azhdarchid jaw and neck anatomy are nothing like those of modern skimming birds (which, you'll recall from chapter 24, are extremely specialized for skim feeding), and the flight energetics of even relatively small azhdarchids are insufficient to shove their chunky jaw tips through water. The latter point applies tenfold to the 10 m wingspan azhdarchids. For all their power, their available energy output is only *5 percent* of that required for effective skim feeding at their enormous sizes (Humphries et al. 2007). Dip feeding has also been criticized on anatomical grounds, with the long, relatively inflexible neck cited as the main (but not only) problem with this hypothesis (Witton and Naish 2008).

The notable terrestrial abilities of azhdarchids lead several authors to conclude that they may have been proficient waders, patrolling water courses and picking up bite-sized animals from the shallows (e.g., Paul 1987; Bennett 2001; Witton 2007). Such a lifestyle explains the extensive jaws of many azhdarchids, along with the long limbs that characterize the group. Their lengthy, relatively inflexible necks are also conducive to this proposal, and would have helped them reach out over deep water or forage in areas that their wading limbs had not disturbed. This hypothesis works very well save for one problem: the tiny appendages azhdarchids possess at the end of their stilt-like legs. Unlike the large, splayed feet common to ctenochasmatoids (chapter 19), the tiny hands and feet of azhdarchids would probably do little to spread their weight over the soft, muddy margins of water bodies, leaving them either unable to access many prime foraging spots, or at risk of miring should they try to reach them.

Small feet are predicted, however, for animals that spend much of their time on firm ground. Short feet minimize the effort needed to push off from the ground when striding forward, thus making them more efficient for walking than long, splayed extremities. Along with a suite of other characters including their long limbs and unspecialized jaw structure, these appendages suggest that azhdarchids were "terrestrial stalkers," ground-foraging pterosaurs akin to modern ground hornbills and some storks (Witton and Naish 2008). In such a hypothesis, long limbs are ideal for efficiently striding through vegetated settings, while a long skull and jaws, naturally downturned because of the posteriorly rotated occipital face, are well suited to reaching the ground. The long, stiffened neck, problematic in many other foraging hypotheses, becomes potentially beneficial by amplifying motions at the neck base to assist lowering the head to reach the ground, as well as providing an elevated vantage point to search for prey. Perhaps, as with the wading hypothesis, long necks also allowed prey to be grabbed before the azhdarchid's footfalls could scare it away. The generalized structure of azhdarchid jaws may have allowed access to a broad omnivorous diet of small animals and fruiting bodies (Ősi et al. 2005), but it seems likely that larger species were mostly carnivorous. Not only would relatively small vegetative matter be difficult to grasp with their enormous jaws, but the net energy gain for such large animals might negate feeding on it in the first place. As such, the larger azhdarchids may have dined exclusively on smaller dinosaurs, eggs, and other small to midsized tetrapods. Making matters even worse for potential azhdarchid foodstuffs is that the abundant and occasionally associated remains of some azhdarchids (e.g., Langston 1981; Cai and Wei 1994; Ősi et al. 2005) suggest some species may have been socially tolerant and foraged in groups. The prospect of several giant, stilt-legged pterosaurs with sharp, elevated visual fields and long, quick jaws must have been pretty frightening for many small animals of the Upper Cretaceous. Interestingly, the latest terrestrial Cretaceous communities containing the largest azhdarchids were devoid of midsized dinosaur predators (e.g., Farlow and Pianka 2002), a situation that contrasts markedly with earlier dinosaur ecosystems. A speculative, but nonetheless intriguing proposal is that giant azhdarchids may have at least partially filled the roles of midsized predators in these ecosystems toward the end of the Cretaceous.

26
The Rise and Fall of the Pterosaur Empire

The preceding chapters have hopefully made a case for pterosaurs as diverse, successful animals that are as far away from their stereotype of Mesozoic gargoyles as could be imagined. But an obvious question remains unanswered: If they were so successful, why are they extinct? Like the nonavian dinosaurs, pterosaurs persisted until the very end of the Cretaceous (Buffetaut et al. 1996), suggesting the last of their number died in the catastrophic Cretaceous/Tertiary (or "KT") extinction event which wiped out 75 percent of all species on the planet at the end of the Mesozoic (fig. 26.1). The cause of this, one of the largest extinction events in Earth's history, remains controversial, but the conspiring effects of an enormous meteorite impact, extensive volcanic activity, and shifting climates in the last few million years of the Mesozoic all probably played some part. But were pterosaurs already on their way out before this, or were they cut off in their prime?

As with most aspects of pterosaur science, the answer to this question is complex. The pterosaur fossil record suggests that by the end of the Cretaceous, pterosaurs were represented by relatively few, morphologically similar species (Butler et al. 2012). There is only evidence for two groups of pterosaurs in uppermost Cretaceous rocks (fig. 26.2): azhdarchids (in some abundance) and nyctosaurids (ultrarare, represented by a solitary humerus [Price 1953]). At face value, this suggests that pterosaur stocks were already broadly diminished when the Mesozoic world ended 65 Ma, which would mean the KT event merely polished off a dying lineage. However, we must be cautious with this interpretation because the pterosaur fossil record may be too incomplete to reveal the overall "health" of the pterosaur clan at any time in their history. This can be assessed by noting major gaps in pterosaurian temporal or spatial distribution, testing for significant biases in their fossil record, and identifying whether certain species or specimens are extremely different, morphologically speaking, from all other pterosaurs. Such specimens would indicate that we have large gaps in our phylogenetic data sets, implying incompleteness of the pterosaur evolutionary record.

Geographically and stratigraphically speaking, the pterosaur fossil record is very patchy (fig. 26.3). Though their remains occur on every continent except Antarctica (Barrett et al. 2008), vast swathes of the planet have yet to yield pterosaur fossils. Africa and Australia, for example, have revealed very few pterosaurs indeed, and vast chunks of Mesozoic strata around the world, representing millions of years have generated few pterosaur fossils, or none at all. Their Triassic and Jurassic records are particularly scrappy, and although the Cretaceous record is generally better, it is still very far from complete. This is not always the case of course, with a wealth of pterosaur fossils known from fossil Lagerstätten, such as the Solnhofen Limestone or the several excellent fossil beds of China's Liaoning region. Noting this, Buffetaut et al. (1996) and Kellner (2003) suggested that the pterosaur fossil record is severely affected by the so-called Lagerstätten effect, a common bias in paleontological data where our knowledge of the taxonomic and spatial distribution of a fossil group is skewed by the abundant fossils of these deposits. Butler, Barrett, Nowbath, and Upchurch (2009), and Butler et al. (2012) provided quantitative evidence of this, showing that pterosaur diversity though time largely correlates with the number of rocks yielding their fossils (fig. 26.3). Thus, the diversity curves we are currently drawing for pterosaurs mainly reflect sampling biases rather than reality, and particularly so for the Triassic and Jurassic. Perhaps it is significant, however, that the latest Cretaceous has a higher than average number of pterosaur-bearing rocks, and yet pterosaur diversity remains low.

Contrasting with these findings is the fact that most newly discovered pterosaur fossils share enough common features with previously known forms that only a

Fig. 26.1. The end. Any pterosaurs alive in the last weekend of the Mesozoic would have seen the Earth plunged into a short-lived nuclear winter, thanks to the vast quantities of ash and dust thrown into the atmosphere by the infamous Chicxulub meterorite impact in Mexico. Though perhaps only lasting a month or two, temperatures at even midlatitudes were subfreezing during this interval, collapsing ecosystems around the globe. Pterosaurs were already on their last dance by this stage, and for the large, energy-demanding animals left carrying their torch in the late Maastrichtian, this winter spelled their doom.

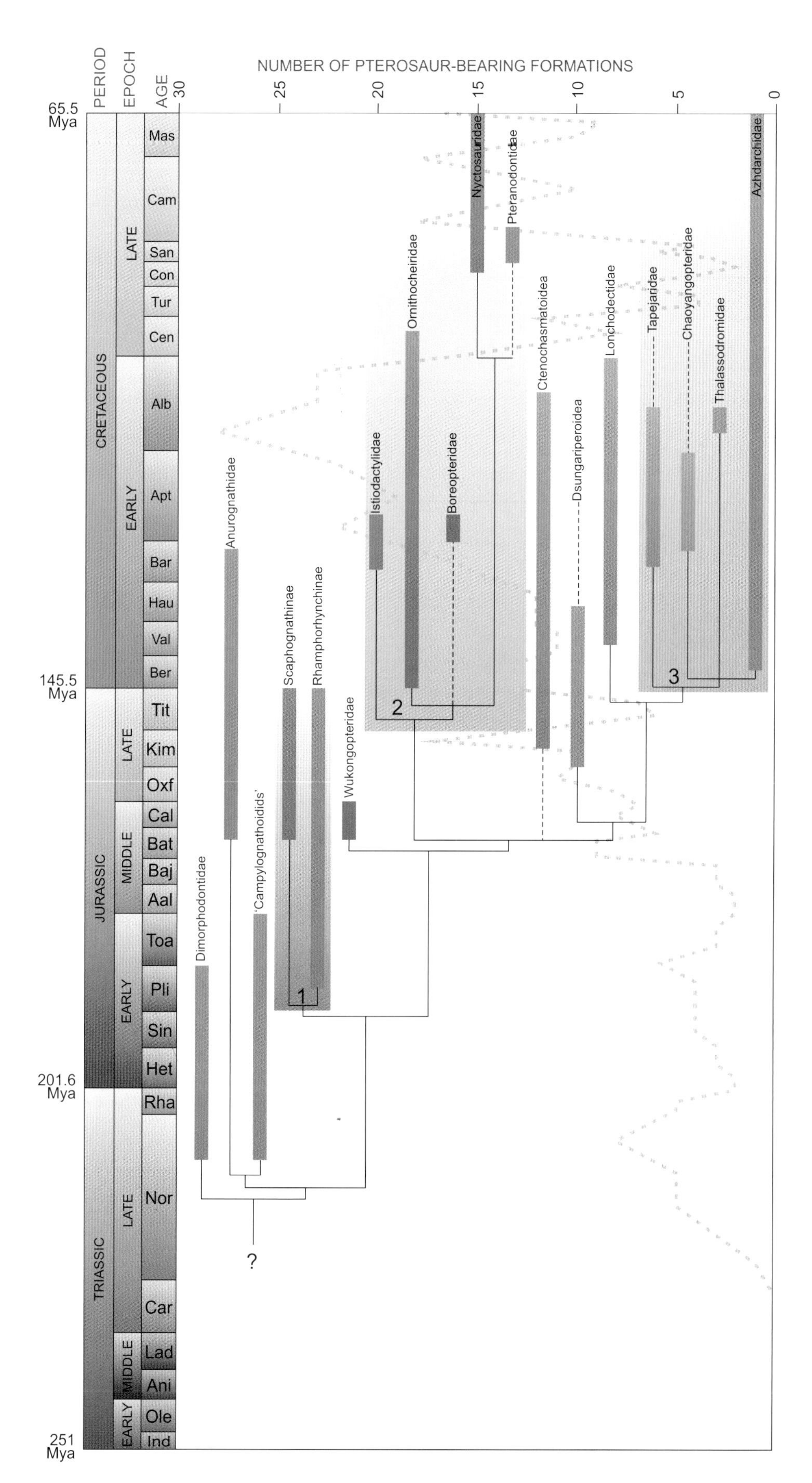

FIG. 26.2. A phylogenetic tree of pterosaur evolution, based on the phylogeny of Lü, Unwin, et al. (2010), plotted against the number of pterosaur-bearing formations through time (the *blue zigzag line*; data from Butler, Barrett, Nowbath, and Upchurch 2009). Numbers denote major pterosaur groups: 1, Rhamphorhynchidae; 2, Ornithocheiroidea; 3, Azhdarchoidea.

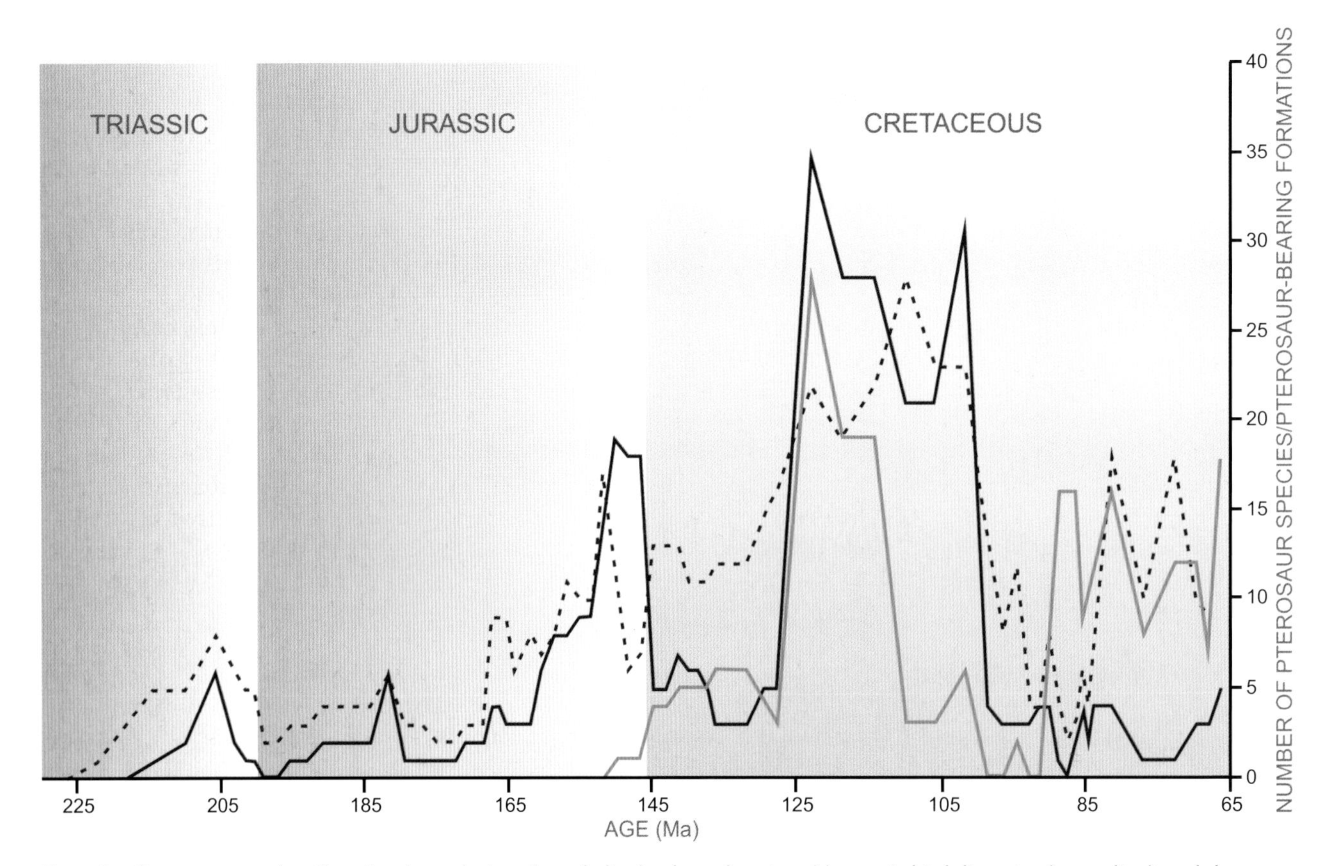

FIG. 26.3. Pterosaur species diversity through time (purple line), plotted against Mesozoic bird diversity (green line) and the number of pterosaur-bearing formations (dotted line) (data from Butler, Barrett, Nowbath, and Upchurch 2009). Note how pterosaur diversity generally matches rock availability closely, but appears unaffected by the number of avian taxa.

handful seem truly out of place in modern phylogenies. Most pterosaur species discovered this millennium have slotted neatly into phylogenetic space without causing too much disruption (e.g., Lü, Unwin, et al. 2008; Lü, Unwin, et al. 2010), suggesting that we have discovered most of the major branches of the pterosaur tree. In addition, there is a rough correlation between pterosaur phylogenies and their distribution in the rock record, indicating that we have a vague idea when each pterosaur clan appeared (though Dyke et al. [2009] suggest an alternative view). This may mean that although we know very little about actual pterosaur species counts across time and space, we may have a rough handle on their overall morphological range. Consequently, the dominance of azhdarchids in the latest Cretaceous may be a genuine feature of pterosaur evolution, because we should be able to easily identify any nonazhdarchid fossils from this time. Of course, the discovery of a pterosaur-bearing Late Cretaceous Lagerstätten could change this perception overnight, but the current data seems to indicate that azhdarchids really did rule the pterosaur roost at the end of the Mesozoic.

If so, the question of pterosaur extinction becomes not one of "why did *pterosaurs* go extinct?" but instead: "Why did *azhdarchids* go extinct?" This, at one level, is not too difficult to answer. Unwin (2005) noted that azhdarchids may have been generally more vulnerable to extinction than other pterosaurs because of their large size. Larger animals are relatively slow to reproduce and adapt, and can only sustain relatively low populations. When ecosystems worldwide fell into chaos as the KT event wreaked its havoc, azhdarchids, and all the pterosaurness they embodied, were doomed. Azhdarchids may not have even needed the chaos of a mass extinction to have died out, as far more mundane catalysts for extinction eradicate species continually. Azhdarchids would be just as prone to these causes of "background extinctions" as any other species, and their departure at the KT boundary may be a mere coincidence. Whichever scenario actually played out is somewhat irrelevant; the important point is that Late Cretaceous pterosaur stock was almost entirely invested in one, slowly adapting lineage and was therefore very vulnerable. Hence, the nature of our question changes again, because the real mystery of pterosaur extinction does not lie at the end of the Mesozoic, but with how the Late Cretaceous became devoid of other pterosaur types.

The patchiness of the pterosaur fossil record becomes a real issue when attempting to unravel this part of the problem. Were pterosaurs dying out at atypically high rates in the Late Cretaceous, or merely not diversifying and replacing themselves against normal "background" rates of extinction? The limited data we have suggests the latter is more likely. After an evolutionary burst in the Early Cretaceous, pterosaurs simply seem to have stopped diversifying to any significant degree. The pterosaurs we see in the middle and latest Cretaceous are really not very different from their lower Cretaceous ancestors. The cause of this evolutionary stasis is not clear. It has been proposed that birds may have gradually displaced pterosaurs ecologically throughout the Cretaceous, preventing the evolution of new forms and leaving only specialized, giant pterosaurs alive for the last 15 million years before the KT event (Unwin 2005). This notion has received little support under testing however. There is no correlation between pterosaur decline and avian diversification (fig. 26.3; Slack et al. 2006; Butler, Barrett, Nowbath, and Upchurch 2009; Dyke et al. 2009), and it is unlikely that early birds and pterosaurs overlapped ecologically to the extent that they directly competed for resources (Slack et al. 2006). Likewise, there is nothing to suggest that azhdarchids outcompeted their own brethren and bullied their way to dominance. Azhdarchid anatomy does not appear more adaptable than that of other pterosaurs, nor in any way "superior." Recent ecomorphological analyses have also found no indication of direct competition between azhdarchids and other clades (Prentice et al. 2011). Pterosaur extinction does not appear, therefore, to have been caused by an azhdarchid coup d'état.

We are left, then, with no clear indication of why pterosaur diversity dwindled so dangerously low in the Late Cretaceous. It is notable that all pterosaurs, not just those of the Cretaceous, seem to have been evolutionarily conservative. Many groups existed for tens of millions of years in disparate regions of the globe without developing overwhelming differences from their ancestors or contemporaries (e.g., Unwin, Lü, and Bakhurina 2000; Lü, Unwin, et al. 2012). Perhaps factors other than the evolution of birds denied pterosaurs the opportunity or ability to continue diversifying in changing Cretaceous ecosystems. Whatever the cause, the lack of new pterosaurs to replace the old meant that their diversity could only decline to progressively lower levels as they approached the end of the Mesozoic, increasing the chance that the extinction of the next lineage would represent the final curtain call for Pterosauria as a whole.

The Difficult Last Paragraph

The true nature of the pterosaur demise is, of course, only one of many unanswered questions about our leathery-winged chums. That we will answer them is not certain, but I suspect we stand a good chance. There are more pterosaur researchers now than at any other time in the last 228 years of pterosaurology, and new discoveries are being unearthed, described, and analyzed at record rates. Novel ways of processing and analyzing this data means that even the hottest controversies of previous years have been resolved with well-constructed, nearly incontrovertible arguments, and the current debates raging in pterosaur research will inevitably be settled in a similar fashion one day. With all this going on, it's hard not to be optimistic about the future of pterosaur studies. I hope this book has convinced you to look forward to these forthcoming breakthroughs, and to the discovery of more crazy-looking new species, just as much as I do.

That's your pterosaurs for now, folks. Thanks very much for reading.

References

Abel, O. 1925. *Geschichte und Methode der Rekonstruktion vorzeitlicher Wirbeltiere.* Jena, G. Fischer.

Andres, B. 2007. Pterosaur systematics. Abstract. In *Flugsaurier: The Wellnhofer pterosaur meeting,* edited by D. Hone, 7. Abstract Volume. Munich, Bavarian State Collection for Palaeontology.

Andres, B., J. M. Clark, and X. Xu. 2010. A new rhamphorhynchid pterosaur from the Upper Jurassic of Xinjiang, and the phylogenetic relationships of basal pterosaurs. *Journal of Vertebrate Paleontology* 30: 163–187.

Andres, B., L. Howard, and L. Steel. 2011. Owen's pterosaurs, old fossils shedding light on new clades. Abstract. In *Programme and Abstracts,* Symposium of Vertebrate Palaeontology and Comparative Anatomy, 59th Annual Meeting, 2011, 4. Lyme Regis, U.K.

Andres, B., and Q. Ji. 2003. Two new pterosaur species from Liaoning, China, and the relationships of the Pterodactyloidea. Abstract. *Journal of Vertebrate Paleontology* 23: 29A.

———. 2006. A new species of *Istiodactylus* (Pterosauria, Pterodactyloidea) from the Lower Cretaceous of Liaoning, China. *Journal of Vertebrate Paleontology* 26: 70–78.

———. 2008. A new pterosaur from the Liaoning Province of China, the phylogeny of the Pterodactyloidea, and the convergence in their cervical vertebrae. *Palaeontology* 51; 453–469.

Arambourg, C. 1954. Sur la présence d'un ptérosaurien gigantesue dans les phosphates de Jordanie. *Comptes rendus de l'Académie des Sciences* 283: 133–134.

———. 1959. *Titanopteryx phildelphiae* nov. gen., nov. sp., pterosaurien géant. *Notes et Mémoires du Moyen Orient* 7: 229–234.

Arbour, V. M., and P. J. Currie. 2010. An istiodactylid pterosaur from the Upper Cretaceous Nanaimo Group, Hornby Island, British Columbia, Canada. *Canadian Journal of Earth Sciences* 48: 63–69.

Arthaber, G. von. 1919. Studien fiber Flugsaurier auf Grund der Bearbeitung des Wiener Exemplares von *Dorygnathus banthensis* Theod. sp. *Denkschrift der Akademie der Wissenschaften Wien, Mathematisch-Naturwissenschaftliche Klasse* 97: 391–464.

Arthaber, G. von. 1922. Über Entwicklung, Ausbildung und Absterben der Flugsaurier. *Paläontologische Zeitschrift* 4: 1–47.

Atanassov, M. 2000. Morphometric analysis of late Jurassic pterodactyloids from Germany and France. *Journal of Vertebrate Paleontology* 20: 27A.

Averianov, A. O. 2007. New records of azhdarchids (Pterosauria, Azhdarchidae) from the Late Cretaceous of Russia, Kazakhstan, and Central Asia. *Paleontological Journal* 41: 73–79.

———. 2010. The osteology of *Azhdarcho lancicollis* Nessov, 1984 (Pterosauria, Azhdarchidae) from the Late Cretaceous of Uzbekistan. *Proceedings of the Zoological Institute RAS* 314: 264–317.

———. 2012. *Ornithostoma sedgwicki*—valid taxon of azhdarchoid pterosaurs. *Proceedings of the Zoological Institute RAS* 316: 40–49.

Averianov, A. O., M. S. Arkhangelsky, E. M. Pervushov, and A. V. Ivanov. 2005. A new record of an azhdarchid (Pterosauria: Azhdarchidae) from the Upper Cretaceous of the Volga Region. *Paleontological Journal* 39: 433–439.

Averianov, A. O., M. S. Arkhangelsky, and E. M. Pervushov. 2008. A new late Cretaceous azhdarchid (Pterosauria, Azhdarchidae) from the Volga Region. *Paleontological Journal* 42: 634–642.

Bakhurina, N. A. 1986. Flying reptile. [In Russian.] *Priroda* 7: 27–36.

Bakhurina, N. A., and D. M. Unwin. 1992. *Sordes pilosus* and the function of the fifth toe in pterosaurs. *Journal of Vertebrate Paleontology* 12: 18A.

———. 1995a. A preliminary report on the evidence for 'hair' in *Sordes pilosus*, an Upper Jurassic pterosaur from Middle Asia. In *Sixth Symposium on Mesozoic Terrestrial Ecosystems and Biota, Short Papers*, edited by A. Sun and Y. Wang, 79–82. Beijing, China Ocean Press.

———. 1995b. A survey of pterosaurs from the Jurassic and Cretaceous of the former Soviet Union and Mongolia. *Historical Biology* 10: 197–245.

———. 2003. Reconstructing the flight apparatus of *Eudimorphodon. Rivista del Museo Civico di Scienze Naturali "E. Caffi" di Bergamo* 22: 5–8.

Bakker, R. T. 1986. *The Dinosaur Heresies.* New York, Citadel Press.

———. 1998. Dinosaur mid-life crisis: The Jurassic-Cretaceous transition in Wyoming and Colorado. In *Lower and Middle Cretaceous Terrestrial Ecosystems, New Mexico Museum of Natural History and Science Bulletin*, edited by S. G. Lucas, J. I. Kirkland, and J. W. Estep. Vol. 14, pp. 67–77.

Barrett, P. M., Butler, R. J., Edwards, N. P., and A. R. Milner. 2008. Pterosaur distribution in space and time: An atlas. *Zitteliana* B28: 61–107.

Bell, C. M. and K. Padian. 1995. Pterosaur fossils from the Cretaceous of Chile: Evidence for a pterosaur colony on an inland desert plain. *Geological Magazine* 132: 31–38.

Bennett, S. C. 1989. A pteranodontid pterosaur from the Early Cretaceous of Peru, with comments on the relationships of Cretaceous pterosaurs. *Journal of Paleontology* 63: 422–434.

———. 1990. A pterodactyloid pterosaur from the Santana Formation of Brazil: Implications for terrestrial locomotion. *Journal of Vertebrate Paleontology* 10: 80–85.

———. 1992. Sexual dimorphism of *Pteranodon* and other pterosaurs with comments on cranial crests. *Journal of Vertebrate Paleontology* 12: 422–434.

———. 1993. The ontogeny of *Pteranodon* and other pterosaurs. *Paleobiology* 19: 92–106.

———. 1994. Taxonomy and systematics of the Late Cretaceous pterosaur *Pteranodon* (Pterosauria, Pterodactyloidea). *Occasional Papers of the Natural History Museum, University of Kansas* 169: 1–70.

———. 1995. A statistical study of *Rhamphorhynchus* from the Solnhofen Limestone of Germany: Year-classes of a single large species. *Journal of Paleontology* 69: 569–580.

———. 1996a. The phylogenetic position of the Pterosauria within the Archosauromorpha. *Zoological Journal of the Linnaean Society* 118: 261–308.

———. 1996b. Year-classes of pterosaurs from the Solnhofen Limestone of Germany: Taxonomic and systematic implications. *Journal of Vertebrate Paleontology* 16: 432–444.

———. 1996c. On the taxonomic status of *Cycnorhamphus* and *Gallodactylus* (Pterosauria: Pterodactyloidea). *Journal of Paleontology* 70: 335–338.

———. 1997a. The arboreal leaping theory of the origin of pterosaur flight. *Historical Biology* 12: 265–290.

———. 1997b. Terrestrial locomotion of pterosaurs: A reconstruction based on *Pteraichnus* trackways. *Journal of Vertebrate Paleontology* 17: 104–113.

———. 2000. Pterosaur flight: The role of actinofibrils in wing function. *Historical Biology* 14: 255–284.

———. 2001. The osteology and functional morphology of the Late Cretaceous pterosaur *Pteranodon*. *Palaeontographica Abteilung A* 260: 1–153.

———. 2002. Soft-tissue preservation of the cranial crest of the pterosaur *Germanodactylus* from Solnhofen. *Journal of Vertebrate Paleontology* 22: 43–48.

———. 2003a. Morphological evolution of the pectoral girdle in pterosaurs: Myology and function. In *Evolution and Palaeobiology of Pterosaurs*, edited by E. Buffetaut and J. M. Mazin, 191–215. Geological Society Special Publication 217. London, The Geological Society.

———. 2003b. New crested specimens of the Late Cretaceous pterosaur *Nyctosaurus*. *Paläontologische Zeitschrift* 77: 61–75.

———. 2003c. A survey of pathologies in large pterodactyloid pterosaurs. *Palaeontology* 46: 195–196.

———. 2005. Pterosaur science or pterosaur fantasy? *Prehistoric Times* 70: 21–23.

———. 2006. Juvenile specimens of the pterosaur *Germanodactylus cristatus*, with a review of the genus. *Journal of Vertebrate Paleontology* 26: 872–878.

———. 2007a. Articulation and function of the pteroid bone of pterosaurs. *Journal of Vertebrate Paleontology* 27: 881–891.

———. 2007b. A second specimen of the pterosaur *Anurognathus ammoni*. *Paläontologische Zeitschrift*, 81, 376-398.

———. 2007c. A review of the pterosaur *Ctenochasma*: Taxonomy and ontogeny. *Neues Jahrbuch für Geologie und Paläontologie* 245: 23–31.

———. 2008. Morphological evolution of the wing of pterosaurs: Myology and function. *Zitteliana* B28: 127–141.

———. 2010. The morphology and taxonomy of *Cycnorhamphus*. Abstract. *Acta Geoscientica Sinica* 31(1): 4.

Benton, M. J. 1990. *The Reign of the Reptiles*. New York, Crescent.

———. 1999. *Scleromochlus taylori* and the origin of dinosaurs and pterosaurs. *Philosophical Transactions of the Royal Society, London B* 354: 1423–1446.

———. 2005. *Vertebrate Palaeontology*, 3rd ed. Oxford, Wiley-Blackwell.

Billon-Bruyat, J. P., and J. M. Mazin. 2003. The systematic problems of tetrapod ichnotaxa: The case study of *Pteraichnus* Stokes, 1957 (Pterosauria, Pterodactyloidea). In *Evolution and Palaeobiology of Pterosaurs*, edited by E. Buffetaut and J. M. Mazin, 315–324. Geological Society Special Publication 217. London, Geological Society.

Bock, W. J. 1960. Secondary articulation of the avian mandible. *The Auk* 77: 19–55.

Bogolubov, N. N. 1914. On the pterodactyl vertebra from the Upper Cretaceous deposits of Saratov Government. [In Russian.] *Ezhegodnik po Geologii i Mineralogii Rossii* 16: 1–7.

Bonaparte, J. F. 1970. *Pterodaustro guinazui* gen. et sp. nov. Pterosaurio de la Formacion Lagarcito, Provincia de San Luis, Argentina y su significado en la geologia regional (Pterodactylidae). *Acta Geologica Lilloana* 10: 209–225.

Bonaparte, J. F., E. H. Colbert, P. Currie, A. de Ricqlés, Z. Kielan Jaworowska, G. Leonardi, N. Morello, and P. Taquet. 1984. *Sulle Orme dei Dinosauri*. Venezia, Erizzo.

Bonaparte, J. F., and T. M. Sánchez. 1975. Restos de un pterosaurio *Puntanipterus globosus* de la formación La Cruz provincia San Luis, Argentina. *Actas Primo Congresso Argentino de Paleontologia e Biostratigraphica* 2: 105–113.

Bonaparte, J. F., C. L. Schultz, and M. B. Soares. 2010. Pterosauria from the Late Triassic of southern Brazil. In *New Aspects of Mesozoic Biodiversity*, edited by S. Bandyopadhyay, 63–71. Lecture Notes in Earth Sciences 132. Berlin Heidelberg, Springer.

Bonde, N., and P. Christiansen. 2003. The detailed anatomy of *Rhamphorhynchus*: Axial pneumaticity and its implications. In *Evolution and Palaeobiology of Ptero-*

saurs, edited by E. Buffetaut and J. M. Mazin, 217–232. Geological Society Special Publication 217. London, Geological Society.

Bowerbank, J. S. 1851. On the pterodactyles of the Chalk formation. *Proceedings of the Zoological Society of London* 1:14–20.

Bramwell, C. D., and G. R. Whitfield. 1974. Biomechanics of *Pteranodon*. *Philosophical Transactions of the Royal Society of London* 267: 503–581.

Broili, F. 1927. Ein Exemplar von *Rhamphorhynchus* mit Resten von Schwimmhaut. *Sitzungsberichte der Bayerischen Akademie der Wissenschaften, Mathematisch-Naturwissenschaftlichen* 1927: 29–48.

———. 1938. Beobachtungen an *Pterodactylus*. *Sitzungsberichte der Bayerische Akademie der Wissenschaften, Mathematisch-Naturwissenschaftlichen* 1938: 139–154.

Brower, J. C. 1983. The aerodynamics of *Pteranodon* and *Nyctosaurus*, two large pterosaurs from the upper Cretaceous of Kansas. *Journal of Vertebrate Paleontology* 3: 84–124.

Brower, J. C., and J. Veinus. 1981. Allometry in pterosaurs. *University of Kansas Paleontological Contributions* 105: 1–32.

Brown, B. 1943. Flying reptiles. *Natural History* 52: 104–111.

Brown, R. E., and A. C. Cogley. 1996. Contributions of the propatagium to avian flight. *Journal of Experimental Zoology* 276: 112–124.

Buckland, W. 1829. On the discovery of a new species of Pterodactyle in the Lias at Lyme Regis. *Transactions of the Geological Society, London* 3: 217–222.

Buffetaut, E., J. B. Clarke, and J. Le Lœuff. 1996. A terminal Cretaceous pterosaur from the Corbiéres (southern France) and the problem of pterosaur extinction. *Bulletin de la Société Geologique de France* 167: 753–759.

Buffetaut, E., D. Grigorescu, and Z. Csiki. 2002. A new giant pterosaur with a robust skull from the latest Cretaceous of Romania. *Naturwissenschaften* 89: 180–184.

———. 2003. Giant azhdarchid pterosaurs from the terminal Cretaceous of Transylvania (western Romania). In *Evolution and Palaeobiology of Pterosaurs*, edited by E. Buffetaut and J. M. Mazine, 91–104. Geological Society Special Publication 217. London, Geological Society.

Buffetaut, E., and J. P. Guibert. 2001. An early pterodactyloid pterosaur from the Oxfordian of Normandy (northwestern France). *Comptes Rendus de l'Académie des Sciences* 333: 405–409.

Buffetaut, E., and P. Jeffery. 2012. A ctenochasmatid pterosaur from the Stonesfield Slate (Bathonian, Middle Jurassic) of Oxfordshire, England. *Geological Magazine* 49: 552–556.

Buffetaut, E., and X. Kuang. 2010. The *tuba vertebralis* in dsungaripterid pterosaurs. Abstract. *Acta Geoscientica Sinica* 31 (1): 10–11.

Buffetaut, E., Y. Laurent, J. Le Lœuff, and M. Bilotte. 1997. A terminal Cretaceous giant pterosaur from the French Pyrenees. *Geological Magazine* 134: 553–556.

Buffetaut, E., J. Lepage, and G. Lepage. 1998. A new pterodactyloid from the Kimmeridgian of the Cape de la Héve (Normandy, France). *Geological Magazine* 135: 719 722.

Buffetaut, E., D. M. Martill, and F. Escuillié. 2004. Pterosaurs as part of a spinosaur diet. *Nature* 430: 33.

Butler, R. J., P. M. Barrett, and D. J. Gower. 2009. Postcranial skeletal pneumaticity and air-sacs in the earliest pterosaurs. *Biology Letters* 5: 557–560.

Butler, R. J., P. M. Barrett, S. Nowbath, and P. Upchurch. 2009. Estimating the effects of the rock record on pterosaur diversity patterns: Implications for hypotheses of bird/pterosaur competitive replacement. *Paleobiology* 35: 432–446.

Butler, R. J., S. L. Brusatte, B. Andres, and R.B.J. Benson. 2012. How do geological sampling biases affect studies of morphological evolution in Deep Time? A case study of pterosaur (Reptilia: Archosauria) disparity. *Evolution* 66: 147–162.

Cai, Z., and F. Wei. 1994. *Zhejiangopterus linhaiensis* (Pterosauria) from the Upper Cretaceous of Linhai, Zhejiang, China. *Vertebrata PalAsiatica* 32: 181–194.

Calvo, J. O., and M. G. Lockley. 2001. The first pterosaur tracks from Gondwana. *Cretaceous Research* 22: 585–590.

Campos, D. A., and A.W.A. Kellner. 1985. Panorama of the flying reptiles study in Brazil and South America. *Anais Academia Brasileira, Ciencias* 57: 453–466.

———. 1997. Short note on the first occurrence of Tapejaridae in the Crato Member (Aptian), Santana Formation, Araripe Basin, Northeast Brazil. *Anais Academia Brasileira, Ciencias* 69: 83–87.

Carpenter, K., D. Unwin, K. Cloward, C. Miles, and C. Miles. 2003. A new scaphognathine from the upper Jurassic Morrison Formation of Wyoming, USA. In *Evolution and Palaeobiology of Pterosaurs*, edited by E. Buffetaut and J. M. Mazin, 45–54. Geological Society Special Publication 217.London, Geological Society.

Casamiquela, R. M. 1975. *Herbstosaurus pigmaeus* (Coeluria, Compsognathidae) n. gen., n. sp. del Jurassico medio del Nequén (Patagonia septentrional): Uno de los más pequenos dinosaurios conocidos. *Acta Primero Congresso Argentino Paleontologia et Biostratigrafia* 2: 87–102.

Chatterjee, S., and R. J. Templin. 2004. Posture, locomotion and palaeoecology of pterosaurs.Geological Society of America Special Publication 376: 1–64.

Chiappe, L. M., and A. Chinsamy. 1996. *Pterodaustro*'s true teeth. *Nature* 379: 211–212.

Chiappe, L. M., L. Codorniú, G. Grellet-Tinner, and D. Rivarola. 2004. Argentinian unhatched pterosaur fossil. *Nature* 432: 571.

Chiappe, L. M., A.W.A. Kellner, D. Rivarola, S. Davila, and M. Fox. 2000. Cranial morphology of *Pterodaustro guinazui* (Pterosauria: Pterodactyloidea) from the Lower Cretaceous of Argentina. *Contributions in Science* 483: 1–19.

Chiappe, L. M., D. Rivarola, A. Cione, M. Fregenal-Martínez, H. Sozzi, L. Buatois, O. Gallego, et al. 1998. Biotic association and palaeoenvironmental reconstruction of the "Loma del *Pterodaustro*" fossil site (Early Cretaceous, Argentina). *Geobios* 31: 349–369.

Chinsamy, A., L. Codorniú, and L. M. Chiappe. 2008. Developmental growth patterns of the filter-feeder pterosaur, *Pterodaustro guinazui*. *Biology Letters* 23: 282–285.

———. 2009. Palaeobiological implications of the bone histology of *Pterodaustro guinazui*. *The Anatomical Record* 292: 1462–1477.

Claessens, L.P.A.M., P. M. O'Connor, and D. M. Unwin. 2009. Respiratory evolution facilitated the origin of pterosaur flight and aerial gigantism. *PLoS ONE* 4: e4497. http://www.plosone.org/article/info:doi/10.1371/journal.pone.0004497.

Clark, J. M., J. A. Hopson, R. Hernández, D. E. Fastovsky, and M. Montellano. 1998. Foot posture in a primitive pterosaur. *Nature* 391: 886–889.

Codorniú, L. 2005. Morfología caudal de *Pterodaustro guinazui* (Pterosauria: Ctenochasmatidae) del Cretácico de Argentina. *Ameghiniana* 42: 505–509. http://www.scielo.org.ar/scielo.php?script=sci_arttext&pid=S0002-70142005000200021&lng=en&nrm=iso.

Codorniú, L., and L. M. Chiappe. 2004. Early juvenile pterosaurs (Pterodactyloidea: *Pterodaustro guinazui*) from the Lower Cretaceous of central Argentina. *Canadian Journal of Earth Sciences* 41: 9–18.

Codorniú, L., L. M. Chiappe, A. Arcucci, and A. Ortiz-Suárez. 2009. First occurrence of gastroliths in Pterosauria (Early Cretaceous, Argentina). *Ameghiniana* 46: 15R–16R.

Colbert, E. H. 1969. A Jurassic pterosaur from Cuba. *American Museum Novitates* 2370: 1–26.

Company, J., J. I. Ruiz-Omeñaca, and X. Pereda Suberbiola. 1999. A long-necked pterosaur (Pterodactyloidea, Azhdarchidae) from the Upper Cretaceous of Valencia, Spain. *Geologie en Mijnbouw* 78: 319–333.

Coombs, W. P. 1978. Theoretical aspects of cursorial adaptations in dinosaurs. *Quarterly Review of Biology* 53: 393–418.

Cope, E. D. 1866. Communication in regard to the Mesozoic sandstone of Pennsylvania. *Proceedings of the Academy of Natural Sciences of Philadelphia* 1866: 290–291.

Cunningham, J. R., and M. Gerritsen. 2003. The flight of the *Nyctosaurus*. *Bulletin of the American Physical Society: Program of the 56th Annual Meeting of the Division of Fluid Dynamics* 48: 24–25.

Currie, P. J., and A. R. Jacobsen. 1995. An azhdarchid pterosaur eaten by a velociraptorine theropod. *Canadian Journal of Earth Sciences* 32: 922–925.

Cuvier, G. 1801. Reptile volant. *Magasin Encyclopédique* 9: 60–82.

———. 1809. Mémoire sur le squelette fossile d'un Reptil volant des environs d'Aichstedt, que quelques naturalistes ont pris pour un oiseau, et done nous formons un genre de Sauriens, sons le nom de *Ptero-Dactyle*. *Annales du Muséum d'histoire naturelle* 13: 424.

Czerkas, S. A., and Q. Ji. 2002. A new rhamphorhynchoid with a head crest and complex integumentary structures. In *Feathered Dinosaurs and the Origin of Flight*. The Dinosaur Museum Journal 1: 15–41.

Dal Sasso, C., S. Maganuco, E. Buffetaut, and M. A. Mendez. 2005. New information on the skull of the enigmatic theropod *Spinosaurus*, with remarks on its size and affinities. *Journal of Vertebrate Paleontology* 25: 888–896.

Dalla Vecchia, F. M. 1993. *Cearadactylus? ligabuei* nov. sp., a new early Cretaceous (Aptian) pterosaur from Chapada do Araripe (Northeastern Brazil). *Bollettino della Societá Paleontologica Italiana* 32: 401–409.

———. 1995. A new pterosaur (Reptilia, Pterosauria) from the Norian (Late Triassic) of Friuli (Northeastern Italy). Preliminary note. *Gortania - Atti del Museo Friulano di Storia Naturale* 16: 59–66.

———. 1998. New observations on the osteology and taxonomic status of *Preondactylus buffarinii* Wild, 1984 (Reptilia, Pterosauria). *Bollettino della Societá Paleontologica Italiana* 36: 355–366.

———. 2003a. A review of the Triassic pterosaur fossil record. *Rivista del Museo Civico di Scienze Naturali "E. Caffi" di Bergamo* 22: 13–29.

———. 2003b. New morphological observations on Triassic pterosaurs. In *Evolution and Palaeobiology of Pterosaurs*, edited by E. Buffetaut and J. M. Mazin, 23–44. Geological Society Special Publication 217. London, Geological Society.

———. 2009a. Anatomy and systematics of the pterosaur *Carniadactylus* gen. n. *rosenfieldi* (Dalla Vecchia, 1995). *Rivista Italiana di Paleontologia e Stratigrafia* 115: 159–188.

———. 2009b. The first Italian specimen of *Austriadactylus cristatus* (Diapsida, Pterosauria) from the Norian (Upper Triassic) of the Carnic Prealps. *Rivista Italiana di Paleontologia e Stratigrafia* 115: 291–304.

Dalla Vecchia, F. M., and G. Ligabue. 1993. On the presence of a giant pterosaur in the Lower Cretaceous (Aptian) of Chapada of Araripe (northeastern Brazil). *Bollettino della Societá Paleontologica Italiana* 32: 131–136.

Dalla Vecchia, F. M., G. Muscio, and R. Wild. 1989. Pterosaur remains in a gastric pellet from the upper Triassic (Norian) of Rio Seazza Valley (Udine, Italy). *Gortania—Atti Museo Friulano di Storia Naturale* 10: 121–132.

Dalla Vecchia, F. M., R. Wild, H. Hopf, and J. Reitner. 2002. A crested rhamphorhynchoid pterosaur from the Late Triassic of Austria. *Journal of Vertebrate Paleontology* 22: 196–199.

Delair, J. B. 1963. Notes on Purbeck fossil footprints, with descriptions of two hitherto unknown forms from Dorset. *Proceedings of the Dorset Natural History and Archaeological Society* 84: 92–100.

Dilkes, D. W. 1998. The Early Triassic rhynchosaur *Mesosu-*

chus browni and the interrelationships of basal archosauromorph reptiles. *Philosophical Transactions of the Royal Society of London B* 353: 501–541.

Döderlein, L. 1923. *Anurognathus ammoni* ein neuer Flugsaurier. *Sitzungsberichte der Bayerischen Akademie der Wissenschaften, Mathematisch-Naturwissenschaftlichen* 1923: 117–164.

Dong, Z. 1982. A new pterosaur (*Huanhepterus quingyangensis* gen. et sp. nov.) from Ordos, China. *Vertebrata PalAsiatica* 20: 115–121.

Dong, Z., and J. Lü. 2005. A new ctenochasmatid pterosaur from the Early Cretaceous of Liaoning Province. *Acta Geologica Sinica* 79: 164–167.

Dong, Z., Y. Sun, and S. Wu. 2003. On a new pterosaur from the Lower Cretaceous of Chaoyang Basin, Western Liaoning, China. *Global Geology* 22: 1–7.

Dyke, G. J., M. J. Benton, E. Posmosanu, and D. Naish. 2010. Early Cretaceous (Berriasian) birds and pterosaurs from the Cornet Bauxite Mine, Romania. *Palaeontology* 54: 79–95.

Dyke, G. J., A. J. McGowan, R, L. Nudds, and D. Smith. 2009. The shape of pterosaur evolution: Evidence from the fossil record. *Journal of Evolutionary Biology* 22: 890–898.

Dyke, G. J., R. L. Nudds, and J.M.V. Rayner. 2006. Flight of *Sharovipteryx mirabilis*: the world's first delta-winged glider. *Journal of Evolutionary Biology* 19: 1040–1043.

Eaton, G. F. 1910. Osteology of *Pteranodon*. *Memoirs of the Connecticut Academy of Arts and Sciences* 2: 1–38.

Eck, K., R. A. Elgin, and E. Frey. 2011. On the osteology of *Tapejara wellnhoferi* Kellner 1989 and the first occurrence of a multiple specimen assemblage from the Santana Formation, Araripe Basin, NE-Brazil. *Swiss Journal of Palaeontology* 130: 277–296.

Edinger, T. 1941. The brain of *Pterodactylus*. *American Journal of Science* 239: 665–682.

Elgin, R. A., and E. Frey. 2011a. A new ornithocheirid, *Barbosania gracilirostris* gen. et. sp. nov. (Pterosauria, Pterodactyloidea) from the Santana Formation (Cretaceous) of NE Brazil. *Swiss Journal of Palaeontology* 130: 259–275.

Elgin, R. A., and E. Frey. 2011b. A new azhdarchoid pterosaur from the Cenomanian (Late Cretaceous) of Lebanon. *Swiss Journal of Geosciences* 104: 21–33.

Elgin, R. A., C. A. Grau, C. Palmer, D.W.E. Hone, D. Greenwell, and M. J. Benton. 2008. Aerodynamic characters of the cranial crest of *Pteranodon*. *Zitteliana*, B28: 167–174.

Elgin, R. A., D.W.E. Hone, and E. Frey, E. 2011. The extent of the pterosaur flight membrane. *Acta Palaeontologica Polonica* 56: 99–111.

Engel, M. S., and D. A. Grimaldi. 2004. New light shed on the oldest insect. *Nature* 6975: 627–630.

Evans, S. E., and M. S. King. 1993. A new specimen of *Protorosaurus* (Reptilia: Diapsida) from the Marl Slate (late Permian) of Britain. *Proceedings of the Yorkshire Geological Society* 49: 229–234.

Everhart, M. J. 2005. *Oceans of Kansas: A Natural History of the Western Interior Sea*. Bloomington, Indiana University Press.

Fabre, J. 1976. Un nouveau Pterodactylidae sur le gisement "Portlandian" de Canjuers (Var): *Gallodactylus canjuerensis* nov. gen., nov. sp. *Comptes Rendus de l'Académie des Sciences* 279: 2011–2014.

Fajardo, R. J., E. Hernandez, and P. M. O'Connor. 2007. Postcranial skeletal pneumaticity: A case study in the use of quantitative microCT to assess vertebral structure in birds. *Journal of Anatomy* 211: 138–147.

Farlow, J. O., and E. R. Pianka. 2002. Body size overlap, habitat partitioning and living space requirements of terrestrial vertebrate predators: Implications for the paleoecology of large theropod dinosaurs. *Historical Biology* 16: 21–40.

Fastnacht, M. 2001. First record of *Coloborhynchus* (Pterosauria) from the Santana Formation (Lower Cretaceous) of the Chapada do Araripe, Brazil. *Paläontologische Zeitschrift* 75: 23–36.

———. 2005a. Jaw mechanics of the pterosaur skull construction and the evolution of toothlessness. PhD thesis, Johannes Gutenberg-Universität.

———. 2005b. The first dsungaripterid pterosaur from the Kimmeridgian of Germany and the biomechanics of pterosaur long bones. *Acta Palaeontologica Polonica* 50: 273–288

———. 2008. Tooth replacement pattern of *Coloborhynchus robustus* (Pterosauria) from the Lower Cretaceous of Brazil. *Journal of Morphology* 269: 332–348.

Fisher, D. C. 1981a. Crocodilian scatology, microvertebrate concentrations, and enamel-less teeth. *Paleobiology* 7: 262–275.

———. 1981b. Taphonomic interpretation of enamel-less teeth in the Shotgun Local Fauna (Paleocene, Wyoming). *Contributions from the Museum of Paleontology, the University of Michigan* 25: 259–275.

Fleming, T. H., and K. R. Lips. 1991. Angiosperm endozoochory: Were pterosaurs Cretaceous seed dispersers? *The American Naturalist* 138: 1058–1065.

Forrest, R. 2003. Evidence for scavenging by the marine crocodile *Metriorhynchus* on the carcass of a plesiosaur. *Proceedings of the Geologists' Association* 114: 363–366.

Fowler, D. W., E. A. Freedman, and J. B. Scannella. 2009. Predatory functional morphology in raptors: Interdigital variation in talon size is related to prey restraint and immobilisation technique. *PLoS ONE* 4: e7999. http://www.plosone.org/article/info%3Adoi%2F10.1371%2Fjournal.pone.0028964.

Frey, E., M. C. Buchy, and D. M. Martill. 2003. Middle- and bottom- decker Cretaceous pterosaurs: Unique designs in active flying vertebrates. In *Evolution and Palaeobiology of Pterosaurs*, edited by E. Buffetaut and J. M. Mazin, 267–274. Geological Society Special Publication 217. London, Geological Society.

Frey, E., M. C. Buchy, W. Stinnesbeck, A. G. González, and A. Stefano. 2006. *Muzquizopteryx coahuilensis*, n. g.,

n. sp., a nyctosaurid pterosaur with soft tissue preservation from the Coniacian (Late Cretaceous) of northeast Mexico (Coahuila). *Oryctos* 6: 19–40.

Frey, E., and D. M. Martill. 1994. A new pterosaur from the Crato Formation (Lower Cretaceous, Aptian) of Brazil. *Neues Jahrbuch für Geologie und Paläontologie* 194: 421–441.

———. 1996. A reappraisal of *Arambourgiania* (Pterosauria, Pterodactyloidea): One of the world's largest flying animals. *Neues Jahrbuch für Geologie und Paläontologie* 199: 221–247.

———. 1998. Soft tissue preservation in a specimen of *Pterodactylus kochi* (Wagner) from the Upper Jurassic of Germany. *Neues Jahrbuch für Geologie und Paläontologie* 210: 421–441.

Frey, E., D. M. Martill, and M. C. Buchy. 2003a. A new species of tapejarid pterosaur with soft tissue head crest. In *Evolution and Palaeobiology of Pterosaurs*, edited by E. Buffetaut and J. M. Mazin, 65–72. Geological Society Special Publication 217. London, Geological Society.

———. 2003*b*. A new crested ornithocheirid from the Lower Cretaceous of northeastern Brazil and the unusual death of an unusual pterosaur. In *Evolution and Palaeobiology of Pterosaurs*, edited by E. Buffetaut and J. M. Mazin, 55–63. Geological Society Special Publication 217. London, Geological Society.

Frey, E., C. A. Meyer, and H. Tischlinger. 2011. The oldest azhdarchoid pterosaur from the Late Jurassic Solnhofen Limestone (Early Tithonian) of Southern Germany. *Swiss Journal of Geoscience* 104: 35–55.

Frey, E., and J. Riess. 1981. A new reconstruction of the pterosaur wing. *Neues Jahrbuch für Geologie und Paläontologie* 194: 379–412.

Frey, E., and H. Tischlinger, H. 2000. Weichteilanatomie der Flugsaurierfüße und Bau der Scheitelkämme: Neue Pterosaurierfunde aus der Solnhofener Schichten (Bayern) und der Crato-Formation (Brasilien). *Archaeopteryx* 18: 1–16.

———. 2012. The Late Jurassic pterosaur *Rhamphorhynchus*, a frequent victim of the ganoid fish *Aspidorhynchus*? *PLoS ONE* 7: e31945. http://www.plosone.org/article/info%3Adoi%2F10.1371%2Fjournal.pone.0031945.

Frey, E., H. Tischlinger, M. C. Buchy, and D. M. Martill. 2003. New specimens of Pterosauria (Reptilia) with soft parts with implications for pterosaurian anatomy and locomotion. In *Evolution and Palaeobiology of Pterosaurs*, edited by E. Buffetaut and J. M. Mazin, 233–266. Geological Society Special Publication 217. London, Geological Society.

Fröbisch, N. B., and J. Fröbisch. 2006. A new basal pterosaur genus from the upper Triassic of the northern Calcareous Alps of Switzerland. *Palaeontology* 49: 1081–1090.

Gans, C., I. Darevski, and L. P. Tatarinov. 1987. *Sharovipteryx*, A reptilian glider? *Paleobiology* 13: 415–426.

Gao, K. Q., Q. Li, M. Wei, H. Pak, and I. Pak. 2009. Early Cretaceous birds and pterosaurs from the Sinuiju Series, and geographic extension of the Jehol Biota into the Korean Peninsula. *Journal of the Palaeontological Society of Korea* 1: 57–61.

Gasparini, Z., M. Fernàndez, and M. de la Fuente. 2004. A new pterosaur from the Jurassic of Cuba. *Palaeontology* 47: 919–927.

Gauthier, J. A. 1986. Saurischian monophyly and the origin of birds. *Memoirs of the Californian Academy of Science* 8: 1–55.

Gilmore, C. W. 1928. A new pterosaurian reptile from the marine Cretaceous of Oregon. *Proceedings of the U. S. National Museum* 73: 1–5.

Godfrey, S. J., and P. J. Currie. 2005. Pterosaurs. In *Dinosaur Provincial Park: A Spectacular Ancient Ecosystem Revealed,* edited by P. J. Currie and E. B. Koppelhus, 292–311. Bloomington, Indiana University Press.

Goldfuss, A. 1831. Beiträge zur Kenntnis verschiedener Reptilien der Vorwelt. *Nova Acta Academiae Leopoldinae* 15: 61–128.

Gower, D. J., and M. Wilkinson. 1996. Is there any consensus on basal archosaur phylogeny? *Proceedings of the Royal Society B* 263: 1399–1406.

Graf, W., C. de Waele, and P. P. Vidal. 1995. Functional anatomy of the head-neck movement system of quadrupedal and bipedal animals. *Journal of Anatomy* 186: 55–74.

Grellet-Tinner, G., S. Wroe, M. B. Thompson, and Q. Ji. 2007. A note on pterosaur nesting behaviour. *Historical Biology* 19: 273–277.

Grimaldi, D. A., and M. S. Engel. 2005. *Evolution of the Insects*. Cambridge, UK, Cambridge University Press.

Habib, M. B. 2007. Structural characteristics of the humerus of *Bennettazhia oregonensis* and their implications for specimen diagnosis and azhdarchoid biomechanics. Abstract. In *Flugsaurier: The Wellnhofer Pterosaur Meeting,* edited by D. Hone, 20. Abstract Volume 20. Munich, Bavarian State Collection for Palaeontology.

———. 2008. Comparative evidence for quadrupedal launch in pterosaurs. *Zitteliana* B28: 161–168.

———. 2010. 10,000 miles: Maximum range and soaring efficiency of azhdarchid pterosaurs. *Journal of Paleontology* 30: 99A–100A.

———. 2011. Functional morphology of anurognathid pterosaurs. Abstract. *Geological Society of America Abstracts with Programs* 43: 118. Northeastern (46th Annual) and North-Central (45th Annual) Joint Meeting, Pittsburgh, Pennsylvania, March 2011.

Habib, M. B., and J. Cunningham. 2010. Capacity for water launch in *Anhanguera* and *Quetzalcoatlus*. Abstract. *Acta Geoscientica Sinica* 31: 24–25.

Habib, M. B., and S. J. Godfrey. 2010. On the hypertrophied opisthotic processes in *Dsungaripterus weii* Young (Pterodactyloidea, Pterosauria). Abstract. *Acta Geoscientica Sinica* 31: 26.

Halstead, L. B. 1975. *The Evolution and Ecology of the Dinosaurs*. London, Peter Lowe.

Hankin, E. H., and D.M.S. Watson. 1914. On the flight of pterodactyls. *Aeronautical Journal* 18: 324–335.

Harksen, J. C. 1966. *Pteranodon sternbergi*, a new fossil pterodactyl from the Niobrara Cretaceous of Kansas. *Proceedings of the South Dakota Academy of Sciences* 45: 74–77.

Harris, J. D., and K. Carpenter. 1996. A large pterodactyloid from the Morrison Formation (Late Jurassic) of Garden Park, Colorado. *Neues Jahrbuch für Geologie und Paläontologie* 8: 473–484.

Hazlehurst, G. A.. and J.M.V. Rayner. 1992. Flight characteristics of Triassic and Jurassic Pterosauria: An appraisal based on wing shape. *Paleobiology* 18 447–463.

He, X., K. Li, and K. J. Cai. 1983. A new pterosaur from the middle Jurassic of Dashanpu, Zigong, Sichuan. *Journal of Chengdu College* 1: 27–33.

Hedenström, A., and M. Rosén. 2001. Predator versus prey: On aerial hunting and escape strategies in birds. *Behavioural Ecology* 12: 150–156.

Henderson, D. M. 2002. The eyes have it: The sizes, shapes, and orientations of theropod orbits as indicators of skull strength and bite force. *Journal of Vertebrate Paleontology* 22: 766–778.

———. 2010. Pterosaur body mass estimates from three-dimensional mathematical slicing. *Journal of Vertebrate Paleontology* 30: 768–785.

Henderson, M. D., and J. E. Peterson. 2006. An azhdarchid pterosaur cervical vertebrae from the Hell Creek Formation (Maastrichtian) of southeastern Montana. *Journal of Vertebrate Paleontology* 26: 192–195.

Henderson, D. M., and D. M. Unwin. 1999. Mathematical and computational model of a walking pterosaur. Abstract. *Journal of Vertebrate Paleontology* 19: 50A.

Hertel, F. 1995. Ecomorphological indicators of feeding behaviour in recent and fossil raptors. *The Auk* 112: 890–903.

Hildebrand, M. 1995. *Analysis of Vertebrate Structure*, 4th ed. New York, John Wiley and Sons.

Hing, R. 2011. A re-examination of a specimen of pterosaur soft-tissue from the Cretaceous Santana Formation of Brazil. MPhil thesis, University of Portsmouth, U.K.

Hone, D.W.E., and M. J. Benton. 2007. An evaluation of the phylogenetic relationships of the pterosaurs among archosauromorph reptiles. *Journal of Systematic Palaeontology* 5: 465–469.

———. 2008. Contrasting supertrees and total-evidence methods: Pterosaur origins. *Zitteliana* B28: 35–60.

Hone, D.W.E., D. M. Henderson, and C. Palmer. 2011. Investigating the buoyancy and floating posture of pterosaurs. Abstract. In *Programme and Abstracts,* Symposium of Vertebrate Palaeontology and Comparative Anatomy, 59th Annual Meeting, 2011, 26. Lyme Regis, U.K.

Hone, D.W.E., and J. Lü. 2010. A new specimen of *Dendrorhynchoides* (Pterosauria: Anurognathidae) with a long tail and the evolution of the pterosaurian tail. Abstract. *Acta Geoscientica Sinica* 31: 29–30.

Hone, D.W.E., D. Naish, and I. Cuthill. 2011. Does mutual sexual selection explain the evolution of head crests in pterosaurs and dinosaurs? *Lethaia* 45: 139–156.

Hone, D.W.E., H. Tischlinger, E. Frey, and M. Röper. 2012. A new non-pterodactyloid pterosaur from the Late Jurassic of Southern Germany. *PLoS ONE* 7: e39312. http://www.plosone.org/article/info%3Adoi%2F10.1371%2Fjournal.pone.0039312.

Hone, D.W.E., T, Tsuhiji, M. Watabe, and K. Tsogbataar. 2012. Pterosaurs as a food source for small dromaeosaurs. *Palaeogeography, Palaeoclimatology, Palaeoecology* 331–332: 27–30.

Hooley, R. W. 1913. On the skeleton of *Ornithodesmus latidens*; an ornithosaur from the Wealden Shales of Atherfield (Isle of Wight). *Quarterly Journal of the Geological Society* 96: 372–422.

———. 1914. On the ornithosaurian genus *Ornithocheirus*, with a review of the specimens of the Cambridge Greensand in the Sedgwick Museum. *Annals and Magazine of Natural History* 8: 529–557.

Howse, S.C.B. 1986. On the cervical vertebrae of the Pterodactyloidea (Reptilia: Archosauria). *Zoological Journal of the Linnaean Society, London* 88: 307–328.

Howse, S.C.B., and A. R. Milner. 1993. *Ornithodesmus*—a maniraptoran theropod dinosaur from the Lower Cretaceous of the Isle of Wight, England. *Palaeontology* 36: 425–437.

———. 1995. The pterodactyloids from the Purbeck Limestone Formation of Dorset. *Bulletin of the Natural History Museum, London* 51: 73–88.

Howse, S.C.B., A. R. Milner, and D. M. Martill. 2001. Pterosaurs. In *Dinosaurs of the Isle of Wight, Field Guide to Fossils*10, edited by D. M. Martill and D. Naish, 324–335. .London, The Palaeontological Association.

Humphries, S., R.H.C. Bonser, M. P. Witton, and D. M. Martill. 2007. Did pterosaurs feed by skimming? Physical modelling and anatomical evaluation of an unusual feeding method. *PLoS Biology* 5: e204. http://www.plosbiology.org/article/info%3Adoi%2F10.1371%2Fjournal.pbio.0050204.

Hwang, K. G., M. Huh, M. G. Lockley, D. M. Unwin, and J. L. Wright. 2002. New pterosaur tracks (Pteraichnidae) from the Late Cretaceous Uhangri Formation, S. W. Korea. *Geological Magazine* 139: 421–435.

Hyder, E., M. P. Witton, and D. M. Martill. In press. Evolution of the pterosaur pelvis. *Acta Palaeontologica Polonica*.

Ibáñez , C., J. Juste, J. L. García-Mudarra, and P. T. Agirre-Mendi. 2001. Bat predation on nocturnally migrating birds. *Proceedings of the National Academy of Sciences* 98: 9700–9792. http://www.pnas.org/content/98/17/9700.full.

Ibrahim, N., D. M. Unwin, D. M. Martill, L. Baidder, and S. Zouhri. 2010. A New Pterosaur (Pterodactyloidea: Azhdarchidae) from the Upper Cretaceous of Morocco. *PLoS ONE* 5: e.10875. http://www.plosone.org/article/info:doi/10.1371/journal.pone.0010875.

Jain, S. L. 1974. Jurassic pterosaur from India. *Journal of the Geological Society of India* 15: 330–335.

Jell, P. A., and P. M. Duncan. 1986. Invertebrates, mainly insects, from the freshwater, Lower Cretaceous, Koonwarra Fossil Bed (Korumburra Group), south Gippsland, Victoria. *Memoir of the Association of Australasian Palaeontologists* 3: 111–205.

Jenkins, F. A., Jr., N. H. Shubin, S. M. Gatesy, and K. Padian. 2001. A diminutive species of pterosaur (Pterosauria: Eudimorphodontidae) from the Greenlandic Triassic. *Bulletin of the Museum of Comparative Zoology, Harvard* 156: 151–170.

Jensen, J. A., and K. Padian. 1989. Small pterosaurs and dinosaurs from the Uncompahgre Fauna (Brusy Basin Member, Morrison Formation: ?Tithonian), Late Jurassic, Western Colorado. *Journal of Paleontology* 63: 364–373.

———. 1998. A new fossil pterosaur (Rhamphorhynchoidea) from Liaoning. [In Chinese.] *Jiangsu Geology* 22:199–206.

Ji, Q., S. A. Ji, Y. N. Cheng, H. L. You, J. Lü, Y. Q. Liu, and C. X. Yuan. 2004. Pterosaur egg with a leathery shell. *Nature* 432: 572.

Ji, Q., and C. Yuan. 2002. Discovery of two kinds of protofeathered pterosaur in the Mesozoic Daohugou Biota in the Ningcheng Region and its stratigraphic and biologic significances. *Geological Review* 48: 221–224.

Ji, S., and Q. Ji. 1997. Discovery of a new pterosaur in Western Liaoning, China. *Acta Geologica Sinica* 71: 115.

Jiang, S., and X. Wang. 2011. Important features of *Gegepterus change* (Pterosauria: Archaeopterodactyloidea, Ctenochasmatidae) from a new specimen. *Acta PalAsiatica* 49: 172–184.

Jouve, S. 2004. Description of the skull of a *Ctenochasma* (Pterosauria) from the latest Jurassic of Eastern France, with a taxonomic revision of European Tithonian Pterodactyloidea. *Journal of Vertebrate Paleontology* 24: 542–554.

Kellner, A.W.A. 1984. Occurrência de uma mandíbula de Pterosauria (*Brasileodactylus araripensis*, nov. gen, nov. sp.) na Formação Santana, Cretáceo da Chapada do Araripe, Ceará, Brasil. *33° Anais Congresso Brasileiro de Geologia* 2: 578–590.

———. 1989. A new edentate pterosaur of the Lower Cretaceous of the Araripe Basin, Northeast Brazil. *Anais da Academia Brasileira de Ciências* 61: 439–446.

———. 1995. The relationships of the Tapejaridae (Pterodactyloidea) with comments on pterosaur phylogeny. In *Sixth Symposium on Mesozoic Terrestrial Ecosystems and Biota, Short Papers*, edited by A. Sun and Y. Wang, 73–77. Beijing, China Ocean Press.

———. 1996a. Description of the braincase of two early Cretaceous pterosaurs (Pterodactyloidea) from Brazil. *American Museum Novitates* 3175: 1–34.

———. 1996b. Reinterpretation of a remarkably well preserved pterosaur soft tissue from the early Cretaceous of Brazil. *Journal of Vertebrate Paleontology* 16: 718–722.

———. 2003. Pterosaur phylogeny and comments on the evolutionary history of the group. In *Evolution and Palaeobiology of Pterosaurs*, edited by E. Buffetaut and J. M. Mazin, 105–137. Geological Society Special Publication 217. London, Geological Society.

———. 2004. New information on the Tapejaridae (Pterosauria, Pterodactyloidea) and discussion of the relationships of this clade. *Ameghiniana* 41: 521–534.

———. 2010. Comments on the Pteranodontidae (Pterosauria, Pterodactyloidea) with the description of two new species. *Anais da Academia Brasileira de Ciências* 82: 1063–1084.

Kellner, A.W.A., and D. A. Campos. 1988. Sobre um novo pterossauro com crista sagital da Bracia do Araripe, Cretáceo Inferior do Nordeste do Brasil. *Anais da Academia Brasileira, Ciências* 60: 459–469.

———. 1990. Preliminary description of an unusual pterosaur skull of the Lower Cretaceous from the Araripe Basin. In *Atas do I Simpósio sobre a Bacia do Araripe e Bacias Interiores do Nordeste*, edited by D. A. Campos, M.S.S. Viana, P. M. Brito, and G. Beurlen, 401–406. Conference Proceedings. Crato, Ceara, Brazil.

———. 1994. A new species of *Tupuxuara* (Pterosauria, Tapejaridae) from the Early Cretaceous of Brazil. *Anais da Academia Brasileira, Ciências* 66: 467–473.

———. 2002. The function of the cranial crest and jaws of a unique pterosaur from the Early Cretaceous of Brazil. *Science* 297: 389–392.

———. 2007. Short note on the ingroup relationships of the Tapejaridae (Pterosauria, Pterodactyloidea). *Boletim do Museu Nacional, Nova Série, Rio de Janeiro—Brasil. Geologia* 75: 1–14.

Kellner, A.W.A., and Y. Hasagawa. 1993. Postcranial skeleton of *Tupuxuara* (Pterosauria, Pterodactyloidea, Tapejaridae) from the Lower Cretaceous of Brazil. *Journal of Vertebrate Paleontology* 13: 44A.

Kellner, A.W.A., and W. Langston, Jr. 1996. Cranial remains of *Quetzalcoatlus* (Pterosauria, Azhdarchidae) from Late Cretaceous sediments of Big Bend National Park. *Journal of Vertebrate Paleontology* 16: 222–231.

Kellner, A.W.A., T. H. Rich, F. R. Costa, P. Vickers-Rich, B. P. Kear, M. Walters, and L. Kool. 2010. New isolated pterodactyloid bones from the Albian Toolebuc Formation (western Queensland, Australia) with comments on the Australian pterosaur fauna. *Alcheringa* 34: 219–230.

Kellner, A.W.A., T. Rodrigues, and F. R. Costa. 2011. Short note on a pteranodontid pterosaur (Pterodactyloidea) from western Queensland, Australia. *Anais da Academia Brasileira de Ciências* 83: 301–308.

Kellner, A.W.A. and Y. Tomida. 2000. Description of a new species of Anhangueridae (Pterodactyloidea) with comments on the pterosaur fauna from the Santana Formation (Aptian -Albian), Northeastern

Brazil. *National Science Museum, Tokyo, Monographs* 17: 1–135.

Kellner, A.W.A., X. Wang, H. Tischlinger, D. A. Campos, D.W.E. Hone, and X. Meng. 2009. The soft tissue of *Jeholopterus* (Pterosauria, Anurognathidae, Batrachognathinae) and the structure of the pterosaur wing membrane. *Proceedings of the Royal Society B* 277: 321–329.

Kripp, D. 1943. Ein Lebensbild von *Pteranodon ingens* auf flugtechnicsher Grundlage. *Luftwissen* 8: 217–246.

Kuhn, O. 1967. Die fossil Wirbeltierklasse Pterosauria. Krailling, Munich, Verlag Oeben.

Langston, W., Jr. 1981. Pterosaurs. *Scientific American* 244: 92–102.

Lawson, D. A. 1975a. Pterosaur from the Latest Cretaceous of West Texas: Discovery of the largest flying creature. *Science* 185: 947–948.

Lawson, D. A. 1975b. Could pterosaurs fly? *Science* 188: 676.

Lee, Y-N. 1994. The Early Cretaceous pterodactyloid pterosaur *Coloborhynchus* from North America. *Palaeontology* 37: 755–763.

Lehman, T. M. 1998. A gigantic skull and skeleton of the horned dinosaur *Pentaceratops sternbergi* from New Mexico. *Journal of Paleontology* 72: 894–906.

Leonardi, G., and G. Borgomanero, G. 1983. *Cearadactylus atrox*, nov. gen. nov. sp.; novo Pterosauria (Pterodactyloidea) da Chapada do Araripe, Ceará, Brasil. Abstract. *Congresso Brasileiro de Paleontologia, resumos*, 17.

———. 1985. *Cearadactylus atrox*, nov. gen. nov. sp.; novo Pterosauria (Pterodactyloidea) da Chapada do Araripe, Ceará, Brasil. *Coletânea de Trabalhos Paleontológicos, Série Geologia, Brasilia* 27: 75–80.

Li, J., J. Lü, and B. K. Zhang. 2003. A new lower Cretaceous sinopterid pterosaur from Western Liaoning, China. *Acta Palaeontologica Sinica* 42: 442–447.

Liem, K. F., W. E. Bemis, W. F. Walker, Jr., and L. Grande. 2001. *Functional Anatomy of the Vertebrates: An Evolutionary Perspective*, 3rd ed. Belmont, CA, Thomson Brooks/Cole.

Lockley, M. G., T. J. Logue, J. J. Moratalla, A.P.P. Hunt, J. Schultz, and J. M. Robinson. 1995. The fossil trackway *Pteraichnus* is pterosaurian, not crocodilian: Implications for the global distribution of pterosaur tracks. *Ichnos* 4: 7–20.

Lockley, M. G., and J. L. Wright. 2003. Pterosaur swim tracks and other ichnological evidence of behaviour and ecology. In *Evolution and Palaeobiology of Pterosaurs*, edited by E. Buffetaut and J. M. Mazin, 297–313. Geological Society Special Publication 217. London, Geological Society.

Logue, T. J. 1977. Preliminary investigation of pterodactyl tracks at Alcora, Wyoming. *Wyoming Geological Association Earth Science Bulletin* 10: 29–30.

Lü, J. 2002. Soft tissue in an Early Cretaceous pterosaur from Liaoning province, China. *Memoir of the Fukui Prefectural Dinosaur Museum* 1: 19–28.

———. 2003. A new pterosaur: *Beipiaopterus chenianus*, gen. et sp. nov. (Reptilia: Pterosauria) from Western Liaoning Province of China. *Memoir of the Fukui Prefectural Dinosaur Museum* 2: 153–160.

———. 2009a. A baby pterodactyloid pterosaur from the Yixian Formation of Ningcheng, Inner Mongolia, China. *Acta Geologica Sinica* 83: 1–8.

———. 2009b. A new non-pterodactyloid pterosaur from Qinglong County, Hebei Province of China. *Acta Geologica Sinica* 83: 189–199.

———. 2009c. New material of dsungaripterid pterosaurs (Pterosauria: Pterodactyloidea) from Western Mongolia and its palaeoecological implications. *Geological Magazine* 287: 283–389.

———. 2010a. An overview of the pterosaur fossil record in China. Abstract. *Acta Geoscientica Sinica* 31 (1): 49–51.

———. 2010b. A new boreopterid pterodactyloid pterosaur from the Early Cretaceous Yixian Formation of Liaoning Province, Northeastern China. *Acta Geologica Sinica* 84: 49–51.

Lü, J., and X. Bo. 2011. A new rhamphorhynchid pterosaur (Pterosauria) from the Middle Jurassic Tiaojishan Formation of Western Liaoning, China. *Acta Geologica Sinica* 85: 977–983.

Lü , J., X. Du, Q. Zhu, X. Cheng, and D. Luo. 1997. Computed tomography (CT) of braincase of *Dsungaripterus weii* (Pterosauria). *Chinese Science Bulletin* 42: 1125–1129.

Lü, J., and X. Fucha. 2010. A new pterosaur (Pterosauria) from the Middle Jurassic Tiaojishan Formation of western Liaoning, China. *Global Geology* 13: 113–118.

Lü, J., X. Fucha, and J. Chen. 2010. A new scaphognathine pterosaur from the Middle Jurassic of Western Liaoning, China. *Acta Geoscientica Sinica* 31: 263–266.

Lü, J., C. Gao, J. Liu, Q. Meng, and Q. Ji. 2006. New material of the pterosaur *Eopteranodon* from the Early Cretaceous Yixian Formation, western Liaoning, China. *Geological Bulletin of China* 25: 555–571.

Lü, J., Y. Gao, L. Xing, Z. Li, and Q. Ji. 2007. A new species of *Huaxiapterus* (Pterosauria: Tapejaridae) from the Early Cretaceous of Western Liaoning, China. *Acta Geologica Sinica* 81: 693–687.

Lü, J., and Q. Ji. 2005a. A new ornithocheirid from the Early Cretaceous of Liaoning Province, China. *Acta Geologica Sinica* 79: 157–163.

———. 2005b. New azhdarchid pterosaur from the Early Cretaceous of Western Liaoning. *Acta Geologica Sinica* 79: 301–307.

Lü, J., Q. Ji, X. Wei, and Y. Liu. 2012. A new ctenochasmatoid pterosaur from the Early Cretaceous Yixian Formation of western Liaoning, China. *Cretaceous Research* 34: 26–30.

Lü J., S. Ji,, C. Yuan, Y. Gao, Z. Sun, and Q. Ji. 2006. New pterodactyloid pterosaur from the Lower Cretaceous Yixian Formation of Western Liaoning. In *Papers from the 2005 Heyuan International Dinosaur Symposium*, edited by J. Lü, Y. Kobayashi, D. Huang, and Y. N. Lee, 195–203. Beijing, Geological Publishing House.

Lü, J., S. Ji, C. Yuan, and Q. Ji. 2006. *Pterosaurs from China*. Beijing, Geological Publishing House.

Lü, J., X. Jin, D. M. Unwin, L. Zhao, Y. Azuma, and Q. Ji. 2006. A new species of *Huaxiapterus* (Pterosauria: Pterodactyloidea) from the Lower Cretaceous of Western Liaoning, China with comments on the systematics of tapejarid pterosaurs. *Acta Geologica Sinica* 80: 315–326.

Lü, J., J. Liu, X. Wang, C. Gao, Q. Meng, and Q. Ji. 2006. New material of the pterosaur *Sinopterus* (Reptilia: Pterosauria) from the Early Cretaceous Jiufotang Formation, Western Liaoning, China. *Acta Geologica Sinica* 80: 783–789.

Lü, J., H. Pu, L. Xu, Y. Wu., and X. Wei. 2012. Largest toothed pterosaur skull from the Early Cretaceous Yixian Formation of western Liaoning, China, with comments on the family Boreopteridae. *Acta Geologica Sinica* 86: 287–293.

Lü, J., D. M. Unwin, D. C. Deeming, X. Jin, Y. Liu, and Q. Ji. 2011. An egg-adult association, gender, and reproduction in pterosaurs. *Science* 331: 321–324.

Lü, J., D. M. Unwin, X. Jin, Y. Liu, and Q. Ji. 2010. Evidence for modular evolution in a long-tailed pterosaur with a pterodactyloid skull. *Proceedings of the Royal Society B* 277: 383–389.

Lü, J., D. M. Unwin, L. Xu, and X. Zhang, X. 2008. A new azhdarchoid pterosaur from the Lower Cretaceous of China and its implications for pterosaur phylogeny and evolution. *Naturwissenschaften* 95: 891–897.

Lü, J., D. M. Unwin, B. Zhou, G. Chunling, and C. Shen. 2012. A new rhamphorhynchid (Pterosauria: Rhamphorhynchidae) from the Middle/Upper Jurassic of Qinglong, Hebei Province, China. *Zootaxa* 3158: 1–19.

Lü, J., L. Xu, H. Chang, and X. Zhang. 2011. A new darwinopterid pterosaur from the Middle Jurassic of Western Liaoning, Northeastern China and its ecological implications. *Acta Geologica Sinica* 85: 507–514.

Lü, J., L. Xu, and Q. Ji. 2008. Restudy of *Liaoxipterus* (Istiodactylidae, Pterosauria), with comments on the Chinese istiodactylid pterosaurs. *Zitteliana* B28: 229–242.

Lü, J., and C. Yuan. 2005. New tapejarid pterosaur from Western Liaoning, China. *Acta Geologica Sinica* 79: 453–458.

Lü, J., and B. K. Zhang. 2005. New pterodactyloid pterosaur from the Yixian Formation of western Liaoning. *Geological Review* 51: 458–462.

Lydekker, R. 1888. *Catalogue of the Fossil Amphibia and Reptilia in the British Museum (Natural History)*, Part 1. London, Trustees of the BM(NH).

Mader, B., and A.W.A. Kellner. 1999. A new anhanguerid pterosaur from the Cretaceous of Morocco. *Boletim do Museu Nacional, Geologia, Nova Série* 45: 1–11.

Maisch, M. W., A. T. Matzke, and G. Sun. 2004. A new dsungaripteroid pterosaur from the Lower Cretaceous of the southern Junggar Basin, north-west China. *Cretaceous Research* 25: 625–634.

Marden, J. H. 1994. From damselflies to pterosaurs: How burst and sustainable flight performance scale with size. *American Journal of Physiology* 266: 1077–1084.

Marsh, O. C. 1871. Note on a new and gigantic species of Pterodactyle. *American Journal of Science* Ser. 3, 1: 472.

———. 1876. Notice of a new sub-order of Pterosauria. *American Journal of Science* Ser. 3, 11: 507–509.

———. 1881. Note on American pterodactyls. *American Journal of Science* Ser. 3, 21: 342–343.

Martill, D. M. 1986. The diet of *Metriorhynchus*, a Mesozoic marine crocodile. *Neues Jahrbuch für Geologie und Paläontologie* H10: 621–625.

———. 1997. From hypothesis to fact in a flight of fancy: The responsibility of the popular scientific media. *Geology Today* 13: 71–73.

———. 2007. The age of the Cretaceous Santana Formation fossil Konservat Lagerstätte of north-east Brazil: A historical review and an appraisal of the biochronostratigraphic utility of its biota. *Cretaceous Research* 29: 895–920.

———. 2010. The early history of pterosaur discovery in Great Britain. In *Dinosaurs and Other Extinct Saurians: A Historical Perspective,* edited by R.T.J. Moody, E. Buffetaut, D. Naish, and D. M. Martill, 287–311. Geological Society Special Publication 343. London, Geological Society.

———. 2011. A new pterodactyloid pterosaur from the Santana Formation (Cretaceous) of Brazil. *Cretaceous Research* 32: 236–243.

Martill, D. M., and S. Etches. In press. A monofenestratan pterosaur from the Kimmeridge Clay Formation (Upper Jurassic, Kimmeridgian) of Dorset, England. *Acta Palaeontologica Polonica*.

Martill, D. M., E. Frey, R. M. Sadaqah, and H. N. Khoury. 1998. Discovery of the holotype of the giant pterosaur *Titanopteryx philadelphiae* Arambourg, 1959 and the status of *Arambourgiania* and *Quetzalcoatlus*. *Neues Jahrbuch für Geologie and Paläeontologie* 207: 57–76.

Martill, D. M., E. Frey, G. C. Diaz, and C. M. Bell. 2000. Reinterpretation of a Chilean pterosaur and the occurrence of Dsungaripteridae in South America. *Geological Magazine* 137: 19–25.

Martill, D. M. and D. Naish. 2006. Cranial crest development in the azhdarchoid pterosaur *Tupuxuara*, with a review of the genus and tapejarid monophyly. *Palaeontology* 49: 925–941.

Martill, D. M. and D. M. Unwin. 1989. Exceptionally well preserved pterosaur wing membrane from the Cretaceous of Brazil. *Nature* 340: 138–140.

———. 2012. The world's largest toothed pterosaur, NHMUK R481, an incomplete rostrum of *Coloborhynchus capito* (Seeley, 1870) from the Cambridge Greensand of England. *Cretaceous Research* 34: 1–9.

Martill, D. M., P. Wilby, and D. M. Unwin. 1990. Stripes on a pterosaur wing. *Nature* 346: 116.

Martill, D. M., and M. P. Witton. 2008. Catastrophic failure in a pterosaur skull from the Cretaceous Santana Formation of Brazil. *Zitteliana* B28: 175–183.

Mateer, N. J. 1975. A study of *Pteranodon*. *Bulletin of the Geological Institute of the University of Uppsala* 6: 23–53.

Mayr, F. X. 1964. Die naturwissenschaftlichen Sammlungen der Philosophisch-theologischen Hochschule, Eichstätt. In *Festschrift zur 400 Jahre Coll. Willibald*, 302–334. Eichstätt, Brönner and Daentler.

Mazin, J. M., P. Hantzpergue, G. Lafaurie, and P. Vignaud. 1995. Des pistes de pterosaurs dans le Tithonien de Crayssac (Quercy, France). *Comptes rendus de l'Académie des Sciences* 321: 417–424.

Mazin, J. M., J. Billon-Bruyat, P. Hantzepergue, and G. Larauire. 2003. Ichnological evidence for quadrupedal locomotion in pterodactyloid pterosaurs: Trackways from the Late Jurassic of Crayssac. In *Evolution and Palaeobiology of Pterosaurs*, edited by E. Buffetaut and J. M. Mazin, 283–296. Geological Society Special Publication 217. London, Geological Society.

Mazin, J. M., J. P. Billon-Bruyat, and K. Padian. 2009. First record of a pterosaur landing trackway. *Proceedings of the Royal Society B* 276: 3881–3886.

McGowen, M. R., K. Padian, M. A. de Sosa, and R. W. Harmon. 2002. Description of *Montanazhdarcho minor*, an azhdarchid pterosaur from the Two Medicine Formation (Campanian) of Montana. *Paleobios* 22: 1–9.

Meyer, H. von. 1834. *Gnathosaurus subulatus*, ein saurus aus dem lithographischen Schiefen von Solnhofen. *Museum Senckenbergianum* 1: 3.

Meyer, H. von. 1847. Homoeosaurus maximiliani *und* Rhamphorhynchus (Pterodactylus) longicaudus, *zwei fossile Reptilien aus dem Kalkscheifer von Solnhofen*. Frankfurt am Main, Schmerber Verlag.

Meyer, H. von. 1851. *Ctenochasma roemeri*. *Palaeontographica* 2: 82–84.

Meyers, T. S. 2010. A new ornithocheirid pterosaur from the Upper Cretaceous (Cenomanian-Turonian) Eagle Ford Group of Texas. *Journal of Vertebrate Paleontology* 30: 280–287.

Miller, H. W. 1972. The taxonomy of *Pteranodon* species from Kansas. *Transactions of the Kansas Academy of Science* 74: 1–19.

Modesto, S. P., and H. D. Sues. 2004. The skull of the Early Triassic Archosauromorph reptile *Prolacerta broomi* and its phylogenetic significance. *Zoological Journal of the Linnaean Society* 140: 335–351.

Molnar, R. E., and R. A. Thulborn. 1980. First pterosaur from Australia. *Nature* 288: 361–363.

———. 2007. An incomplete pterosaur skull from the Cretaceous of North-Central Queensland, Australia. *Arquivos do Museu Nacional, Rio de Janeiro* 65: 461–470.

Naish, D. 2010. *Tetrapod Zoology Book One*. Bideford, North Devon, U.K., CFZ Press.

Nesbitt, S. J. 2011. The early evolution of archosaurs: Relationships and the origin of major clades. *Bulletin of the American Museum of Natural History* 352: 1–292.

Nesbitt, S. J., and D.W.E. Hone. 2010. An external mandibular fenestra and other archosauriform character states in basal pterosaurs. *Palaeodiveristy* 3: 223–231.

Nesbitt, S. J., A. C. Sidor, R. B. Irmis, K. D. Angielczyk, R.M.H. Smith, and L. A. Tsuji. 2010. Ecologically distinct dinosaurian sister group shows early diversification of Ornithodira. *Nature* 464: 95–98.

Nessov, L. A. 1984. Pterosaurs and birds of the Late Cretaceous of Central Asia. *Paläontologische Zeitschrift* 1: 47–57.

Nessov, L. A. 1991. Gigantskiye lyetayushchiye yashchyeryi semyeistva Azhdarchidae. I. Morfologiya, sistematika. *Vestnik Leningradskogo Gosudarstvennogo Universiteta. Seriya 7* 2: 14–23.

Nessov, L. A., L. F. Kanznyshkina, and G. O. Cherepanov. 1987. Dinosaurs, crocodiles and other archosaurs from the Late Mesozoic of Central Asia and their place in ecosystem. Abstract. In *Abstracts of 33*rd *session of the All-Union Palaeontological Society, Leningrad*, 46–47.

Newman, E. 1843. Note on the Pterodactyle Tribe considered as marsupial bats. *Zoologist* 1: 129–131.

Newton, E. T. 1888. On the skull, brain and auditory organ of a new species of pterosaurian (*Scaphognathus purdoni*) from the Upper Lias near Whitby, Yorkshire. *Proceedings of the Royal Society, London* 43: 436–440.

Norberg, U.M.L., and J.M.V. Rayner. 1987. Ecological morphology and flight in bats (Mammalia; Chiroptera): wing adaptations, flight performance, foraging strategy and echolocation. *Philosophical Transactions of the Royal Society, London B* 316: 335–427.

O'Connor, P. M. 2004. Pulmonary pneumaticity in the postcranial skeleton of extant Aves: A case study examining Anseriformes. *Journal of Morphology* 261: 141–161.

———. 2006. Postcranial pneumaticity: An evaluation of soft-tissue influences on the postcranial skeleton and the reconstruction of pulmonary anatomy in archosaurs. *Journal of Morphology* 10: 1199–1226.

O'Connor, P. M., and L.P.A.M. Claessens. 2005. Basic avian pulmonary design and flow-through ventilation in non-avian theropod dinosaurs. *Nature* 436: 253–256.

Ősi, A. 2010. Feeding-related characters in basal pterosaurs: Implications for jaw mechanism, dental function and diet. *Lethaia* 44: 136–152.

Ősi, A., E. Buffetaut, and E. Prondvai. 2011. New pterosaurian remains from the Late Cretaceous (Santonian) of Hungary (Iharkút, Csehbánya Formation). *Cretaceous Research* 32: 4556–463.

Ősi, A., D. B. Weishampel, and C. M. Jianu. 2005. First evidence of azhdarchid pterosaurs from the Late Cretaceous of Hungary. *Acta Palaeontologica Polonica* 50: 777–787.

Owen, R. 1851. *Monograph on the fossil Reptilia of the Cretaceous Formations. Part I.* Chelonia, Lacertilia, *etc.*, 80–104. London, The Palaeontographical Society.

———. 1861. *Monograph on the fossil Reptilia of the Cretaceous Formations Supplement III. Pterosauria* (Pterodactylus), 1–19. London, The Palaeontographical Society.

———. 1870. *Monograph on the Fossil Reptilia of the Liassic Formations. I. Part III. Plesiosaurus,* Dimorphodon, *and Ichthyosaurus*, 1–81. London, The Palaeontographical Society.

———. 1874. *A Monograph on the Fossil Reptilia of the Mesozoic formations. Monograph on the Order Pterosauria*, 1–14. London, The Palaeontographical Society.

Padian, K. 1983a. A functional analysis of flying and walking in pterosaurs. *Palaeobiology* 9: 218–239.

———. 1983b. Osteology and functional morphology of *Dimorphodon macronyx* (Buckland) (Pterosauria: Rhamphorhynchoidea) based on new material in the Yale Peabody Museum. *Postilla* 189: 1–44.

———. 1984a. The origin of pterosaurs. In *Third Symposium on Mesozoic Terrestrial Ecosystems*, edited by W.-E. Reif and F. Westphal, 163–168. Short papers. Tübingen, Attempto.

———. 1984b. A large pterodactyloid pterosaur from the Two Medicine Formation (Campanian) of Montana. *Journal of Vertebrate Paleontology* 4: 516–524.

———. 2003. Pterosaur stance and gait and the interpretation of trackways. *Ichnos* 10: 115–126.

———. 2008a. The Early Jurassic pterosaur *Dorygnathus banthensis* (Theodori, 1830). *Special Papers in Palaeontology* 80: 1–64.

———. 2008b. The Early Jurassic pterosaur *Campylognathoides* Strand, 1928. *Special Papers in Palaeontology* 80: 65–107.

Padian, K., and P. E. Olsen. 1984. The fossil trackway *Pteraichnus*: Not pterosaurian, but crocodilian. *Journal of Paleontology* 58: 178–184.

Padian, K. and J.M.V. Rayner. 1993. The wings of pterosaurs. *American Journal of Science* 293: 91–166.

Padian, K., A. J. Ricqlés, and J. R. Horner. 1995. Bone histology determines a new fossil taxon of pterosaur (Reptilia: Archosauria). *Comptes Rendus de l'Académie des Sciences* 320: 77–84.

Padian, K., and M. Smith. 1992. New light on Late Cretaceous pterosaur material from Montana. *Journal of Vertebrate Paleontology* 12: 87–92.

Padian, K., and R. Wild. 1992. Studies of the Liassic Pterosauria, I. The holotype and referred specimens of the Liassic pterosaur *Dorygnathus banthensis* (Theodori) in the Petrefaktensammlung Banz, northern Bavaria. *Palaeontographica Abteilung A* 235: 59–77.

Paganoni, A. 2003. *Eudimorphodon* after 30 years: History of the finding and perspectives. *Rivista del Museo Civico di Scienze Naturali 'E. Caffi'* 22: 47–51.

Palmer, C. and G. J. Dyke. 2010. Biomechanics of the unique pterosaur pteroid. *Proceedings of the Royal Society B* 277: 1121–1127.

Parker, L. R. and J. K. Balsley. 1989. Coal mines as localities for studying dinosaur trace fossils. In *Dinosaur Tracks and Traces*, edited by D. D. Gillette and M. G. Lockley, 353–359. Cambridge, U.K., Cambridge University Press.

Paul, G. S. 1987. Pterodactyl habits – real and radio controlled. *Nature* 328: 481.

———. 1991. The many myths, some old, some new, of dinosaurology. *Modern Geology* 16: 69–99.

———. 2002. *Dinosaurs of the Air: The Evolution and Loss of Flight in Dinosaurs and Birds*. Baltimore, John Hopkins University Press.

Pennycuick, C. J. 1971. Gliding flight of the white-backed vulture *Gyps africanus*. *Journal of Experimental Biology* 55: 13–38.

———. 1990. Predicting wingbeat frequency and wavelength of birds. *Journal of Experimental Biology* 150: 171–185.

———. 1998. Computer simulation of fat and muscle burn in long-distance bird migration. *Journal of Theoretical Biology* 191: 47–61.

Pereda Suberbiola, X., N. Bardet, S. Jouve, M. Iarochéne, B. Bouya, and M. Amaghzaz. 2003. A new azhdarchid pterosaur from the Late Cretaceous phosphates of Morocco. In *Evolution and Palaeobiology of Pterosaurs*, edited by E. Buffetaut and J. M. Mazin, 79–90. Geological Society Special Publication 217. London, Geological Society.

Persons, W. S., IV. 2010. Convergent evolution of prezygapophysis/chevron rods in the tails of pterosaurs and dromaeosaurs and the parallel pattern of caudal muscle reduction in pterosaurs and birds. *Acta Geoscientica Sinica* 31(1): 54.

Peters, D. 2000. A reexamination of four prolacertiforms with implications for pterosaur phylogenies. *Rivista Italiana di Paleontologia e Stratigrafia* 106: 293–336.

———. 2008. The origin and radiation of the Pterosauria. In *Flugsaurier: The Wellnhofer Pterosaur Meeting, Munich, Abstract Volume* edited by D. Hone, 27–28. Munich, Bavarian State Collection for Palaeontology.

———. 2009. A reinterpretation of pteroid articulation in pterosaurs. *Journal of Vertebrate Paleontology* 29: 1327–1330.

Pinheiro, F. L., D. C. Fortier, C. L. Schultz, J.A.F.G. Andrade, and R.A.M. Bantim. 2011. New information on the pterosaur *Tupandactylus imperator*, with comments on the relationships of Tapejaridae. *Acta Palaeontologica Polonica* 56: 567–580.

Plieninger, F. 1901. Beiträge zur Kenntnis der Flugsaurier. *Palaeontographica* 48: 65–90.

Ponomarenko, A. G. 1976. The new insect from the Cretaceous of Transbaicalia a probable parasite of pterosaurien. [In Russian.] *Paleontologicheskii Zhurnal* 3: 102–106.

Prange, H. D., J. F. Anderson, and H. Rahn. 1979. Scaling of skeletal mass to body mass in birds and mammals. *American Naturalist* 113: 103–122.

Prentice, K. C., M. Ruta, and M. J. Benton. 2011. Evolution of morphological disparity in pterosaurs. *Journal of Systematic Palaeontology* 9: 337–353.

Price, L. I. 1953. A presença de Pterosáuria no Cretáceo superior do Estada da Paraiba. *Divisão de Geologia e Mineralogia Notas Preliminares e Estudos* 71: 1–10.

———. 1971. A presença de Pterosauria no Cretêceo Inferior de Chapada do Araripe, Brasil. *Anais da Academia Brasileira de Ciências* 43: 452–461.

Prieto, I. R. 1998. Morfología funcional y hábitos alimentarios de *Quetzalcoatlus* (Pterosauria) [Functional morphology and feeding habits of Quetzalcoatlus (Pterosauria)]. *Coloquios de Paleontologia* 49: 129–144.

Prondvai, E., and D.W.E. Hone. 2009. New models for the wing extension in pterosaurs. *Historical Biology* 20: 237–254.

Quenstedt, F. A. 1855. *Über* Pterodactylus suevicus *im lithographischen Schiefer Wüttembergs*. Tübingen, Universität Tübingen.

Rayner, J.M.V. 1988. Form and function in avian flight. *Current Ornithology* 5: 1–66.

Rayner, J.M.V., P. W. Viscardi, S. Ward, and J. R. Speakman. 2001. Aerodynamics and energetics of intermittent flight in birds. *American Zoologist* 41: 188–204.

Reily, S. M., L. D. McBrayer, and T. D. White. 2001. Prey processing in amniotes: Biomechanical and behavioural patterns of food reduction. *Comparative Biochemistry and Physiology Part A* 128: 397–415.

Riabinin, A. N. 1948. Remarks on a flying reptile from the Jurassic of Kara-Tau. [In Russian.] *Akademii Nauk, Paleontological Institute, Trudy* 15: 86–93.

Rieppel, O. 2008. The relationships of turtles within amniotes. In *Biology of Turtles*, edited by J. Wyneken, M. H. Godfrey, and V. Bels, 345–354. Boca Raton, FL, CRC Press.

Ricqlés, A. de, K. Padian, J. R. Horner, and H. Francillon-Veillot. 2000. Palaeohistology of the bones of pterosaurs (Reptilia, Archosauria): Anatomy, ontogeny, and biomechanical implications. *Zoological Journal of the Linnaean Society* 129: 349–385.

Rodrigues, T., and A.W.A. Kellner. 2008. Review of the pterodactyloid pterosaur *Coloborhynchus*. *Zitteliana* B28: 219–228.

Rodriguez-de la Rosa, R. A. 2003. Pterosaur tracks from the latest Campanian Cerro del Pueblo Formation of southeastern Coahuila, Mexico. In *Evolution and Palaeobiology of Pterosaurs*, edited by E. Buffetaut and J. M. Mazin, 275–282. Geological Society Special Publication, 217. London, Geological Society.

Sánchez-Hernández, B., M. J. Benton, and D. Naish. 2007. Dinosaurs and other fossil vertebrates from the Late Jurassic and Early Cretaceous of the Galve area, NE Spain. *Palaeogeography, Palaeoclimatology, Palaeoecology* 249: 180–215.

Sanz, J. L., L. M. Chiappe, Y. Fernández-Jalvo, F. Ortega, B. Sánchez-Chillón, F. J. Poyato-Ariza, and B. P. Pérez-Moreno. 2001. An Early Cretaceous pellet. *Nature* 409: 998–999.

Sato, K., K. Sakamoto, Y. Watanuki, A. Takahashi, N. Katsumata, C. Bost, and H. Weimerskirch. 2009. Scaling of soaring seabirds and implications for flight abilities of giant pterosaurs. *PLoS ONE* 4: e5400. http://www.plosone.org/article/info%3Adoi%2F10.1371%2Fjournal.pone.0005400.

Sayão, J. M., and A.W.A. Kellner. 2000. Description of a pterosaur rostrum from the Crato Member, Santana Formation (Aptian-Albian) northeastern Brazil. *Boletim do Museu Nacional, Nova Série Geologia* 54: 1–8.

Schutt, W. A., Jr., J. S. Altenbach, H. C. Young, D. M. Cullinane, J. W. Hermanson, F. Muradli, and J.E.A. Bertram. 1997. The dynamics of flight-initiating jumps in the common vampire bat *Desmodus rotundus*. *Journal of Experimental Biology* 200: 3003–3012.

Schweigert, G., G. Dietl, and R. Wild. 2001. Miscellanea aus dem Nusplinger Plattenkalk (Ober-Kimmeridgium, Schwäbische Alb) 3. Ein Speiballen mit Flugsaurierresten. *Jahresberichte und Mitteilungen des Oberrheinischen Geologischen Vereins* 83: 357–364.

Seeley, H. G. 1864. On the pterodactyle as evidence of a new subclass of Vertebrata (Saurornia). *Reports of the British Association of Scientists, 34th Meeting* 1864: 69.

———. 1870. *The Ornithosauria: An Elementary Study of the Bones of Pterodactyles*. Deighton, U.K., Bell.

———. 1887. On a sacrum apparently indicating a new type of bird, *Ornithodesmus cluniculus* Seeley from the Wealden of Brook. *Quarterly Journal of the Geological Society, London* 43: 206–211.

———. 1901. *Dragons of the Air*. London, Meuthuen.

Sereno, P. C. 1991. Basal archosaurs: Phylogenetic relationships and functional implications. *Journal of Vertebrate Paleontology Memoir* 2: 1–53.

Sereno, P. C., and A. B. Arcucci. 1993. Dinosaur precursors from the Middle Triassic of Argentina: *Lagerpeton chanarensis*. *Journal of Vertebrate Paleontology* 13: 385–399.

———. 1994. Dinosaurian precursors from the Middle Triassic of Argentina: *Marasuchus lilloensis*, gen. nov. *Journal of Vertebrate Paleontology* 14: 53–73.

Sharov, A. G. 1971. New flying Mesozoic reptiles from Kazakhstan and Kirgizia. [In Russian.] *Transactions of the Paleontological Institute, Academy of Sciences, USSR* 130: 104–113.

Slack, K. E., C. M. Jones, T. Ando, G.L.(A.) Harrison, R. E. Fordyce, U. Arnason, and D, Penny. 2006. Early penguin fossils, plus mitochondrial genomes, calibrate avian evolution. *Molecular Biology and Evolution* 23: 1144–1155.

Smith, D. K., R. K. Sanders, and K. L. Stadtman. 2004. New material of *Mesadactylus ornithosphyos*, a primitive pterodactyloid pterosaur from the Upper Jurassic of Colorado. *Journal of Vertebrate Paleontology* 24: 850–856.

Soemmerring, S. T. 1812. Uber einen *Ornithocephalus*. *Denkschriften der Akademie der Wissenschaften München, Mathematisch-Physik Klasse* 3: 89–158.

———. 1817. Uber einen *Ornithocephalus brevirostris* der Vorwelt. *Denkschriften der Akademie der Wissenschaften München, Mathematisch-Physik Klasse* 6: 89–104.

Spoor, C. F., and D. M. Badoux. 1986. Descriptive and functional myology of the neck and forelimb of the striped hyena (*Hyaena hyaena*, L. 1758). *Anatomischer Anzeiger* 161: 375–387.

Stecher, R. 2008. A new Triassic pterosaur from Switzerland (Central Austroalpine; Grisons), *Raeticodactylus filisurensis* gen. et sp. nov. *Swiss Journal of Geosciences* 101: 185–201.

Steel, L. 2008. The palaeohistology of pterosaur bone: An overview. *Zitteliana* B28: 109–125.

———. 2010. The pterosaur collection at the Natural History Museum, London, UK: History, overview, recent curatorial developments and exciting new finds. Abstract. *Acta Geoscientica Sinica* 31(1): 59–61.

Steel, L., D. M. Martill, D. M. Unwin, and J. D. Winch. 2005. A new pterodactyloid pterosaur from the Wessex Formation (Lower Cretaceous) of the Isle of Wight, England. *Cretaceous Research* 26: 686–698.

Stein, R. S. 1975. Dynamic analysis of *Pteranodon ingens*: A reptilian adaptation to flight. *Journal of Paleontology* 49: 534–548.

Stein, B. R. 1989. Bone density and adaptation in semiaquatic mammals. *Journal of Mammalogy* 70: 467–476.

Stieler, C. 1922. Neuer Rekonstrucktionsversuch eines liassischen Flugsaueriers. *Naturwissenschaftliche Wochenschrift* 20: 273–280.

Stokes, W. L. 1957. Pterodactyl tracks from the Morrison Formation. *Journal of Paleontology* 31: 952–954.

Stokes, W. L., and J. H. Madsen, Jr. 1979. Environmental significance of pterosaur tracks in the Navajo Sandstone (Jurassic) Grand County, Utah. *Brigham Young University Studies* 26: 21–26.

Sullivan, R. M., and D. Fowler. 2011. *Navajodactylus boerei*, n. gen., n. sp. (Pterosauria, ?Azhdarchidae) from the Upper Cretaceous Kirtland Formation (Upper Campanian) of New Mexico. *New Mexico Museum of Natural History and Science Bulletin* 53: 393–404.

Sweetman, S. C., and D. M. Martill. 2010. Pterosaurs of the Wessex Formation (Early Cretaceous, Barremian) of the Isle of Wight, southern England: A review with new data. *Journal of Iberian Geology* 36: 225–242.

Swennen, C., and T. Yu. 2004. Notes on feeding structures of the black-faced spoonbill *Platalea minor*. *Ornithological Science* 3: 119–124.

Taquet, P. 1972. Un crâne de *Ctenochasma* (Pterodactyloidea) du Portlandien inférieur de la Haute-Marne, dans les collections du Musée de St-Dizier. *Comptes Rendus de l'Académie des Sciences* 174: 362–364.

Taquet, P., and K. Padian. 2004. The earliest known restoration of a pterosaur and the philosophical origins of Cuvier's *Ossemens Fossiles*. *Comptes Rendus. Palaevol* 3: 157–175.

Taylor, M. P., M. J. Wedel, and D. Naish. 2009. Head and neck posture in sauropod dinosaurs inferred from extant animals. *Acta Palaeontologica Polonica* 54: 213–220.

Tischlinger, H. 2001. Bemerkungen zur Insekten-Taphonomie der Solnhofen Plattenkalke. *Archaeopteryx* 19: 29–44.

———. 2010. Pterosaurs of the "Solnhofen" Limestone: New discoveries and the impact of changing quarrying practises. *Acta Geoscientica Sinica* 31(1): 76–78.

Tischlinger, H., and E. Frey. 2010. Multilayered is not enough! New soft tissue structures in the *Rhamphorhynchus* flight membrane. Abstract. *Acta Geoscientica Sinica* 31(1): 64.

Tomkins, J. L., N. R. LeBas, M. P. Witton, D. M. Martill, and S. Humphries. 2010. Positive allometry and the prehistory of sexual selection. *American Naturalist* 176: 141–148.

Tütken, T. and D.W.E. Hone. 2010. The ecology of pterosaurs based on carbon and oxygen isotope analysis. *Acta Geoscientica Sinica* 31(1): 65–67.

Unwin, D. M. 1987. Pterosaur locomotion: Joggers or waddlers? *Nature* 327: 13–14.

———. 1988a. New pterosaurs from Brazil. *Nature* 332: 398–399.

———. 1988b. New remains of the pterosaur *Dimorphodon* (Pterosauria: Rhamphorhynchoidea) and the terrestrial ability of early pterosaurs. *Modern Geology* 13: 57–68.

———. 1988c. A new pterosaur from the Kimmeridge Clay of Kimmeridge, Dorset. *Proceedings of the Dorset Natural History and Archaeological Society* 109: 150–153.

———. 1989. A predictive method for the identification of vertebrate ichnites and its application to pterosaur tracks. In *Dinosaur Tracks and Traces*, edited by D. D. Gillette and M. G. Lockley, 259–274. Cambridge, U.K., Cambridge University Press.

———. 1992. The phylogeny of the Pterosauria. *Journal of Vertebrate Paleontology* 12: 57A.

———. 1996. The fossil record of Middle Jurassic Pterosaurs. In *The Continental Jurassic*, edited by M. Morales,. *Museum of Northern Arizona Bulletin* 60: 291–304.

———. 1997. Pterosaur tracks and the terrestrial ability of pterosaurs. *Lethaia* 29: 373–386.

———. 1999. Pterosaurs: Back to the traditional model? *Trends in Ecology and Evolution* 14: 263–268.

———. 2001. An overview of the pterosaur assemblage from the Cambridge Greensand (Cretaceous) of Eastern England. *Mitteilungen Museum für Naturkunde Berlin, Geowissenschaftliche* 4: 189–221.

———. 2002. On the systematic relationships of *Cearadactylus atrox*, an enigmatic Early Cretaceous pterosaur from the Santana Formation of Brazil. *Mitteilungen Museum für Naturkunde Berlin, Geowissenschaftliche* 5: 239–263.

———. 2003. On the phylogeny and evolutionary history of pterosaurs. In *Evolution and Palaeobiology of Pterosaurs*, edited by E. Buffetaut and J. M. Mazin, 139–190. Geological Society Special Publication 217. London, Geological Society.

———. 2005. *The Pterosaurs from Deep Time*. New York, Pi Press.

———. 2011. A new dimorphodontid pterosaur from the Lower Jurassic of Dorset, southern England. Abstract. In *Programme and Abstracts,* Symposium of Vertebrate Palaeontology and Comparative Anatomy, 59th Annual Meeting, 2011, 4. Lyme Regis, U.K.

Unwin, D. M., V. R. Alifanov, and M. J. Benton. 2000. Enigmatic small reptiles from the Middle-Late Triassic of Kirgizstan. In *The Age of Dinosaurs in Russia and Mongolia*, edited by M. J. Benton, M. A. Shishkin, D. M. Unwin, and E. N. Kurochkin, 177–186. Cambridge, U.K., Cambridge University Press.

Unwin, D. M., and N. N. Bakhurina. 1994. *Sordes pilosus* and the nature of the pterosaur flight apparatus. *Nature* 371: 62–64.

———. 2000. Pterosaurs from Russia, Middle Asia and Mongolia. In *The Age of Dinosaurs in Russia and Mongolia*, edited by M. J. Benton, M. A. Shiskin, D. M. Unwin, and E. N. Kurochkin, 420–433. Cambridge, U.K., Cambridge University Press.

Unwin, D. M., and D. C. Deeming. 2008. Pterosaur eggshell structure and its implications for pterosaur reproductive biology. *Zitteliana* B28: 19–207.

Unwin, D. M., E. Frey, D. M. Martill, J. B. Clarke, and J. Reiss. 1996. On the nature of the pteroid in pterosaurs. *Proceedings of the Royal Society B* 263: 45–52.

Unwin, D. M., and W. D. Heinrich. 1999. On a pterosaur jaw from the Upper Jurassic of Tendaguru (Tanzania). *Mitteilungen aus dem Museum für Naturkunde in Berlin Geowissenschaftliche* 2: 121–134.

Unwin, D. M., and D. Henderson. 1999. Testing the terrestrial ability of pterosaurs with computer-based methods. *Journal of Vertebrate Paleontology* 19: 81A.

Unwin, D. M., and J. Lü. 1997. On *Zhejiangopterus* and the relationships of pterodactyloid pterosaurs. *Historical Biology* 12: 199–210.

———. 2010. *Darwinopterus* and its implications for pterosaur phylogeny. *Acta Geoscientica Sinica* 31(1): 68–69.

Unwin, D. M., J. Lü, and N. N. Bakhurina. 2000. On the systematic and stratigraphic significance of pterosaurs from the Lower Cretaceous Yixian Formation (Jehol Group) of Liaoning, China. *Mitteilungen aus dem Museum für Naturkunde in Berlin Geowissenschaftliche* 3: 181–206.

Unwin, D. M., and D. M. Martill. 2007. Pterosaurs from the Crato Formation. In *Window into an Ancient World: The Crato Fossil Beds of Brazil*, edited by D. M. Martill, G. Bechly, and R. F. Loveridge. Cambridge, U.K., Cambridge University Press.

Unwin, D. M., X. Wang, and M. Xi. 2008. How the Moon Goddess, Chang-e, helped us to understand pterosaur evolutionary history. Abstract. In *Programme and Abstracts,* Symposium of Vertebrate Palaeontology and Comparative Anatomy, 56th Annual Meeting, 2008, edited by G. Dyke, D. Naish, and M. Parkes, 55–56. Dublin.

Veldmeijer, A. J. 2003. Description of *Coloborhynchus spielbergi* sp. nov. (Pterodactyloidea) from the Albian (Lower Cretaceous) of Brazil. *Scripta Geologica* 125: 35–139.

Veldmeijer, A. J., M. Signore, and H.J.M. Meijer. 2005. Description of two pterosaur (Pterodactyloidea) mandibles from the Lower Cretaceous Santana Formation, Brazil. *Deinsea* 11: 67–86.

Veldmeijer, A. J., M. Signore, and E. Bucci. 2006. Predator-prey interaction of Brazilian Cretaceous toothed pterosaurs: A case example. In *Predation in Organisms – A Distinct Phenomenon*, edited by A.M.T. Elewa, 295–308. Berlin, Springer-Verlag.

Veldmeijer, A.J., H.J.M. Meijer, and M. Signore. 2009. Description of pterosaurian (Pterodactyloidea: Anhangueridae, *Brasileodactylus*) remains from the Lower Cretaceous of Brazil. *Deinsea* 13: 9–40.

Vidal, P. P., W. Graf, and A. Berthoz. 1986. The orientation of the cervical vertebral column in unrestrained awake animals. *Experimental Brain Research* 61: 549–559.

Vidarte, C. F., and M. Calvo. 2010. Un nuevo pterosaurio (Pterodactyloidea) en el Cretácico Inferior de La Rioja (España). *Boletín Geológico y Minero* 121: 311–328.

Vila Nova, B. C., A.W.A. Kellner, and J. M. Sayão. 2010. Short note on the phylogenetic position of *Cearadactylus atrox,* and comments regarding its relationships to other pterosaurs. *Acta Geoscientica Sinica* 31(1): 73–75.

Vršanský, P., D. Ren, and C. Shih. 2010. Nakridletia ord. n. – enigmatic insect parasites support sociality and endothermy of pterosaurs. *Amba Projekty* 8: 1–16.

Vull, R., E. Buffetaut, and M. J. Everhart. 2012. Reappraisal of *Gwawinapterus beardi* from the Late Cretaceous of Canada: A saurodontid fish, not a pterosaur. *Journal of Vertebrate Paleontology* 32:1198–1201.

Vullo, R., J. Marugán-Lobónm, A.W.A. Kellner, A. Buscalioni, M. Fuente, and J. J. Moratalla. 2012. A new crested pterosaur from the Early Cretaceous of Spain: The first European tapejarid (Pterodactyloidea: Azhdarchoidea). *PLoS ONE* 7: e38900. http://www.plosone.org/article/info%3Adoi%2F10.1371%2Fjournal.pone.0038900.

Wagler, J. G. 1830. *Natürliches System der Amphibien*. Munich, Stuttgart, and Tübingen, J. G. Cotta.

Wagner, A. 1861. Neue Beiträge zur Kenntnis d. urwltl. Fauna des lithograph. Schiefers. 2. Abt. Schildkröten und Saurier. *Abhandlungen der königlichen bayerischen Akademie der Wissenschaften* 9: 67–124.

Wall, W. P. 1983. The correlation between high limb-bone density and aquatic habits in recent mammals. *Journal of Paleontology* 57: 197–207.

Wang, X., D. A. Campos, Z. Zhou, and A.W.A. Kellner. 2008. A primitive istiodactylid pterosaur (Pterodactyloidea) from the Jiufotang Formation (Early Cretaceous), northeast China. *Zootaxa* 18: 1–18.

Wang, S., A.W.A. Kellner, S. Jiang, and X. Cheng. 2012. New toothed flying reptile from Asia: Close similarities between early Cretaceous pterosaur faunas from China and Brazil. *Naturwissenschaften* 99: 249–257.

Wang, X., A.W.A. Kellner, S. Jiang, X. Cheng, X. Meng,

and T. Rodrigues. 2010. New long-tailed pterosaurs (Wukongopteridae) from western Liaoning, China. *Anais da Academia Brasileira de Ciências* 82: 1045–1062.

Wang, X., A.W.A. Kellner, S. Jiang, and X. Meng. 2009. An unusual long-tailed pterosaur with elongated neck from western Liaoning of China. *Anais da Academia Brasileira de Ciências* 81: 793–812.

Wang, X., A.W.A. Kellner, Z. Zhou, and D. A. Campos. 2005. Pterosaur diversity and faunal turnover in Cretaceous terrestrial ecosystems in China. *Nature* 437: 875–879.

———. 2007. A new pterosaur (Ctenochasmatidae, Archaeopterodactyloidea) from the Lower Cretaceous Yixian Formation of China. *Cretaceous Research* 28: 245–260.

———. 2008. Discovery of a rare arboreal forest-dwelling flying reptile (Pterosauria, Pterodactyloidea) from China. *Proceedings of the National Academy of Sciences*105: 1983–1987.

Wang, L., L. Li, Y. Duan, and S. L. Cheng. 2006. A new iodactylid [*sic*] pterosaur from western Liaoning, China. *Geological Bulletin of China* 25: 737–740.

Wang, X., and J. Lü. 2001. Discovery of a pterodactylid pterosaur from the Yixian Formation of western Liaoning, China. *Chinese Science Bulletin* 46: 1112–1117.

Wang, X., and Z. Zhou. 2003a. Two new pterodactyloid pterosaurs from the early Cretaceous Jiufotang Formation of Western Liaoning, China. *Vertebrata PalAsiatica* 41: 34–41.

———. 2003b. A new pterosaur (Pterodactyloidea, Tapejaridae) from the Early Cretaceous Jiufotang Formation of Western Liaoning, China and its implications for biostratigraphy. *Chinese Science Bulletin* 47: 15–21.

———. 2004. Pterosaur embryo from the Early Cretaceous. *Nature* 429: 621.

———. 2006a. Pterosaur assemblages of the Jehol Biota and their implication for the Early Cretaceous pterosaur radiation. *Geological Journal* 41: 405–418.

———. 2006b. Pterosaur adaptive radiation of the Early Cretaceous Jehol Biota. In *Originations, Radiations and Biodiversity changes – Evidences [sic] from the Chinese Fossil Record*, edited by J. Rong, Z. Fang, Z. Zhou, R. Zhan, X. Wang, and X. Yuan, 665–689, 937–938. Beijing, Beijing Science Press.

Wang, X., Z. Zhou, F. Zhang, and X. Xu. 2002. A nearly completely articulated rhamphorhynchoid pterosaur with exceptionally well-preserved wing membranes and "hairs" from Inner Mongolia, northeast China. *Chinese Science Bulletin* 47: 226–230.

Wedel, M. J. 2003. Vertebral pneumaticity, air sacs, and the physiology of sauropod dinosaurs. *Paleobiology* 29: 243–255.

Wellnhofer, P. 1970. Die Pterodactyloidea (Pterosauria) der Oberjura-Plattenkalke Süddeutschlands. *Bayerische Akademie der Wissenschaften, Mathematisch- Wissenschaftlichen Klasse, Abhandlungen* 141: 1–133.

———. 1974. *Campylognathoides liasicus* (Quenstedt), an Upper Jurassic pterosaur from Holzmaden – the Pittsburgh specimen. *Annals of the Carnegie Museum* 45: 5–34.

———. 1975. Die Rhamphorhynchoidea (Pterosauria) der Oberjura-Plattenkalke Süddeutschlands. *Palaeontographica A* 148: 1–33, 132–186; 149: 1–30.

———. 1978. *Handbuch der Paläoherpetologie. Teil 19: Pterosauria*. Stuttgart, Gustav Fischer Verlag.

———. 1980. Flugaurierreste aus der Gosau-Kriede von Muthmannsdorf (Niederösterreich – ein Beitrag zue Kiefermechanik der Pterosaurier. *Mitteilungen der Bayerischen Staatssammlung für Paläontologie und Historische Geologie* 20: 20–112.

———. 1985. Neue pterosaurier aus der Santana-Formation (Apt) der Chapada do Araripe, Brasilien. *Palaeontographica. Abteilung A* 187: 105–182.

———. 1987a. New crested pterosaurs from the Lower Cretaceous of Brazil. *Mitteilungen der Bayerischen Staatssammlung für Paläontologie und Historische Geologie* 27: 175–186.

———. 1987b. Die Flughaut von *Pterodactylus* (Reptilia, Pterosauria) am Beispiel des Weiner Exemplares von *Pterodactylus kochi* (Wagner). *Annalen des Naturhistorischen Museums in Wein* 88: 149–162.

———. 1988. Terrestrial locomotion in pterosaurs. *Historical Biology* 1: 3–16.

———. 1991a. *The Illustrated Encyclopaedia of Pterosaurs*. London, Salamander Books.

———. 1991b. Weitere pterosaurierfunde aus der Santana-Formation (Apt) der Chapada do Araripe, Brasilien [Additional pterosaur remains from the Santana Formation (Aptian) of the Chapada do Araripe, Brazil]. *Palaeontographica Abteilung A* 215: 43–101.

———. 2003. A Late Triassic pterosaur from the Northern Calcareous Alps (Tyrol, Austria). In *Evolution and Palaeobiology of Pterosaurs*, edited by E. Buffetaut and J. M. Mazin, 5–22. Geological Society Special Publication 217. London, Geological Society.

———. 2008. A short history of pterosaur research. *Zitteliana* B28: 7–19.

Wellnhofer, P., and E. Buffetaut. 1999. Pterosaur remains from the Cretaceous of Morocco. *Palaontologische Zeitschrift* 73: 133–142.

Wellnhofer, P., and A.W.A. Kellner. 1991. The skull of *Tapejara wellnhoferi* Kellner (Reptilia, Pterosauria) from the Lower Cretaceous Santana Formation of the Araripe Basin, Northeastern Brazil. *Mitteilungen der Bayerischen Staatssammlung für Paläontologie und Historische Geologie* 31: 89–106.

Wellnhofer, P., and B. W. Vahldiek. 1986. Ein Flugsaurier-Rest aus dem Posidonienschiefer (Unter-Toarcium) von Schandelah bei Braunschweig. *Paläontologische Zeitschrift* 60: 329–340.

Wild, R. 1971. *Dorygnathus mistelgauensis* n. sp., ein neuer Flugsaurier aus fem Lias Epsilon von Mistelgau

(Fränkischer Jura). *Geologisches Blätter, NO-Bayern* 21: 178–195.
———. 1978. Die Flugsaurier (Reptilia, Pterosauria) aus der Oberen Trias von Cene bei Bergamo, Italien. *Bollettino Società Paleontologica Italiana* 17: 176–256.
———. 1983. Über den Ursprung der Flugsaurier. In *Erwin Rutte-Festschrift. , Weltenburger Akademie*, 231–238. Kelheim/Weltenburg.
———. 1984a. Flugsaurier aus der Obertrias von Italien. *Naturwissenschaften* 71: 1–11.
———. 1984b. A new pterosaur (Reptilia, Pterosauria) from the Upper Triassic (Norian) of Friuli, Italy. *Gortania – Atti Museo Friulano di Storia Naturale* 5: 45–62.
———. 1994. A juvenile specimen of *Eudimorphodon ranzii* Zambelli (Reptilia, Pterosauria) from the Upper Triassic (Norian) of Bergamo. *Rivista del Museo Civico di Scienze Naturali 'E. caffi' di Bergamo* 16: 95–120.
Williston, S. W. 1897. Restoration of *Ornithostoma* (*Pteranodon*). *Kansas University Quarterly* 6: 35–51.
———. 1902a. On the skull of *Nyctodactylus*, an Upper Cretaceous pterodactyl. *Journal of Geology* 10: 520–531.
———. 1902b. On the skeleton of *Nyctodactylus*, with restoration. *American Journal of Anatomy* 1: 297–305.
———. 1903. On the osteology of *Nyctosaurus* (*Nyctodactylus*). With notes on American pterosaurs. *Field Columbian Museum Publications, Geological Series* 2: 125–163.
Wilkinson, M. T. 2008. Three dimensional geometry of a pterosaur wing skeleton, and its implications for aerial and terrestrial locomotion. *Zoological Journal of Linnaean Society* 154: 27–69.
Wilkinson, M. T., D. M. Unwin, and C. P. Ellington. 2006. High-lift function of the pteroid bone and forewing of pterosaurs. *Proceedings of the Royal Society B* 273: 119–126.
Wiman, C. 1925. Über *Pterodactylus* Westmani und andere Flugsaurier. *Bulletin of the Geological Institute of the University of Uppsala* 20: 1–38.
Witmer, L. M. 1997. The evolution of the antorbital cavity of archosaurs: A study in soft-tissue reconstruction in the fossil record with an analysis of the function of pneumaticity. *Journal of Vertebrate Paleontology* 17: 1–73.
Witmer, L. M., S. Chatterjee, J. Franzosa, and T. Rowe. 2003. Neuroanatomy of flying reptiles and implications for flight, posture and behaviour. *Nature* 425: 950–953.
Witton, M. P. 2007. Titans of the skies: azhdarchid pterosaurs. *Geology Today* 23: 33–38.
———. 2008a. A new approach to determining pterosaur body mass and its implications for pterosaur flight. *Zitteliana* B28: 143–159.
———. 2008b. The palaeoecology and diversity of pterosaurs. PhD thesis, University of Portsmouth, U.K.
———. 2008c. A new azhdarchoid pterosaur from the Crato Formation (Lower Cretaceous, Aptian?) of Brazil. *Palaeontology* 51: 1289–1300.
———. 2009. A new species of *Tupuxuara* (Thalassodromidae, Azhdarchoidea) from the Lower Cretaceous Santana Formation of Brazil, with a note on the nomenclature of Thalassodromidae. *Cretaceous Research* 30: 1293–1300.
———. 2010. *Pteranodon* and beyond: The history of giant pterosaurs from 1870 onward. In *Dinosaurs and Other Extinct Saurians: A Historical Perspective*, edited by R.T.J. Moody, E. Buffetaut, E. Naish, and D. M. Martill, 313–323. Geological Society Special Publication 310. London, Geological Society.
———. 2011. The pectoral girdle of *Dimorphodon macronyx* (Pterosauria, Dimorphodontidae) and the terrestrial abilities of non-pterodactyloid pterosaurs. Abstract. In *Programme and Abstracts,* Symposium of Vertebrate Palaeontology and Comparative Anatomy, 59th Annual Meeting, 2011, 22. Lyme Regis, U.K.
———. 2012. New insights into the skull of *Istiodactylus latidens* (Ornithocheiroidea, Pterodactyloidea). *PLoS ONE* 7: e33170. http://www.plosone.org/article/info%3Adoi%2F10.1371%2Fjournal.pone.0033170.
Witton, M. P., and M. B. Habib. 2010. On the size and flight diversity of giant pterosaurs, the use of birds as pterosaur analogues and comments on pterosaur flightlessness. *PLoS ONE* 5: e13982. http://www.plosone.org/article/info%3Adoi%2F10.1371%2Fjournal.pone.0013982.
Witton, M. P., D. M. Martill, and M. Green. 2009. On pterodactyloid diversity in the British Wealden (Lower Cretaceous) and a reappraisal of *"Palaeornis" cliftii* Mantell, 1844. *Cretaceous Research* 30: 676–686.
Witton, M. P., D. M. Martill, and R. F. Loveridge. 2010. Clipping the wings of giant pterosaurs: Comments on wingspan estimations and diversity. *Acta Geoscientica Sinica* 31(1): 79–81.
Witton, M. P. and D. Naish. 2008. A reappraisal of azhdarchid pterosaur functional morphology and paleoecology. *PLoS ONE* 3: e2271. http://www.plosone.org/article/info%3Adoi%2F10.1371%2Fjournal.pone.0002271.
Wolff, E.D.S., S. W. Salisbury, J. R. Horner, and D. J. Varricchio. 2009. Common avian infection plagued the tyrant dinosaurs. *PLoS ONE* 4: e7288. http://www.plosone.org/article/info%3Adoi%2F10.1371%2Fjournal.pone.0007288.
Wourms, M. K., and F. E. Wasserman. 1985. Bird predation on Lepidoptera and the reliability of beak-marks in determining predation pressure. *Journal of the Lepidopterists' Society* 39: 239–261.
Wright, J. L., D. M. Unwin, M. G. Lockley, and E. C. Rainforth. 1997. Pterosaur tracks from the Purbeck Limestone Formation of Dorset, England. *Proceedings of the Royal Society* 108: 39–48.

Xing, L., J. Wu, Y. Ly, J. Lü, and Q. Ji. 2009. Aerodynamic characteristics of the crest with membrane attachment on Cretaceous pterodactyloid *Nyctosaurus*. *Acta Geologica Sinica* 83: 25–32.

Young, C. C. 1964. On a new pterosaurian from Sinkiang, China. *Vertebrate Palasiatica* 8: 221–225.

———. 1973. Pterosaurian fauna from Wuerho, Sinkiang. *Reports of Paleontological Expedition to Sinkiang II*. 18–34. Nanjing, Kexue Chubanshe.

Zambelli, R. 1973. *Eudimorphodon ranzii* gen. nov., sp. nov., uno Pterosauro Triassico (nota preliminare). *Rendiconti Istituto Lombardo Accademia B* 107: 27–32.

Zhang, J., D. Li, M. Li, M. G. Lockley, and Z. Bai. 2006. Diverse dinosaur-, pterosaur-, and bird-track assemblages from the Hakou Formation, Lower Cretaceous of Gansu Province, northwest China. *Cretaceous Research* 27: 44–55.

Zheng, X. T., L. H. Hou, X. Xu, and Z. Dong. 2009. An Early Cretaceous heterodontosaurid dinosaur with filamentous integumentary structures. *Nature* 458: 333–336.

Zusi, R. L. 1962. *Structural Adaptations of the Head and Neck in the Black Skimmer*, Rynchops nigra *Linnaeus*. Publications of the Nuttal Ornithological Club 3. Cambridge, MA, Nuttal Ornithological Club.

Zweers, G., F. de Long, H. Berkhoudt, and J. C. Vanden Berge. 1995. Filter feeding in flamingos (*Phoenicopterus ruber*). *The Condor* 97: 297–324.

Index

Page numbers in italics refer to images.

acetabulum, *30*, 35
aerial hawking, 19, 56, 61–62, *62*; 84–86, *85*; in anurognathids, 86, 110, 111–12; in azhdarchids, 257; in dimorphodontids, 103; in scaphognathines, 134, in wukongopterids, 141–42
Aetodactylus, *154*, 157
Africa, 259
Agadirichnus, 68
Age of Enlightenment, 7
air sacs, 39, *41*. *See also* skeletal pneumaticity
aktinofibrils, *52*, 53, *53*. *See also* wing membrane
Alanqa saharica, *246*, 248, 251; wingspan, 249
albatross, as analogues of pterosaurs, 132, 179, 255–56
Alberta, 247
Amniota, 13
anapsid, 14
ancestry, of pterosaurs, 12–22
Angustinaripterus longicephalus, *125*, 127
Anhanguera blittersdorfi, *156*
Anhanguera cuvieri, 152, *155*
Anhanguera santanae, *11*; endocast, *11*, *44*; flight, *62*; hindlimb anatomy of, *36*; terrestrial locomotion, *66*, 67;
Anhanguera, *153*, 154–55, *154*, 158, *158*, *162*, 163; wing shape, *61*
Anhangueridae, 152
ankle. *See* tarsals
Anning, Mary, 98
Antarctica, 259
antorbital fenestra, 23, *26*; use in reptile phylogeny, 17
antungual sesamoid, *100*, 101, 119, 122, 130, 133
Anurognathidae, *92*, 104–12, *106*, 140, 148, *261*; as denizens of continental environments, 104; flight, 110; osteology, 107–9; paleoecology, 110, 111–12; position within Pterosauria, 105; resting posture, 110–11, *111*; soft tissues, 109–10; terrestrial competence, 110; wingspans, 107–8
Anurognathus ammoni, 104, *105*, *106*, 107, *107*; anatomy, *108*, 109, 110; diet, 86; flight, *62*, 110; muscle preservation, 44; wing shape, *61*; wingspan, 107
Aralazhdarcho bostobensis, *246*, 248
Arambourgiania philadelphiae, 244, *246*, 254; wingspan, 250–51
Araripe Group, 234
"*Araripesaurus*," 155
Archaeoistiodactylus linglongtaensis, *137*, 138
Archaeopterodactyloidea, 185
Archaeopteryx, 5, 86
Archegetes neuropterum, 84–86, *85*
Archosauria, 14, 16, 18
Archosauriformes, 14–15, 18
Archosauromorpha, 15
Argentina, 187, 205; fossil of, *188*
Arthaber, G. von, 154, *155*
arthritis, 87
arthrology: of acetabulum, 35, 69, 121–22; of ankle, 36; of knee, 35; cervical vertebrae, 28–29, 178, 241; of elbow, 32–33; of glenoid, 31, 69, 71, 122, 162; of wing finger, 35; of wrist, 33–34, 162
Arthurdactylus conandoylei, *154*, 157
aspect ratio: definition, 61; of flying vertebrates, *62*
Asphidorhynchus, 88, 133–34
"*Auroazhdarcho primordius*," 186
Aussiedraco, *154*, 157
Australia, 157, 259
Austria, 116
Austriadactylus cristatus, *115*, 116, *116*; anatomy, 116–19; flight, 121; wingspan, 116
Avemetatarsalia. *See* Ornithodira
Azhdarchidae, *77*, *92*, 158, 179, 195, 228, 232, 234, 241–42, 244–58, *246*, 259, *260*, *261*, 262–63; anatomy, 249–55; as denizens of continental environments, 58, 248–49, 256; flight, *57*, 255–56; osteology, 251–55; pneumaticity, 254; size, 244, 249–51, *250*; terrestrial competence, *65*, *69*, 256–57; paleoecology, 243, 257–58; wingspans, 249–51
Azhdarcho lancicollis, *246*, 248, 251; wingspan, 249
Azhdarchoidea, *92*, 170, 186, 214–15, 208, 216–58, *261*; characteristics of, 216

background extinction, 262–263
Bactracognathus volans, 104, *106*; flight, 110; wingspan, 107–8
Bakonydraco guinazui, *246*, *247*, 248, 251–52; wingspan, 249
Barbosania, *154*, 155
bats, as pterosaur analogues, 1, 3, 61–63, 150; flight, *62*; launch strategy, 59–60
Bavaria, 104
beaks. *See* rhamphothecae
Beipiaopterus chenianus, *185*, 187, 197; embryo, *76*
Bellubrunnus rothgaengeri *125*, 128, 129
Bennett, Christopher, 66–67, 81, 174
Bennettazhia oregonensis, 248
Big Bend National Park, 244
birds: as catalyst for pterosaur extinction, 263; as pterosaur analogues, 1, 3, 61–63; as pterosaur prey, 42, *43*; flight, 60, *62*; launch strategy, 59–60
blood vessels, cranial, 49, 117, 242; wing membranes, *52*, 53, *53*, 132
body mass, of pterosaurs, 56, 58–63 *59*, *62*, 100, 141
"*Bogolubovia orientalis*," 248

bone histology, 37–38, *37*, 77–78
bone strength, 38
bone walls, reinforced, 201, 204; function, 205, 209–10. *See also* expanded joints
Boreopteridae, *92*, 143, 164–169, *166*, 179, 188, *261*; anatomy, 166–67; locomotion, 167, 169; palaeoecology, 169; wingspans, 166
Boreopterus cuiae, 164, *166*, 166; wingspan, 166
bounding, 20, 70–71, 163
brachiopatagium, 52, *52*, 196–197, *196*; attachment site, 54, 66
braincase. *See* endocast
Brasileodactylus, *154*, 155, 157–58
Brazil, 8, 37, 52, 88, 95, 155, 175, 211, 216, 218–19, 227, 228, 230, 234, 237; fossils of, *48*, *50*, *156*, *160*, *218*, *222*, *226*, *237*, *239*, *231*
Britain, 68, 95, 125, 143, 152, 157, 186, 204, 211, 238; fossils of, *84*, *146*, *155*, *194*, *213*
British Columbia, 146
Brown, Barnum, 127
Buckland, Reverend William, 98
burst fliers, 63, 101

Cacibupteryx caribensis, *125*, 128, 130; wingspan, 130
Cambridge Greensand, *93*, 152, 154–155, 157, 172, 211; fossils of, *6*, *155*, *214*
Campylognathoides, *115*; anatomy, 116, 117–21; diet, 122; flight, 121; terrestrial competence, 121–22
"*Campylognathoides indicus*," 113
Campylognathoides liasicus, 113, *115*, *119*; wingspan, 116
Campylognathoides zitteli, 113; wingspan, 116
"campylognathoidids," *92*, 113–22, *115*, 183, 224, 242, *261*; flight, 121; definition, 113; osteology, 116–21; paleoecology, 122; soft tissues, 121; taxonomy of, 113; terrestrial compctence, 121–22; wingspans, 116
"*Campylognathus*," 114, 132
Canada, 88, 143
Capillaria, 87
Carniadactylus rosenfieldi, *115*, 116; anatomy, 116–19; flight, 121; terrestrial competence, 122; wingspan, 116
carpus, anatomy of, 33–35, *33*, *34*
carrion, as food source for pterosaurs, 122, *144*, 150–151, 257
cassowary, 216
Cathayopterus grabaui, *185*, 187
caudal rods, *177*, 178
caudal vertebrae, *177*, anatomy of, 30–31, *31*
Caulkicephalus, *154*, 157–58
Caviramus filisurensis, *114*, 116, *117*, *118*, *119*, *122*
Caviramus schesaplanensis, 116
Caviramus, *115*; anatomy, 116–21; diet, 122; flight, 121; terrestrial competence, 122; wingspan, 116
Cearadactylus, *154*, 155–58
cervical vertebrae, *29*, *210*, *214*, *246*, *253*; anatomy of, 28–29, *29*; ribs, 28, 31; use in reptile phylogeny, 17
Chalk, the, 152, 211; fossils of, *155*, *213*
Changchengopterus pani, 135, *137*, wingspan, 138
"Chang-e," 211
Chaoyangopteridae, *92*, 216, 228–233, *230*, 244, *261*; anatomy, 230–33; locomotion, 233; paleoecology, 233; wingspans, 231
Chaoyangopterus zhangi, 228, 230–231, *230*
chevrons. *See* caudal vertebrae
chewing, 28, *118*, 122
Chile, 205
Chimera, 107
China, 9, *9*, 106, 127, 128, 135, 145–46, 157, 164, 187, 201, 204–5, 211, 216, 219, 222, 227, 228, 230, 248, 259; fossils of, *8*, *10*, *50*, *76*, *140*, *146*, *167*, *208*, *225*, *231*, *247*, *249*
claws, anatomy of, 35–36; sheaths, 48, *48*, 109, 226
climbing, 67, 71–72, *72*, 101–3, 227
Collini, Cosimo, 6–7, 186
Coloborhynchus, 154–55, *154*, 157–58; wingspan, 157
Coloborhynchus clavirostris, *155*
Coloborhynchus spielbergi, *156*, *160*
color, 49, *50*, 81, 128; of anurognathids, 111
Colorado, 128, 205
competitive displacement, 263
computed tomography (CT), 10, *11*
Conan Doyle, Arthur, 157. See also *Lost World, The*
continental soaring, 62–63
Cookie Monster, 110
Cope, Edward Drinker, 95, 172
coracoid. *See* scapulocoracoid
cranial crests: anatomy, 24, *26*, *27*; antler crests, 174–75, *174*, 178; as aerodynamic devices, 79, 81, 179, 197, 226–27, 241; as thermoregulatory devices, 79, 81, 227, 242; as display devices, 79, 227, 242; discovery, 170; function, 79, *80*, 81–82, *82*, 226–27, 242; growth, 79, *80*, 175, 182, 210, 227, 242; soft tissues, 24, *27*, 48, 49, *50*, 128, 131, 139, 196, 225–26; taxonomic use of, 81, 183
Crato Formation, 9, *9*, 82, 157, 218–20, 224–25, 230, 234, 237, 239; fossils of, *48*, *50*, *93*, *226*, *231*
Crayssac, 68
Cretaceous Period: duration of, 3, *93*; extinction event of, 3, 175, 244, 259; and pterosaur record, 259, *261*, *262*. *See also* extinction
"*Criorhynchus*," 154
crossbills, as analogues of pterosaurs, 200
crows, 226
crows, as pterosaur analogues, 101, 134
Ctenochasma, *185*; anatomy, 193–95; diet, 200; flight, *62*, 197; growth, 199; wing shape, *61*
Ctenochasma elegans, 186, *187*, *191*, *194*, *198*
"*Ctenochasma porocristata*," 186
Ctenochasma roemeri, 186
Ctenochasma taqueti, 186
Ctenochasmatidae, 192, 185, *185*
Ctenochasmatoidea, *92*, 127, 145, 156, 164, 166, 183–200, *185*, 214, 224, 251, 257, *261*; diet, 199–200; flight, 197; growth, 197, 199; pneumaticity, 189–190; osteology, 188–95; soft tissues, 195–197; taxonomy, 185, *185*; terrestrial competence, 71, 197; wingspans, 190–91
Cuba, 127, 128

Cuspicephalus scarfi, *137*, 138; wingspan, 138
Cuvier, Baron Georges, 6–7, 13–14, 64, 186
Cycnorhamphus suevicus, *184*, 185, *185*, 186–87, *193*; anatomy, *189*, 194–95; diet, 200; flight, 197; wingspan, 191
Cymatophlebia longialata, 84–86, *85*

Daohugou beds, 106
Darwinopterus, 135, *137*
Darwinopterus linglongtaensis, 135
Darwinopterus modularis, *10*, 135, *136*, *138*, *139*, *140*; reproduction, 135; sexual dimorphism, 81, *82*
Darwinopterus robustodens, 135–36, 140, *140*, 142
Dawndraco kanzai, 174
Dendrorhynchoides curvidentatus, 106, *106*; anatomy, 109; wingspan, 107
dentition, multicusped, 113, 116, *116*, 117–18, *118*; ontogeny, 79; overview, 27–28; overgrown by jaw bones, 28, 207–8; replacement of, 27, 207–8
diapsid, 14
digitigrady, 18, 67
Dimorphodon macronyx, *7*, 95–98, *96*, *97*, *98*, 120; anatomy, 99–101, *99*, *100*; climbing adaptations, *72*; diet, *96*, 103; flight, *62*, 101, 103; terrestrial competence, 66, *66*, 71, 101–3; wing shape, *61*; wingspan, 99
"*Dimorphodon weintraubi*," *97*, 98
Dimorphodontidae, *92*, 95–103, *261*; anatomy, 99–101; discovery, 95, 98–99, *97*; taxonomy, 98; flight of, 101, *102*; paleoecology, 103; terrestrial competence, 101–2; wingspans, 99
dinosaurs, 1, 14, 17, *56*, 78, 98, 170, 227, 248, 254, 259; as pterosaur predators, *75*, 88, *89*, *102*, 151; as pterosaur prey, *235*, 258
directional terms, 23
disease, 87. *See also* pathologies
diversification rates, of pterosaurs, 263
diving, *171*, 182
Döderlein, Ludwig, 104
Domeykodactylus ceciliae, *203*, 205
dorsal vertebrae, anatomy of, 29–30, *30*
Dorygnathus, 113, *125*; anatomy, 128–32; endocast, 131; wingspan, 128–30
Dorygnathus banthensis, 125, *127*
Dorygnathus mistelgauensis, 125
Dorygnathus purdoni, 125
dragonfly, 84–86, *85*
dromaeosaurs, 88
Dsungaripteridae, 201, 205
Dsungaripteroidea, *92*, 185, 191, 201–10, *203*, 224, 254, *261*; anatomy, 205–9; flight, 209; terrestrial competence, *69*, 209–10; nomenclature of, 201; palaeoecology, 210; pneumaticity, 205, 208; wingspans, 205
Dsungaripterus weii, *8*, 201, *202*, *203*, 204–205; anatomy, 205–8; diet, 210; flight, *62*, 209; skull, 205–6, *24*, *208*; terrestrial competence, 209; wingspan, 205; wing shape, *61*
duck-billed pterosaurs, 143, 150
durophagy, 210

eggs, 10, 74, 76–77, *76*, *77*, 135, 141, 157, 187, 197; as pterosaur prey, 258
eggshell histology, 74
Elanodactylus prolatus, *185*, 187, 195; wingspan, 191
embryos, 74, 76–77, *76*, 157, 197, 219
endocast, *11*, 24, 42, 158, 163
Eoazhdarcho liaoxiensis, 230, *230*, 233
Eopteranodon lii, 230–31, *230*, 233
Eosipterus yangi, *185*, 187
esophagus, 41, *42*
Euctenochasmatia, 185
"*Eudimorphodon*" *cromptonellus*, *115*, 116
Eudimorphodon ranzii, *71*, 113, *115*, 116, *116*, *118*, anatomy, 116–19; flight, *62*, 121; gut content, 82; diet, 122; terrestrial competence, 122; wing shape, *61*; wingspan, 116
Euparkeria capensis, 15, *16*
Europejara olcadesorum, *218*, 219, 221
evolution, of pterosaurs, *261*,
evolutionary stasis, 107, 263
excrement, 42
expanded joints, 205, 209–10
extinction, 6–7, 175, 216, 244, 259–63
eye socket. *See* orbit

falcons, as analogues of pterosaurs, 110, 121
Faxinalipterus minima, 95, *97*
feathers, 51
feeding traces, 83–84, *84*, *85*
Feilongus youngi, 164, 166, *185*, 187–88, 192, 199; wingspan, 191
femur, anatomy of, 35, *36*; of dsungaripteroids, *8*, 209
Fenghuangopterus lii, *125*, 128, 130, 135; wingspan, 130
fibrolamellar bone, 38, 77
fifth pedal digit, anatomy, 36–37, *36*; early development and function, 22; as membrane support, 54–55
filter feeding, 42, 150, 199
fish, as pterosaur prey, 1, 41, *42*, 82, *83*,103, 122, 133–134, 150, 163, 169, *171*, 182, 233, 257; as pterosaur predators, 87–88
flap-gliding, 60
flapping. *See* powered flight
flight muscle, use in launch, 58–60
flight, *2*, 56–63, 101, 103, 121, 132–33, 140–41, 149–50, 160–62, 179–80, 197, 209, 215, 226, 241, 255–56; evolution of, 18–22, 56; research history, 9, 56, 62, 179; diversity of, 62–63, *62*
flightless pterosaurs, 59–60, 179, 255
flocculus, 42, 163
foot pads, 47–49, *48*, 226, 256
footprints. *See* trackways
forelimb, anatomy of, 32–25, *32*, *33*, *34*; growth, 79, *79*
forgery, of fossils, 106, 135, 156, 187–88
fossil record, quality of pterosaur, 259, 262–63, *262*
fossilization potential of pterosaur bone, 38
France, 68, 186–87, 205

frigatebirds, as analogues of pterosaurs, 132, 160, 180, 182
frogmouths, as analogues of pterosaurs, 112
fur. *See* pycnofibers

"*Gallodactylus canjuerensis*," 187. See also *Cycnorhamphus suevicus*
gastralia, 30, *30*, 39
gastroliths, 42, 199. See also *Pterodaustro*
geese, as analogues of pterosaurs, 255
Gegepterus changi, *185*, 187, 196, 199
Geosternbergia maiseyi, 174
Geosternbergia sternbergi, 174
Germanodactylus cristatus, *68*, 201, *203*, 205, *206*, *255*; growth, 210, *210*; wingspan, 205
Germanodactylus rhamphastinus, 201, 205; wingspan, 205
Germanodactylus, 185, 201, *203*, 204; anatomy, 205, 207–9;
Germany, 5, 7, 84, 88, *107*, 114, 113, 123, 125, 127–28, 183, 186, 205; fossils of, *10*, *53*, *85*, *115*, *126*, *127*, *187*, *194*, *198*, *203*
giant pterosaurs, 58, 170, 177, 179, 228, 244, 249–51, *250*; flight, 255–56; evolution of, 251; maximum size, 251; propensity to extinction, 262; skulls, 251–53; terrestrial locomotion, *67*, 67, 69
giraffe, 250
gizzard, 42. *See also* gastroliths
Gladocephaloideus jingangshanensis, *185*, 187
glenoid, anatomy, 31
gliding flight, 19, 22
Gnathosaurus macrurus, 186
Gnathosaurus subulatus, 186, *194*
Gnathosaurus, *185*; anatomy, 192–94; diet, 199–200
Greenland, 116
griffins, 13
ground hornbills, as analogues of pterosaurs, 258
grouse, as pterosaur analogues, 101
growth, 77–79, *78*, *80*, 130, *132*, 133–34, 163, 164, 166, 174, 181–82, 197, 199, 210, 219, 227; determinate, 78; allometry, 78
guano, 42
Guidraco, *154*, 157–58
gular pouches, 196
gulls, as pterosaur analogues, 106, 123, 132
gut content of pterosaurs: birds, 42, *43*; fish, 41, *42*, 82, *83*, 122, 133, 182; plants, 82, *156*
Gwawinapterus beardi, 147

Haenamichnus, 256
Haenamichnus sp., *65*, *67*, 68
Haenamichnus uhangrensis, *256*
hanging, from feet, *8*, 64, 72, 150
Haopterus gracilis, 145, *145*; anatomy, 148–49
Harpactognathus gentryii, *125*, 128, 131; wingspan, 130
Hatzegopteryx thambema, *245*, *246*, *247*, 248, *250*; anatomy, 251–53, *252*; wingspan, 250–51
headcrest. *See* cranial crest
Herbstosaurus pigmaeus, *203*, 204
heterothermy, 51
high speed wings, 121
hindlimb, anatomy of, 35–37, *36*
homeothermy, 51
Hongshanopterus lacustris, *145*, 146, 148
Hooley, Reginald Walter, 145, 155, 211
horned dinosaurs, 253
Huanhepterus quingyangensis, *185*, 187–88, 192, 195, 199; flight, *62*; wing shape, *61*; wingspan, 191
"*Huaxiapterus*," *218*, 219, 224–225; flight, *62*, 226
"*Huaxiapterus*" *benxiensis*, 219, *220*, *222*, *255*
"*Huaxiapterus*" *corallatus*, 219, *220*; wingspan, 221
"*Huaxiapterus*" jii. See *Sinopterus jii*
humerus, *79*, *160*, *213*, *246*; anatomy of, 32–33; deltopectoral crest, definition, 32
Hungary, 248; fossils of, *247*
Hypothetical Pterosaur Ancestor (HyPtA), *12*, 19–22, *20*; HyPtA A, 20; HyPtA B, 20, 22; HyPtA C, *12*, 22; HyPtA D, *21*, 22.

ilium, anatomy of, *30*, 35
injuries. *See* pathologies
insects: development of insect flight, 56; as pterosaur prey, 19, 56, 61–62, 84–86, *85*, *105*, 110, 111–12, 122, 134, 142; role in evolution of pterosaur flight, 19, 56
integument, 46–51
Irritator, *75*, 88
ischium, anatomy of, *30*, 35
Istiodactylidae, *91*, *92*, 138, 143–51, *145*, 152, 167, 179, *261*; anatomy, 147–49; as denizens of continental environments, 150; flight, *91*, 149–51; terrestrial competence, 150; paleoecology, 151; pneumaticity, 148; taxonomy, 147; wingspans, 147
Istiodactylus, 145, *145*, wing shape, *61*
Istiodactylus latidens, *144*, 145, *146*, 147; anatomy, 147–48, *149*; diet, 151, *151*; flight, 149; terrestrial competence, 150; wingspan, 147
Istiodactylus sinensis, 146–47, *146*; wingspan, 147
Italy, *9*, 18, 87, 95, 113, 116; fossils of *115*, 116

Japan, 205
"Javelina azhdarchid," 247, *247*, 251, 238; wingspan, 249
Javelina Formation, 244, 246–48
jaw joint, 26
Jeholopterus ningchenensis, 106–7; flight, 110; soft-tissue anatomy, 107, 110; wingspan, 108
Jianchangopterus zhaoianus, *125*, 128, 130, 138
Jidapterus edentus, 230–31, *230*
Jiufotang Formation, *9*, 145–47, 157, 211, 219, 228, 230, 233; fossils of, *93*, *146*, *147*, *225*, *231*
Jordan, 244, fossils of, *246*
Jurassic Coast, 98
Jurassic Period, duration, 3, *93*; pterosaur record, 259, *261*, *262*

Kansas, 173; fossils of, *33, 34, 173, 174, 177*
Kazakhstan, 104, 128
Kellner, Alexander, 216
Kem Kem beds, 219, 248
"Kem Kem tapejarid," 219; wingspan, 221

Kepodactylus insperatus, *203*, 205
K/T extinction. *See* Cretaceous extinction.
Kungpengopterus sinensis, 135, *137*

lacewing, 84–68, *85*
Lacusovagus magnificens, *229*, 230–31, *230*, *231*
Lagerstätten, *9*, 9, 39, 82, *93*, 155; Lagerstätten effect, 259
landing, 63–64, *63*, 209
Las Hoyas, 42, 82, 219; fossils of, *43*
launch: bipedal, 58–59, 64, *255*; environmentally assisted, 58, 64; quadrupedal, 22, 59–60, *60*, 101, 110, 121, 133, 141, 150, 167, 168, 179, 197, 209, 255; taxiing, 59–60. *See also* water launch
Lebanon, 230
leks, 81, 182
Liaoningopterus, *154*, 157–58
Liaoxipterus brachyognathus, 145, *145*, 147
lines of Arrested Growth (LAGs), 77–78
Loma del Pterodaustro, *93*, 187, *200*
Lonchodectes *79*, 211, *212*, *213*
Lonchodectes compressirostris, 211
Lonchodectes giganteus, 211, *213*, 214
Lonchodectes machaerorhynchus, 211, 214
Lonchodectes microdon, 211
Lonchodectes platystomus, 211
Lonchodectes sagittirostris, 211, *213*, 214
Lonchodectidae, *92*, 195, 211–15, *213*, 224, *261*; anatomy, 214–15, *214*; classification, 213–14; palaeoecology, 215; wingspans, 214
Lonchognathosaurus acutirostris, *203*, 205; wingspan, 205
Longchengpterus zhoai, *145*, 146–47, *147*
Lophocratia, *92*, 143, 183n; characterizing features of, 183
Lost World, *The* (Conan Doyle), 3, 157
Ludodactylus sibbicki, *154*, *156*, 157–58; with deadly plant frond, 82, *156*
lungs, 39, 41, *41*

mammalian identification of pterosaurs, 13, 123
mandible, *83*, *115*, *127*, *140*, *146*, *188*, *193*, *194*, *208*, *210*, *213*, *222*, *236*, *237*, *247*; mandibular symphysis, 26; mandibular fenestra, *26*, 27, 99, 117; overview of anatomy, 26–27
manus, *100*, anatomy of, *32*, *33*, *34*, 34–35; *See also* wing metacarpal; wing finger
manus-only trackways, 69–70
marine reptiles, 1, 14; as pterosaur predators, 88
Marsh, Othniel Charles, 170, 172–73
mastiff bats, as analogues of pterosaurs, 121
megapodes, 74, 74n
"*Mesadactylus ornithosphyos*," 107, *185*, 187
Mesozoic Era: duration, 1, *93*; divisions of time, 90, *93*
metabolism, 19, 29, 51
metacarpals, anatomy of, *32*, *33*, *34*, 35
metatarsals, anatomy of, 36–37, *36*
meteorite impact, 3, 259
Mexico, 68, 175, 256; fossils of, *84*
Microtuban altivolans, 230, *230*
migration, 197
modular evolution, 135
Mongolia, 88, 104, 106, 204, *204*
Monofenestrata, *92*, *137*, 139–40
Montana, 247
Montanazhdarcho minor, *246*, 248, 254; wingspan, 248–49
"Moon Goddess," 211, 213–15
Morganopterus zhuiana, 166, *185*, 187–88, 192, 199; wingspan, 190–91
Mörnsheim limestones, 201
Morocco, 68, 216, 219, 248; fossils of, *253*
Morrison Formation, 107, 128, 145, 187, 205
Muppets, 104, *108*
muscles. *See* myology.
Muzquizopteryx coahuilensis, *172*, 175, 178–79, wingspan, 177
myology, 43–46, *107*, 109–10; of forelimbs, 46, *47*, 160; of hindlimbs, 46, *48*, 68–69, 148, 160, 162–63; of jaws, 44–45, *45*, 117, *122*, 151, *151*, 199, 243; of neck, 45–46, *45*, 151, 169, 210, 226; of tail, 46; of wing tissues, 52–53, *52*
Mythunga, *154*, 157

nasal septum, 225
nasoantorbital fenestra, 23, *26*
Navajodactylus boerei, *246*, 248
Nemicolopterus crypticus, *218*, 219, *220*, 224–25, 227; wingspan, 219
Neoazhdarchia, *92*, 216, 228, 230–31, 234, 238, 240–41, 244, 251; characteristics of, 216
Nesodactylus hesperius, *125*, 127–30
nests, *77*, 141. *See also* reproductive strategy
New Mexico, 248
niche partitioning, 133, 181, 199, 210, 239, 243
nightjars, as analogues of pterosaurs, 111
Ningchengopterus liuae, *185*, 187, 197
Niobrara Chalk (Niobrara Formation), *9*, *93*, 170, 172, 174; fossils of, *33, 34, 173, 174, 177*
Noripterus complicidens, 201, *203*, 204; anatomy, 205, 207–9; wingspan, 205
Normanognathus wellnhoferi, *203*, 205–6
North America, 7, 88, 170–71
North Korea, 107
notarium, 30–31, *30*, *160*; evolution of, 201
Nurhachius ignaciobritoi, 145–47, *145*; anatomy, 147–49; flight, *62*, 149; wingspan, 147
Nusplingen Limestone, 183, 186
Nyctosauridae, 171, 174–75, 177–82, 228, 248, 259, *261*; wing shape, *150*
Nyctosaurus, *172*, 174–75, *174*, *177*, 226; anatomy, *176*, 177–79; crest function, 81; diet, 182; flight, *62*, 180; growth, 181–82; terrestrial competence, 180, *180*; wing shape, *61*; wingspan, 177
Nyctosaurus bonneri, 174
Nyctosaurus gracilis, 174
"*Nyctosaurus lamegoi*," *172*, 175; wingspan, 177
Nyctosaurus nanus, 174

occipital face, anatomy, 24, *26*; occipital condyle, 26
olfactory bulbs, 43
omnivory, 122, 134, 227, 233, 258
open-billed storks, as analogues of pterosaurs,200

opisthotic processes, 204, 205, 210
optic lobe, 42–43. *See also* visual acuity
orbit, anatomy, 23, *26*; closed, 148, 151, 205, 210
Oregon, 248
"*Ornithocephalus*," 98n, 186
Ornithocheiridae, 75, *92*, 143, 152–63, *154*, 164, 179, 242, 256, *261*; anatomy, 157–60; as denizens of marine settings, 153; flight, 160–62; paleoecology, 163; pneumaticity, 158; taxonomic controversy of, 152, 163; terrestrial competence, 162–63, *162*; wing shapes, *150*; wingspans, 157
Ornithocheiroidea, *92*, 143–82, 195, 211, 214, 241, *261*; characterizing features of, 143, 148; wing shapes, *150*
Ornithocheirus mesembrinus, *73*, *156*, *159*, *255*
Ornithocheirus simus, *6*, *155*
Ornithocheirus, 154–55, *154*, 158, *158*, *160*, *162*; wingspan, 157
Ornithodesmus cluniculus, 143, 145
"*Ornithodesmus*" *latidens*, 143. See also *Istiodactylus latidens*
Ornithodira, 17–18
ornithosaurs, 13
Ornithostoma sedgwicki, 172–73, *172*
ossified tendons, 179–80
Owen, Sir Richard, *6*, 98, 154, 173

Padian, Kevin, 66, 102
palatal surface of jaw, 26, *239*, *240*
paleoecology, 74–89, 103, 111–12, 122, 133–34, 141–42, 150–51, 163, 169, 180–82, 197, 199–200, 210, 215, 227, 233, 242–43, 257–58
Parapsicephalus purdoni. See *Dorygnathus purdoni*
parasites, 87, *88*
parrots, as pterosaur analogues, 101, 226
pathologies, 86–87, *86*, 174
peck marks. *See* feeding traces
pectoral girdle, anatomy of *30*, 31–32; "bottom decker" configuration, 224; of ornithocheiroids, 148, *160*, 224, 241
pellet, gut regurgitate, 42, 82, 87–88
pelvic girdle, *8*, *160*; anatomy of *30*, 35; of ornithocheiroids, 159, *160*, 162–63
Pentaceratops sternbergi, 253
pes, *48*, *100*; anatomy of, 36–37, *36*. *See also* fifth pedal digit
Peteinosaurus zambelli, 95, *97*, *113*; terrestrial competence, 101–3; wingspan, 99
petrels, as pterosaur analogues, 179
phalanges, *100*; anatomy of, 35–37,
"*Phobetor*" *parvus*, *203*, 204, 205, *207*; wingspan, 205
Phosphatodraco mauritanicus, *246*, 248, *253*; wingspan, 249
phylogeny: of pterosaurs, 90, *92*, 106, *261*, 262; of reptiles, 15; of tetrapods, *13*
pigeons, as pterosaur analogues, 101
plantigrady, 18, 68
Plataleorhynchus streptorophodon, *185*, 193, *194*, 199
pneumatic foramina, 32, 39. *See also* skeletal pneumaticity
pneumaticity. *See* skeletal pneumaticity
Posidonienschiefer, *93*, 113, 125; fossils of, *115*, *127*
postacetabular process, *30*, 35. *See also* ilium
posture: of head, 26, 43, of limbs, 66–71 *67*, *68*, 69–71, *71*, 98, 121–22, 133, 122, 162, 256
potoos, as analogues of pterosaurs, 112
powered flight, 19, 22, 60
preacetabular process, *30*, 35, 46. *See also* ilium
Prejanopterus curvirostra, 211
premaxillary ridge, 240–41, *240*
Preondactylus buffarinii, 95, *97*, 120, *137*; anatomy, *97*, 99–101; flight, *62*, 101, 121; terrestrial capacity, 101–3; wing shape, *61*; wingspan, 99
prepubes, *30*, 35; use in respiration, 39
probing, foraging strategy, 83, 199, 210, 257
"prolacertiforms." *See* protorosaurs
propatagium, 52, *52*, 54, 140–41; importance to flight, 54. *See also* pteroid
protopterosaur, 18–19, *19*, 56
"*Protopterosaurus*," 18–19, *19*
protorosaurs, 14–15, 14n2, 17–18
Pteraichnus saltwashensis, *67*
Pteraichnus, 64–68, *84*
Pteranodon, 172–74, *172*, *173*, 195, 224, 243, 254; anatomy, 177–79, *177*; diet, 182; flight, *62*, 179–180; growth, 180–182; gut content, 82, *83*; pathologies, *86*, 87; as prey for *Squalicorax*, 88; pulmonary system, *41*; research history, 62; sexual dimorphism, 81, 141, 177, 180–82, *181*; terrestrial competence, 180, *180*; trunk anatomy, *30*; wrist and hand anatomy, *33*, *34*; wing membrane distribution, *52*; wing shape, *61*, wingspan, 177
Pteranodon longiceps, *8*, 173, *173*, *175*, 178; forelimb myology, *47*; hindlimb myology, *48*; launching, *60*; neck myology, *45*; sexual dimorphism, *80*; skeletal anatomy, *25*; terrestrial locomotion, *66*
Pteranodon sternbergi, *171*, 173, *173*, 178, *181*
Pteranodontia, 92, 143, 152, 158, 170–82, *172*, 241–42; discovery, 172–73; anatomy, 177–79; as denizens of marine settings, 179; diet, 182; flight, 179–80; pneumaticity, 178; sexual dimorphism, *80*, 81, 171, 177, 180–82, *181*; terrestrial competence, 180; wingspans, 177
Pteranodontidae, 171, 172–74, 177–82, *261*; wing shape, *150*
Pteranodontoidea, 143n
Ptero-dactyle, 186
Pterodactyloidea, 92–94, *92*, 135, 139, 143, 183, 201
Pterodactylus, 98, 98n; 114, *137*, 145, *162*, 172, 183, 185, *185*, *198*, 201; diet, 199; flight, *62*, 197; gut content, 82; osteology, 191–92, 195; sexual dimorphism, 199; soft tissues, *49*; 195–97, *196*; wing shape, *61*; wingspan, 191
Pterodactylus antiquus, *4*, *5*, 186, *190*, *255*; caudal vertebrae, *31*; discovery of, *5*, 5–6, 186; humerus, *79*
Pterodactylus kochi, 186, *193*, *196*, 201; "Vienna" specimen, 54, 196–97, *196*
"*Pterodactylus*" *longicollum*, *185*, 186
"*Pterodactylus*" *micronyx*, 186
Pterodaustro guinazui, *185*, 187, *188*, *255*; anatomy, *192*, 194–95; diet, 199, *200*; embryo, 76; flight, *62*, 197; gastroliths, 42, 199; growth, 78, *78*, 197, 199; pneumaticity, 190; wing shape, *61*; wingspan, 191
pteroid: articulation of, *33*, 34–35, *34*; evolution of, 22, 34; orientation of, 34–35, 54

Pterorhynchus wellnhoferi, 50, *125*, 128, 135; anatomy, 130, 132; wingspan, 130
pterosaur research, origin of, 5, *5*, 183, 186
pubis, anatomy of, *30*, 35
pulmonary system, 37–38, 39, 41, *41*
Purbeck Limestone, 186–87
Purbeckopus pentadactylus, 68, *84*
"*Puntanipterus globosus*," 187
pycnofibers, *40*; 51, 110, 128, 132, 196; evolution of, 20, 51
pygostyle, 104,109

Qinglongopterus guoi, *125*, 127, 135
Quetzalcoatlus, 62, 243, 247–48, 250, 255
Quetzalcoatlus northropi, *79*, 245–46, *246*; wingspan, 250–51
Quetzalcoatlus sp., *29*, 246–47, *247*, *253*, *254*, 257; anatomy, 249, 251–52; flight, *62*; jaw myology, *45*; wingspan, 249; wing shape, *61*

radius, anatomy of, *32*, 33, *33*, *34*
"*Raeticodactylus filisurensis*." See *Caviramus filisurensis*
raptors, 243
reproductive strategies, 74, 76–77, 141, 180–82. *See also* sexual maturity
reptilian identification of pterosaurs, 13–14
respiration. 39, 41
Rhabdopelix, 95
Rhamphocephalus, 125, *125*; wingspan, 128–130
Rhamphocephalus bucklandi, 127
Rhamphocephalus depressirostris, 127
Rhamphorhynchidae, *92*, 106, 123–34, *124*, *125*, *261*; characterizing features, 128; flight, 131–33; osteology, 128–30; paleoecology, 133–34; palaeoenvironmental biases, 132; skeletal pneumaticity, 129; research history, 123, 125, 127–28; soft tissues, 131–32; terrestrial competence, 133; wingspans, 128–29, 130
Rhamphorhynchinae, 123, 125, *125*, 127–34, 242, *261*. *See also* Rhamphorhynchidae
"Rhamphorhynchoidea," 92
Rhamphorhynchus muensteri, *2*, *11*, 41, *42*, *126*; anatomy, *25*, *29*, *31*, 33, *36*, *48*, 128–32, *129*; as prey, 88, 133–34; "Darkwing" specimen, *53*; diet, 133; discovery, 123, 125, *125*; endocast, *44*, 131; flight, *62*; growth, 78, *80*, *130*, 133–34; gut content, 82, *83*; sexual dimorphism, 133; soft tissues, *49*, *52*, *53*, 131–32; wing shape, *61*; wingspan, 128–30; "Zittel wing" specimen, *53*
rhamphothecae, 27, 49, *49*, 131, 196, 225
ribs, dorsal, anatomy of, *30*, 31,
Romania, 205, 248–49; fossils of, *246*, *247*
Russia, 157, 248
Rynchops, 133, *240*, 242–43

sacral vertebrae, *160*; anatomy of, 30, *30*; synsacrum, *30*
sail crests, 178, 226
Santana Formation, 9, *9*, 67, 82, *93*, 155, 157, 216, 218, 234, 237, 241; fossils of, *37*, *156*, *160*, *222*, *237*, *239*
"*Santanadactylus*," *155*
"*Santadactylus*" *spixi*, 237
"*Santanadactylus pricei*," *32*
Saurichthys, 88
Sauropsida, 14, *15*
Saurornitholestes, 88, *89*
scales, 47–48, *48*
Scanning Electron Microscope (SEM), 10, *118*
Scaphognathinae, 123, *124*, 125, *125*, 128, 130–34, *261*. *See also* Rhamphorhynchidae
Scaphognathus crassirostris, *126*, 128; anatomy, 130, *131*; discovery, 123, 125, *125*; growth, 134; wingspan, 130
scapula, elongate, 195, 197, 224. *See also* scapulocoracoid
scapulocoracoid, *254*; anatomy of, *30*, 31–32. *See also* pectoral girdle
scavenging, *144*, 150–51, 257
Scleromochlus taylori, *16*, 17–18
sclerotic ring, 24
seabirds, as pterosaur analogues, 63, 133, 160
seed dispersal, 227
Seeley, Harry, 7, *7*, 13, 64, 143, 155, 172; *Dragons of the Air*, 7
semicircular canals, 42–43
senses, 42–43
Sericipterus wucaiwanensis, *125*, 127; wingspan, 129
sexual dimorphism, 81, 139, 141, *171*, 174, 177, 180–82, *181*
sexual maturity, 78
sexual selection, 182
sharks, as pterosaur predators, 88
Sharovipteryx mirabilis, 14, *16*, 17
shellfish, as pterosaur prey, 210
Shenzhoupterus chaoyangensis, 230–32, *230*, *231*, *232*
Sinopterus, *218*, 224, 226, 225, 227, 241; flight, *62*, 226; wing shape, *61*; wingspan, 221
Sinopterus dongi, 219, *220*, *222*, *225*
"*Sinopterus gui*," 219, *220*
Sinopterus jii, 219, *220*, *225*
Siroccopteryx, *154*, 157
skeletal pneumaticity, 23, 37–38, 58, 100, 119, 129, 148, 158, 178, 189–90, 205, 241, 254; function, 37–38;
skim feeding, 122, 133, 163, 182, 242–243, 257. *See also* *Rynchops*
skin, 48
skulls, *115*, *127*, *130*, *140*, *146*, *167*, 173, *174*, *187*, *188*, *193*, *194*, *208*, *210*, *218*, *220*, *222*, *231*, *236*, *237*, *239*, *240*, *247*; anatomy of, 23–24, 26, *26*; proportions in giant pterosaurs, 251–253
soaring flight, 62, 121, 132, 149, 151, 160, 179–80, 255
social lives, 81–82, 258
Soemmerring, Samuel Thomas van, 64, 123
soft tissues, 23–24, 39–55, 109–10, 125, 131–32, 195–97, 219, 225–26, 241
Solnhofen Limestone, 5, 9, *9*, 82, 84, 104, 107, 123, 128, 133, 183, 186, 201; fossils of, *10*, *53*, *85*, *93*, *107*, *126*, *187*, *193*, *196*, *198*, *203*
songbirds, as analogues of pterosaurs, 142
Sordes pilosus, *124*, *125*, 128; anatomy, 130, 132, *132*; flight, *62*; growth, 134; wing shape, *61*
South America, 42
South Dakota, 173
South Korea, *67*, 68, 256; fossils of, *256*
Spain, 42, 68, 146, 211, 216, 219; fossils of, *43*

species diversity: of birds through time, *262*, 263; of pterosaurs, *3*, 94; of pterosaurs through time, 259, 262–63, *261*, *262*
spinosaurs, *75*, 88, *235*
Spinosaurus aegyptiacus, 253
spongiose bone, *37*, 38
spoonbills, as analogues of pterosaurs, 199
sprawling. *See* posture
Squalicorax, 88
Squamata, 14, 17
sternum, *173*; anatomy of, *30*, 31–32; orientation, 32
Stokes, William Lee, 64
stomach, 41–42
Stonesfield Slate, 125, 127, 138, 141, 186
"Stonesfield wukongopterid," *137*, 138
storks, as analogues of pterosaurs, 243, 258
Strashila incredibilis (pterosaur parasite), *88*
Stratigraphic column of the Mesozoic, *93*
supraneural plate, *160*; anatomy of, 30,
sutures, definition, 23
swans, 255
swifts, as analogues of pterosaurs, 110, 111
swimming, 6–7, 72, *73*, 103, 163, 169, *171*, 179, 210, 256, 257
Switzerland, 116
synapsid, 14
syncarpals, proximal and distal: anatomy of, 33–34; articulation with pteroid, 34

tactile-feeding, 199–200
tail vanes, 52n, 132, *132*
takeoff. *See* launch
Tanzania, 205
Tapejara wellnhoferi, 216, 218–19, *218*; anatomy, 220–21, *222*, 224; wingspan, 221
Tapejaridae, *48*, *92*, 148, 216–27, *218*, 228, 231, 233, 234, *261*; classification, 216; flight, 226, osteology, 220–25; pneumaticity, 224; soft tissues, 225–226, *226*; terrestrial competence, 227; palaeoecology, 227
tarsals, anatomy of, 36, *36*
teeth. *See* dentition
temporal fenestra: definition, 24; use in reptile phylogeny, 14
Tendaguripterus recki, *203*, 205
terns, as analogues of pterosaurs, 160
terrestrial foraging, 227, 233, 243, 256–58
terrestrial locomotion, 64–72, 101–3, 110–111, 121–22, 133, 141, 150, 162–63, *162*, 169, 180; 197, 209–10, 215, 227, 233, 241, 256–57; bipedal, 64, 66–69, 102, 180; folding of wing, *70*; gait, *65*, 68–70, *68*; quadrupedal, 64, 67–62; running, 69, *69*, 162; walk cycle, 69. *See also* trackways; manus-only trackways
terrestrial stalkers, 258
Testudines, 14, 14n1
Tetrapoda, 13, *13*
Texas, 157, 238, 244; fossils of, *246*, *247*, *253*, *254*
Thalassodromeus sethi, 81, *85*, *235*, *236*, 237, *237*, *240*; anatomy, 239–41; crest function, 242; etymology, 237n; foraging habits, 242–43; wingspan, 239
Thalassodromidae, *92*, 216, 228, 230–31, 234–43, *236*, 254, *261*; anatomy, 239–41; flight, 241; foraging strategies, 242–43; growth, 242; soft tissue, 241; terrestrial competence, 241; wingspan, 239
throat. *See* esophagus
Tiaojishan Formation, *93*, 128, 135, 141; fossils of, *10*, *50*, *146*
tibiotarsus, anatomy of, 35–36, *35*
tinamous, as pterosaur analogues, 101
Tischlinger, Helmut, 39. *See also* ultraviolet (UV) light fossil photography
Titanopterygiidae, 248
Titanopteryx. See *Arambourgiania*
toothlessness. *See* beaks
trabeculae, *37*, 38
trackways, caiman, 66
trackways, pterosaur, 63, *63*, 64, *65*, 66–72, *84*, 255, *256*; models of, 67–68; in space, 68; tail drags, 83–84; in time, 70–71. *See also* footprints; *Pteraichnus*
trailing edge structure, 53, 121, 132
Triassic Period: duration, 3, *93*; pterosaur record, 259, *261*, *262*
Trichomonas, 87
"*Tropeognathus*," 155
Tulugu Group, 204
Tupandactylus, *218*, 219, 222; wingspan, 221
Tupandactylus imperator, *217*, 218–19, *218*, 224–25; size, *221*
Tupandactylus navigans, *40*, *49*, *50*, *218*, 219, 222, 225
Tupuxuara, 234, *236*, 247; anatomy, 239–241; foraging habits, 243; wingspan, 239
Tupuxuara deliradamus, *80*, 234, *236*
Tupuxuara leonardii, *79*, 234, *236*, 238, *239*, *240*, 241, *255*; anatomy, 238–39
Tupuxuara longicristatus, 234, *236*, 237, *239*, 241–42

Uktenadactylus, *154*, 157
ulna, anatomy of, *32*, 33, *33*, *34*
ultraviolet (UV) light fossil photography, 10, *11*, 39, *43*, *107*, 183, *187*, *196*
United States of America, 68, 95, *173*; discovery of American pterosaurs, 170, 172–73
Unwin, David, 154, 211
Unwindia trigonus, 211, *213*, 214
uropatagium, 52, *52*, 54–55, 121, 128, 133, 196–97, *196*
Utah, *67*, *73*, *84*
Uzbekistan, 248

Velociraptor, 88
vertebral column, *8*, *160*, *210*, *213*, *246*, *253*; anatomy of, 28–31, *29*, *30*, *31*
Vertebrata, 13
visual acuity, 42–43; of anurognathids, 111; of azhdarchids, 258; of boreopterids, 169, of ornithocheirids, 163
Volgadraco bogolubovi, *246*, 248
vomit. *See* pellet
vultures, as pterosaur analogues, 150–51

wading, 71, 183, 197, 199–200, *200*, 210, 257
water launch, *60*, 150, 160–62, *161*, 167, 179–80, 182, 226, 256–57. *See also* launch
Wealden, 143, 146, 157, 211; fossils of, *146*, *213*
webbing, between digits, 48, *196*, 226
weight, of pterosaurs. *See* body mass
Wellnhofer, Peter, 5, 7, 67, 183; *Encyclopedia of Pterosaurs*, 9; *Pterosaur Handbook*, 9
Western Interior Seaway, 88, 170, 179, 181
Wild, Rupert, 18, 116
Williston, Samuel, 174
wing finger: anatomy of, *32*, *33*, 35; evolution of, 22, *32*
wing loading: definition, 61; of flying vertebrates, *62*
wing membranes, 51–55, *52*, *53*, *107*, 109, 121, 125, 128, 132, 196–97, 226, *226*, 241; anchorage to wing bones, 53; distribution of, *52*, 54–55; evolution of, 22; fossils, 11; histology, 52–53, *52*, *53*; structural fibers, 17; *52*, 53, *53*, *70*, 132
wing metacarpal: anatomy of, *32*, *33*, *34*, 35; evolution of, 22
wing shapes, *61*, *150*
wingspan, 130
woodpeckers, as pterosaur analogues, 101
wrist. *See* carpus
Wukongopteridae, *92*, 127–28, 135–42, 183, 191, *261*; anatomy, 138–40; evolution of, *137*; flight, 140–41; paleoecology, 141–42; sexual dimorphism, 139, 141; terrestrial competence, 141; wingspans, 138
Wukongopterus lii, 135, *137*, 140,
Wyoming, 65, *73*, 173

Yixian Formation, 9, *9*, *76*, *93*, 106, 145, 157, 164, 166, *167*, 187, 228, 230, 233
Yixianopterus, *154*, 157
Young, Chung-Chien, 8, 204

Zhejiangopterus linhaiensis, *246*, *247*, 248, *249*; anatomy, 249, 254–55, *255*; flight, *62*; wingspan, 249
Zhenyuanopterus longirostris, 164, *165*, 166–67, *166*, *167*, *168*, 195